I0071991

Protein Engineering and Design

Protein Engineering and Design

Edited by Anton Torres

SYRAWOOD
PUBLISHING HOUSE

New York

Published by Syrawood Publishing House,
750 Third Avenue, 9th Floor,
New York, NY 10017, USA
www.syrawoodpublishinghouse.com

Protein Engineering and Design
Edited by Anton Torres

© 2017 Syrawood Publishing House

International Standard Book Number: 978-1-68286-402-9 (Hardback)

This book contains information obtained from authentic and highly regarded sources. Copyright for all individual chapters remain with the respective authors as indicated. All chapters are published with permission under the Creative Commons Attribution License or equivalent. A wide variety of references are listed. Permission and sources are indicated; for detailed attributions, please refer to the permissions page and list of contributors. Reasonable efforts have been made to publish reliable data and information, but the authors, editors and publisher cannot assume any responsibility for the validity of all materials or the consequences of their use.

The publisher's policy is to use permanent paper from mills that operate a sustainable forestry policy. Furthermore, the publisher ensures that the text paper and cover boards used have met acceptable environmental accreditation standards.

Trademark Notice: Registered trademark of products or corporate names are used only for explanation and identification without intent to infringe.

Cataloging-in-publication Data

Protein engineering and design / edited by Anton Torres.
    p. cm.
Includes bibliographical references and index.
ISBN 978-1-68286-402-9
1. Protein engineering. 2. Proteins. I. Torres, Anton.
TP248.65.P76 P76 2017
660.63--dc23

Printed in the United States of America.

TABLE OF CONTENTS

Permissions

List of Contributors

Index

PREFACE

Protein Engineering is a nascent yet rapidly expanding field of study concerned with the useful development of proteins. This book elucidates new techniques and their applications in a multidisciplinary approach. The discipline of protein engineering is a growing field of study. This book traces the progress of this field and highlights some of its key concepts and applications. This field of study is understood through two major strategies that is - rational protein design and the directed evolution. The text has combined both these strategies so as to give it a holistic approach. Those in search of information on this field to further their knowledge will be greatly assisted by this book. It includes contributions of experts and scientists from across the globe which will provide innovative insights into this field.

This book aims to highlight the current researches and provides a platform to further the scope of innovations in this area. This book is a product of the combined efforts of many researchers and scientists from different parts of the world. The objective of this book is to provide the readers with the latest information in the field.

I would like to express my sincere thanks to the authors for their dedicated efforts in the completion of this book. I acknowledge the efforts of the publisher for providing constant support. Lastly, I would like to thank my family for their support in all academic endeavors.

Editor

The Core Proteome and Pan Proteome of *Salmonella* Paratyphi A Epidemic Strains

Li Zhang[1], Di Xiao[1], Bo Pang[1,2], Qian Zhang[1], Haijian Zhou[1], Lijuan Zhang[1], Jianzhong Zhang[1,2], Biao Kan[1,2]*

1 State Key Laboratory for Infectious Disease Prevention and Control, National Institute for Communicable Disease Control and Prevention, Chinese Center for Disease Control and Prevention, Beijing, P. R. China, **2** Collaborative Innovation Center for Diagnosis and Treatment of Infectious Diseases, Hangzhou, P.R.China

Abstract

Comparative proteomics of the multiple strains within the same species can reveal the genetic variation and relationships among strains without the need to assess the genomic data. Similar to comparative genomics, core proteome and pan proteome can also be obtained within multiple strains under the same culture conditions. In this study we present the core proteome and pan proteome of four epidemic *Salmonella* Paratyphi A strains cultured under laboratory culture conditions. The proteomic information was obtained using a Two-dimensional gel electrophoresis (2-DE) technique. The expression profiles of these strains were conservative, similar to the monomorphic genome of *S.* Paratyphi A. Few strain-specific proteins were found in these strains. Interestingly, non-core proteins were found in similar categories as core proteins. However, significant fluctuations in the abundance of some core proteins were also observed, suggesting that there is elaborate regulation of core proteins in the different strains even when they are cultured in the same environment. Therefore, core proteome and pan proteome analysis of the multiple strains can demonstrate the core pathways of metabolism of the species under specific culture conditions, and further the specific responses and adaptations of the strains to the growth environment.

Editor: Dongsheng Zhou, State Key Laboratory of Pathogen and Biosecurity, Beijing Institute of Microbiology and Epidemiology, China

Funding: This study was supported by the Priority Project on Infectious Disease Control and Prevention (2012ZX10004215). The funders had no role in study design, data collection and analysis, decision to publish, or preparation of the manuscript.

Competing Interests: The authors have declared that no competing interests exist.

* E-mail: kanbiao@icdc.cn

Introduction

Over 2500 serotypes have been reported in *Salmonella*, and most of them result in diarrhea. Within these serotypes, *Salmonella enterica* serovar Typhi and Paratyphi, can lead to systemic infections in humans, known as typhoid and paratyphoid fever. These diseases cause epidemics in Asia, Africa and Latin America [1,2]. Before the 1990s, *S.* Typhi was the main causative agent of enteric fever in southeast Asia and in China, but in the mid-1990s, the number of cases caused by *S.* Paratyphi A started to increase, and paratyphoid fever subsequently became the major enteric fever [3,4,5,6].

The whole genomes of some *S.* Typhi and *S.* Paratyphi A strains have been sequenced [7,8,9,10]. Genetically monomorphic genomes and relatively low sequence diversity were found, which may be the result of a high restriction of host adaption [11]. Multilocus sequence typing (MLST) and pulsed-field gel electrophoresis (PFGE) [12] were used to generate phylogenetic information and obtain a population variance analysis, and for *S.* Typhi and *S.* Paratyphi A genotyping. Genomic sequencing and a single nucleotide polymorphism (SNP) analysis provided high-throughput and high-resolution genome variation methodology [13], and were applied for the epidemic analysis of *S.* Typhi strains [14,15,16,17]. All of the results showed a low level of genetic variation in *S.* Paratyphi A, and a high clonality of strains involved in epidemics.

A genome comparison among different strains is used to identify the core genome and pan genome [18]. The core genome includes the core, conserved genes and surviving characteristics which keep the microorganism evolving. In contrast, the pan genome includes newly transferred genes, and demonstrates the diversity of the organism. Genome comparisons help investigators discover the divergence of the same genes between different organisms. However, a genome analysis cannot show the differences in the protein levels, which are the actual determinants of the growth and survival of the organism. Proteomic studies can illustrate the expression levels of various gene products under given culture conditions, discover the responses to different biological systems and uncover protein modifications and protein-protein interactions [19,20]. A comparison of the proteomes of different strains can indicate their shared and unique features. Besides the shared proteins, it may also help identify newly acquired gene products.

Many technologies for proteome analysis are in use [21,22]. In this study, we conducted a comparative proteomics analysis for four strains with different geospatial and temporal characteristics by performing 2-DE, and obtained their core and pan proteomes. We found that the proteome was highly conserved for the four *S.* Paratyphi A strains, consistent with the conservative genomes of *S.* Paratyphi A. However, some of the core proteins had significant differences in abundance among the strains, suggesting that there are variations in the protein expression in different strains, even though the strains have strict convergence in their genomes.

Materials and Methods

1. Strains

Among the strains collected during the surveillance of typhoid and paratyphoid fever in China, and from the PFGE (*Xba*I) subtyping database, we selected the *S.* Paratyphi A strains from patients in 2-DE analysis: YN07077 (isolated in Yunnan province in 2007) and GZ9A05036 (isolated in Guizhou province in 2005), which have the predominant PFGE subtype, and ZJ98053 (isolated in Zhejiang province in 1998), with the nondominant subtype, for the 2-DE analysis. Strain ATCC 9150, which was isolated in Malaysia in 1993 clinically, was also included for comparison, it has a different PFGE subtype from the other three strains.

2. PFGE

We performed PFGE according to the method previously conducted in the paper [23].

3. Protein Extraction

The protein samples used for 2-DE were prepared according to the protocol described in a previous study [24]. In brief, the strains were cultured in Colombia blood agar for 16–18 hours, then the cells were scraped from four plates (9 cm in diameter) and washed four times in ice-cold low salt PBS. The cells were resuspended in deionized water and urea (7 M), thiourea (2 M), CHAPS (4%) and IPG buffer (1%), then DTT (1%) was added respectively, in a final volume of 5 ml. A protease inhibitor cocktail tablet (Roche applied science) was added to each sample. The samples were sonicated to lyse the cells, then 125 μg RNase A and 50 U DNase were added. The samples were kept at ambient temperature for 1 hour to make proteins sufficient dissolution, centrifuged at 40,000×g for 1 hour, then the supernatant was collected and the protein content was quantified with the PlusOne Quant Kit. The samples (800 μg protein) were aliquotted and either directly used for IEF or frozen at −80°C until use.

4. 2-DE and Image Scanning

Isoelectric focusing (IEF; 17 cm, pH 4–7, Bio-Rad; 18 cm, pH 6–11, Amersham Biosciences) and 12.5% sodium dodecyl sulfate polyacrylamide gel electrophoresis (SDS-PAGE) were performed according to the manufacturer's instructions (Bio-Rad, PROTEAN IEF CELL, Protean II Xi apparatus). Briefly, passive rehydration was performed for 4 hours, and active rehydration was performed for 8 hours at 50 V, and IEF was conducted using the following conditions: 300 V linear for 1 hour, 600 V linear for 1 hour, 1000 V linear for 1 hour, 8000 V linear for 1 hour and 8000 V rapid for 8 hours. After the IEF and equilibration, the proteins were transferred by SDS-PAGE, using 10 mA for the electrophoresis of each strip for 30 minutes, which was then increased to 30 mA until the bromophenol blue line just shifted off of the lower edge of the gel. The procedure was then stopped, and the gel was dyed with Coomassie blue G-250. The gels were scanned with a UMAX2100XL device (Umax Technologies Inc.). All the samples were replicated the same procedure for three times.

5. In-gel Protein Digestion and Identification

The Coomassie-stained protein spots were cut and in-gel protein digestion was conducted as the previously described protocol [25]. Protein identification was carried out by using tandem matrix-assisted laser desorption/ionization time-of-flight (MALDI-TOF/TOF) mass spectrometry (MS, 4700 MALDI-TOF/TOF Mass Spectrometer, Applied Biosystems) as described previously [26]. The spectrum of every sample was acquired in the mass range between 800 and 4000 Da by using 1500 laser shots. MS/MS spectra were acquired by using 2000 laser shots with air as the collision gas. The single charged peaks were analyzed by using an interpretation method provided in the 4000 Series ExplorerTM software version 3.0, which selected the five most intense peaks and automatically generated the MS/MS spectra by excluding the peaks associated with the matrix and those were formed due to trypsin autolysis. The spectra were processed and analyzed by the Global Protein Server Workstation (GPS Applied Biosystems, Foster City, CA, USA), which uses internal Mascot v2.1 software for searching the peptide mass fingerprints. The searches were performed by using the NCBI non-redundant protein database (ftp://ftp.ncbi.nih.gov/blast/db/FAST/nr.gz, updated in 2011) with the following criteria: NCBI bacteria database; trypsin digestion; Moxidation and iodoacetamide alkylation as the variable modifications; missed digestion site of 1; and the MS mass error of 0.1 Da. Identifications with a GPS confidence interval greater than 95% were accepted. The inversion database was used to remove false positives (Protein identification was listed in Table S1 and Table S2, MS map of some proteins was listed in Attachment S2).

6. Data Analysis

An analysis of the proteomic data was performed using the PDQuest^{TM} Advanced 2-DE Analysis software program. We used the basic model and default parameters (Attachement S1). After matching the spots using the software program, we revised the protein spot identification manually. Each spot displayed in all four gels was allocated to the core proteins, while spots displayed in only one strain were considered to be specific proteins. The data could be output using the following steps within the same window: File, Export, Export (Text) Experiment, Spot data by gel. We selected the center position option, so the (X, Y) values for each protein could be obtained. To normalize the coordinate values, all of the core proteins in each strain were designated to use the same coordinate value as ATCC9150, while the other shared proteins (minus the core proteins) were normalized using the same coordinate values as ATCC 9150, ZJ98053 or YN07077. For example, when a spot was found for ATCC 9150, its coordinate value in all strains that displayed the spot was designated to be the same as in ATCC 9150. Spots that were not present in ATCC 9150, but were found for ZJ98053, were designated to be the same as in ZJ98053. If spots were not found in either ATCC 9150 or ZJ98053, but were displayed in the gels for YN07077, its coordinate value would be designated to be the same as that in YN07077. Specific proteins for each strain were assigned an original coordinate value.

A scatter plot for pan proteins was generated using the Origin software program (Origin Lab), since each protein in each strain has a specific coordinate value (X, Y). Red represented the core proteins shared by all four strains, blue represented ATCC 9150-specific proteins, green represented ZJ98053-specific proteins, dark green represented YN07077-specific proteins, cyan represented GZ9A05036-specific proteins and black was used to indicate proteins other than the core and specific proteins.

Core protein and pan protein trend lines were generated using the Origin software program. A similarity matrix was generated according to the r values produced by the PDQuest software, version 8.0.1 (Bio-Rad).

Functional protein assignments were based on notation and classification on Tigr website (http://cmr.jcvi.org/tigr-scripts/CMR/CmrHomePage.cgi).

Results

1. The Core Proteome and Pan Proteome of the Epidemic Strains

The 2-DE was performed within two pH ranges for the four strains of *S. Paratyphi* A, and the scanned patterns were analyzed using the PDQuest software program (Fig. S1–S8). Within the range of pH 4–7, 849, 858, 857 and 860 spots were detected in the strains ATCC 9150, ZJ98053, YN07077 and GZ9A05036 respectively, and 380, 389, 366 and 355 spots were detected within the range of pH 6–11 in these strains. Any spot detected in all four strains was considered to be a core protein, and the total number of core proteins identified was 739 and 318 within the ranges of pH 4–7 and pH 6–11. The core proteins covered from 85.9%–87.0% of the spots within the range of pH 4–7 and 81.8%–89.6% of the spots within the range of pH 6–11 in each strain, which suggested a high similarity in the protein expression among *S. Paratyphi* A strains, indicating that the proteome was highly conserved.

Within the ranges of pH 4–7 and pH 6–11, there were 946 and 435 pan proteins for the four strains. Core proteins covered a proportion of 78.1% and 73.1% of the pan proteins, confirming their conservation.

To display the proportions of core proteins and strain-specific proteins in pan proteins, we drew scatter diagrams to show the pan proteome within the two pH ranges. The principle and process have already been described above. In brief, in the scatter diagrams, specific proteins in the four strains were represented by four different colors (there were no specific proteins for ZJ98053 within the pH range of 4–7). Core proteins are presented in red. The proteins other than the core proteins and specific proteins are shown in black (Fig. 1). We also presented a constitution map to show the proportion of core proteins and strain-specific proteins within the pan proteins.

The trend lines for the core and pan proteins exhibited the amount of protein change for each of the four *S. Paratyphi* A strains (Fig. 2). From strain ATCC 9150 to ZJ98053, which were isolated in 1993 and 1998, respectively, the number of pan proteins significantly increased. During this period, the incidence of paratyphoid fever increased dramatically in Southeast Asia and China. After adding strains YN07077 and ZJ98053, the slope of the increase slowed down, indicating that the proteome did not change very much. As far as the core protein trend line was concerned, it decreased quickly at the beginning and then slowed down, but the core proteins still covered a large proportion of the total proteins in each strain, suggesting that *S. Paratyphi* A has a conservative proteome.

The above data showed the expression level of proteins included in the core proteome, which include crucial proteins involved in the normal biological processes occurring within cells, which maintain the cells' survival and basic physiological processes. The core proteome was distinguished from the core genome, because the latter is only theoretically crucial, and the gene transcription has not been confirmed.

The two pairing proteome comparisons among these four strains displayed various similarities, which were somewhat consistent with the PFGE clustering. However, there were also many differences among the strains (Fig. 3). Strains YN07077 and GZ9A05036 were the closest (with similarity of 80.4%) in terms of the protein pattern, and they had the same PFGE pattern. Compared to ATCC 9150, ZJ98053 was more similar to YN07077 and GZ9A05036 in terms of the proteome pattern, with similarity values of 79.44% and 78.16% respectively. Strain ATCC 9150 showed less similarity to YN07077 and GZ9A05036

(74.3% and 71.7%) than strain ZJ98053. Since strain ATCC 9150 was isolated in Malaysia in 1993, while strains ZJ98053, YN07077 and GZ9A05036 were from adjacent provinces in China, this suggests that the geospatial and temporal characteristics of the strains influence their proteomic pattern. In terms of the PFGE subtyping, strain ZJ98053 showed a nondominant pattern, ATCC 9150 showed a subdominant pattern and strains YN07077 and GZ9A05036 showed a predominant pattern. Strain ATCC 9150 was closer to YN07077 and GZ9A05036 than to ZJ98053 in terms of PFGE clustering. The differences in the proteomic and genomic patterns were likely due to the fact that the proteomic studies explored the more rapid proteomic response in cells when they were adapting to the environment around them, while the genome may take a longer time to show changes.

2. Constitution of the Expressed Proteins

Among the core proteins, the largest functional category was energy metabolism, then protein fate, protein synthesis, cellular processes, transport and binding proteins, central intermediary metabolism, etc. (Fig. 4). The functional constitution of the pan proteins other than the core proteins was slightly different from that of the core proteins. Energy metabolism was still the main category, but transport and binding proteins was the second most common functional category (Fig. 5).

3. Diverse Expression Levels of the Core Proteins

Although these four *S. Paratyphi* A strains had a conserved proteome and they shared over 80% of their proteins, differences in the abundance of some protein spots were observed among the strains. Fig. 6 showed that some spots had a higher abundance in ATCC9150 than in the other three strains. Fig. 7 showed that other spots had a lower abundance in ATCC9150 than in the other three strains. Of these differentially-expressed spots, strain ZJ98053 had a more consistent protein expression level with YN07077 and GZ9A05036 compared to ATCC 9150, however, its proteome had a higher regression value with strain ATCC 9150 than with the other two strains (Fig. 3).

4. Strain-specific Proteins

We blasted all the coding genes for the MS identified proteins to ATCC9150 genome, but did not find new acquired gene products. All proteins were variants of the core proteins and non-core proteins.

Discussion

In this study, the core proteome and pan proteome of four *S. Paratyphi* A strains cultured under laboratory conditions were compared, based on the core genome and pan genome comparison method. The previous genome comparisons revealed that *S. Paratyphi* A was highly clonal [10,17]. We also found that there was limited genetic diversity in terms of the level of protein expression when strains were cultured under the same conditions. In the four tested strains, the core proteins covered a large proportion (>70%) of the pan proteomes. For each strain, the core proteins covered a proportion from 81.8% to 89.6% of the global proteins. Thus, the proteome of *S. Paratyphi* A was also highly conserved, which was consistent with the highly clonal genome.

The PFGE cluster analysis showed that strain YN07077 had the same pattern as GZ9A05036, less similarity with ATCC9150 and much less similarity with ZJ98053. Nevertheless, based on the regression matrix derived from the proteomic analysis, strain ZJ98053 was approaching YN07077 and GZ9A05036 in similarity, with less in common with ATCC9150. In terms of the amount

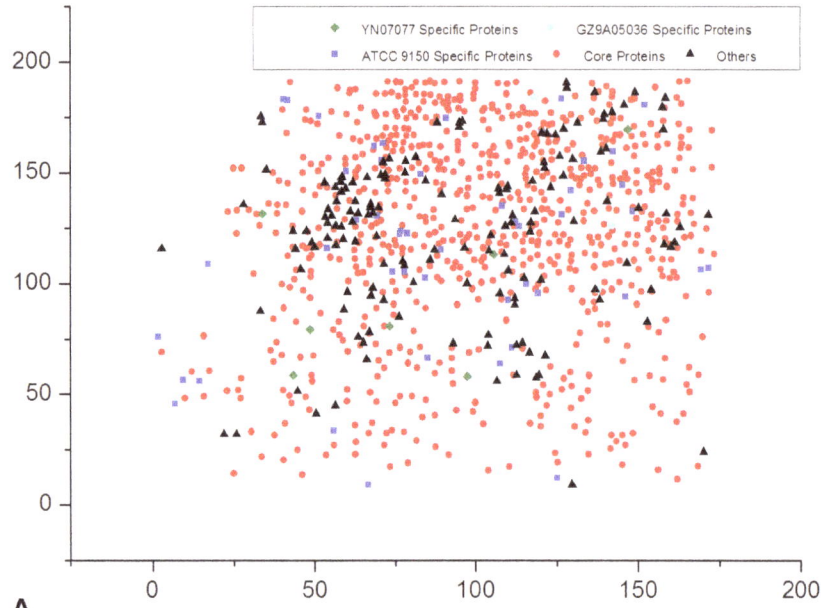

A

Scatter Plot for global proteins (pH 4-7)

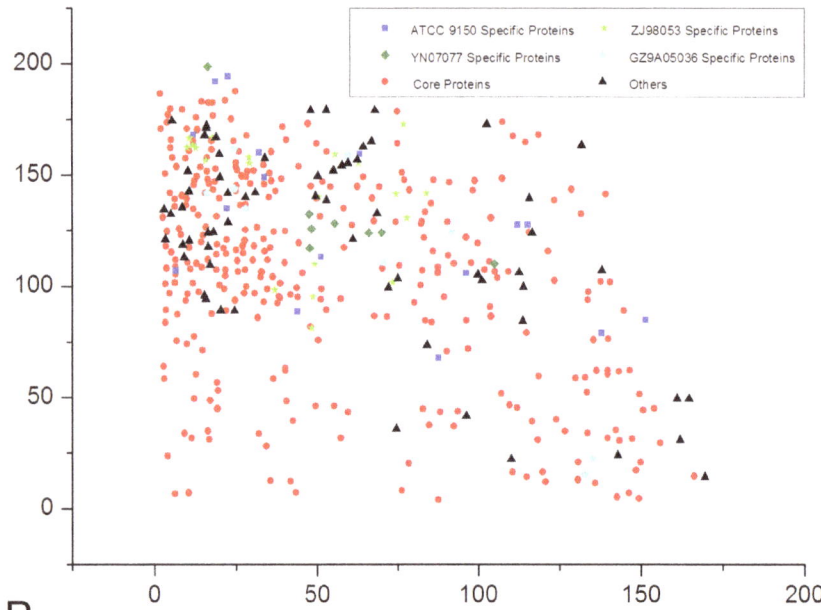

B

Scatte plot for global proteins (pH 6-11)

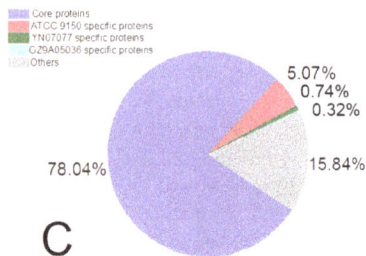

C

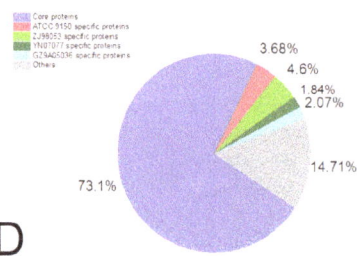

D

Figure 1. Scatter plot for the four strains of *S.* Paratyphi A pan proteins within pH range 4–7 and 6–11. A,B:Red spots represented core proteins, blue spots represented ATCC 9150 specific proteins, dark green spots represented YN07077 specific proteins, cyan spots represented GZ9A05036 specific proteins, black spots represented the other proteins except core or specific proteins in each strain; C,D represented proportion of the above proteins covered in the pan proteins within pH ranges 4–7 and 6–11 respectively.

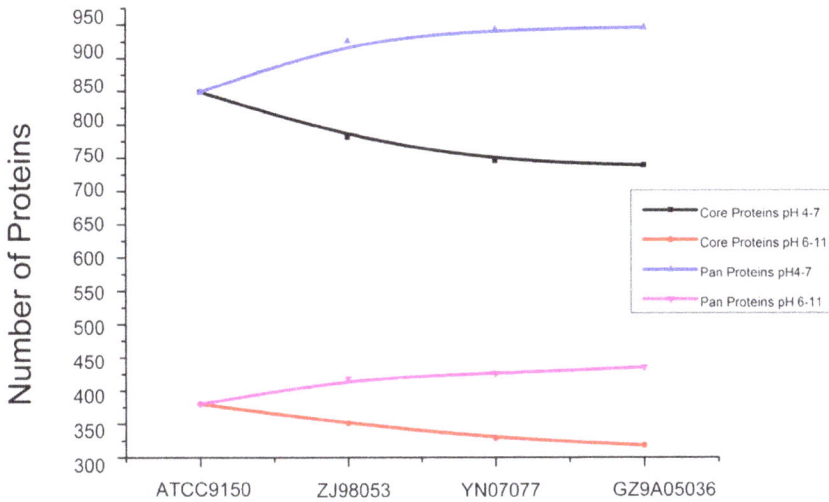

Figure 2. Number treadline for core and pan proteins when introducing more strains. Black and red lines showed core proteins tread lines within range pH 4–7 and pH 6–11; Blue and pink lines showed pan proteins tread lines within range pH –7 and pH 6–11.

of proteins, strains YN07077, GZ9A05036 and ZJ98053 had 813 (pH 4–7) core protein spots, which decreased to 739 (pH 4–7) after adding strain ATCC 9150, which indicated that similar genomes do not necessarily result in similar proteomes. Although the *S.* Paratyphi A strains had both conservative proteomes and genomes, they actively displayed distinct metabolic and other characteristics, which were not apparent at the genome level. Moreover, strains YN07077, GZ9A05036 and ZJ98053 were

isolated from very close geographical regions, which might be the epidemiological basis for their high similarity in terms of the proteome, and their trend lines for core proteins and pan proteins exhibited no big changes and there were not significant differences between their proteomes, suggesting that the genomes and expression profiles of these strains were quite conservative, and that they had undergone stable evolution.

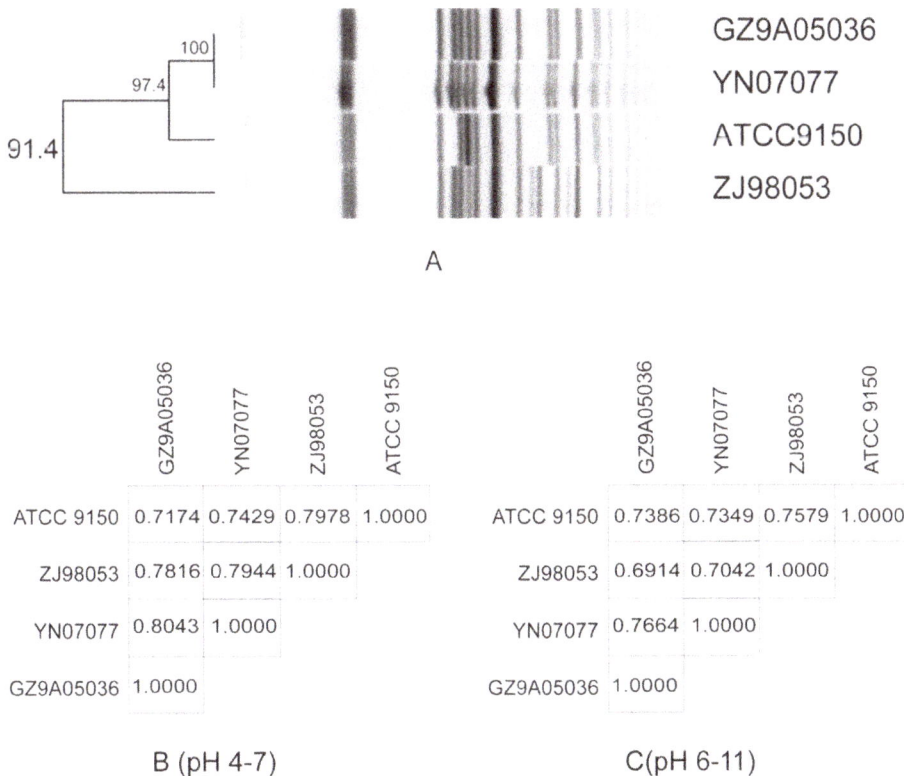

A

	GZ9A05036	YN07077	ZJ98053	ATCC 9150
ATCC 9150	0.7174	0.7429	0.7978	1.0000
ZJ98053	0.7816	0.7944	1.0000	
YN07077	0.8043	1.0000		
GZ9A05036	1.0000			

B (pH 4-7)

	GZ9A05036	YN07077	ZJ98053	ATCC 9150
ATCC 9150	0.7386	0.7349	0.7579	1.0000
ZJ98053	0.6914	0.7042	1.0000	
YN07077	0.7664	1.0000		
GZ9A05036	1.0000			

C(pH 6-11)

Figure 3. PFGE clustering and matrix for the four strains of *S.* Paratyphi A. A, Strain YN07077 and GZ9A05036 were the predominant PFGE type, and strain ATCC 9150 was subdominant PFGE type, strain ZJ98053 was the non dominant PFGE type; C,D was the proteome matrix.

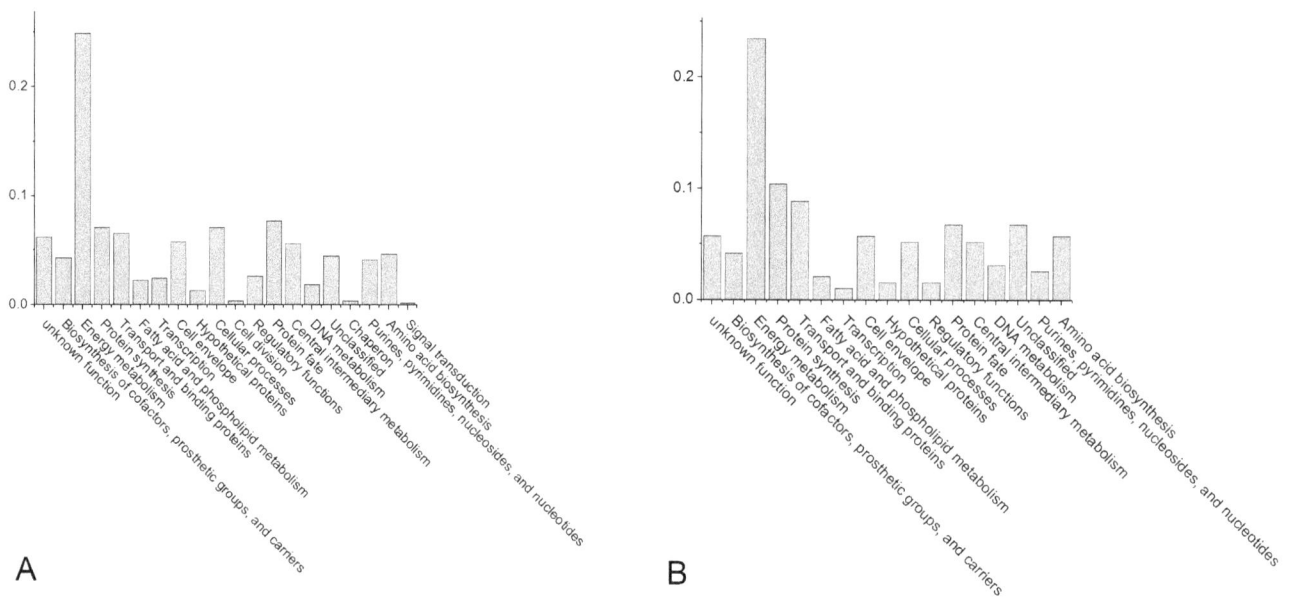

Figure 4. Functional categories of the core proteins within pH ranges 4–7 and 6–11. Each column represented the proportion of the protein number in this category to the total number of core proteins. A represents pH 4–7 and B represents pH 6–11.

According to the functional classification of core proteins, the function of most of the core proteins was mainly focused on the survival of the organisms. Interestingly, some of the pan proteins (excluding the core proteins) fit in similar functional categories, which may reflect the high concordance of the expression profiles of these strains based on their conservative genomes. When the bacteria were grown under nutrient-rich conditions, the spread of the functional classification was nonspecific, because they were mainly experiencing routine metabolism that did not require new adaptations to improve survival.

Although *S.* Paratyphi A had a highly conserved proteome in terms of the protein species, some core proteins had significant fluctuations with regard to their abundance between strains. Strains YN07077, GZ9A05036 and ZJ98053 had some protein

spots that were expressed at a similar abundance, such as spots SSP 3403, SSP 3806, SSP 4302, SSP 6304 and SSP 7806 at pH 4–7 and SSP 7117 at pH 6–11, which were expressed at a much higher abundance in these three strains than in the ATCC9150 strain. Both spots SSP 3403 and SSP 3806 were identified as outer membrane protein A, the surface-exposed porin proteins in high-copy number [27], which may play an important role in the structural stability and in the maintenance of the cell morphology, but has low-efficiency porin activity [28,29,30,31]. It exposes to and interacts with outside circumstance factors. Their variants with subtle difference on modifications might adapt to diverse environments and host immunity, which might subsequently develop to inherited and characterized phenotypes. SSP 4302, SSP 7802 and SSP 7806 were correlated to the central

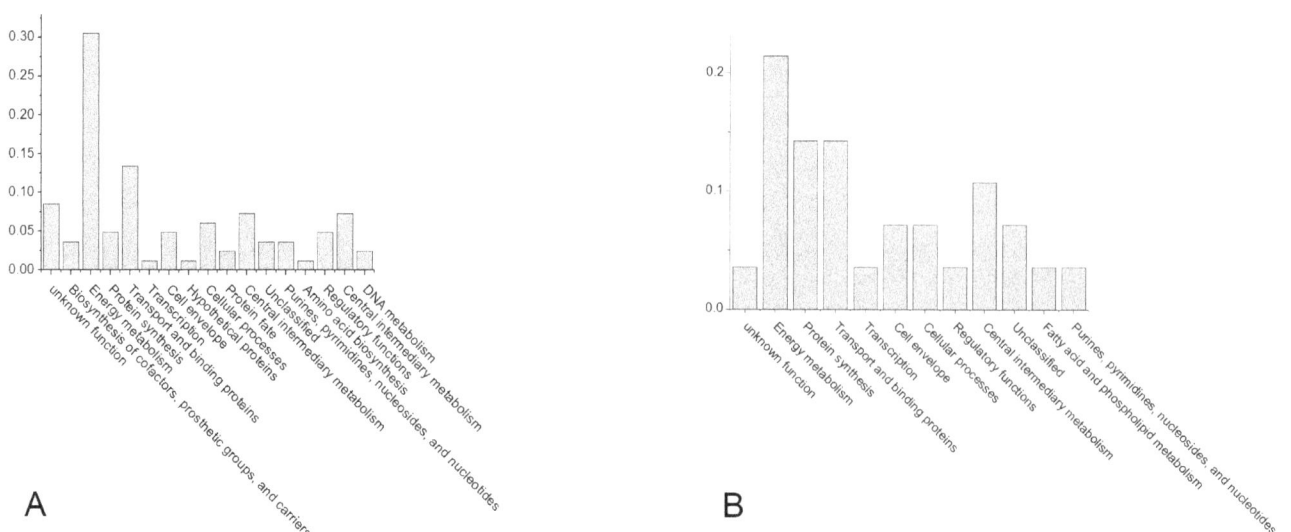

Figure 5. Functional categories of the non core proteins within pH range 4–7 and 6–11. Each column represented the proportion of the protein number in this category to the total number of non core proteins. A represents pH 4–7 and B represents pH 6–11.

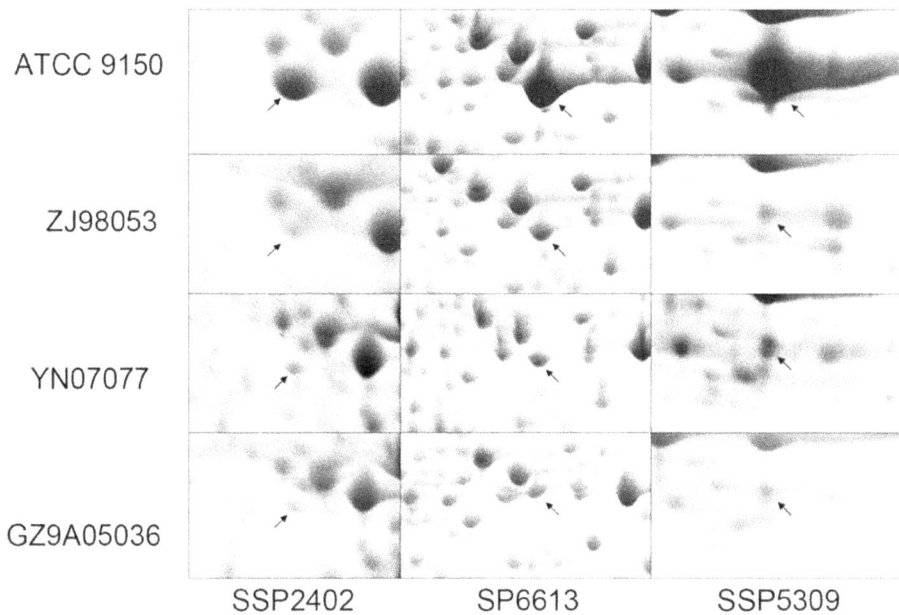

Figure 6. Spots with a higher protein expression level in ATCC 9150 than in the other *S.* Paratyphi A strains. The lines from the left to the right were the different expression level of core proteins SSP2402 (pH 4–7), SSP6613 (pH 4–7), SSP5309 (pH 6–11).

stationary-phase-specific sigma subunit of RNA polymerase σ[s] [32,33], SSP 4302 (*arcA*) is a negative regulator for *rpoS* [34], SSP 7802 and SSP 7806 were positively regulated by *rpoS* [35]. It has been proved that *rpoS* was essential for *Salmonella* virulence, *rpoS* mutant of serovar Typhi is less cytotoxic for macrophages than the parental strain, therefore *rpoS* maybe involved in the virulence of serovar Typhi [36]. *S.* Paratyphi A has similar infection mechanism to *S.* Typhi, we could speculate that different growth status and cytotoxicity of bacteria might result in diverse expression of response factors in the regulative cascade.

However, some core proteins had a higher abundance in the ATCC9150 strain than in the other three strains, such as SSP 2402, SSP 5309 and SSP 6613. Genes of SSP 2402 (*rbsK*) and SSP 5309 (*rbsB*) locate in the same operon, which participate D-Ribose

transportation and utilization [37]. This operon is transposable [38]. Up to now the real role of the higher expression is still unknown, but it may imply their biological roles in vary degrees in different strains and need further studies in detail.

ZJ98053 was in the middle in terms of its year of isolation (1998) compared with the other three strains (1993 for ATCC9150, 2005 for GZ9A05036 and 2007 for YN07077), but it was geographically close to strains GZ9A05036 and YN07077, and it exhibited high genomic similarity to ATCC9150 and high proteomic similarity to YN07077 and GZ9A05036. It also showed independent characteristics from all the other strains. For example, spots SSP 3204, SSP 6703 and SSP 1405 were more abundant in strain ZJ98053 than in the other three strains, which suggests that ZJ98053 might have evolved separately from the other three strains.

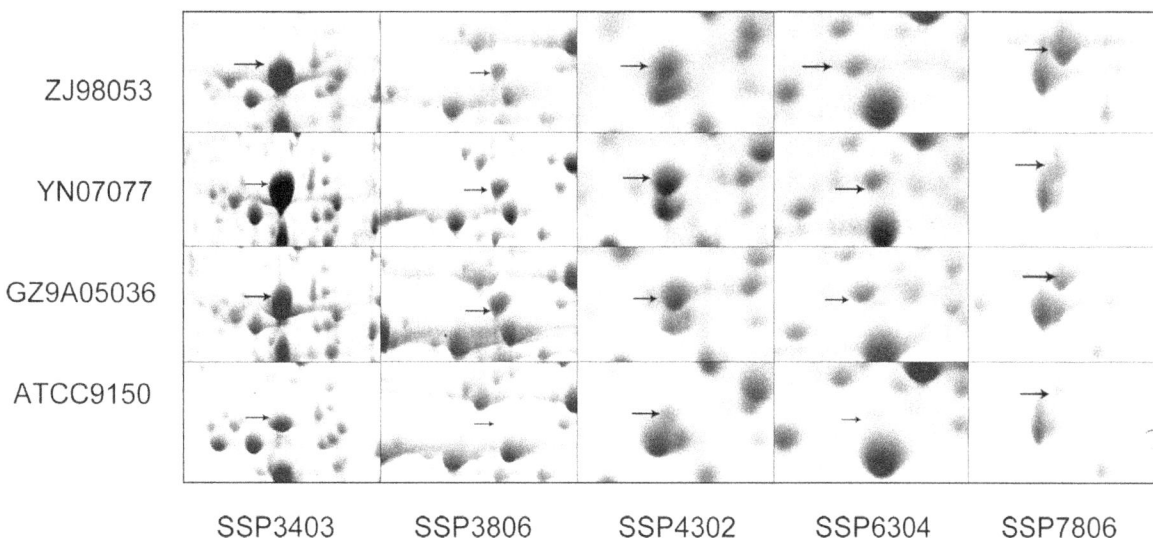

Figure 7. Spots with a lower protein expression level in ATCC 9150 than in the other *S.* Paratyphi A strains in pH 4–7.

The above differentially-expressed core protein spots were spread throughout various metabolic pathways. The variable expression levels of core proteins revealed the metabolic diversity present in the different strains. Thus, even core proteins produced under the same culture conditions can display diverse expression levels and different modifications to exert different functions, which eventually become a characteristic genetic phenotype [39,40]. Such phenotypes were common in this study, and may have been connected to the function of the individual proteins.

With regard to the specific spots, we blasted (http://blast.ncbi.nlm.nih.gov/Blast.cgi) the gene sequence for the ATCC 9150 genome, and found that there were limited differences caused by the differences in the genome or pseudogenes. Most differentially-expressed spots were considered to have been caused by differences in the transcription level or post-translational modifications.

A high-throughput genome comparison can provide a detailed gene map, including the genes, their arrangement, recombination, pseudogene accumulation and similarity between strains, whereas information about the gene expression, protein modification and regulatory network cannot be obtained from such studies. Different expression profiles (including the protein species and their abundance) can be observed even when strains have the same or similar gene clusters, since large differences can arise due to differences in the gene expression, regulatory networks and protein modifications. Thus, biological studies, and interpreting the results of such studies, remain challenging even when the whole genome sequences are known. Proteomic studies can provide information about the true expression of the genes under the studied culture condition, and the core proteome reveals the conservative expression of the genomes of different strains under this condition. Further, proteomic comparisons may show the genome-based differences, and even evolutionary relationships, among the strains, even when the genome sequences are unknown.

In summary, we herein compared the core proteome and pan proteome of *S.* Paratyphi A strains isolated during recent epidemics. Our results may provide a new approach to analyzing the expression profiles of strains at the species level, which can help to understand their genetic differences, without requiring the genomic sequence, and can facilitate understanding their common biological processes under specific conditions, which will provide information about their fundamental metabolism and survival strategies. In addition, more sensitive and high-throughput technology, such as iTRAQ-based LC-MS/MS analyses, may make it possible to perform large scale analyses of proteomic data, and may also provide information for a powerful database that can be used to assess newly-identified or emerging strains.

Supporting Information

Figure S1   Two-dimensional electrophoresis and identified spots of whole-cell proteins for ATCC 9150 within pH range 4–7.

Figure S2   Two-dimensional electrophoresis and identified spots of whole-cell proteins for ATCC 9150 within pH range 6–11.

Figure S3   Two-dimensional electrophoresis and identified spots of whole-cell proteins for ZJ98053 within pH range 4–7.

Figure S4   Two-dimensional electrophoresis and identified spots of whole-cell proteins for ZJ98053 within pH range 6–11.

Figure S5   Two-dimensional electrophoresis and identified spots of whole-cell proteins for YN07077 within pH range 4–7.

Figure S6   Two-dimensional electrophoresis and identified spots of whole-cell proteins for YN07077 within pH range 6–11.

Figure S7   Two-dimensional electrophoresis and identified spots of whole-cell proteins for GZ9A05036 within pH range 4–7.

Figure S8   Two-dimensional electrophoresis and identified spots of whole-cell proteins for GZ9A05036 within pH range 6–11.

Table S1   Protein identification for the global spots of strain ATCC9150 and differential spots of strain ZJ98053, YN07077 and GZ9AO5036 within pH range 4–7.

Table S2   Protein identification for the global spots of strain ATCC9150 and differential spots of strain ZJ98053, YN07077 and GZ9AO5036 within pH range 6–11.

Attachment S1   Spots counting parameter for 2-DE map using PDQuest software.

Attachment S2   Mass spectrum identification for some proteins of *S.* Paratyphi A.

Acknowledgments

We are grateful for the reviewing and advisement from Dr. Henliang Wang.

Author Contributions

Conceived and designed the experiments: BK JZZ. Performed the experiments: LZ DX QZ LJZ. Analyzed the data: LZ BP HJZ. Contributed reagents/materials/analysis tools: LZ. Wrote the paper: LZ BK. Preparing bacteria: ZL ZLJ. Freeze dryer using and keeping: ZL ZQ. Software analysis: ZL PB KB. Obtained permission for MS identification: XD.

References

1. Crump JA, Luby SP, Mintz ED (2004) The global burden of typhoid fever. Bull World Health Organ 82: 346–353.
2. Crump JA, Mintz ED (2010) Global trends in typhoid and paratyphoid Fever. Clin Infect Dis 50: 241–246.
3. Ochiai RL, Wang X, von Seidlein L, Yang J, Bhutta ZA, et al. (2005) Salmonella paratyphi A rates, Asia. Emerg Infect Dis 11: 1764–1766.
4. Dong BQ, Yang J, Wang XY, Gong J, von Seidlein L, et al. (2010) Trends and disease burden of enteric fever in Guangxi province, China, 1994–2004. Bull World Health Organ 88: 689–696.
5. Yan M, Liang W, Li W, Kan B (2005) Epidemics of Typhoid and Paratyphoid Fever From 1995 Through 2004 in China. DISEASE SURVEILLANCE 20: 401–403.

6. Sood S, Kapil A, Dash N, Das BK, Goel V, et al. (1999) Paratyphoid fever in India: An emerging problem. Emerg Infect Dis 5: 483–484.

7. Liang W, Zhao Y, Chen C, Cui X, Yu J, et al. (2012) Pan-genomic analysis provides insights into the genomic variation and evolution of Salmonella Paratyphi A. PLoS One 7: e45346.

8. Parkhill J, Dougan G, James KD, Thomson NR, Pickard D, et al. (2001) Complete genome sequence of a multiple drug resistant Salmonella enterica serovar Typhi CT18. Nature 413: 848–852.

9. Deng W, Liou SR, Plunkett G 3rd, Mayhew GF, Rose DJ, et al. (2003) Comparative genomics of Salmonella enterica serovar Typhi strains Ty2 and CT18. J Bacteriol 185: 2330–2337.

10. McClelland M, Sanderson KE, Clifton SW, Latreille P, Porwollik S, et al. (2004) Comparison of genome degradation in Paratyphi A and Typhi, human-restricted serovars of Salmonella enterica that cause typhoid. Nat Genet 36: 1268–1274.

11. Kidgell C, Reichard U, Wain J, Linz B, Torpdahl M, et al. (2002) Salmonella typhi, the causative agent of typhoid fever, is approximately 50,000 years old. Infect Genet Evol 2: 39–45.

12. Chen C, Zhao Y, Han H, Pang B, Zhang J, et al. (2012) Optimization of pulsed-field gel electrophoresis protocols for Salmonella Paratyphi A subtyping. Foodborne Pathog Dis 9: 325–330.

13. Holt KE, Parkhill J, Mazzoni CJ, Roumagnac P, Weill FX, et al. (2008) High-throughput sequencing provides insights into genome variation and evolution in Salmonella Typhi. Nat Genet 40: 987–993.

14. Baker S, Holt K, van de Vosse E, Roumagnac P, Whitehead S, et al. (2008) High-throughput genotyping of Salmonella enterica serovar Typhi allowing geographical assignment of haplotypes and pathotypes within an urban District of Jakarta, Indonesia. J Clin Microbiol 46: 1741–1746.

15. Holt KE, Baker S, Dongol S, Basnyat B, Adhikari N, et al. (2010) High-throughput bacterial SNP typing identifies distinct clusters of Salmonella Typhi causing typhoid in Nepalese children. BMC Infect Dis 10: 144.

16. Holt KE, Dolecek C, Chau TT, Duy PT, La TT, et al. (2011) Temporal fluctuation of multidrug resistant salmonella typhi haplotypes in the mekong river delta region of Vietnam. PLoS Negl Trop Dis 5: e929.

17. Baker S, Holt KE, Clements AC, Karkey A, Arjyal A, et al. (2011) Combined high-resolution genotyping and geospatial analysis reveals modes of endemic urban typhoid fever transmission. Open Biol 1: 110008.

18. Jacobsen A, Hendriksen RS, Aaresturp FM, Ussery DW, Friis C (2011) The Salmonella enterica pan-genome. Microb Ecol 62: 487–504.

19. Pandey A, Mann M (2000) Proteomics to study genes and genomes. Nature 405: 837–846.

20. Naaby-Hansen S, Waterfield MD, Cramer R (2001) Proteomics-post-genomic cartography to understand gene function. Trends Pharmacol Sci 22: 376–384.

21. Encheva V, Wait R, Begum S, Gharbia SE, Shah HN (2007) Protein expression diversity amongst serovars of Salmonella enterica. Microbiology 153: 4183–4193.

22. Steel LF, Haab BB, Hanash SM (2005) Methods of comparative proteomic profiling for disease diagnostics. J Chromatogr B Analyt Technol Biomed Life Sci 815: 275–284.

23. Ribot EM, Fair MA, Gautom R, Cameron DN, Hunter SB, et al. (2006) Standardization of pulsed-field gel electrophoresis protocols for the subtyping of Escherichia coli O157:H7, Salmonella, and Shigella for PulseNet. Foodborne Pathog Dis 3: 59–67.

24. Yuan J, Zhu L, Liu X, Li T, Zhang Y, et al. (2006) A proteome reference map and proteomic analysis of Bifidobacterium longum NCC2705. Mol Cell Proteomics 5: 1105–1118.

25. Ying T, Wang H, Li M, Wang J, Shi Z, et al. (2005) Immunoproteomics of outer membrane proteins and extracellular proteins of Shigella flexneri 2a 2457T. Proteomics 5: 4777–4793.

26. Zhang MJ, Zhao F, Xiao D, Gu YX, Meng FL, et al. (2009) Comparative proteomic analysis of passaged Helicobacter pylori. J Basic Microbiol 49: 482–490.

27. Confer AW, Ayalew S (2013) The OmpA family of proteins: roles in bacterial pathogenesis and immunity. Vet Microbiol 163: 207–222.

28. Jap BK, Walian PJ (1990) Biophysics of the structure and function of porins. Q Rev Biophys 23: 367–403.

29. Singh SP, Williams YU, Miller S, Nikaido H (2003) The C-terminal domain of Salmonella enterica serovar typhimurium OmpA is an immunodominant antigen in mice but appears to be only partially exposed on the bacterial cell surface. Infect Immun 71: 3937–3946.

30. Sugawara E, Nikaido H (1992) Pore-forming activity of OmpA protein of Escherichia coli. J Biol Chem 267: 2507–2511.

31. Sugawara E, Nikaido H (1994) OmpA protein of Escherichia coli outer membrane occurs in open and closed channel forms. J Biol Chem 269: 17981–17987.

32. O'Neal CR, Gabriel WM, Turk AK, Libby SJ, Fang FC, et al. (1994) RpoS is necessary for both the positive and negative regulation of starvation survival genes during phosphate, carbon, and nitrogen starvation in Salmonella typhimurium. J Bacteriol 176: 4610–4616.

33. Talukder AA, Yanai S, Nitta T, Kato A, Yamada M (1996) RpoS-dependent regulation of genes expressed at late stationary phase in Escherichia coli. FEBS Lett 386: 177–180.

34. Sevcik M, Sebkova A, Volf J, Rychlik I (2001) Transcription of arcA and rpoS during growth of Salmonella typhimurium under aerobic and microaerobic conditions. Microbiology 147: 701–708.

35. Ibanez-Ruiz M, Robbe-Saule V, Hermant D, Labrude S, Norel F (2000) Identification of RpoS (sigma(S))-regulated genes in Salmonella enterica serovar typhimurium. J Bacteriol 182: 5749–5756.

36. Khan AQ, Zhao L, Hirose K, Miyake M, Li T, et al. (1998) Salmonella typhi rpoS mutant is less cytotoxic than the parent strain but survives inside resting THP-1 macrophages. FEMS Microbiol Lett 161: 201–208.

37. Iida A, Harayama S, Iino T, Hazelbauer GL (1984) Molecular cloning and characterization of genes required for ribose transport and utilization in Escherichia coli K-12. J Bacteriol 158: 674–682.

38. Abou-Sabe M, Pilla J, Hazuda D, Ninfa A (1982) Evolution of the D-ribose operon on Escherichia coli B/r. J Bacteriol 150: 762–769.

39. Meysman P, Sanchez-Rodriguez A, Fu Q, Marchal K, Engelen K (2013) Expression divergence between Escherichia coli and Salmonella enterica serovar Typhimurium reflects their lifestyles. Mol Biol Evol 30: 1302–1314.

40. Leekitcharoenphon P, Lukjancenko O, Friis C, Aarestrup FM, Ussery DW (2012) Genomic variation in Salmonella enterica core genes for epidemiological typing. BMC Genomics 13: 88.

2

Fine Epitope Mapping of the Central Immunodominant Region of Nucleoprotein from Crimean-Congo Hemorrhagic Fever Virus (CCHFV)

Dongliang Liu[1], Yang Li[1], Jing Zhao[1], Fei Deng[2], Xiaomei Duan[1], Chun Kou[1], Ting Wu[1], Yijie Li[1], Yongxing Wang[1], Ji Ma[1], Jianhua Yang[1,4], Zhihong Hu[2], Fuchun Zhang[1], Yujiang Zhang[3]*, Surong Sun[1]*

1 Xinjiang Key Laboratory of Biological Resources and Genetic Engineering, College of Life Science and Technology, Xinjiang University, Urumqi, Xinjiang, China, 2 State Key Laboratory of Virology, Chinese Academy of Sciences, Wuhan, Hubei, China, 3 Center for Disease Control and Prevention of the Xinjiang Uyghur Autonomous Region, Urumqi, Xinjiang, China, 4 Texas Children's Cancer Center, Department of Pediatrics, Dan L. Duncan Cancer Center, Baylor College of Medicine, Houston, Texas, United States of America

Abstract

Crimean-Congo hemorrhagic fever (CCHF), a severe viral disease known to have occurred in over 30 countries and distinct regions, is caused by the tick-borne CCHF virus (CCHFV). Nucleocapsid protein (NP), which is encoded by the S gene, is the primary antigen detectable in infected cells. The goal of the present study was to map the minimal motifs of B-cell epitopes (BCEs) on NP. Five precise BCEs (E1, ^{247}FDEAKK^{252}; E2a, ^{254}VEAL^{257}; E2b, ^{258}NGYLNKH^{264}; E3, ^{267}EVDKA^{271}; and E4, ^{274}DSMITN^{279}) identified through the use of rabbit antiserum, and one BCE (E5, ^{258}NGYL^{261}) recognized using a mouse monoclonal antibody, were confirmed to be within the central region of NP and were partially represented among the predicted epitopes. Notably, the five BCEs identified using the rabbit sera were able to react with positive serum mixtures from five sheep which had been infected naturally with CCHFV. The multiple sequence alignment (MSA) revealed high conservation of the identified BCEs among ten CCHFV strains from different areas. Interestingly, the identified BCEs with only one residue variation can apparently be recognized by the positive sera of sheep naturally infected with CCHFV. Computer-generated three-dimensional structural models indicated that all the antigenic motifs are located on the surface of the NP stalk domain. This report represents the first identification and mapping of the minimal BCEs of CCHFV-NP along with an analysis of their primary and structural properties. Our identification of the minimal linear BCEs of CCHFV-NP may provide fundamental data for developing rapid diagnostic reagents and illuminating the pathogenic mechanism of CCHF.

Editor: Jens H. Kuhn, Division of Clinical Research, United States of America

Funding: This work was supported partly by Science and Technology Basic Work Program 2013FY113500 from Ministry of Science and Technology of China, the National Science Foundation of China (grant no. 81460303, 30860225), and the Open Research Fund Program of Xinjiang Key Laboratory of Biological Resources and Genetic Engineering (grant no. XJDX020I-2014-04). The funders had no role in study design, data collection and analysis, decision to publish, or preparation of the manuscript.

Competing Interests: The authors have declared that no competing interests exist.

* Email: xjsyzhang@163.com (YZ); sr_sun2005@163.com (SS)

Introduction

The Crimean-Congo hemorrhagic fever virus (CCHFV) is a human pathogenic agent that causes Crimean-Congo hemorrhagic fever (CCHF), a severe disease with case-fatality rates up to 30% [1–3]. CCHFV is broadly distributed across much of the Middle East, Africa, and Asia as well and has also been found in parts of Eastern Europe [4–6]. Humans are generally infected through tick bites, direct contact with blood or tissue of infected livestock, or through nosocomial infections [7–9]. In China, the first CCHF cases were reported in 1965 when the CCHFV strain BA66019 was isolated in a patient living in Bachu County of the Xinjiang Autonomous Region, which is now known to have the highest occurrences of CCHF in the country [10]. Despite the high mortality associated with CCHF, the biology and pathogenesis of the disease remain poorly understood for several reasons: CCHF outbreaks are sporadic and have been generally restricted to a relatively small number of cases, limited animal model development, and the handling of the infectious virus requires the highest

level of laboratory containment (BSL-4) [11]. Thus, early diagnosis and vaccine development are critical for both patient survival and for the prevention of potential nosocomial infection and transmission in China.

CCHFV belongs to the *Nairovirus* genus within the family Bunyaviridae [2,12]. The genome consists of three negative-stranded RNAs, designated as small (S), medium (M) and large (L) in accordance with their relative nucleotide length, and which encode the viral nucleocapsid protein (NP), the glycoprotein precursor (GP) and the putative RNA-dependent polymerase, respectively [13]. Studies have indicated that NP is the predominant protein which is present in high levels early after infection, thereby inducing a high immune response that can be detected in infected cells [14-17]. As a major protein primarily detected during the viral invasion phase, NP has been increasingly regarded as an important target of antivirus and clinical diagnosis [2]. In previous studies, complete NP expressed in bacteria has been used to detect CCHFV immunoglobulin G (IgG) and IgM antibodies; however, the instability of the protein has limited its application for routine use [18–20]. Thus there is a need to develop truncated NP or a multi-epitope peptide for CCHF diagnosis. In a prior study, Saijo et al. [21] reported that high titer sera of CCHF patients reacted only with amino acid residues 201 to 306 (NP$^{201-306}$) of the NP central fragment, a highly conserved region among various isolates. In our previous study, the NP region containing amino acid residues 237 to 305 (NP$^{237-305}$) was found to have remarkable reactivity both with a rabbit polyclonal antibody (pAb) against CCHFV-NP and with a mouse monoclonal antibody (mAb) 14B7 in Western blotting analysis [22].

Advances have made epitope mapping much easier today than it was before. Many approaches and technologies, including recombinant DNA [23], peptide synthesis [24], and peptide [25] or protein display [26] have highlighted the need for epitope mapping and raised the possibility of mapping to a sufficient level the epitopes of certain antigens of interest [27]. Biosynthetic peptide technology is often used to express several 15–25mer peptide segments covering a certain target protein to determine the presence of an antigenic region or regions for a mAb or pAb by the use of Western blotting. Epitope mapping can be subsequently performed with a set of synthetic overlapping 8mer peptides for the positive segment(s) detected by immunoblotting [28–30]. Herein, based on the findings of a previous study, we describe the fine epitope mapping of immunodominant region NP$^{237-305}$ of the CCHFV using an improved biosynthetic peptide method [28,29].

In this paper, a total of six overlapping 16–22mer peptides (Y1–Y6) and forty-one 8mer peptides (P1–P41), both fused with a truncated carrier protein, were biosynthesized and expressed for minimal epitope mapping of the antigenic properties of NP$^{237-305}$. Five potential pAb BCEs and one potential mAb BCE were identified and mapped on the stalk region of CCHFV-NP for the first time.

Materials and Methods

Ethics Statement

The study was approved by the Research Ethics Committee (Animal Ethics Committee of Xinjiang University) and the procedures that followed were in accordance with the policies and regulations of experimental animals of China. The field studies in Bachu County were permitted by Xinjiang Wildlife Conservation Association (XJWCA). The serum samples were collected using method of random sampling and this process was not involving sacrifice.

Plasmids, Antibodies and Strains

The plasmids pGEX-KG and pXXGST-1 [28] were used to express biosynthetic peptides. The prokaryotic expression plasmid pGEX-KG was maintained by the Xinjiang Key Laboratory of Biological Resources and Genetic Engineering, and pXXGST-1 was donated by Professor Wanxiang Xu of the Shanghai Institute of Planned Parenthood Research. Rabbit polyclonal antibody against CCHFV-NP (pAb) was prepared as previously described [31]. A mouse monoclonal antibody cell line (14B7) secreting IgM type monoclonal antibody 14B7 against CCHFV was obtained from Xinjiang Centers for Disease Control and Prevention (XJCDC). A high titer of mAb was separated from mice ascites [32]. A pooled sheep serum of five samples collected from Bachu County with a confirmed history of CCHFV infection was included in the study and used for reconfirming the antigenicity of identified BCEs of CCHFV-NP in Western blotting assay. Serum sample of one healthy sheep with no history of CCHFV infection was used as negative control. All the sheep sera used in

the study were collected in 2005 and kindly provided by Professor Zhang Yujiang of XJCDC [33]. The serum samples of sheep infected with CCHFV were previously identified by using indirect immunofluorescent assay (IFA) and reverse transcription polymerase chain reaction (RT-PCR) [33]. *Escherichia coli* (*E. coli*) BL21 (DE3) competent cells, used for expression of recombinant plasmids, were purchased from Beijing TransGen Biotech Co., Ltd.

Epitope Prediction

To predict the B-cell epitopes (BCEs) on the NP$^{237-305}$ fragment of CCHFV, the corresponding amino acid sequence was analyzed using the DNAStar Protean system. Secondary structure prediction of the truncated protein was performed by using the methods of Garnier and Robson [34] and Chou and Fasman [35]. The surface properties of the structural proteins, namely, hydrophilicity, flexibility, accessibility and antigenicity, were analyzed using the methods of Kyte and Doolittle [36], Karplus and Schulz [37], Emini [38] and Jameson and Wolf [39], respectively. According to the results obtained using these methods, peptides with good hydrophilicity, high accessibility, high flexibility and strong antigenicity were selected as epitope candidates. In general, peptides located in α-spiral and β-sheet regions, which do not readily form epitope regions, were excluded [40].

Biosynthesis of 8-22mer Peptides and Recombinant Plasmid Construction

Six biosynthetic 16–22mer peptides (designated Y1–Y6) spanning the NP$^{237-305}$ segment and overlapping 6 ~ 9 amino acid residues each other, which all fused with GST or a truncated GST188 carrier protein were expressed in *E. coli*, respectively [30]. The sequences of the peptides were as follows: Y1 (KLAETEGKGVFDEAKKTVEA), Y2 (AKKTVEALNGYLN-KHK), Y3 (YLNKHKDEVDKASADSM), Y4 (DKASADSMI-TNLLKHI), Y5 (TNLLKHIAKAQELYK) and Y6 (IAKAQE-LYKNSSALRAQGAQID), which correspond to NP$^{237-256}$, NP$^{250-265}$, NP$^{260-276}$, NP$^{269-284}$, NP$^{277-292}$ and NP$^{284-305}$, respectively. For the immunodominant peptides identified (Y1–Y4), four additional sets of 8mer peptides spanning the Y1 to Y4 fragments overlapping 7 residues each other were generated (totaling 41 biosynthetic peptides designated P1 to P41) for fine epitope mapping (Fig. 1c). Briefly, the synthesized DNA fragments encoding the Y1–Y6 and P1–P41 peptides (based on the S gene sequence [41] were flanked by *Bam*H I and TTA-*Sal* I sites at the 5′ and 3′ ends, respectively, then inserted into the *Bam*H I and *Sal* I sites downstream of the GST or GST188 encoding gene in the pGEX-KG or pXXGST-1 plasmid.

Expression of Fusion Proteins

The resultant recombinant plasmids expressing each 8–22mer peptide fused with GST or GST188 were transformed into *E. coli* BL21 (DE3) competent cells. Each recombinant clone was cultivated in 3 mL LB medium containing 100 μg/mL ampicillin at 30°C with continuous shaking at 200 rpm overnight. The next day, 30 μL of cell suspension was added to 3 mL fresh LB medium and grown for 4 h until reaching a bacterial density of 0.6–0.8 at OD600. The cells were grown for an additional 4 h with 0.8 mM IPTG (Y1, Y3 and Y6 fusion peptides) or without IPTG (all other fusion peptides) at 42°C to induce the expression of the recombinant proteins. For the screening of positive recombinant clones, an SDS-PAGE gel was run for each harvested cell pellet, with the pellet corresponding to GST or GST188 protein

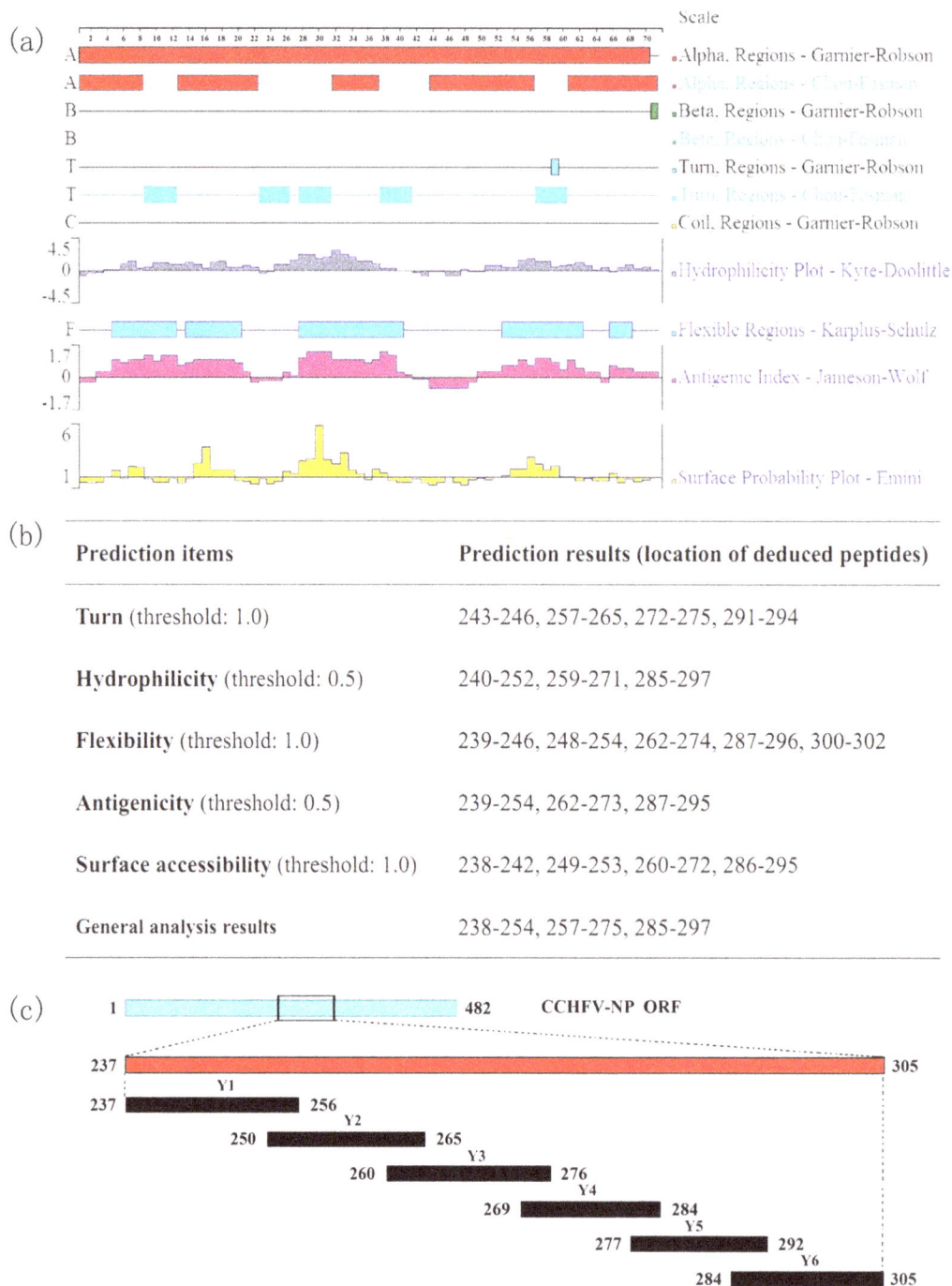

Figure 1. Prediction and mapping strategy of the epitopes in the central region of CCHFV-NP. (a) Epitope prediction for amino acid residues 237–305 of the NP sequence of the YL04057 strain using DNAStar Protean software. The secondary structure, flexibility plot, hydrophilicity, surface probability, and antigenicity index for NP$^{237-305}$ were taken into consideration. (b) Epitope predictions of the NP$^{237-305}$ fragment of the YL04057 strain based on various principles. (c) Schematic of the BCE mapping strategy with six 16–22mer overlapping biosynthetic peptides spanning the NP$^{237-305}$ fragment. The blue band represents the full length nucleoprotein. The red band represents the immunodominant fragments of YL04057 NP.

expressed by pGEX-KG or pXXGST-1 as a negative control. All recombinant clones were subsequently sent out for sequencing determination. The cell pellets containing the short peptide fusion proteins were stored at −20°C.

SDS-PAGE and Western Blotting

The cell pellets obtained from 3 mL medium were boiled in 400 μL of 1×SDS-PAGE loading buffer for 10 min, and the proteins were resolved by SDS-PAGE under reducing conditions using 15% gels [42]. Gels were either stained with Coomassie brilliant blue R-250 to analyze the bands corresponding to the fusion proteins or processed for Western blotting by electrotransfer of the proteins onto a 0.2 μm nitrocellulose membrane (Whatman GmbH, Dossel, Germany) [43]. Complete transfer of proteins was ensured by staining the nitrocellulose membrane with 0.1% (w/v) Ponceau S dye liquor. After cleaning and blocking, the nitrocellulose membrane was subsequently treated with pAb (1:1000 dilution in PBS containing 0.05% Tween 20 and 1% skim milk powder), mAb 14B7 (1:500 dilution) or a pooled serum (1:100 dilution) collected from sheep with confirmed CCHFV infection. A serum sample from a known healthy sheep with no history of CCHFV infection was used as a negative control. Specific antigen-antibody reactions on the membrane were visualized using goat anti-rabbit IgG, goat anti-mouse IgM, or rabbit anti-sheep IgG conjugated to horseradish peroxidase (HRP) (Proteintech Group, Chicago, USA) at a 1:1000 dilution. The blot was performed using ECL plus Western blotting detection reagent (GE Healthcare, Buckinghamshire, UK) according to the manufacturer's instructions.

Sequence Conservation Analysis and Three-Dimensional Modeling

To assess the sequence conservation of the identified epitopes, nine NP amino acid sequences collected from strains in different countries were obtained from GenBank. Amino acid residues 170–305 of the NP sequences from the nine virus strains were selected for multiple alignment analysis against the corresponding sequence of the YL04057 strain (GenBank code: ACM78470.1) using the ClustalW program (http://www.ebi.ac.uk/services) [44].

Three-dimensional structures of the immunodominant epitopes identified using pAb and 14B7 were simulated using PyMOL™ software [45].

Results

BCE Prediction and Mapping Strategy

Accessibility, variability, fragment mobility, charge distribution and hydrophilicity are important features of antigenic epitopes. The presence of flexible regions, such as coil and turn regions, provide further evidence for epitope identification. In this study, the secondary structure of $NP^{237-305}$ was predicted using the methods of Garnier and Robson [34] and Chou and Fasman [35] and the NP gene sequence of YL04057 CCHFV. A hydrophilicity plot, flexibility plot, surface probability plot and antigenic index for the truncated protein were obtained using the methods of Kyte and Doolittle [36], Karplus and Schulz [37], Emini [38] and Jameson and Wolf [39], respectively (Fig. 1a). The potential BCEs on $NP^{237-305}$ were predicted (Fig. 1b) based on the methods mentioned above. The finding that the secondary structure of the $NP^{237-305}$ fragment consists of five main turn motifs suggested the presence of multiple significant BCEs in this region. In our previous study, the $NP^{237-305}$ truncated fragment was found to exhibit remarkable antigen-antibody reactivity when either pAb or mAb was used in Western blotting analysis [22]. However, other unpredicted amino acids in this same region should also be considered because they may also contain BCEs, some of which may be predominant BCEs. To identify how many epitopes there are in the fragment of NP, we therefore designed a feasible strategy for BCE mapping of the $NP^{237-305}$ (Fig. 1c). Briefly, six truncated polypeptides (Y1–Y6) spanning $NP^{237-305}$ were incorporated into prokaryotic expression plasmids. Based on the results of the Western blotting analysis, sets of 8mer peptides were constructed for each of the immunodominant polypeptides identified for further BCE mapping.

Mapping Epitopes on CCHFV-NP Using pAb

All 16–22mer Peptides fused with a GST or GST188 carrier were expressed through constructing short peptide fusion expression plasmids using each synthesized encoding DNA fragments [28]. To define the fine epitopes on the $NP^{237-305}$ fragment of CCHFV, epitope mapping was performed in two steps. For the first round of antigenic peptide mapping, $NP^{237-305}$ was divided into six overlapping fragments (Y1–Y6), which were fused with GST/GST188 and expressed in E. coli, respectively. As determined by SDS-PAGE, bands corresponding to the GST-fused proteins (Lane 2, 4 and 7) were approximately 33 kD, and those corresponding to the GST188-fused proteins (Lane 3, 5 and 6) were approximately 25 kD (Fig. 2a). Western blot analysis showed that pAb reacted with polypeptides Y1–Y4 (Fig. 2b).

To further map the epitopes on $NP^{237-305}$, four sets of 8mer peptides spanning Y1 to Y4 were constructed, which have an overlap of seven amino acid residues each other in second round of fine epitope mapping. A total of 41 recombinant 8mer peptides (designated P1 to P41) were constructed and expressed in E. coli (Fig. 3). Among the 13 recombinant clones corresponding to Y1 (Fig. 3a), Western blot analysis showed that 8mer peptides P9 (GVFDEAKK), P10 (VFDEAKKT) and P11 (FDEAKKTV) were recognized by pAb against NP, suggesting that the epitope minimal motif within Y1 was the FDEAKK (named as epitope 1, E1) according to their shared residues number (Fig. 4). Three antigenic peptides Y2–Y4 were similarly identified and analyzed (Fig. 3 and 4): the fine epitopes were the VEAL (E2a) and NGYLNKH (E2b) in Y2, EVDKA (E3) in Y3 and DSMITN (E4) in Y4, respectively. Thus, five specific BCE motifs within the $NP^{237-305}$ segment were found using rabbit pAb to NP.

Epitope Mapping Using mAb 14B7 Against CCHFV-NP

In our previous study, $NP^{237-305}$ was also found to exhibit antigen-antibody reactivity with 14B7. To reveal its antibody-reactive epitope motif, using same strategy described above to map its fine epitope motif. That is, the mAb 14B7 was identified to recognize antigenic peptide Y2 in the first round of mapping (Fig. 2c) and then its epitope motif was confirm as NGYL (designated as E5, amino acid residues 258–261) in the second round of fine mapping (Fig. 5). Interestingly, its epitope motif was located in the E2b identified by the rabbit pAb, suggesting the diversity of antibody production in mouse and rabbit.

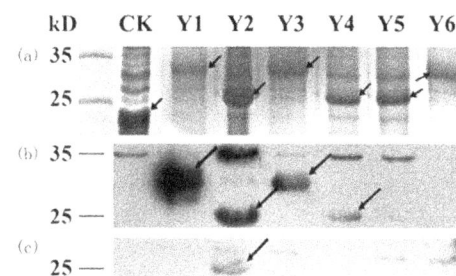

Figure 2. Prokaryotic expression and immunoblotting analysis of Y1–Y6 fused proteins. (a) SDS-PAGE analysis of expressed pXXGST-1 (CK) and Y1-Y6 peptides fused with a GST (Y1, Y3 and Y6) or GST188 tag (Y2, Y4 and Y5). (b) Western blotting of fusion proteins Y1–Y6 using the rabbit polyclonal antibody against CCHFV-NP. (c) Western blotting of fusion proteins Y1–Y6 using the mouse IgM-type monoclonal antibody 14B7 against CCHFV. The arrows represent expressed target peptides in SDS-PAGE and the corresponding positive antigenic-peptides in Western Blotting analysis.

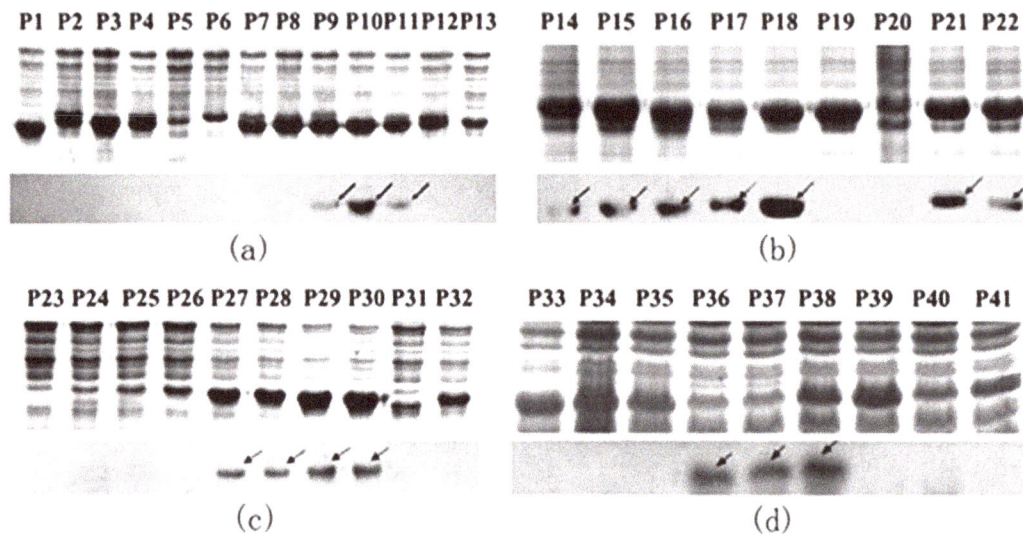

Figure 3. SDS-PAGE identification and Western blotting analysis of the minimal epitopes on NP$^{237-305}$ using pAb. (a) Thirteen 8mer peptides (P1–P13) corresponding to the Y1 protein. (**b**) Nine 8mer peptides (P14–P22) corresponding to the Y2 protein. (**c**) Ten 8mer peptides (P23–P32) corresponding to the Y3 protein. (**d**) Nine 8mer peptides (P33–P41) corresponding to the Y4 protein. The arrows stand for 8mer peptides which display a positive antigen-antibody reaction in Western Blotting analysis.

Determination of the Antigenicity of Identified BCEs by CCHFV Antibody-positive Sheep Sera

To determine whether the BCEs identified are rabbit/mouse specific or also recognizable by the immune systems of other host species, five randomly selected 8-mer peptides, each of which containing one of the five pAb-identified BCEs, were carried out Western blot test by using sheep sera with or without CCHFV infection (Fig. 6a). As shown in this study, the serum samples of sheep with a confirmed history of CCHFV infection could react with all the five 8-mer peptides to varying degrees, while the

Peptide items	Amino acids	Position in NP	Peptide items	Amino acids	Position in NP
P1	KLAETEGK	237-244	P23	YLNKHKDE	260-267
P2	LAETEGKG	238-245	P24	LNKHKDEV	261-268
P3	AETEGKGV	239-246	P25	NKHKDEVD	262-269
P4	ETEGKGVF	240-247	P26	KHKDEVDK	263-270
P5	TEGKGVFD	241-248	P27	HKDEVDKA	264-271
P6	EGKGVFDE	242-249	P28	KDEVDKAS	265-272
P7	GKGVFDEA	243-250	P29	DEVDKASA	266-273
P8	KGVFDEAK	244-251	P30	EVDKASAD	267-274
P9	GVFDEAKK	245-252	P31	VDKASADS	268-275
P10	VFDEAKKT	246-253	P32	DKASADSM	269-276
P11	FDEAKKTV	247-254			
P12	DEAKKTVE	248-255			
P13	EAKKTVEA	249-256			
P14	AKKTVEAL	250-257	P33	DKASADSM	269-276
P15	KKTVEALN	251-258	P34	KASADSMI	270-277
P16	KTVEALNG	252-259	P35	ASADSMIT	271-278
P17	TVEALNGY	253-260	P36	SADSMITN	272-279
P18	VEALNGYL	254-261	P37	ADSMITNL	273-280
P19	EALNGYLN	255-262	P38	DSMITNLL	274-281
P20	ALNGYLNK	256-263	P39	SMITNLLK	275-282
P21	LNGYLNKH	257-264	P40	MITNLLKH	276-283
P22	NGYLNKHK	258-265	P41	ITNLLKHI	277-284

E1 (P9–P11); E2a (P14–P18); E2b (P21–P22); E3 (P27–P30); E4 (P36–P38); E5 (P18–P22)

Figure 4. The synthetic 8mer peptide sequences derived from a span of the immunodominant peptides Y1, Y2, Y3, and Y4 respectively. The yellow and magenta highlighting represents the common sequences among peptides which react with pAb or mAb using Western blotting analysis.

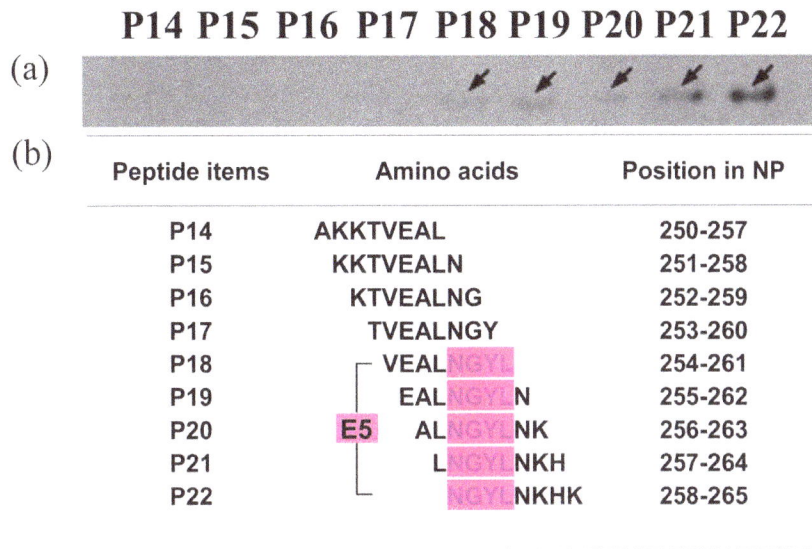

Figure 5. Minimal epitope identification on CCHFV-NP using mAb 14B7. (a) A reactivity profile of the 8mer peptides P14–P22 corresponding to Y2 using Western blotting analysis. (b) Sequences of the 8mer peptides and their positions in NP. Magenta highlighting indicates the common sequence identified as the minimal BCE (E5) on NP using 14B7. The arrows represent 8mer peptides which display positive antigen-antibody reactions using Western Blotting analysis.

CCHFV antibody-negative sera could not react with anyone of them. Of the five peptides, P10 (containing E1) and P18 (containing E2a) showed the strongest antigen-antibody reaction activities with CCHFV-infected sheep sera; Meanwhile, P29 (containing E3) and P38 (containing E4) displayed the weakest reaction intensity among the five 8-mer peptides.

Sequence Conservation Analysis and Three-Dimensional Modeling

To analyze primary structural properties of identified each BCE, the sequence corresponding to amino acid residues 170–305, which contains the identified BCEs and flanking sequences, was used to conduct multiple sequence alignment (MSA) (Fig. 7). The analysis revealed that this region (using the sequence from Chinese strain YL04057, ACM78470.1) is highly conserved when compared to the corresponding region from other CCHFV strains,

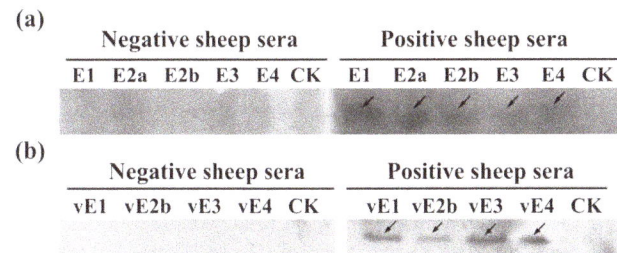

Figure 6. Western blot of five 8-mer peptides containing identified BCEs with or without one residue variation performed using positive sheep sera with a confirmed history of CCHFV infection. Five randomly selected 8-mer peptides containing identified BCEs (a) or BCEs with one residue variation (b) expressed as GST188 fusion protein in *E. coli*. A serum sample of healthy sheep with no history of CCHFV infection was used as a negative control. CK was a GST188 protein tag. The arrows represent 8mer peptides displaying positive antigen-antibody reactions based upon Western Blotting analysis.

with 90.4% sequence identity. Notably, an even higher sequence identity (92.8%) was found for $NP^{237-305}$ compared to the other nine strains. The five BCEs (E5 epitope motif was included within E2b) identified were also found to be highly conserved. Only single amino acid differences were found for four BCEs (E1, E2b, E3 and E4) compared with the other strains. To further determine whether the epitope peptides with single residue difference revealed in Fig. 7 could be used as a or all universal diagnostic reagent(s), we prepared four biosynthetic peptides (vE1, FDEAKR; vE2b, NGYLDKH; vE3, EVDRA; vE4, DNMITN) using methods mentioned above and explored their antigenic properties. Specifically, a single amino acid substitution was made within E1 (K252R), E2b (N262D), E3 (K270R), and E4 (S275N). Our investigation demonstrated that the four biosynthetic peptides with one variable residue (vE1, vE2a, vE3, vE4) can remarkably react with CCHFV antibody-positive sheep sera compared with the negative sheep sera panel (Fig. 6b), suggesting that BCEs E1–E4 derived from different CCHFV strains shared conservation in antigenicity aspect. Intriguingly, we found the E2a epitope (VEAL) and E5 epitope (NGYL) identified using mAb 14B7 were highly conserved in a majority of CCHFV isolates (Fig. 7).

Computer modeling using PyMOL™ software indicated that all of the antigenic motifs are located on the stalk domain of CCHFV-NP (Fig. 8a). Furthermore, the five BCEs are located in a flexible "helix-turn-helix" (HTH) structure (Fig. 8b). According to the surface representations (Fig. 8c and 8d), all of the identified BCEs are located on the surface of NP, which is consistent with the antigenic principles of surface accessibility and hydrophilicity.

Discussion

Nucleocapsid research is an important branch of viral study, as virus nucleocapsids may stimulate human immune responses, most of which are of the humoral immunity type [46]. Thus, the identification and mapping of minimal BCEs on NP represent significant steps in the development of novel diagnostic tools and multi-epitope peptide vaccines. In a previous study, a bacterially expressed recombinant NP antigen was used to detect IgG

```
ACM78470.1      (China)     LSDMIRRRNLILNRGGDENPRGPVSREHVEWCREFVKGKYIMAFNPPPWGDINKSGRSGIA    229
AAL28095.2   (Yugoslavia)   ............................................................    229
ABB72472.1      (Russia)    ............................................................    229
AAP46054.2     (Bulgaria)   ...................S...................H....................    229
AAZ38665.1(South Africa)    ...............Q......................................D.....    229
ADD64468.1      (China)     ...............................S............................    229
AAQ23152.2   (Tajikistan)   ......................F.....................................    229
AE072054.1      (India)     ............................................................    229
AAK40124.1      (China)     ..................S.........................................    229
ABD98123.1       (Iran)     ..................K.........................................    229
```

```
ACM78470.1      (China)     LVATGLAKLAETEGKGVFDEAKKTVEALNGYLNKHKDEVDKASADSMITNLLKHIAKAQE    289
AAL28095.2   (Yugoslavia)   ................FDEAKK.VEALNGYLDKHR.EVDKA..DSMITN...........    289
ABB72472.1      (Russia)    ................FDEAKK.VEALNGYLDKHR.EVDKA..DSMITN...........    289
AAP46054.2     (Bulgaria)   ................FDEAKK.VEALNGYLDKHR.EVDKA..DSMITN...........    289
AAZ38665.1(South Africa)    ................FDEAKK.VEALNGYLNKH..EVDKA..DSMITN...........    289
ADD64468.1      (China)     ................FDEAKK.VEALNGYLNKH..EVDKA..DSMITN...........    289
AAQ23152.2   (Tajikistan)   ................FDEAKR.VEALNGYLDKH..EVDKA..DNMITN...........    289
AE072054.1      (India)     .I..............FDEAKK.VEALNGYLDKH..EVDKA..DNMITN...........    289
AAK40124.1      (China)     ...............G....FDEAKK.VEALNGYLDKH..EVDKA..DNMITN.......    289
ABD98123.1       (Iran)     ................FDEAKK.VEALNGYLDKH..EVDKA..DSMITN...........    289
                                            E1       E2a  E2b       E3    E4
```

```
ACM78470.1      (China)     LYKNSSALRAQGAQID    305
AAL28095.2   (Yugoslavia)   ................    305
ABB72472.1      (Russia)    ................    305
AAP46054.2     (Bulgaria)   ................    305
AAZ38665.1(South Africa)    ................    305
ADD64468.1      (China)     ................    305
AAQ23152.2   (Tajikistan)   ................    305
AE072054.1      (India)     ................    305
AAK40124.1      (China)     ................    305
ABD98123.1       (Iran)     ................    305
```

Figure 7. Amino acid sequence comparison of the NP[170−305] fragment from the YL04057 strain (ACM78470. 1) and other CCHFV strains using the ClustalW program. The GenBank codes and sources are shown at left. The five minimal epitopes E1, E2a, E2b, E3, and E4 recognized by pAb are highlighted in yellow, and the variable amino acids within the minimal epitopes are highlighted in red. Dots (.) indicate identical amino acids within the ten strains.

antibodies against CCHFV; the instability however, of the protein in soluble expression as well as serological diagnosis restricted the application of this protein [47,48]. The use of non-complete NP or multi-epitope peptides for CCHF diagnosis has attracted increasing attention, along with CCHF studies in general. At the same time, there have been increased efforts related to the epitope mapping of CCHFV-NP. For instance, Saijo et al. reported that in Western blotting analysis, high titer sera of CCHF patients reacted only with the highly conserved NP fragment which contained the amino acid residues 201 to 306 (NP[201−306]) [21]. Similarly, Burt et al. found that NP[123−396] of CCHFV includes a highly antigenic region with application toward the development of antibody detection assays [48]. Previously, our group showed that NP[237−305] is an immunogenic region of CCHFV-NP using a polyclonal antibody and two monoclonal antibodies against CCHFV with Western blot analysis [22]. It is worth noting that the NP[237−305] region is smaller and more detailed and completely encompassed by the NP[201−306] and NP[123−396] regions. The consistent finding from these independent research groups suggests that the high antigenicity region of CCHFV-NP is located in the central region of NP rather than the N- or C-terminal regions. Although several antigenic peptides have been mapped on CCHFV-NP, to our knowledge, no minimal motifs have been previously identified, due to methodological limitations.

The biosynthetic peptide method has been successfully used by several research groups to identify the minimal epitopes on human zona pellucida protein [28–30]. In the present study, we used two prokaryotic plasmids for expressing 8–22mer peptides fused with a GST or GST188 tag to avoid the influence of different expression systems on the stability and antigenicity of the recombinant peptides generated. The simplicity, cost effectiveness, reliability, and adaptability of this approach are highly suitable for minimal motif identification [28]. Herein, we demonstrate the use of this method in mapping the minimal motifs of the BCEs of CCHFV-NP. Thus, this methodology may accelerate research requiring the minimal motif mapping of known viral antigenic epitope fragments. In the present study, we mapped six minimal BCEs on NP, five of which were identified by pAb (E1, [247]FDEAKK[252]; E2a, [254]VEAL[257]; E2b, [258]NGYLNKH[264]; E3, [267]EVDKA[271]; and E4, [274]DSMITN[279]) and one by mAb (E5, [258]NGYL[261]). Herein, the antigenicity of the five pAb-identified BCEs was reconfirmed by utilizing natural sera from the sheep with CCHFV infection history, indicating that the identified BCEs may have significant potential in acting as a diagnostic tool to identify whether certain wild animals or even human beings were infected by CCHFV in natural conditions. Additionally, four of the six BCEs were identified in our antigen prediction analysis (Fig. 1b), demonstrating that the epitope prediction tool combined with the biosynthetic peptide method is a reliable approach for epitope mapping and may reduce the experimental effort and expense of identifying and mapping epitopes for immunodiagnostics.

The five minimal BCEs found on NP[237−305] span amino acid residues 247 to 279 of CCHFV-NP, and all of them were found to have high sequence similarity among different CCHFV strains according to MSA analysis (88.57% for E2b, 95% for E4, 98% for E3, 98.33% for E1, and 100% for E2a and E5) (Fig. 7). To give specifics, the lysine252 was replaced by arginine within E1 (K252R) in one strain. In certain strains, a single amino acid

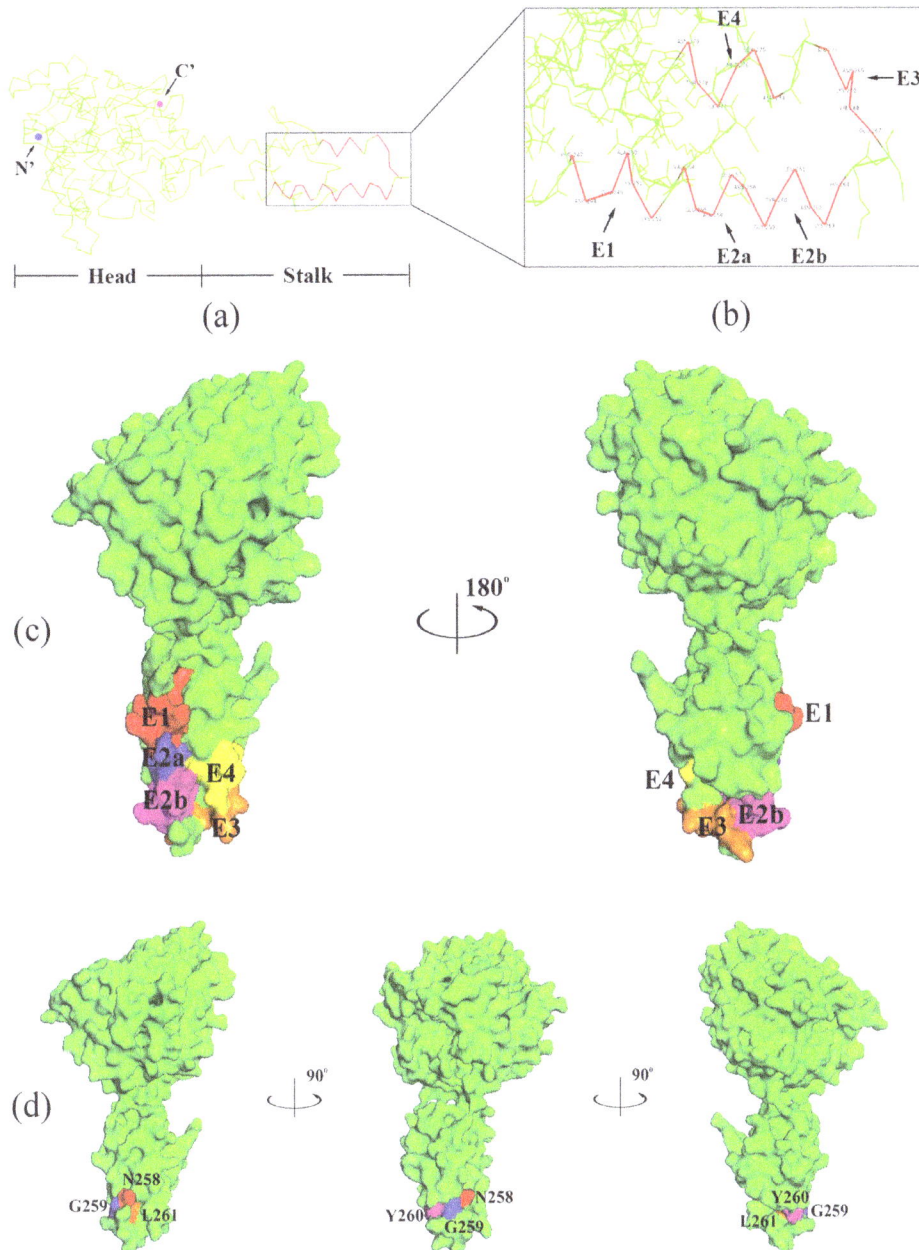

Figure 8. Location and three-dimensional structure of the epitopes identified using pAb and mAb 14B7 on CCHFV-NP. (a) The ribbon diagram shows the overall secondary structure of CCHFV-NP from strain YL04057 (PDB code: 3U3I). The motifs within the frame indicate the five minimal epitopes E1–E4. **(b)** E1–E4 sites on the CCHFV-NP stalk domain. **(c)** Surface properties of CCHFV-NP. The molecular surfaces of E1 (red), E2a (blue), E2b (magenta), E3 (orange) and E4 (yellow) are shown. **(d)** The structural representations show the location and spatial conformation of epitope E5 (tetrapeptide "NGYL") identified by Amb 14B7. Residues N258, G259, Y260, and L261 are shown in different colors. The figures were generated using the PyMOL molecular graphics system.

substitution was also found within E2b (N262D), E3 (K270R), and E4 (S275N). Despite one residue difference, the antigen-antibody reaction was still obvious when using positive sera of sheep naturally infected with CCHFV (Fig. 6b), reflecting highly antigenic conservation. However, ideally, the sera of CCHF patients should be utilized to verify the conservation and specificity of BCEs, which is crucial for future applications in CCHF diagnosis and prevention. In this study, we only provided the fundamental data that the antigenicity of CCHFV-NP was researched using rabbit polyclonal antiserum against the CCHFV-NP, using the mouse monoclonal antibody against the CCHFV, and using the sera of sheep naturally infected with CCHFV. Thereby, the properties, structure, antigenicity, and immunogenicity of the NP protein, and in particular, identification of human sera infected with CCHFV will be further studied in order to be applied to CCHF diagnosis and therapy in the future. To our knowledge, the ten CCHFV strains used were isolated in countries which were directly affected by at least one of the five CCHF-epidemic areas, namely countries in Europe, Africa, Central Asia, South Asia, and the Middle East (Table S1). As depicted, epitopes E2a and E5 showed 100% conservative properties among all strains from the five CCHF-epidemic areas.

Epitope E1 was fully conserved in the countries of Europe, Africa, South Asia, the Middle East, and part of Central Asia. Similarly, epitope E3 also displayed complete homology in the different CCHF-epidemic areas, with the exception of Africa. It is worth noting that the epitope E2b from isolate YL04057, though it was not consistent with the other eight strains from different CCHF-epidemic areas, had merely one amino acid difference among ten CCHFV strains. As far as we know, eleven complete NP sequences of CCHFV strains isolated within China have been registered in the GenBank database. To further confirm whether the epitope E2b (^{258}NGYLNKH^{264}) showed high homology among strains from China, the sequences of the eleven Chinese strains corresponding to amino acid residues 241 to 300 of NP were retrieved from the GenBank for sequence alignment using the ClustalW program (Figure S1). Our study indicated that there was only a single difference, which of asparagine changing to aspartic acid (N262D) within epitope E2b in four strains of the eleven, suggesting that asparagine262 may well exist only in the Chinese CCHFV strains.

It has been previously reported that the Dugbe virus, another member of the *Nairovirus* genus, shares some antigenic and genetic properties with CCHFV [49]. Based on our findings, however, the amino acid sequences of NP$^{247-279}$ of the two viruses (GenBank codes: ACM78470.1 and AAL73396.1) display only a 9.1% similarity, despite a 57.4% sequence similarity between the two complete NP sequences (data not given). These findings suggest that the identified BCEs may be unique to CCHFV and thus highly species specific.

To further investigate the structural aspects of the minimal BCEs, the three-dimensional structure of YL04057 NP was retrieved from the Protein Databank (PDB code: 3U3I). The five pAb BCEs and one mAb BCE were all found to be located on the NP stalk domain (BCE surface properties shown in different colors in Fig. 8c and 8d). Structural analyses, particularly of the surface structure of epitopes, provide a good foundation in the search for and creation of structurally complementary drugs with clinical applications. The flexible "helix-turn-helix" structure containing the five BCEs may form a discontinuous epitope and easily react with antibodies or drugs. Human MxA protein has been shown to inhibit the CCHFV replication process [50]. In a protein-protein docking study of MxA with CCHFV-NP emphasizing epitope-based immunoinformatics, Srinivasan et al. [51] showed a complementary wrapping of the NP stalk around the MxA model. Together with the present findings, these results suggest that the CCHFV-NP stalk domain may play a critical role in immune system processes and virus interaction. Intrabodies, or intracellular antibodies, are powerful tools for cell biology studies as well as therapeutic applications [52]. They are commonly used to either block the intracellular antibody target or to image endogenous target dynamics [53,54]. It is reported that intrabodies induced cell death via activation of the caspase-3-mediated apoptotic pathway [55]. In a related vein, recent structural studies of CCHFV-NP revealed that the amino acid residues DEVD at positions 266 to 269 on the NP stalk domain comprise a putative cleavage site of caspase-3, indicating that caspase-3 cleavage of NP

may represent a host defense mechanism against lytic CCHFV infection [56–59]. Interestingly, the occurrence of the DEVD motif is within the epitope-rich region of amino acid residues 247 to 279. Our study also raises the possibility of a combination of caspase-3-dependent apoptosis and intrabody therapy in fighting a CCHFV infection in the future.

In the present study, we identified five fine linear BCEs on the stalk region of CCHFV-NP using a peptide biosynthesis strategy, thereby demonstrating the utility of this approach in peptide-based assays aimed at antibody detection. However, it remains a topic of further research whether the antigenic activities, consisting of specificity and sensitivity, can be enhanced by linearly fusing the five BCEs so that they would be more easily and more effectively applied in clinical diagnosis and epidemiological investigation.

Conclusion

In this study, the five highly conserved or 100% conserved B-cell epitopes E1, E2a, E2b (which fully overlaps E5), E3 and E4 do not only react with a prepared polyclonal antibody, but also with the positive sera of sheep naturally infected with CCHFV. More importantly, the four epitope mutants vE1, vE2b, vE3, and vE4 are distinctly recognizable through the use of naturally infected sheep sera. Our discovery has demonstrated a high antigenic-conservation of these identified minimal epitopes, which might be useable as universal epitopes in CCHF diagnosis. It is of great importance that human sera infected with CCHFV be used to test these identified epitopes in future study. Furthermore, these BCEs were determined to be located on the surface of the NP stalk region, suggesting they very well may play significant roles in the process of interaction with the host immune system, being easily recognized by antibodies. These findings would provide fundamental data for the development of novel diagnostic reagents and the illumination of the pathogenic mechanism of CCHFV.

Acknowledgments

We are very grateful to Professor Wanxiang Xu for his technical support and proofreading work.

Author Contributions

Conceived and designed the experiments: SRS DLL YJZ FCZ. Performed the experiments: DLL YL JZ XMD CK TW. Analyzed the data: SRS DLL YJL JM YXW. Contributed reagents/materials/analysis tools: DLL SRS YJZ ZHH FD. Wrote the paper: DLL SRS JHY YJZ ZHH.

References

1. Elliott RM (1990) Molecular biology of the Bunyaviridae. J Gen Virol 71: 501–522.
2. Ergönül Ö (2006) Crimean-Congo haemorrhagic fever. Lancet Infect Dis 6: 203–214.
3. World Health Organization (2001) Crimean-Congo haemorrhagic fever. Fact sheet 208.
4. Deyde VM, Khristova ML, Rollin PE, Ksiazek TG, Nichol ST (2006) Crimean-Congo hemorrhagic fever virus genomics and global diversity. J Virol 80: 8834–8842.
5. Hoogstraal H (1979) The epidemiology of tick-borne Crimean-Congo hemorrhagic fever in Asia, Europe, and Africa. J Med Entomol 15: 307–417.
6. Maltezou HC, Papa A (2010) Crimean-Congo hemorrhagic fever: Risk for emergence of new endemic foci in Europe? Travel Med Infect Di 8: 139–143.

7. Swanepoel R, Shepherd AJ, Leman PA, Shepherd SP, McGillivray GM, et al. (1987) Epidemiologic and clinical features of Crimean-Congo hemorrhagic fever in southern Africa. Am J Trop Med Hyg 36: 120–132.

8. Chinikar S (2009) An overview of Crimean-Congo hemorrhagic fever in Iran. Iran. J Microbiol 1: 7–12.

9. Gürbüz Y, Sencan I, Öztürk B, Tütüncü E (2009) A case of nosocomial transmission of Crimean–Congo hemorrhagic fever from patient to patient. Int J Infect Dis 13: e105–e107.

10. Sun SR, Dai X, Aishan M, Wang XH, Meng WW, et al. (2009). J Clin Microbiol 47: 2536–2543.

11. Bergeron É, Albariño CG, Khristova ML, Nichol ST (2010) Crimean-Congo hemorrhagic fever virus-encoded ovarian tumor protease activity is dispensable for virus RNA polymerase function. J Virol 84: 216–226.

12. Whitehouse CA (2004) Crimean–Congo hemorrhagic fever. Antivir Res 64: 145–160.

13. Walter CT, Barr JN (2011) Recent advances in the molecular and cellular biology of bunyaviruses. J Gen Virol 92: 2467–2484.

14. Magurano F, Nicoletti L (1999) Humoral response in Toscana virus acute neurologic disease investigated by viral-protein-specific immunoassays. Clin Vaccine Immunol 6: 55–60.

15. Schwarz TF, Gilch S, Pauli C, Jäger G (1996) Immunoblot detection of antibodies to Toscana virus. J Med Virol 49: 83–86.

16. Vapalahti O, Kallio-Kokko H, Närvänen A, Julkunen I, Lundkvist Å, et al. (1995) Human B-cell epitopes of puumala virus nucleocapsid protein, the major antigen in early serological response. J Med Virol 46: 293–303.

17. Dowall SD, Richards KS, Graham VA, Chamberlain J, Hewson R (2012) Development of an indirect ELISA method for the parallel measurement of IgG and IgM antibodies against Crimean-Congo haemorrhagic fever (CCHF) virus using recombinant nucleoprotein as antigen. J Virol Methods 179: 335–341.

18. Tang Q, Saijo M, Zhang Y, Asiguma M, Tianshu D, et al. (2003) A patient with Crimean-Congo hemorrhagic fever serologically diagnosed by recombinant nucleoprotein-based antibody detection systems. Clin Vaccine Immunol 10: 489–491.

19. Saijo M, Tang Q, Shimayi B, Han L, Zhang Y, et al. (2005). Recombinant nucleoprotein-based serological diagnosis of Crimean-Congo hemorrhagic fever virus infections. J Med Virol 75: 295–299.

20. Garcia S, Chinikar S, Coudrier D, Billecocq A, Hooshmand B, et al. (2006) Evaluation of a Crimean-Congo hemorrhagic fever virus recombinant antigen expressed by Semliki Forest suicide virus for IgM and IgG antibody detection in human and animal sera collected in Iran. J Clin Virol 35: 154–159.

21. Saijo M, Tang Q, Niikura M, Maeda A, Ikegami T, et al. (2002) Recombinant nucleoprotein-based enzyme-linked immunosorbent assay for detection of immunoglobulin G antibodies to Crimean-Congo hemorrhagic fever virus. J Clin Microbiol 40: 1587–1591.

22. Wei PF, Luo YJ, Li TX, Wang HL, Hu ZH, et al. (2010) Serial expression of the truncated fragments of the nucleocapsid protein of CCHFV and identification of the epitope region. Virol Sin 25: 45–51.

23. Morrow JF, Cohen SN, Chang AC, Boyer HW, Goodman HM, et al. (1974) Replication and Transcription of Eukaryotic DNA in Esherichia coli. Proc Natl Acad Sci 71: 1743–1747.

24. Merrifield RB (1968) Solid-phase peptide synthesis. Adv Enzymol Relat Areas Mol Biol 32: 221–296.

25. Cwirla SE, Peters EA, Barrett RW, Dower WJ (1990) Peptides on phage: a vast library of peptides for identifying ligands. Proc Natl Acad Sci 87: 6378–6382.

26. Roberts BL, Markland W, Ley AC, Kent RB, White DW, et al. (1992) Directed evolution of a protein: selection of potent neutrophil elastase inhibitors displayed on M13 fusion phage. Proc Natl Acad Sci 89: 2429–2433.

27. Ladner RC (2007) Mapping the epitopes of antibodies. Biotechnol Genet Eng 24: 1–30.

28. Xu WX, He YP, Tang HP, Jia XF, Ji CN, et al. (2009) Minimal motif mapping of a known epitope on human zona pellucida protein-4 using a peptide biosynthesis strategy. J Reprod Immunol 81: 9–16.

29. Xu WX, Bhandari B, He YP, Tang HP, Chaudhary S, et al. (2012) Mapping of Epitopes Relevant for Induction of Acrosome Reaction on Human Zona Pellucida Glycoprotein-4 Using Monoclonal Antibodies. Am. J Reprod Immunol 68: 465–475.

30. Xu WX, He YP, Wang J, Tang HP, Shi HJ, et al. (2012) Mapping of Minimal Motifs of B-Cell Epitopes on Human Zona Pellucida Glycoprotein-3. Available: http://www.hindawi.com/journals/jir/2012/831010/abs/. Accepted 2 September 2011.

31. Liu DL, Li Y, Zhao J, Wu T, Sun SR (2013) Preparation of polyclonal antibody to nucleoprotein from Xinjiang hemorrhagic fever virus and its immunological evaluation. Chinese J Cell Mol Immunol 29: 838–841. (In Chinese)

32. Kints JP, Manouvriez P, Bazin H (1989) Rat monoclonal antibodies VII. Enhancement of ascites production and yield of monoclonal antibodies in rats following pretreatment with pristane and Freund's adjuvant. J Immunol Methods 119: 241–245.

33. Dai X, Muhtar, Feng CH, Sun SR, Tai XP, et al. (2006) Geography and host distribution of Crimean - Congo hemorrhagic fever in the Tarim Basin. Chin J Epidemiol 27: 1048–1052. (In Chinese)

34. Garnier J, Robson B (1989) The GOR method for predicting secondary structures in proteins. In Prediction of protein structure and the principles of protein conformation. pp. 417–465.

35. Chou PY, Fasman GD (1978) Prediction of the secondary structure of proteins from their amino acid sequence. Adv Enzymol Relat Areas Mol Biol 47: 45–148.

36. Kyte J, Doolittle RF (1982) A simple method for displaying the hydropathic character of a protein. J Mol Biol 157: 105–132.

37. Karplus PA, Schulz GE (1985) Prediction of chain flexibility in proteins. Naturwissenschaften 72: 212–213.

38. Emini EA, Hughes JV, Perlow D, Boger J (1985) Induction of hepatitis A virus-neutralizing antibody by a virus-specific synthetic peptide. J Virol 55: 836–839.

39. Jameson BA, Wolf H (1988) The antigenic index: a novel algorithm for predicting antigenic determinants. Comput Appl Biosci 4: 181–186.

40. Zhang ZW, Zhang YG, Wang YL, Pan L, Fang YZ, et al. (2010) Screening and identification of B cell epitopes of structural proteins of foot-and-mouth disease virus serotype Asia1. Vet Microbiol 140: 25–33.

41. Zhou Z, Meng W, Deng F, Xia H, Li T, et al. (2013) Complete genome sequences of two Crimean-Congo hemorrhagic fever viruses isolated in China. Genome Announc 1: e00571-13.

42. Laemmli UK (1970) Cleavage of structural proteins during the assembly of the head of bacteriophage T4. Nature 227: 680–685.

43. Towbin H, Staehelin T, Gordon J (1979) Electrophoretic transfer of proteins from polyacrylamide gels to nitrocellulose sheets: procedure and some applications. Proc Natl Acad Sci 76: 4350–4354.

44. Chenna R, Sugawara H, Koike T, Lopez R, Gibson TJ, et al. (2003) Multiple sequence alignment with the Clustal series of programs. Nucleic Acids Res 13: 3497–3500.

45. DeLano WL (2002) The PyMOL molecular graphics system.

46. Lundkvist Å, Meisel H, Koletzki D, Lankinen H, Cifire F, et al. (2002) Mapping of B-cell epitopes in the nucleocapsid protein of Puumala hantavirus. Viral Immunol 1: 177–192.

47. Samudzi RR, Leman PA, Paweska JT, Swanepoel R, Burt FJ (2012) Bacterial expression of Crimean-Congo hemorrhagic fever virus nucleoprotein and its evaluation as a diagnostic reagent in an indirect ELISA. J Virol Methods 179: 70–76.

48. Burt FJ, Samudzi RR, Randall C, Pieters D, Vermeulen J, et al. (2013) Human defined antigenic region on the nucleoprotein of Crimean-Congo haemorrhagic fever virus identified using truncated proteins and a bioinformatics approach. J Virol Methods 193: 706–712.

49. Papa A, Ma B, Kouidou S, Tang Q, Hang C, et al. (2002) Genetic characterization of the M RNA segment of Crimean Congo hemorrhagic fever virus strains, China. Emerg Infect Dis 8: 50–53.

50. Andersson I, Bladh L, Mousavi-Jazi M, Magnusson KE, Lundkvist Å, et al. (2004) Human MxA protein inhibits the replication of Crimean-Congo hemorrhagic fever virus. J Virol 78: 4323–4329.

51. Srinivasan P, Kumar SP, Karthikeyan M, Jeyakanthan J, Jasrai YT, et al. (2011) Epitope-based immunoinformatics and molecular docking studies of nucleocapsid protein and ovarian tumor domain of Crimean-Congo hemorrhagic fever virus. Front Gen 2: 72–80.

52. Lo AS, Zhu Q, Marasco WA (2008) Intracellular antibodies (intrabodies) and their therapeutic potential. Handb Exp Pharmacol 181: 343–373.

53. Moutel S, Nizak C, Perez F (2012) Selection and Use of Intracellular Antibodies (Intrabodies). Methods Mol Biol 907: 667–679.

54. Stocks M (2005) Intrabodies as drug discovery tools and therapeutics. Curr Opin Chem Biol 9: 359–365.

55. Tse E, Rabbitts TH (2000) Intracellular antibody-caspase-mediated cell killing: An approach for application in cancer therapy. Proc Natl Acad Sci USA 97: 12266–12271.

56. Karlberg H, Tan YJ, Mirazimi A (2011) Induction of caspase activation and cleavage of the viral nucleocapsid protein in different cell types during Crimean-Congo hemorrhagic fever virus infection. J Biol Chem 286: 3227–3234.

57. Wang Y, Dutta S, Karlberg H, Devignot S, Weber F, et al. (2012) Structure of Crimean-Congo hemorrhagic fever virus nucleoprotein: superhelical homo-oligomers and the role of caspase-3 cleavage. J Virol 86: 12294–12303.

58. Carter SD, Surtees R, Walter CT, Ariza A, Bergeron É, et al. (2012) Structure, function, and evolution of the Crimean-Congo hemorrhagic fever virus nucleocapsid protein. J Virol 86: 10914–10923.

59. Guo Y, Wang WM, Ji W, Deng MP, Sun YN, et al. (2012) Crimean-Congo hemorrhagic fever virus nucleoprotein reveals endonuclease activity in bunyaviruses. Proc Natl Acad Sci 109: 5046–5051.

3

Break CDK2/Cyclin E1 Interface Allosterically with Small Peptides

Hao Chen[2][9], Yunjie Zhao[2][9], Haotian Li[1][9], Dongyan Zhang[1], Yanzhao Huang[1], Qi Shen[3], Rachel Van Duyne[4,5], Fatah Kashanchi[4], Chen Zeng[1,2], Shiyong Liu[1]*

1 Department of Physics, Huazhong University of Science and Technology, Wuhan, Hubei, China, 2 Department of Physics, The George Washington University, Washington, D. C., United States of America, 3 BNLMS, Center for Quantitative Biology, Peking University, Beijing, China, 4 George Mason University, National Center for Biodefense & Infectious Diseases, Manassas, Virginia, United States of America, 5 The George Washington University Medical Center, Department of Microbiology, Immunology, and Tropical Medicine, Washington, D. C., United States of America

Abstract

Most inhibitors of Cyclin-dependent kinase 2 (CDK2) target its ATP-binding pocket. It is difficult, however, to use this pocket to design very specific inhibitors because this catalytic pocket is highly conserved in the protein family of CDKs. Here we report some short peptides targeting a noncatalytic pocket near the interface of the CDK2/Cyclin complex. Docking and molecular dynamics simulations were used to select the peptides, and detailed dynamical network analysis revealed that these peptides weaken the complex formation via allosteric interactions. Our experiments showed that upon binding to the noncatalytic pocket, these peptides break the CDK2/Cyclin complex partially and diminish its kinase activity in vitro. The binding affinity of these peptides measured by Surface Plasmon Resonance can reach as low as 0.5 µM.

Editor: Chandra Verma, Bioinformatics Institute, Singapore

Funding: SYL is supported by the National Natural Science Foundation of China [31100522]; and the National High Technology Research and Development Program of China [2012AA020402]; and Specialized Research Fund for the Doctoral Program of Higher Education [20110142120038]; and the Fundamental Research Funds for the Central Universities, HUST: 2013QN019. YZH is supported by NSFC [11174093]. FK is supported by George Mason University funds and NIH grant AI043894. The funders had no role in study design, data collection and analysis, decision to publish, or preparation of the manuscript.

* Email: liushiyong@gmail.com

[9] These authors contributed equally to this work.

Introduction

Protein-protein interactions play critical roles in many biological processes, and therefore may become the targets for drug design [1–4]. In this approach, functional proteins [5,6] and small inhibitors [7–10] are successfully designed by grafting, docking and high-throughput NMR screening.

The main strategies for designing effective peptide inhibitors fall into three categories: 1) Cutting peptide sequence [11,12] from native protein-protein interface; 2) Phage display [13–16]; and 3) Computational design, including docking [17–20], molecular dynamics simulation [21–23], normal mode analysis [24], template-based searching [25] and sequence design [26–28]. Peptide inhibitors derived from natural protein-protein interfaces are found to disrupt protein-protein interaction [11,12,29,30]. For example, Schon et al. [11] cut parts of P53 (sequence 15–29) and tested their binding with MDM2. They found that the peptide PMD2 (ETFSDLWKLL, $K_d = 46$ nM) bound MDM2 stronger than peptide PMD1 (SQETFSDLWKLLPEN, $K_d = 580$ nM). However, sometimes this cutting strategy does not work. For example, Gondeau et al. [12] found the peptide C4 derived from Cyclin A with $IC_{50} = 1.8$ µM does not disrupt CDK2/Cyclin A complex. Besides this "cutting" strategy, Hu et al. [13] found a peptide pDI (LTFEHYWAQLTS) with the ability to disrupt P53-MDM2 interaction by phage display. And then, using the same

technology, Pazgier et al. [16] found a novel peptide PMI (TSFAEYWNLLSP, $K_d = 3.4$ nM) bound with MDM2 stronger than the wild p53 peptide (ETFSDLWKLLPE). Later, Li et al. [15] reported that systematic alanine scanning on PMI resulted in a mutant N8A that is the strongest binder with MDM2 ($K_d = 490$ pM). Phage display is a useful method for designing peptide inhibitor of protein-protein interaction, but it is limited to the size of the random library. It cannot cover the entire sequence space. Though the alanine scanning could make up for a number of shortcomings of the phage display technology, the optimized peptide sequence may still not be found without the help of theoretical computational method.

Structure-based computational design of inhibitor has been studied for many years. Protein-peptide docking is one such method [31,32]. London et al. [19] cut the peptide from protein-protein interface in protein-protein docking benchmark 3.0 and CAPRI targets, and docked the peptide to the protein by FlexPepDock [17]. They showed that the derived peptides contributed dominantly to binding free energy, however, it is necessary to validate experimentally if such peptides actually bind their targets. In 2008, Fu et al. [24] successfully designed a 26-mer peptide by modeling backbone flexibility with NMA (normal mode analysis) from Bcl-X_L/Bim-BH3 complex structure. 8 of their 17 designed peptides are validated experimentally to bind well with

Bcl-X$_L$. This approach relies on the knowledge of the protein-peptide structure. In most cases, peptide binding does not induce large conformational changes [33]. However, in the case of CDK2/Cyclin complex, peptide binding may induce large conformational change in its T-loop region. CDK2 inhibition and activation by phosphorylation have been studied by using 1–3 ns molecular dynamics simulations [34], which showed that its glycine-rich loop moves away from the ATP binding pocket. The T-loop is extremely flexible in the unbound state but rigid in any of the CDK2/Cyclin complexes [35]. The previous study shows that the active site cleft is blocked by the T-loop and becomes accessible to the substrate only after activation by Cyclin binding. From its inactive to active conformation, the CDK2 needs to bind Cyclin with a large conformational change in the T-loop region. Finally, its complete activation is achieved by phosphorylation at Thr160 in the T-loop. Within 5 ns MD simulations on CDK2/Cyclin A, it was observed that the T-loop with phosphorylated Thr160 stayed in its active conformation and began to reconfigure with unphosphorylated Thr160 [36].

Recently, a peptide TAALS was found experimentally to break the CDK2/Cyclin interface and inhibit HIV-1 replication [19]. Two key CDK2 residues (Y180 and K178) for the binding between TALLS and CDK2 were identified because 100% and 50% loss in binding were observed for two mutants Y180A and K178A, respectively [19]. These key residues are located at a pocket near the CDK2/Cyclin interface and the T-loop of CDK2. TAALS thus disrupts the complex formation of CDK2/Cyclin by targeting a nearby pocket instead of the interface directly. This indicated that the interface residues can be affected via allosteric interactions upon peptide binding occurred at some distance away from the interface.

In this work, we present a novel strategy to design peptide inhibitors by combining a series of computational methods and experiments, including docking simulation, MD simulation, dynamical network analysis, and SPR assay. The paper is organized as follows. We first described how the peptides were selected by docking simulations. We then identified some peptides that can bind to the nearby pockets and further weaken the CDK2/Cyclin interface using molecular dynamics simulation and dynamical network analysis. Finally, we performed *in vitro* experiments to verify our predictions.

Results

Peptide selection

We constructed the active (PDB ID: 1FIN) and inactive (PDB ID: 1E1X) CDK2 conformations with flexible T-loop (amino acids 150–165) by Rosetta and Morph server totaling 30 models. The peptides used in docking simulation were generated by mutating the two end residues of TAALS yielding 400 double mutants. We focus on the end residues because previous studies [19] on single mutation indicated that the middle residues are conserved. See Materials and Methods for more details. The constructed CDK2 models and peptides were used as starting structures for docking simulation. The final resulting conformations from CDK2-peptide docking simulation were clustered into 10 clusters by lowest binding free energy. One typical structure (decoy) from each cluster was kept, so the ideal number of docking structures should be 30*400*10 = 120,000. However, some cases resulted in fewer than 10 clusters. The actual number of CDK2-peptide decoys turns out to be 115,976. In order to get more accurate information, we have used three different methods to identify the peptides.

Peptide selection according to frequency analysis

We have analyzed the structural occurrence probabilities from the top 1000 protein-peptide decoys with lowest energy calculated by AutoDock. The results show that the top 3 occurrence number of SET2_06, SET3_07, SET3_09 are 528, 110, 92, respectively. So the protein conformations SET2_06, SET3_07 and SET3_09 are favorite conformations to be used to select peptides from top peptide list. Finally, 5 peptides were selected, which are RAALF, RAALG, RAALQ, FAALA, and GAALY, respectively (see Table 1).

Peptide selection according to binding energy calculation

The binding energy describes the strength of the intermolecular interactions. The ranking results show that the peptides of RAALW, RAALQ, GAALY, PAALA, and RAALM are the top 5 peptides with lowest AutoDock binding energy.

Peptide selection according to a knowledge-based potential

The Pmfscore [37] has been used successfully for protein-protein binding energy prediction. Therefore, we apply this knowledge-based potential to re-rank the protein-peptide docking decoy to get more candidate structures. According to this new ranking result, top 5 peptides are KAALE, DAALT, YAALE, YAALQ, and TAALL, respectively.

Considering all results of the three methods above, 13 peptides were finally selected for further MD simulations as shown in Table 2.

MD simulations

There may be some conformational changes of CDK2/Cyclin complex induced by peptide binding that may render the conformations obtained from docking simulations unstable since the protein is held rigid in the simulations. In order to observe the dynamical behavior, we have done MD simulations using two different sets of Van der Waals cut off parameters to analyze the stabilities of peptides and the correlated motions of the CDK2/Cyclin interface.

First, we used a sensitive cut-off 14 Å to analyze the stabilities of the 13 CDK2-peptides (shown in Table 2). As a control, we also checked the stabilities of the peptide-CDK2 complexes of TAALD, TAALS, and LAALS. The three peptides have been investigated computationally and experimentally in previous work [20,38,39]. TAALS and LAALS as inhibitor are found experimentally to be effective; TAALD, while having the highest predicted binding affinity, however, does not show any inhibitory effect [38]. After 5 ns MD simulations, the conformations of CDK2-peptide complex for LAALS, TAALS, DAALT, YAALQ, RAALW, RAALG, FAALA, KAALE were stable with the peptides remaining in the binding pockets. Peptide TAALD was less stable. Moreover, the peptides RAALF, YAALE, and TAALL were moving away. The MD simulations of all CDK2-peptide decoys are summarized in Table 1. For example, TAALS stayed in the binding pocket (Figure 1), however, RAALF moved away from the binding pocket (Figure 2). Finally, we selected six peptides based on these MD simulation results as summarized in Table 3.

It is known that the ATP-binding sites of CDK2 are modified and regulated by Cyclin binding. A stable interface of CDK2/Cyclin complex is required for ATP binding and thus its enzymatic activity. In order to analyze the dynamical motions of the CDK2/Cyclin interface, we applied a method of dynamical

Table 1. MD simulations of CDK2-peptide docking decoys.

RANK	Protein-peptide models	AutoDock Energy (Kcal/mol)	Selected	Methods	MD simulation
49	SET2_RAALF	−12.84	RAALF	Frequency	Swam away
23	SET2_RAALG	−13.11	RAALG	Frequency	Stay
3	SET3_RAALQ	−14.67	RAALQ	Frequency	Blowing up
16	SET2_FAALA	−13.3	FAALA	Frequency	Stay
4	SET2_GAALY	−14.33	GAALY	Frequency	Stay
RANK	**Protein-peptide models**	**Pmfscore (Kcal/mol)**	**Selected**		
7483	SET2_KAALE	−11.34	KAALE	Pmfscore	Stay
26490	SET2_DAALT	−10.37	DAALT	Pmfscore	Stay
73048	SET1_YAALE	−10.34	YAALE	Pmfscore	Swam away
73571	SET1_YAALQ	−9.99	YAALQ	Pmfscore	Stay
40624	SET2_TAALL	−9.87	TAALL	Pmfscore	Swam away
RANK	**Protein-peptide models**	**AutoDock Energy (Kcal/mol)**	**Selected**		
1	SET2_RAALW	−15.89	RAALW	AutoDock Energy	Stay
3	SET3_RAALQ	−14.67	RAALQ	AutoDock Energy	Blowing up
4	SET2_GAALY	−14.33	GAALY	AutoDock Energy	Stay
5	SET2_PAALA	−13.86	PAALA	AutoDock Energy	Stay
6	SET3_RAALM	−13.82	RAALM	AutoDock Energy	Stay
CONTROL	**Protein-peptide models**	**AutoDock Energy (Kcal/mol)**			
	SET2_TAALS	−11.28			Stay
	SET2_LAALS	−10.98			Stay
	SET2_TAALD	−11.58			Swam away & move back

RANK: The rank of the protein-peptide model sorted by AutoDock binding energy. Methods: Frequency, Pmfscore and AutoDock (details see table 2).
SET1, SET2 and SET3 have been defined as CDK2 with different T-loop conformation (see text).
CONTROL: The previous experimental result [20] shows that TAALS and LAALS bound to unphosplorylated form of CDK2, but TAALD not.
Stay: That means that the peptide is staying in the pocket during the MD simulation.

correlation analysis to the CDK2/Cyclin interface based on the MD simulations with Van der Waals cut-off 10 Å.

If any two heavy atoms of two residues were less than 4.5 Å apart for 75% of the snapshots taken at the interval of 100 ps during 20 ns trajectories, the two residues were said to be correlated and the correlation value was computed, otherwise the correlation value was set to zero. If the residues move in the same (opposite) direction in most snapshots, the motions are defined as correlated (anti-correlated) with positive (negative) correlation values. A correlation value close to zero indicates uncorrelated motion. We focused on the residues at the CDK2/Cyclin interface. The average correlation value of the interface residues

Table 2. Designed peptides based on three scoring methods.

Frequency[1]	AutoDock[2]	Pmfscore[3]
FAALA	RAALM	KAALE
RAALF	RAALQ	DAALT
RAALG	RAALW	YAALE
RAALQ	GAALY	YAALQ
GAALY	PAALA	TAALL

[1]Frequency: Top 5 was selected according to the number of the peptide sequence in the top 1000 lowest energy docking decoys.
[2]AutoDock: Top 5 was selected according to the calculated binding energy by AutoDock.
[3]Pmfscore is a statistical potential developed by Jiang et al. [37]. Top 5 was selected according to the Pmfscore.

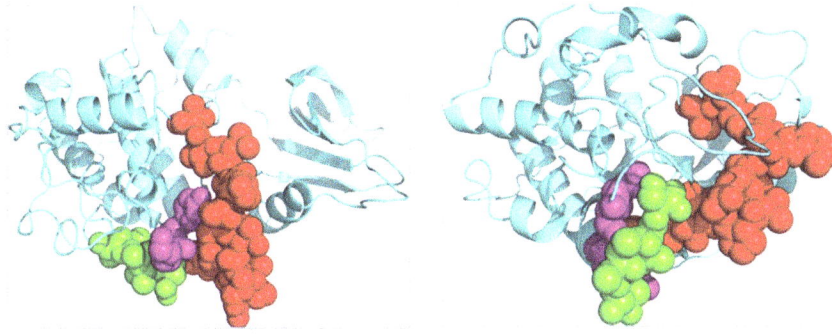

Figure 1. MD simulation of TAALS-CDK2 docking decoy. Left: the docked TAALS and CDK2 complex structure, as an initial structure for MD simulation; Right, after 5 ns MD simulation, the TAALS and CDK2 complex structure is shown. The green represent peptide TAALS, and the purple balls are atoms from the key residues: K178, Y180, and the red is the T-loop of CDK2. The MD simulation shows that after 5 ns, the peptide TAALS (Green) induced the conformational change of the CDK2 and moved to the gap between purple and red.

in the absence of peptide is 0.38. Figure 3 shows the correlation analysis results of the six selected peptides. The interface regions displaying high degree of correlation are marked in white rectangles. The correlation values for the cases of DAALT, YAALQ, RAALG, FAALA, KAALE, and RAALW are 0.31, 0.27, 0.44, 0.39, 0.33, 0.38, respectively. The correlation values reflect the coupled motions between CDK2 and Cyclin in the interface regions, and thus larger correlation values indicate more stable interface.

Therefore, the order of stability of the CDK2/Cyclin interface is YAALQ<DAALT<KAALE<RAALW<FAALA<RAALG. These computational results suggest that the interface regions become less stable if the peptides YAALQ and DAALT bind to CDK2. This prediction is consistent with the experimental results described in the next section. While longer MD simulations would undoubtedly provide a more pronounced correlation map, the short simulations performed here could nonetheless provide an estimate on which peptides may break up the CDK2/Cyclin interface.

Dissociation of CDK2/Cyclin E in vitro in the presence of six designed peptides

To visualize and verify the dissociation of CDK2/Cyclin complex by each of the six designed peptides, immunoprecipitations against Cyclin and IgG, with the latter being a negative control for nonspecific background signal, were performed and

followed by Western blot for CDK2 as shown in Figure 4(A). Comparing the band intensity on Lane 2 for Cyclin pulldown to that on Lane 3 for IgG pulldown, we clearly observed a weaker intensity indicating the dissociation of CDK2 from the CDK2/Cyclin complex in the presence of peptide DAALT. Upon closer inspection, the left band (lane 4) of YAALQ appears slightly wider and darker than the right band (lane 5) indicating a weak complex disassociation. However, additional evidence is needed to differentiate peptide YAALQ from other four peptides (lanes 6–13) that failed to break up the complex. This is discussed below by the kinase activity experiment.

Figure 4(B) further illustrates how the dissociation of CDK2 inhibits the kinase activity of CDK2/Cyclin complex. Here an immunoprecipitation of the CDK2/Cyclin complex was performed as previously described, followed by the kinase reaction with H1 histone being added as the substrate. The levels of phosphorylation of H1 are shown in the presence of the six designed peptides. Again, the two peptides DAALT and YAALQ exhibit a clear loss of kinase activity. To figure out how strong these two effective peptides bind to CDK2, we measured their binding affinities via Surface Plasmon Resonance as described below.

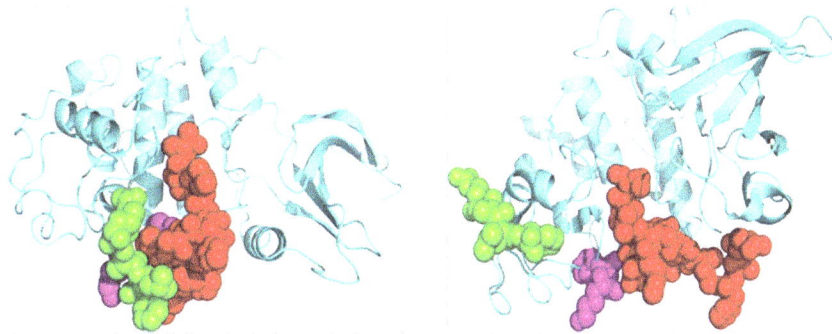

Figure 2. MD simulation of RAALF-CDK2 docking decoy. Left: the docked RAALF and CDK2 complex structure,as an initial structure for MD simulation; Right, after 5 ns MD simulation, the RAALF and CDK2 complex structure is shown. The green represent peptide RAALF, and the purple balls are atoms from the key residues: K178, Y180, and the red is the T-loop of CDK2. The MD simulation shows that after 5 ns, the peptide RAALF (Green) swam away from the key pocket sites of CDK2.

Table 3. Selection based on MD simulation results.

RANK	Protein-peptide models	AutoDock Energy (Kcal/mol)	Selected peptide	Methods	MD simulation
23	SET2_RAALG	−13.11	RAALG	Frequency	Stay[1]
16	SET2_FAALA	−13.30	FAALA	Frequency	Stay between key residues and T-loop[2]
7483	SET2_KAALE	−9.35(−11.34)[3]	KAALE	Pmfscore	Stay between key residues and T-loop[2]
26490	SET2_DAALT	−7.90(−10.37)[3]	DAALT	Pmfscore	Stay[1]
73571	SET1_YAALQ	−6.05(−9.99)[3]	YAALQ	Pmfscore	Stay[1]
1	SET2_RAALW	−15.89	RAALW	AutoDock	Stay[1]

RANK: The rank of the protein-peptide model sorted by AutoDock binding energy. Methods: Frequency, Pmfscore and AutoDock (details see table 2).
SET1 and SET2 have been defined as CDK2 with different T-loop conformation (see text).
[1]Stay: That means that the peptide is staying in the pocket during the MD simulation.
[2]Key residue and T-loop: Key residues are that Y180, K178of CDK2.
[3]The value in brackets is calculated by Pmfscore.

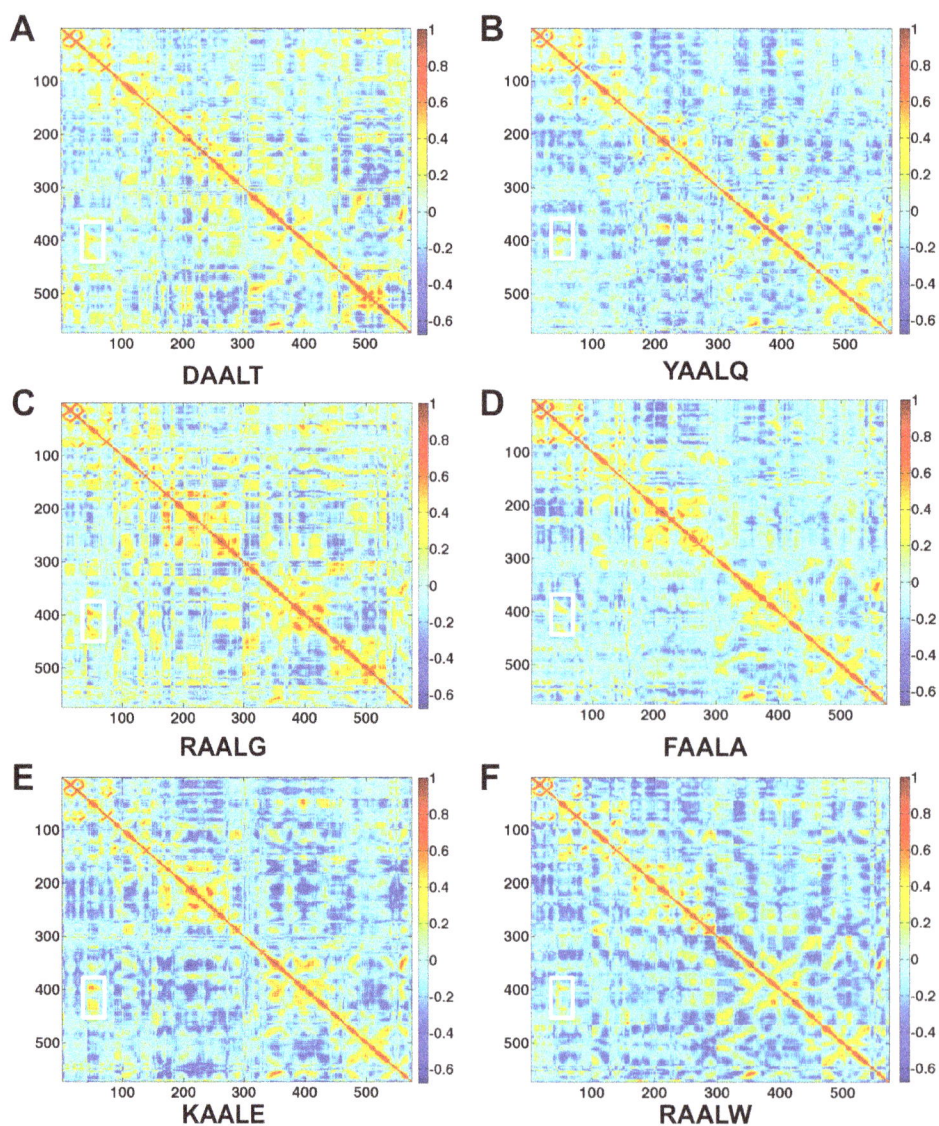

Figure 3. Correlation analysis of the motion during a 20-ns MD simulation of the CDK2/Cyclin/peptide complex structures.
Monomers with highly (anti)correlated motion are orange or red (blue). Interface regions displaying high degree of (anti)correlation are marked in white rectangles.

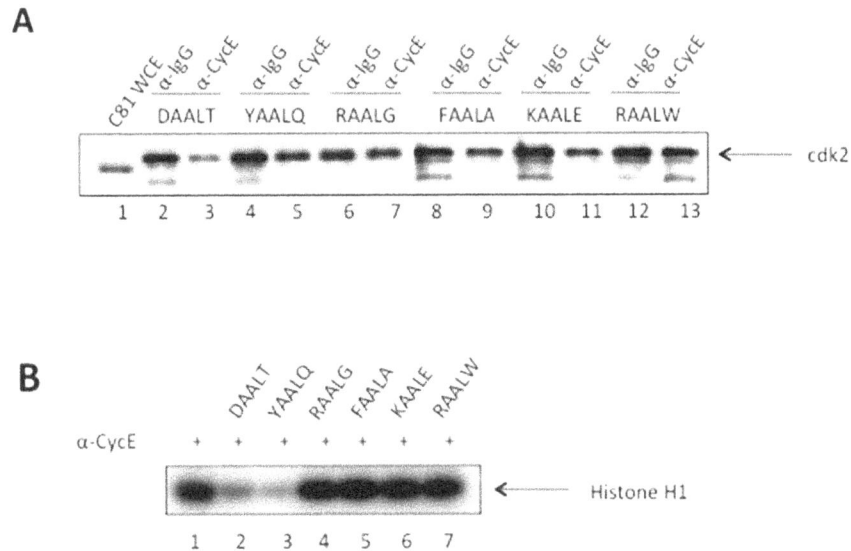

Figure 4. Dissociation of CDK2/Cyclin E in the presence of designed peptides. A) C81 fractionated cell extracts containing cdk2/Cyclin E complex were incubated with α-Cyclin E antibody in the presence of six designed peptides at 10 μM concentration. Following immunoprecipitation of Cyclin E, Western blot for CDK2 was shown here. α-IgG is included as a negative control. B) Immunoprecipitated Cyclin E samples in the presence of peptides were assessed for kinase activity. Histone H1 (1 μg/reaction) was added to each reaction tube along with 2 μl of (γ-32P) ATP (3000 Ci/mmol). Reactions were incubated at 37°C for 30 min and stopped by the addition Laemmli buffer. The samples were separated on a 4–20% Tris–Glycine gel. Samples were ran on a gel, dried, and exposed to a PhosphorImager cassette and analyzed using Molecular Dynamic's ImageQuant Software.

Peptide binding measurement by a Surface Plasmon Resonance (SPR) assay

Two peptides (DAALT and YAALQ) and two positive controls (TAALS and LAALS) are tested in the same condition with CDK2 by T200 (all binding data see Table 4).

The response curves of various analyte concentrations were globally fitted to the two-step binding model described by the following equation [40],

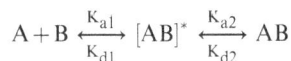

$$A + B \underset{K_{d1}}{\overset{K_{a1}}{\longleftrightarrow}} [AB]^* \underset{K_{d2}}{\overset{K_{a2}}{\longleftrightarrow}} AB$$

Where the equilibrium constants of each binding step are $K_1 = K_{a1}/K_{d1}$ and $K_2 = K_{a2}/K_{d2}$, and the overall equilibrium binding constant is calculated as $K_A = K_1 (1+K_2)$ and $K_D = 1/K_A$. In this model, the analyte (A) binds to the ligand (B) to form an initial complex [AB]* and then undergoes subsequent binding or conformational change to form a more stable complex AB. Data

were fitted globally by using the standard two state models provided by Biacore T200 Software v2.0. The binding affinities K_D to CDK2 for peptides DAALT and YAALQ were measured to be 0.47 μM and 98 μM, respectively. After ATP with a concentration of 60 μM was added, the binding affinities K_D for peptides DAALT and YAALQ were changed to 37 μM and 61 μM, respectively. Given the large uncertainty of the fitting in SPR kinetic assays, we consider a 10-fold change in binding affinity not very significant. For the three peptides, TAALS, LAALS and YAALQ, the changes in K_D in the presence/absence of ATP are all within 10-fold. Thus, ATP does not have a significant effect on the binding of these peptides to CDK2. Therefore, we conclude that YAALQ does not compete directly with ATP for the ATP binding pocket. For DAALT, however, a larger-than-10-fold decrease in the presence of ATP was observed. While it is possible for DAALT to compete directly with ATP by occupying the ATP binding pocket on CDK2, it is more likely that it competes indirectly with ATP. For example, binding of ATP to

Table 4. SPR-derived binding affinities of CDK2 for four peptides with and without 60 μM ATP.

Peptides	K$_{a1}$(1/Ms)	K$_{d1}$(1/s)	K$_{a2}$(1/Ms)	K$_{d2}$(1/s)	K$_D$(M)
TAALS	11.3±0.1	26.0±0.1 E-3	11.1±0.2 E-3	6.8±2.7 E-6	1.4±0.6 E-6
TALLS*	3.7±0.1	28.2±1.4 E-3	9.1±0.2 E-3	3.8±1.0 E-6	3.3±0.8 E-6
LAALS	498.0±8.8	11.9±0.2 E-3	5.8±2.0 E-5	2.9±0.8 E-5	8.0±2.9 E-6
LAALS*	237.4±6.9	51.8±2.3 E-3	7.9±0.5 E-3	3.1±0.3 E-3	6.1±0.6 E-5
DAALT	434.1±6.6	23.2±0.4 E-3	14.1±0.4 E-4	1.3±0.6 E-5	4.7±2.0 E-7
DAALT*	612.9±13.0	101.2±4.3 E-3	34.6±1.0 E-3	10.1±0.1 E-3	3.7±0.3 E-5
YAALQ	165.7±4.0	84.0±2.7 E-3	9.0±0.3 E-3	2.2±0.1 E-3	9.8±0.7 E-5
YAALQ*	100.9±1.5	53.9±1.4 E-3	12.3±0.3 E-3	1.6±0.1 E-3	6.1±0.3 E-5

*With 60 μM ATP.

CDK2 stabilizes a certain CDK2 conformation that is less favorable for DAALT binding. The detailed binding mode between DAALT and CDK2 needs to be resolved by other means beyond SPR experiments. From Figure 5, we can see that the peptide TAALS used a binding mechanism different from that of three other peptides. More precisely, upon injection of peptide, the response curve for TAALS shows a slower increase before reaching a steady value or a horizontal curve as well as a slower decrease after the peak than those for the other three peptides where a sharp jump and a dramatic drop are seen. This indicates that the mechanism of binding of TAALS represents slow binding and slow dissociation. On the contrary, LAALS displays fast binding and fast dissociation, similar to the other two peptides, DAALT and YAALQ. The SPR results confirmed a direct interaction between the peptides and CDK2. The two-state model was a better fit, suggesting that there are two different states in peptide-CDK2 binding processes. We hypothesize that the second state is the induced conformation of CDK2 by peptide binding. This is consistent with our MD simulations. The binding affinities predicted by computational docking simulations and measured by SPR between peptides and CDK2 fall into the same range (0.1 μM~40 μM). Compared to other peptides, the association and dissociation processes of TAALS are very slow.

Discussion

We have performed a series of computational simulations to design and select the effective peptide inhibitors against CDK2/Cyclin complex. In the structural modeling and docking selection steps, all the T-loops of the selected CDK2 were built by Rosetta loop modeling algorithm from inactive (1E1X [41]) and active (1FIN [42], Chain A) conformations. These results suggest that Rosetta loop modeling algorithm may be better for sampling flexible loop conformation than the Morph Server. In the MD simulation step, we have used two sets of Van der Waals cut-off parameters. The larger Van der Waals cut-off value considers more non-bounded interactions and is more sensitive for MD simulations. In the 5 ns MD simulations, we wanted to speed up the observation of potential instabilities of the peptides binding to CDK2/Cyclin complexes, and thus used a larger cut-off value of 14 Å. However, in the subsequent correlation analysis, we wanted to analyze the dynamical motions in detail and used the default cut-off value of 10 Å. It is generally difficult to design peptide or small molecule inhibitor to target the interface directly since typical protein-protein interface is rather diffusive. Most of the known CDK2 inhibitors target the catalytic ATP-binding pocket of CDK2 [43]. However, this pocket for the family of CDK proteins is so conserved that it is difficult to design specific CDK inhibitors for this pocket. It is thus highly desirable to discover non-ATP competitive inhibitors that function through allosteric interactions. Our previous studies [20,38,39] have identified such a binding pocket next to the T-loop of CDKs for designing allosteric inhibitors. A recent study by Betzi et al. [10] discovered yet another binding site near the ATP-binding pocket for designing non-ATP competitive small molecule inhibitors. Here we report a new computational methodology that combined MD simulations and novel dynamic network analysis to uncover subtle correlations not revealed by static structures in a systematic manner. This provides a means to not only identify allosteric binding sites but also understand the mechanism of allosteric interactions to design effective small molecule or peptide inhibitors. In our paper, the stability of its interface requires specific conformations of its flexible T-loop. This provide an alternative strategy to design inhibitors that disrupt the complex formation by targeting nearby pockets that could induce both the conformational changes of the T-loop and the stability of the CDK2/Cyclin interface via allosteric interactions. Equally important is the dynamical network analysis such as correlation analysis on the MD simulations because it could provide a benchmark for selecting effective peptide inhibitors.

Materials and Methods

Preparation of the ensemble of CDK2 and initial peptide structures

In order to model the flexibility of CDK2, T-loop (residues 150–165) conformations are reconstructed by Rosetta (version 3.1) [44] with KIC algorithm, and ten models are kept for inactive (1E1X [41]) and active (1FIN [42], Chain A) conformation, respectively. On the other hand, ten intermediate conformations of CDK2 between inactive (1E1X) and active (1FIN, Chain A) conformations are also generated by the Morph server [45] (http://molmovdb.mbb.yale.edu/). There are thirty conformations for CDK2 in total. The ten models from 1E1X by Rosetta loop prediction are named SET1; the ten models from 1FINA by Rosetta loop prediction are named SET2; and the ten models by Morph server prediction are named SET3.

Single mutation scanning experiment [20] shows that peptide (xAALx with x representing any residue) could break CDK2/Cyclin complex. So, a double mutation on position x is performed. The side-chain conformation of the double mutants are built by SCWRL4 [46]. The backbone template of the peptide used here is 1HS6A_128553_5.pdb (sequence: TAALT), which is downloaded from pepx database [47](http://pepx.switchlab.org/) by query sequence pattern. AAL.

Flexible CDK2-peptide docking based on an ensemble of CDK2

Docking Protocol. The peptides and CDK2 docking were performed using the Lamarckian genetic algorithm with the default parameters by AutoDock [48] (version 4.2). AutoDock-Tools (http://autodock.scripps.edu/) was used to prepare the ligands and the receptor. Mass-centered grid maps were generated by the AutoGrid program using the default parameters. The center of grid is set to be in the local peptide binding pocket (−12.299 28.510 35.091) in receptor (CDK2), and the number of grid points in xyz are set to be 60. The key residues (ARG150, LYS178, TYR180) in the binding pocket of CDK2, which are determined experimentally by point mutation, are set to be flexible when docking. The flexible residues of the receptor are treated in a similar way as the ligand. Hydrogen atoms were added by REDUCE (version 3.14) [49]. The results were clustered using a tolerance of 2.0 Å. Finally, 10 conformations with the lowest binding free energy were kept for further analysis.

MD simulation Protocol. In this study, two separate sets of MD simulations were undertaken by using the software GROMACS [50]. One is to focus on the CDK2 structures with different conformation of its T-loop; the other is for the docked CDK2-peptide structures. Each system was solvated in a cubic box with 10 Å SPC water. Then, each system was first minimized. The protein was constrained with all-bonds and the solvent molecules with counter ions were allowed to move during a 1,000-step minimization and a 100 ps long MD simulation. After relaxation, the system was simulated for 5 ns in total for all studied systems. The temperature was set at 300 K. The force field G53a6 was used in all simulations.

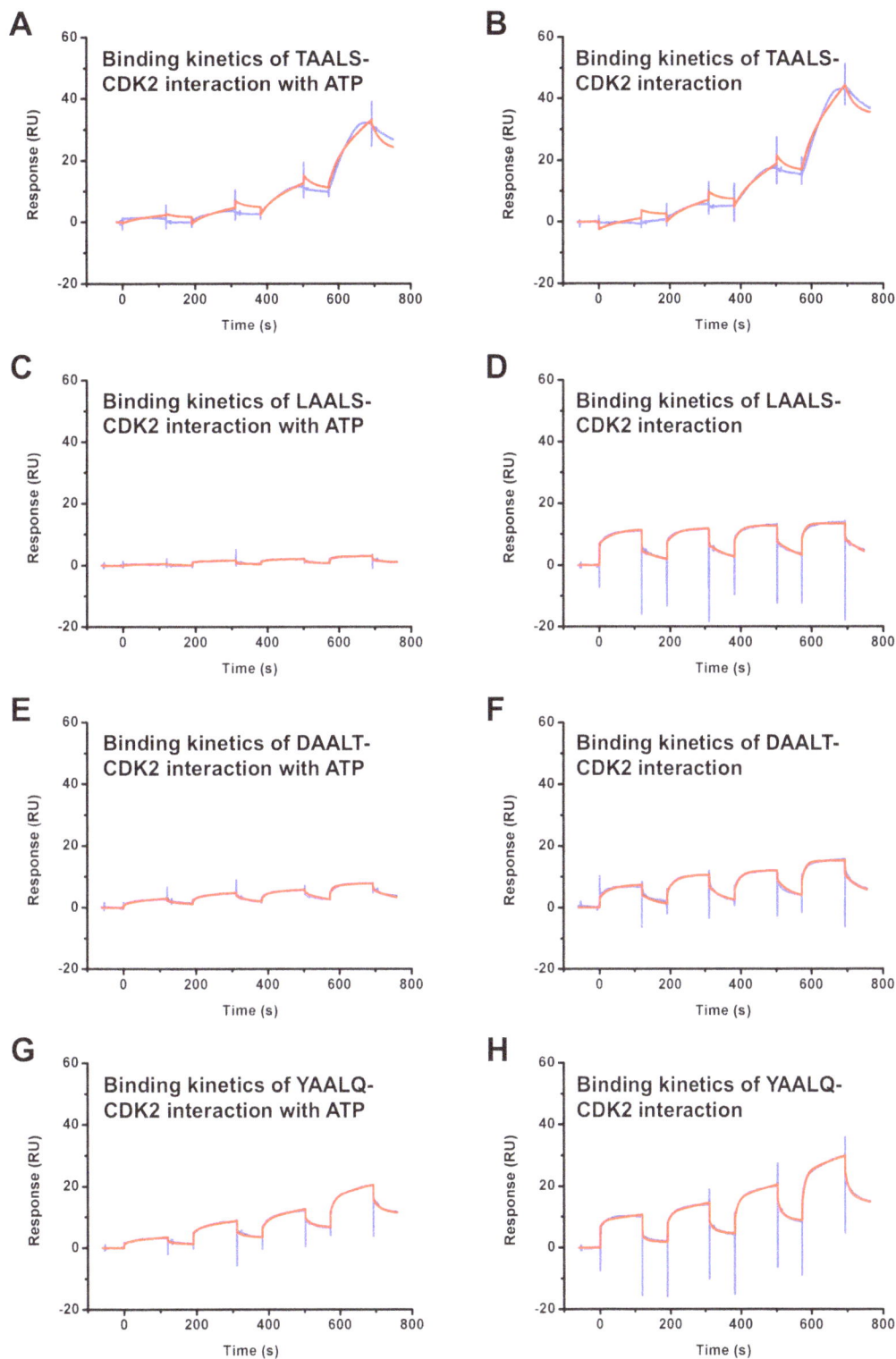

Figure 5. SPR binding assay results of CDK2 and peptides with and without ATP. The blue lines are experimental data, and the red lines are fitted results. The binding affinities (K_D) between CDK2 and TAALS (A) with and (B) without 60 µM ATP are 3.3 µM, 1.4 µM, respectively. The binding affinities (K_D) between CDK2 and LAALS (C) with and (D) without 60 µM ATP are 61 µM, 8.0 µM, respectively. The binding affinities (K_D) between CDK2 and DAALT (E) with and (F) without 60 µM ATP are 37 µM, 0.47 µM, respectively. The binding affinities (K_D) between CDK2 and YAALQ (G) with and (H) without 60 µM ATP are 61 µM, 98 µM, respectively.

Correlation Analysis

In the protein network, a node is defined as one single amino acid. If the distance of any two heavy atoms of a pair of different nodes is less than 4.5 Å for at least 75% of the snapshots, then this pair of nodes were said to form an edge [51]. The neighboring nodes in sequence are not considered to be in contact. We have done 30 ns MD simulations for each different state. The dynamical network is constructed with the final 20 ns of the 30 ns trajectories sampled every 100 ps. Then, we define the pairwise correlations (C_{ij}) as $C_{ij} = \langle \Delta \vec{r}_i(t) \bullet \Delta \vec{r}_j(t) \rangle / (\langle \Delta \vec{r}_i(t)^2 \rangle \langle \Delta \vec{r}_j(t)^2 \rangle)^{1/2}$, $\Delta \vec{r}_i(t) = \vec{r}_i(t) - \langle \vec{r}_i(t) \rangle$, and the $\vec{r}_i(t)$ is the position of the atom corresponding to the ith node. We calculated the correlations from the MD simulation trajectories using the program Carma [52].

Cell culture

C81 is an HTLV-1-infected T-cell line that expresses Tax protein established from patients with T-cell leukemia. These cells are available through AIDS reagent catalog [53–55]. Cells were cultured in RPMI-1640 containing 10% fetal bovine serum, 1% penicillin/streptomycin, and 1% L-glutamine (Quality Biological) and were incubated in a 5% CO_2 incubator at 37°C. Cells were cultured to confluency and pelleted at 4°C for 15 min at 3,000 rpm. The cell pellets were washed twice with 25 ml of phosphate-buffered saline (PBS) without Ca^{2+} and Mg^{2+} (Quality Biological) and centrifuged once more. Cell pellets were resuspended in lysis buffer (50 mM Tris–HCl, pH 7.5, 120 mM NaCl, 5 mM EDTA, 0.5% NP-40, 50 mM NaF, 0.2 mM Na_3VO_4, 1 mM DTT, one complete protease cocktail tablet/ 50 ml) and incubated on ice for 20 min, with a gentle vortexing every 5 min. Cell lysates were transferred to Eppendorf tubes and were centrifuged at 10,000 rpm for 10 min. Supernatants were transferred to a fresh tube where protein concentrations were determined using Bio-Rad protein assay (Bio- Rad, Hercules, CA).

Peptide synthesis

All peptides used for this study were commercially synthesized (RS Sythesis, Lousiville, KY) with the following sequences:

NH2-D-A-A-L-T-OH
NH2-Y-A-A-L-Q-OH
NH2-R-A-A-L-G-OH
NH2-F-A-A-L-A-OH
NH2-K-A-A-L-E-OH
NH2-R-A-A-L-W-OH

The purity of each peptide was analyzed by HPLC to greater than 95%. Mass spectral analysis was also performed to confirm the identity of each peptide as compared to the theoretical mass (Applied Biosystems Voyager System 1042). Peptides were resuspended in dH2O to a concentration of 1 mg/ml and stored at −70C. Peptides were only thawed once prior to use for biochemical experiments.

Size-exclusion chromatography

C81 whole cell lysate (5 mg) was fractionated on a Superose 6 HR 10/30 column (Amersham Biosciences, Piscataway, NJ) in Buffer D (20 mM HEPES (pH 7.9), 0.05 M KCl, 0.2 mM EDTA, 0.5 mM PMSF, 0.05 DTT, and 20% Glycerol). Flow-through was collected at 0.5 ml for 50 fractions. Every 10th fraction was analyzed by immunoblotting for cdk2 in order to determine the elution location of the cdk2/Cyclin E complex.

Immunoprecipitation

Cdk2 containing chromatography fractions (28–31) were pooled together for immunoprecipitation. The pooled C81 extracts were combined with 10 μM of each respective peptide. Cyclin E antibody (Santa Cruz, sc-198) was added to each reaction tube (10 μl, 2 μg), the reaction mixture was brought up to 500 μl with TNE$_{50}$+0.1% NP-40 (100 mM Tris, pH 8.0; 50 mM NaCl; 1 mM EDTA, 0.1% Nonidet P-40) and was allowed to incubate while rotating overnight at 4°C. α-IgG was added to extract as a negative control, and an IP was performed in the absence of competing peptide, acting as a positive control. The following day, 30 μl of a 30% Protein A & G bead slurry (CalBioChem, La Jolla, CA) was added to each reaction tube and allowed to incubate while rotating for 2 h at 4°C. Samples were spun and washed 2× with TNE$_{300}$+0.1% NP-40 (100 mM Tris, pH 8.0; 300 mM NaCl; 1 mM EDTA, 0.1% Nonidet P-40) and 1× with TNE$_{50}$+0.1% NP-40 to remove non-specifically bound proteins. 2× Laemmli buffer was added to each sample and heated at 95°C for 3 min. Samples were loaded and run on a 4–20% Tris–Glycine SDS/PAGE gel to be used for both Western blots and kinase assays.

Western Blot

Immunoprecipitated samples were separated on SDS/PAGE gels and were transferred to a nitrocellulose membrane via a constant current of 70 mA overnight. The membrane was blocked with a 3% BSA solution in PBS containing 0.1% Tween-20, rocking for 2 h at 4°C. A 1:1000 dilution of α-cdk2 antibody (Santa Cruz, sc-163) was added to the blocking solution and incubated rocking overnight at 4°C. The membrane was washed with a fresh PBS+0.1% Tween-20 solution in order to wash off any residual primary antibody solution. A 1:1000 dilution of α-rabbit secondary antibody was added to a fresh 3% BSA solution in PBS+0.1% Tween-20 and incubated with the membrane, rocking for 2 h at 4°C. The membrane was washed 2× with PBS+0.1% Tween-20 and 1× with PBS to remove any residual antibody. The membrane was exposed to chemiluminescence reagent (Pierce) in the dark for 5 min., and was developed using a BioRad Imager.

Kinase assay

Immunoprecipitated samples were assessed for kinase activity. After the final TNE$_{50}$+0.1% NP-40 wash, beads were washed with kinase buffer (50 mM HEPES, 10 mM MgCl2, 5 mM MnCl2, 1 mM DTT, 50 mM NaF, 0.2 mM Na_3VO_4 and one complete tablet of protease cocktail inhibitor/50 ml buffer) to equilibrate the reaction. Histone H1 (1 μg) was added to each reaction tube along with the γ-32P ATP (2 μl at 3000 Ci/mmol). Reactions were incubated at 37°C for 30 min and stopped by the addition of 15 μl Laemmli buffer. The samples were separated by reducing SDS-PAGE on a 4–20% Tris–Glycine gel. Gels were stained with Coomassie blue, destained, and then dried for 2 hours. Following drying, the gels were exposed to a PhosphorImager cassette and analyzed utilizing Molecular Dynamic's ImageQuant Software.

Peptide binding measurement by SPR

The peptides TAALS, LAALS, DAALT, and YAALQ were bought from Sangon Biotech. The purity of the four peptides were greater than 98% and they were stored at −20°C. When they were used for experiments, they are dissolved at 25°C.

Single-cycle kinetics experiments were performed with a T200 apparatus. The experiments were done on S-CM5 sensor chips coated with 6000 RU of CDK2. A flow cell left blank was used for double-referencing of the sensorgrams. Binding experiments were performed in standard PBS-P buffer with 60 μM ATP as running buffer (10 mM NaH2PO4/Na2HPO4, 150 mM NaCl, 60 μM ATP and 0.05% surfactant P20, pH 7.4) at 25°C with a flow rate of 30 µl/min. The peptide samples were prepared in the running buffer, and were injected.

The regeneration of the surface was achieved with a 30 second pulse of 20 mM NaOH. Single-cycle kinetics assay were performed using the standard SCK method implemented by the T200 Control Software. In single-cycle analysis, the analyte is injected with increasing concentrations in a single cycle. The surface is not regenerated between injections. A blank injection of buffer only was subtracted from each curve, and reference sensorgrams were subtracted from experimental sensorgrams to yield curves representing specific binding. Data were fitted globally by using the standard two-state model provided by T200 Software

v2.0. The data shown are representative of at least three independent experiments.

Acknowledgments

We are grateful to Professor Houjin Zhang for providing CDK2 and Professor Luhua Lai for SPR testing.

Author Contributions

Conceived and designed the experiments: CZ SL. Performed the experiments: QS RD DZ FK. Analyzed the data: HC YZ HL SL YH. Contributed reagents/materials/analysis tools: SL QS FK. Wrote the paper: YZ CZ SL.

References

1. Wells JA, McClendon CL (2007) Reaching for high-hanging fruit in drug discovery at protein-protein interfaces. Nature 450: 1001–1009.
2. Arkin MR, Wells JA (2004) Small-molecule inhibitors of protein-protein interactions: progressing towards the dream. Nat Rev Drug Discov 3: 301–317.
3. Arkin M (2005) Protein-protein interactions and cancer: small molecules going in for the kill. Curr Opin Chem Biol 9: 317–324.
4. Bourgeas R, Basse MJ, Morelli X, Roche P (2010) Atomic analysis of protein-protein interfaces with known inhibitors: the 2P2I database. PLoS One 5: e9598.
5. Liu S, Zhu X, Liang H, Cao A, Chang Z, et al. (2007) Nonnatural protein-protein interaction-pair design by key residues grafting. Proceedings of the National Academy of Sciences of the United States of America 104: 5330–5335.
6. Fleishman SJ, Whitehead TA, Ekiert DC, Dreyfus C, Corn JE, et al. (2011) Computational design of proteins targeting the conserved stem region of influenza hemagglutinin. Science 332: 816–821.
7. Winter A, Higueruelo AP, Marsh M, Sigurdardottir A, Pitt WR, et al. (2012) Biophysical and computational fragment-based approaches to targeting protein-protein interactions: applications in structure-guided drug discovery. Q Rev Biophys: 1–44.
8. Vassilev LT, Vu BT, Graves B, Carvajal D, Podlaski F, et al. (2004) In vivo activation of the p53 pathway by small-molecule antagonists of MDM2. Science 303: 844–848.
9. Oltersdorf T, Elmore SW, Shoemaker AR, Armstrong RC, Augeri DJ, et al. (2005) An inhibitor of Bcl-2 family proteins induces regression of solid tumours. Nature 435: 677–681.
10. Betzi S, Alam R, Martin M, Lubbers DJ, Han H, et al. (2011) Discovery of a potential allosteric ligand binding site in CDK2. ACS Chem Biol 6: 492–501.
11. Schon O, Friedler A, Bycroft M, Freund SM, Fersht AR (2002) Molecular mechanism of the interaction between MDM2 and p53. J Mol Biol 323: 491–501.
12. Gondeau C, Gerbal-Chaloin S, Bello P, Aldrian-Herrada G, Morris MC, et al. (2005) Design of a novel class of peptide inhibitors of cyclin-dependent kinase/cyclin activation. J Biol Chem 280: 13793–13800.
13. Hu B, Gilkes DM, Chen J (2007) Efficient p53 activation and apoptosis by simultaneous disruption of binding to MDM2 and MDMX. Cancer Res 67: 8810–8817.
14. Phan J, Li Z, Kasprzak A, Li B, Sebti S, et al. (2010) Structure-based design of high affinity peptides inhibiting the interaction of p53 with MDM2 and MDMX. J Biol Chem 285: 2174–2183.
15. Li C, Pazgier M, Yuan W, Liu M, Wei G, et al. (2010) Systematic mutational analysis of peptide inhibition of the p53-MDM2/MDMX interactions. J Mol Biol 398: 200–213.
16. Pazgier M, Liu M, Zou G, Yuan W, Li C, et al. (2009) Structural basis for high-affinity peptide inhibition of p53 interactions with MDM2 and MDMX. Proc Natl Acad Sci U S A 106: 4665–4670.
17. London N, Raveh B, Cohen E, Fathi G, Schueler-Furman O (2011) Rosetta FlexPepDock web server–high resolution modeling of peptide-protein interactions. Nucleic Acids Res 39: W249–253.
18. Trellet M, Melquiond AS, Bonvin AM (2013) A unified conformational selection and induced fit approach to protein-peptide docking. PLoS One 8: e58769.
19. London N, Raveh B, Movshovitz-Attias D, Schueler-Furman O (2010) Can self-inhibitory peptides be derived from the interfaces of globular protein-protein interactions? Proteins 78: 3140–3149.
20. Chen H, Van Duyne R, Zhang N, Kashanchi F, Zeng C (2009) A novel binding pocket of cyclin-dependent kinase 2. Proteins 74: 122–132.
21. Antes I (2010) DynaDock: A new molecular dynamics-based algorithm for protein-peptide docking including receptor flexibility. Proteins 78: 1084–1104.
22. Dagliyan O, Proctor EA, D'Auria KM, Ding F, Dokholyan NV (2011) Structural and dynamic determinants of protein-peptide recognition. Structure 19: 1837–1845.
23. Zacharias M (2012) Combining coarse-grained nonbonded and atomistic bonded interactions for protein modeling. Proteins.
24. Fu X, Apgar JR, Keating AE (2007) Modeling backbone flexibility to achieve sequence diversity: the design of novel alpha-helical ligands for Bcl-xL. J Mol Biol 371: 1099–1117.

25. Verschueren E, Vanhee P, Rousseau F, Schymkowitz J, Serrano L (2013) Protein-peptide complex prediction through fragment interaction patterns. Structure 21: 789–797.
26. Grigoryan G, Reinke AW, Keating AE (2009) Design of protein-interaction specificity gives selective bZIP-binding peptides. Nature 458: 859–864.
27. Smith CA, Kortemme T (2010) Structure-based prediction of the peptide sequence space recognized by natural and synthetic PDZ domains. J Mol Biol 402: 460–474.
28. Zhang C, Shen Q, Tang B, Lai L (2013) Computational design of helical peptides targeting TNFalpha. Angew Chem Int Ed Engl 52: 11059–11062.
29. Sattler M, Liang H, Nettesheim D, Meadows RP, Harlan JE, et al. (1997) Structure of Bcl-xL-Bak peptide complex: recognition between regulators of apoptosis. Science 275: 983–986.
30. Slivka PF, Shridhar M, Lee GI, Sammond DW, Hutchinson MR, et al. (2009) A peptide antagonist of the TLR4-MD2 interaction. Chembiochem 10: 645–649.
31. Vanhee P, van der Sloot AM, Verschueren E, Serrano L, Rousseau F, et al. (2011) Computational design of peptide ligands. Trends Biotechnol 29: 231–239.
32. London N, Raveh B, Schueler-Furman O (2013) Peptide docking and structure-based characterization of peptide binding: from knowledge to know-how. Curr Opin Struct Biol 23: 894–902.
33. London N, Movshovitz-Attias D, Schueler-Furman O (2010) The structural basis of peptide-protein binding strategies. Structure 18: 188–199.
34. Bartova I, Otyepka M, Kriz Z, Koca J (2004) Activation and inhibition of cyclin-dependent kinase-2 by phosphorylation; a molecular dynamics study reveals the functional importance of the glycine-rich loop. Protein Sci 13: 1449–1457.
35. Bartova I, Koca J, Otyepka M (2008) Functional flexibility of human cyclin-dependent kinase-2 and its evolutionary conservation. Protein Sci 17: 22–33.
36. Barrett CP, Noble ME (2005) Molecular motions of human cyclin-dependent kinase 2. J Biol Chem 280: 13993–14005.
37. Jiang L, Gao Y, Mao F, Liu Z, Lai L (2002) Potential of mean force for protein-protein interaction studies. Proteins 46: 190–196.
38. Van Duyne R, Cardenas J, Easley R, Wu W, Kehn-Hall K, et al. (2008) Effect of transcription peptide inhibitors on HIV-1 replication. Virology 376: 308–322.
39. Agbottah E, Zhang N, Dadgar S, Pumfery A, Wade JD, et al. (2006) Inhibition of HIV-1 virus replication using small soluble Tat peptides. Virology 345: 373–389.
40. Futamura M, Dhanasekaran P, Handa T, Phillips MC, Lund-Katz S, et al. (2005) Two-step mechanism of binding of apolipoprotein E to heparin: implications for the kinetics of apolipoprotein E-heparan sulfate proteoglycan complex formation on cell surfaces. J Biol Chem 280: 5414–5422.
41. Arris CE, Boyle FT, Calvert AH, Curtin NJ, Endicott JA, et al. (2000) Identification of novel purine and pyrimidine cyclin-dependent kinase inhibitors with distinct molecular interactions and tumor cell growth inhibition profiles. J Med Chem 43: 2797–2804.
42. Jeffrey PD, Russo AA, Polyak K, Gibbs E, Hurwitz J, et al. (1995) Mechanism of CDK activation revealed by the structure of a cyclinA-CDK2 complex. Nature 376: 313–320.
43. Echalier A, Endicott JA, Noble ME (2010) Recent developments in cyclin-dependent kinase biochemical and structural studies. Biochim Biophys Acta 1804: 511–519.
44. Mandell DJ, Coutsias EA, Kortemme T (2009) Sub-angstrom accuracy in protein loop reconstruction by robotics-inspired conformational sampling. Nat Methods 6: 551–552.
45. Krebs WG, Gerstein M (2000) The morph server: a standardized system for analyzing and visualizing macromolecular motions in a database framework. Nucleic Acids Res 28: 1665–1675.
46. Krivov GG, Shapovalov MV, Dunbrack RL, Jr. (2009) Improved prediction of protein side-chain conformations with SCWRL4. Proteins 77: 778–795.
47. Vanhee P, Reumers J, Stricher F, Baeten L, Serrano L, et al. (2010) PepX: a structural database of non-redundant protein-peptide complexes. Nucleic Acids Res 38: D545–551.
48. Morris GM, Huey R, Lindstrom W, Sanner MF, Belew RK, et al. (2009) AutoDock4 and AutoDockTools4: Automated docking with selective receptor flexibility. J Comput Chem 30: 2785–2791.

49. Word JM, Lovell SC, Richardson JS, Richardson DC (1999) Asparagine and glutamine: using hydrogen atom contacts in the choice of side-chain amide orientation. J Mol Biol 285: 1735–1747.

50. Van Der Spoel D, Lindahl E, Hess B, Groenhof G, Mark AE, et al. (2005) GROMACS: fast, flexible, and free. J Comput Chem 26: 1701–1718.

51. Sethi A, Eargle J, Black AA, Luthey-Schulten Z (2009) Dynamical networks in tRNA: protein complexes. Proc Natl Acad Sci U S A 106: 6620–6625.

52. Glykos NM (2006) Software news and updates. Carma: a molecular dynamics analysis program. J Comput Chem 27: 1765–1768.

53. Easley R, Carpio L, Guendel I, Klase Z, Choi S, et al. (2010) Human T-lymphotropic virus type 1 transcription and chromatin-remodeling complexes. J Virol 84: 4755–4768.

54. Kehn K, Fuente Cde L, Strouss K, Berro R, Jiang H, et al. (2005) The HTLV-I Tax oncoprotein targets the retinoblastoma protein for proteasomal degradation. Oncogene 24: 525–540.

55. Wu K, Bottazzi ME, de la Fuente C, Deng L, Gitlin SD, et al. (2004) Protein profile of tax-associated complexes. J Biol Chem 279: 495–508.

4

Activity-Modulating Monoclonal Antibodies to the Human Serine Protease HtrA3 Provide Novel Insights into Regulating HtrA Proteolytic Activities

Harmeet Singh[1,2]*, Tracy L. Nero[3], Yao Wang[1,2], Michael W. Parker[3,4], Guiying Nie[1,2]*

1 MIMR-PHI Institute of Medical Research, Clayton, Victoria, Australia, **2** Monash University, Clayton, Victoria, Australia, **3** ACRF Rational Drug Discovery Centre, St Vincent's Institute of Medical Research, Fitzroy, Victoria, Australia, **4** Department of Biochemistry and Molecular Biology, Bio21 Molecular Science and Biotechnology Institute, the University of Melbourne, Parkville, Victoria, Australia

Abstract

Mammalian HtrA (high temperature requirement factor A) proteases, comprising 4 multi-domain members HtrA1-4, play important roles in a number of normal cellular processes as well as pathological conditions such as cancer, arthritis, neurodegenerative diseases and pregnancy disorders. However, how HtrA activities are regulated is not well understood, and to date no inhibitors specific to individual HtrA proteins have been identified. Here we investigated five HtrA3 monoclonal antibodies (mAbs) that we have previously produced, and demonstrated that two of them regulated HtrA3 activity in an opposing fashion: one inhibited while the other stimulated. The inhibitory mAb also blocked HtrA3 activity in trophoblast cells and enhanced migration and invasion, confirming its potential *in vivo* utility. To understand how the binding of these mAbs modulated HtrA3 protease activity, their epitopes were visualized in relation to a 3-dimensional HtrA3 homology model. This model suggests that the inhibitory HtrA3 mAb blocks substrate access to the protease catalytic site, whereas the stimulatory mAb may bind to the PDZ domain alone or in combination with the N-terminal and protease domains. Since HtrA1, HtrA3 and HtrA4 share identical domain organization, our results establish important foundations for developing potential therapeutics to target these HtrA proteins specifically for the treatment of a number of diseases, including cancer and pregnancy disorders.

Editor: Robert B. Sim, Oxford University, United Kingdom

Funding: This work was supported by the National Health and Medical Research Council (NHMRC) of Australia Fellowship #1041835 (to GN) and Program grant #494802, the Bill & Melinda Gates Foundation (OPP1025076 and OPP1086155 to GN) and the Victorian Government's Operational Infrastructure Support Program. MWP is a NHMRC Senior Principal Research Fellow (#1021645). The funders have no role in study design, data collection and analysis, decision to publish, or preparation of the manuscript. Prince Henry's Institute data audit number for this work is 13–19.

Competing Interests: HS and GN are co-inventors of a provisional Australian patent application titled "Antibodies to HtrA3". This patent application is for the monoclonal antibodies described in this paper and their use for diagnostics and therapeutics. This patent only restricts commercial applications and has no impact on academic researchers.

* Email: guiying.nie@mimr-phi.org (GN); harmeet.singh@mimr-phi.org (HS)

Introduction

High-temperature requirement (HtrA) proteins belong to a unique family of oligomeric serine proteases that are conserved from prokaryotes to humans [1]. HtrAs in general are involved in protein quality control through sensing protein folding stress and regulating signal transduction cascades [2]. In contrast to other quality control proteases, some HtrAs such as DegP of *Escherichia coli* also exhibit a chaperone function to stabilize specific proteins [3].

There are four human HtrAs in the genome: HtrA1, HtrA2, HtrA3 and HtrA4 [2,4–9]. These HtrAs play important roles in cell growth, apoptosis, invasion and inflammation; they also control cell fate via regulating protein quality control [2]. The altered expression of human HtrAs is associated with a number of diseases, including cancer, arthritis, neurodegenerative and neuromuscular disorders, age-related macular degeneration, and the pregnancy-specific disease preeclampsia [1,10–16].

HtrA proteases are comprised of a serine protease domain and one or more C-terminal protein-protein interaction domains [1]. They usually form higher order oligomers ranging from ~100 kDa up to 1.2 MDa [17–21]. All HtrA proteases share a common trimeric pyramidal architecture, where each monomer comprises two or three major domains, and exhibit a similar mechanism of activation [22]. Alteration of the N-terminal structural organization leads to functional diversity amongst the HtrA members and amino acid sequence variation gives rise to their substrate specificity [23–26].

HtrA1, HtrA3 and HtrA4 share an identical domain organization, suggesting that they may have similar functions, but they display different tissue expression patterns [7,9]. In contrast, HtrA2 has a completely different N-terminal domain architecture (Figure 1A) [7]. Recent structural and biochemical studies have demonstrated that HtrA1 exists as a trimer and that a conformational change induced by substrate binding is required to stimulate its proteolytic activity [27,28]. It has also been reported for HtrA1 that neither its N-terminal region [an insulin-like growth factor binding domain (IGFB) and a Kazal-type protease inhibitor domain (hereafter referred to as the Kazal domain)] nor the C-terminal postsynaptic density protein 95-Discs large-Zona occludens 1 (PDZ) domain has any involvement in the protease activity [27–29].

Mammalian HtrA3 was initially identified in the developing placenta as a serine protease associated with pregnancy both in the mouse and human [7,30–32]. Two HtrA3 isoforms [long (HtrA3-L) and short (HtrA3-S)], due to alternative mRNA splicing, have been identified in the human placenta [7]. HtrA3-L has four distinct domains including an IGFB, Kazal, trypsin-like serine protease (referred to as the protease domain) and PDZ domain (responsible for protein-protein interaction) [2,7] (Figure 1B). HtrA3-S and HtrA3-L are identical except HtrA3-S lacks the PDZ domain (Figure 1B) [7].

HtrA3 negatively regulates trophoblast invasion during placental development [33,34] and abnormal levels of HtrA3 during early pregnancy in women are associated with risks of developing preeclampsia (a severe pregnancy-specific disorder) [15,35]. HtrA3 is also down-regulated in a number of cancers (eg. ovary, endometrium and lung) and has been suggested to inhibit transforming growth factor-β signaling [10,36–40].

To gain insight into the functional importance of HtrA3, we recently generated five HtrA3 monoclonal antibodies (mAbs), mapped their linear epitopes and established the specificity of these mAbs against HtrA1 and HtrA2 [41]. In this current study, we further establish the specificity of these HtrA3 mAbs against human HtrA4 (the fourth member of the HtrA family), and also investigate the capacity of these mAbs to modulate HtrA3 activity and their modes of action. We show that one of these mAbs inhibits whereas another stimulates HtrA3 activity with high specificity. We also demonstrate that the inhibitory mAb blocks HtrA3 activity in placental trophoblast cells to enhance migration and invasion, demonstrating its potential utility *in vivo*. To understand how the binding of these mAbs may modulate HtrA3 activity, we constructed a 3-dimensional (3D) HtrA3-L (hereafter referred to as HtrA3) homology model using HtrA1 small angle X-ray scattering (SAXS) data and mapped the location of the mAb epitopes onto the homology model. The inhibitory mAb appears to block substrate access to the HtrA3 protease catalytic site; whereas the stimulatory mAb may bind to either the PDZ domain alone or in combination with the N-terminal and protease domains. These results lay the foundation for developing therapeutics to specifically target HtrA3. Furthermore, since mammalian HtrA3 shares a similar domain organization with HtrA1 and HtrA4, our results have important implications in targeting these HtrA proteases.

Materials and Methods

Recombinant human HtrA proteins

C-terminally His-tagged full length recombinant (r) human HtrA1 (amino acids 1–480, Accession # Q92743, Cat # 30600102), HtrA3 (amino acids 1–453, Accession # P83110, Cat # 30600503) and HtrA4 (amino acids 1–476, Accession # P83105, Cat # 30600403) (all produced in insect cells) were obtained from ProteaImmun GmbH (Berlin, Germany), while mature HtrA2 protein (amino acids 134–458, Accession # O43464, Cat # 1458-HT, *Escherichia coli* origin) was from R&D Systems (Minneapolis, MN, USA).

Mouse monoclonal antibody production

HtrA3 mAbs 3E6, 6G6, 10H10 and 9C9 were produced against catalytically inactive HtrA3-L-S305A, constructed by mutating the catalytic site serine residue 305 to alanine [42]; mAb 2C4 was raised against the synthetic peptide TIKIHPKKKL (corresponding to residues 230–239 in both HtrA3-L and HtrA3-S) [41]. The linear epitopes of these mAbs were determined by screening a custom-synthesized peptide library (PepSet, Mimotopes) [41].

Cell culture

The HTR8/SVneo (HTR8) cell line was derived from primary explants cultures of human first trimester placentas (8–10 wk gestation) and immortalized with SV40 virus [43]. The HTR8 trophoblast cells, kindly provided by Dr C.H. Graham (Queen's University, Kingston, ON, Canada), were cultured at 37°C as previously described [34].

Western blot

Human rHtrA1, rHtrA2, rHtrA3 and rHtrA4 proteins (50 ng) were analyzed using standard Western blot (12% reducing SDS-PAGE and PVDF membrane). Primary antibodies included HtrA3 mAbs (50 μg/ml final concentration) and an anti-HtrA4 antibody (200 ng/ml final concentration, affinity-purified rabbit polyclonal, Abcam, Cambridge, UK). The membranes were incubated overnight at 4°C with primary antibodies and probed for 1 hour at room temperature with the following secondary conjugates: rabbit anti-mouse IgG HRP (1:5000, Cell Signaling, Beverley, MA, USA) or goat anti-rabbit IgG HRP (1:5000, DAKO, Carpinteria, CA, USA). Bands were visualized with Pierce ECL Western Blotting Substrate (Thermo Fisher Scientific, Rockford, IL, USA) and ChemiDoc MP Imaging system (Bio-Rad, Hercules, CA, USA).

In vitro protease activity assay

The protease activity of HtrA3 was determined by the cleavage of a custom-made fluorescence-quenched peptide substrate H2-Opt [Mca-IRRVSYSF(Dnp)KK, synthesized by GL Biochem Ltd., Shanghai China] as previously described [27], with minor modifications.

In brief, the activity was determined in a final 50 μl reaction in half-area 96-well clear-bottomed black plates (Sigma). Firstly, a 40 μl reaction mixture containing a final concentration of 1.25 μM rHtrA3 was prepared in 50 mM Tris-HCl (pH 8.0) containing 200 mM NaCl and 0.25% CHAPS. The fluorescence-quenched peptide substrate (10 μl, final concentration 2.5 μM) was then added, and the plates were incubated at 37°C for 30 min during which the real-time kinetic progression of substrate cleavage (increase in fluorescence signal) was monitored every 15 sec at 340 nm/405 nm (Wallac, Victor 2 spectrophotometer, Perkin Elmer, MA). The rate of substrate cleavage (fluorescence increase/min) was calculated from the initial 10 min linear phase of the kinetic progression curve and used as the activity unit. In every assay, the incubation buffer was used as a blank and other controls included PBS substitution for rHtrA3.

To determine the modulatory activity of HtrA3 mAbs on the proteolytic activity of rHtrA3, the *in vitro* activity assay was performed in the presence of HtrA3 mAbs. Firstly, a 40 μl reaction mixture containing rHtrA3 protein (final concentration 1.25 μM), 5 μl HtrA3 mAbs (3E6, 6G6, 2C4, 10H10 or 2C4; final concentration 20 μg/ml) or control mAb (a non-HtrA3 mAb produced/purified similarly to HtrA3 mAbs, also at final concentration of 20 μg/ml) or buffer only, was prepared in 50 mM Tris-HCl (pH 8.0) containing 200 mM NaCl and 0.25%

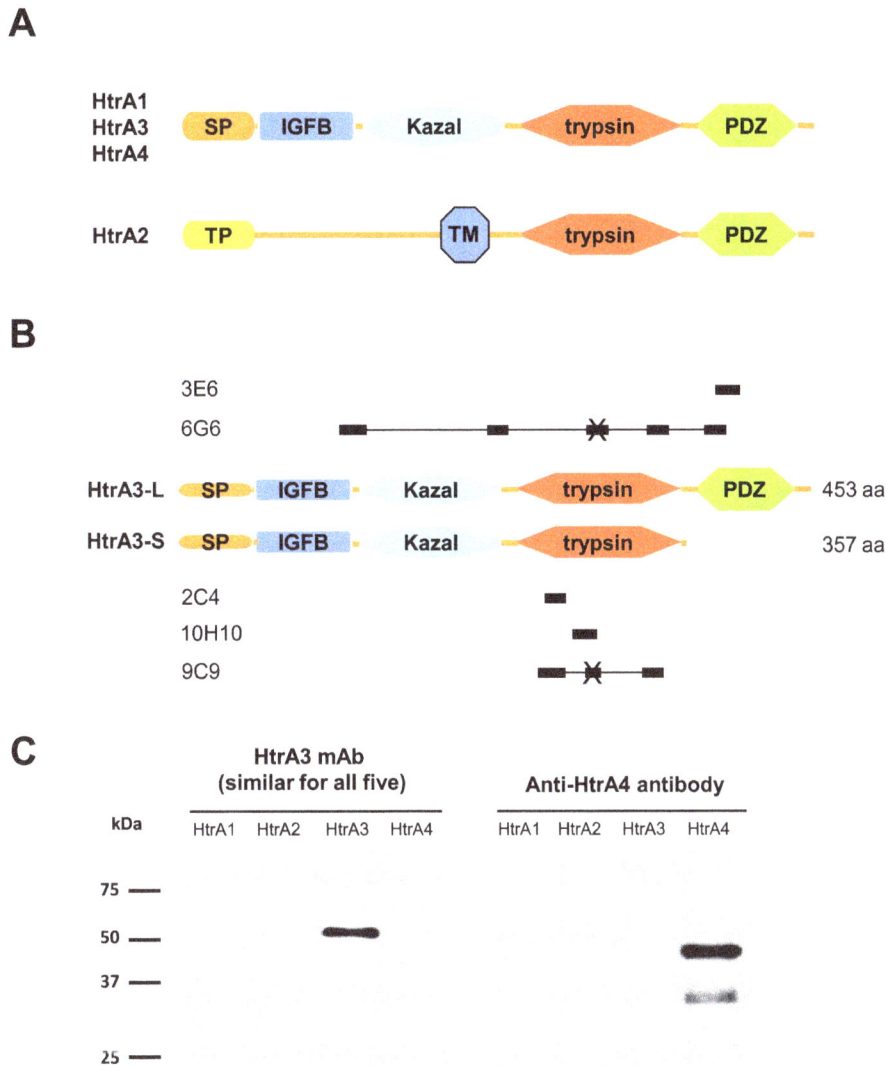

Figure 1. Schematic representation of HtrA3 domain organization, mAb epitope locations and confirmation of mAbs specificity. (A) The domain structure of HtrA1, HtrA3, HtrA4 and HtrA2. (B) The domain structure of HtrA3-L and HtrA3-S. The solid bars above or below the protein domains denote the locations of epitope residues of each mAb identified by the linear peptide library mapping assay. "X" indicates a peptide deemed likely to be a false positive. SP, signal peptide; IGFB, IGF-binding domain; Kazal, Kazal-type S protease inhibitor domain; trypsin, trypsin-like serine protease domain; PDZ, PDZ domain; TM, transmembrane; TP, transient peptide. (C) An equal amount (50 ng) of recombinant human HtrA proteins HtrA1, HtrA2, HtrA3 (HtrA3-L-S305A) and HtrA4 were separated on reducing 12% SDS-PAGE gels and analyzed by Western blot with HtrA3 mAbs (3E6, 6G6, 2C4, 10H10 and 9C9), and an HtrA4-specific antibody.

CHAPS. This mixture was incubated for 1 hour at 37°C, then the fluorescence-quenched peptide substrate (final concentration 2.5 μM) was added and the plates were incubated at 37°C for 30 min to monitor the real-time kinetic progression of substrate cleavage, as described above. In every assay, the incubation buffer was used as a blank and other controls included PBS substitution for HtrA3 protein and control mAb replacing HtrA3 mAbs. Additional control included 6G6 mAb only.

The dose-dependency of the inhibitory (10H10) and stimulatory (6G6) HtrA3 mAbs was further tested at a final concentration of 4 and 20 μg/ml. Furthermore, to test whether HtrA3 following 6G6 stimulation could be inhibited by 10H10, rHtrA3 was first incubated with 6G6 (20 μg/ml) for 1 hour at 37°C, then 10H10 (4 or 20 μg/ml) was added and the reaction was incubated for a further 1 hour at 37°C, before the addition of the fluorescence-

quenched peptide substrate. To test if rHtrA3 following 10H10 inhibition could be activated by 6G6, rHtrA3 was first incubated with 10H10 (20 μg/ml) for 1 hour at 37°C, then 6G6 (4 or 20 μg/ ml) was added and the reaction was incubated for a further 1 hour at 37°C, before the addition of the fluorescence-quenched peptide substrate. Three independent experiments were performed for each condition. Data were expressed as fold changes in the rate of substrate cleavage relative to control.

Trophoblast cell migration and invasion in the presence of HtrA3 activity-modulating mAbs

To determine the effects of HtrA3 activity-modulating mAbs on cellular processes, real-time monitoring of migration and invasion of trophoblast HTR8 cells was carried out in the presence or absence of HtrA3 mAbs using xCELLigence, RTCA DP instrument (Roche Diagnostics GmbH, Germany) that was placed in a humidified incubator and maintained at 37°C with 95% air/ 5% CO_2. For proliferation, growth curves were constructed using 16-well plates (E-plate 16, Roche Diagnostics GmbH). Briefly, HTR8 cells were seeded at 40,000/well in medium containing 1% FCS and monitored once every 2 min for 40 min and then once every hour. Following cell adhesion, HtrA3 mAbs (10H10 or 6G6, final concentration; 5 μg/ml) were added, and the plates were then monitored once every 15 min for 2 hours, then once every hour.

Cell migration and invasion were assessed using specially designed 16-well plates (CIM-plate 16, Roche Diagnostics GmbH) with 8-mm pores. These plates are similar to conventional transwell with the micro-electrodes located on the underside of the membrane of the upper chamber. To measure cell invasion, the upper surface of the transwell was coated with growth factor reduced Matrigel (BD BioSciences, Bedford, MA USA; 1:10 diluted in serum free media). HTR8 cells were incubated with HtrA3 inhibitory (10H10) or stimulatory (6G6) mAbs each at 5 μg/ml for 20 min at 37°C prior to seeding into the upper chamber at 40,000/well in medium containing 1% FCS. In the lower chamber, the media containing 5% FCS was added as a chemo-attractant. Controls included untreated and control-IgG-treated HTR8 cells on culture inserts. The plates were monitored every 2 min for 40 min, then once every 15 min. Data analysis was carried out using the RTCA Software v1.2. Each migration and invasion assay was repeated three times and data were expressed as percent changes (± SEM) relative to untreated control.

Construction of the HtrA3 trimer homology model

HtrA1 and HtrA3 (HtrA3-L) share a similar domain organization (Figure 1A) and 61% sequence identity [7], the sequence alignment is shown in Figure 2. The only available crystal structure of HtrA3 is that of the PDZ domain [44]. However, crystal structures of the IGFB, Kazal and protease domains of HtrA1 are available [27,28] and these were used to model the corresponding HtrA3 domains. The sequence identity between the HtrA1 and HtrA3 IGFB, Kazal and protease domains is 53%, 53% and 76%, respectively [alignments were carried out using the program Muscle, (http://toolkit.tuebingen.mpg.de/muscle)]. The trimeric HtrA3 homology model was constructed in two stages: the first involved the construction of a monomer model and the second, packing three HtrA3 monomers to form a trimer.

The monomeric homology model of HtrA3 consisting of residues 24–453, without the signal peptide (residues 1–23), was constructed as follows. The HtrA1 and HtrA3 amino acid sequences for the (i) N-terminal region (i.e. the IGFB and Kazal domains) and (ii) protease domain were aligned using the default parameters in the program Muscle. The resulting alignment of the N-terminal regions was used to model HtrA3 residues 24–130 using the program Modeller v9.9 (http://toolkit.tuebingen.mpg.de/modeller#, [45]). The crystal structure of the N-terminal region (PDB id: 3TJQ, [27]) of human HtrA1 was the structural template for the HtrA3 N-terminal region (residues 24–130). Likewise, the homology model for HtrA3 residues 131–350 was constructed using the crystal structure of the catalytically active human HtrA1 protease domain (PDB id: 3NZI, A chain [28]) as the template. The SAXS envelope for full length HtrA1, kindly provided by Charles Eigenbrot and Mark Ultsch [27], was used to guide the manual placement of the HtrA3 N-terminal region in relation to the protease domain. Refinement of the orientation of the HtrA3 N-terminal region (residues 24–130) to the protease domain (residues 131–350) was then performed using the

RosettaDock Server (Rosetta Suite 2.1; http://rosettadock.graylab.jhu.edu/). The top 10 solutions obtained from Rosetta-Dock were overlaid onto the HtrA1 SAXS envelope and the one with the best fit was used in the HtrA3 homology model. The C-terminus of the HtrA3 N-terminal region (i.e. residue 130) was attached manually to the N-terminus (i.e. residue 131) of the HtrA3 protease domain. The missing residues (Pro-380 and Glu-381) in the crystal structure of the human HtrA3 PDZ domain (PDB id: 2P3W, A chain) were manually inserted. Using the HtrA1 SAXS envelope as a guide, HtrA3 residues Asp-351, Trp-352 and Lys-353 were manually added to the C-terminus of the HtrA3 protease domain and then the N-terminus of the HtrA3 PDZ domain was manually attached to Lys-353, ultimately generating the monomeric homology model of HtrA3.

To construct the trimeric HtrA3 homology model, the protease domain (residues 131–350) of the monomeric HtrA3 model was aligned to each of the protein chains in the trimeric HtrA1 protease domain crystal structure (PDB id: 3NZI, A, B and C chains [28]). The three HtrA3 monomers packed together with only a few minor steric conflicts between adjacent monomer side-chains; to alleviate these steric problems, the conformation of the side-chains were adjusted using amino acid conformer libraries. The HtrA3 monomeric and trimeric models were geometry optimized after each model building step for at least 2000 iterations (or until the gradient of successive iterations was < 0.05 kcal/mol • Å) using the molecular mechanics Amber02 force field, Amber partial atomic charges and conjugate gradient minimization method (all other parameters were at default values) within the program Sybyl-X 2.0 (Certara, L.P.; http://tripos.com). All manual manipulations were performed using Sybyl-X 2.0. The resulting trimeric HtrA3 homology model was deemed to be a good quality model using PROCHECK [46], with 98.6% of residues in favored (89.8%) or allowed (8.8%) conformations.

Statistics

Data are expressed as mean ± SEM of fold changes relative to control. Statistical analysis was performed on raw data using one-way ANOVA and Tukey's post hoc test using PRISM version 5.00 (GraphPad Software, San Diego, CA), and $P<0.05$ was taken as significant.

Results

Epitopes of HtrA3 mAbs and their specificity

The domain organization of all four human HtrAs (1–4) is schematically illustrated in Figure 1A, and the similarities and differences between human HtrA3-L and HtrA3-S isoforms are shown in Figure 1B. Immunization of mice against a minimal mutant of HtrA3-L (HtrA3-L-S305A) or a synthetic peptide (residues 230–239 of human HtrA3-L) and subsequent cloning resulted in five distinct HtrA3 mAbs: 3E6, 6G6, 2C4, 10H10 and 9C9 [41]. The linear epitopes of these mAbs, schematically shown in Figure 1B, corresponded to the following residues in HtrA3: 3E6, 403–417; 6G6, 73–92, 133–147, 288–302, 313–327 and 398–412; 2C4, 230–239; 10H10, 278–292; and 9C9, 223–242, 283–297 and 308–322 (Figure 2).

All five mAbs were previously shown by Western blot to recognize wild type HtrA3 (both isoforms or HtrA3-L only), but not rHtrA1 or rHtrA2 [41]. While HtrA3-L was recognized by all five mAbs, HtrA3-S was not detected by 3E6 or 6G6, consistent with their epitopes containing HtrA3-L specific sequences in the PDZ domain (Figure 1B) [41]. In this study, we further confirmed that these mAbs were highly specific to HtrA3 against the entire human HtrA family including the

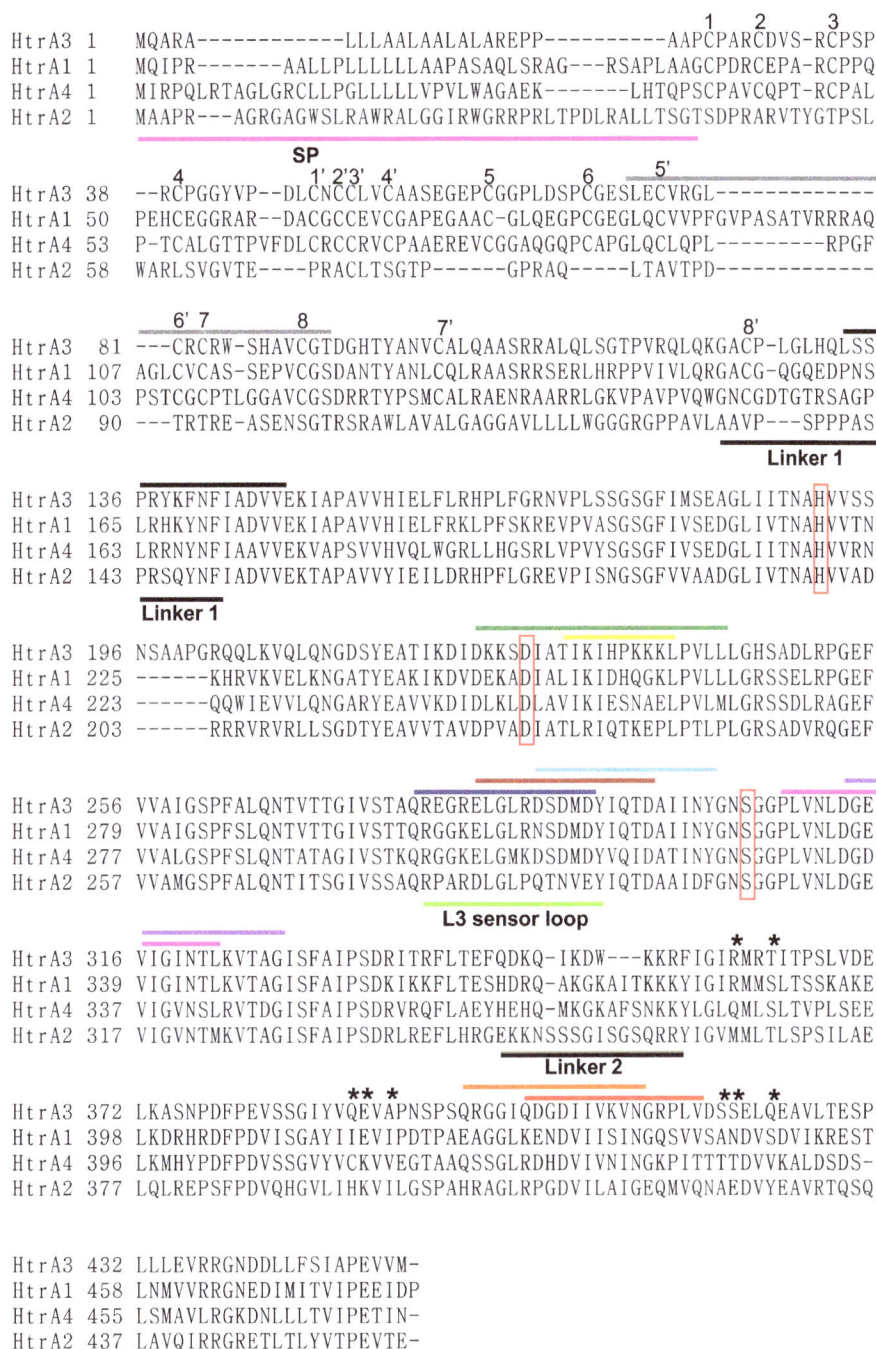

Figure 2. HtrA3 mAb epitope regions and sequence comparison within the human HtrA family members. The HtrA3 signal peptide (SP, residues 1–23) is defined by the pink underline. The eight disulfide bridges in the HtrA1 N-terminal domain are identified by the paired numbers (1 and 1'; 2 and 2' etc) above the Cys residues involved. These 8 disulfide bridges appear to be conserved in both HtrA3 and HtrA4. The location of the flexible linker separating the N-terminal domain from the protease domain is indicated by the black underline labeled Linker 1; the flexible linker connecting the protease domain to the C-terminal PDZ domain is labeled Linker 2. The catalytic triad Ser-His-Asp residues are enclosed in red boxes. The location of the protease L3 sensor loop is indicated by the green underline. The black asterisks denote the HtrA3 PDZ domain residues involved in protein/peptide interactions. The location of the HtrA3 mAb epitopes are indicated by the colored lines above the HtrA3 sequence: residues 73–92 (dark grey, mAb 6G6), residues 133–147 (black, mAb 6G6), residues 223–242 (dark green, mAb 9C9), residues 230–239 (yellow, mAb 2C4), residues 278–292 (dark blue, mAb 10H10), residues 283–297 (brown, mAb 9C9), residues 288–302 (blue, mAb 6G6), residues 308–322 (magenta, mAb 9C9), residues 313–327 (purple, mAb 6G6), residues 398–412 (orange, mAb 6G6) and residues 403–417 (red, mAb 3E6). The sequence alignment was carried out using the program Muscle.

newly discovered HtrA4 (Figure 1C). While rHtrA4 was detected by an HtrA4 antibody, none of the HtrA3 mAbs recognized HtrA4 (Figure 1C).

Identification of HtrA3 activity modulating mAbs

We next assessed whether these HtrA3 mAbs could modulate the proteolytic activity of HtrA3. When pure human HtrA3

(HtrA3-L) was incubated with a fluorescence-quenched peptide substrate, a progressive increase in fluorescence signal resulting from substrate cleavage was detected (Figure 3A). To test the effects of the mAbs on HtrA3 activity, an equal amount of each individual HtrA3 mAbs or control IgG (20 μg/ml) was added into the enzyme reaction and the substrate cleavage kinetics were monitored. Compared to the control mAb, 10H10 reduced whereas 6G6 enhanced the proteolysis, while mAbs 3E6, 2C4 or 9C9 did not significantly affect the HtrA3 activity (Figure 3A). In the presence of mAb 6G6, the peptide substrate cleavage was faster (Figure 3A). These data indicate that mAb 6G6 stimulates whereas 10H10 inhibits HtrA3 activity.

To further investigate dose-dependent effect of mAbs 6G6 and 10H10, the enzyme reaction was carried out with different concentrations of these mAbs (0, 4 and 20 μg/ml). To illustrate the dose-dependency, the rate of substrate cleavage was expressed as a percentage of the control (containing control mAb at 4 or 20 μg/ml). Indeed, 10H10 inhibited (Figure 3B) whereas 6G6 stimulated (Figure 3C) HtrA3 activity in a clear dose-dependent manner. No enzyme activity was detected when 6G6 alone was incubated with the substrate, confirming that 6G6 itself had no peptidase activity (data not shown).

We next investigated whether HtrA3 activity, following mAb 6G6 stimulation, could be inhibited by mAb 10H10. HtrA3 was first incubated with 6G6 (20 μg/ml) and then with 10H10 (4 or 20 μg/ml) before the substrate cleavage was determined. The mAb 10H10 inhibited HtrA3 activity after 6G6 stimulation and the inhibition was also dose-dependent (Figure 3D), further confirming the blocking function of mAb 10H10. Likewise, we examined whether HtrA3, following 10H10 inhibition, could be activated by 6G6. However, no activity was detected when HtrA3 was first incubated with 10H10 (20 μg/ml) then with 6G6 (4 or 20 μg/ml, data not shown).

Neither the 10H10 nor 6G6 had any effect on the proteolytic activity of other human HtrA members – HtrA1, HtrA2 and HtrA4 (data not shown), consistent with these two HtrA3 mAbs recognizing HtrA3 only (Figure 1C).

Confirmation that mAb 10H10 increases trophoblast migration and invasion in vitro

We next determined whether these HtrA3 activity-modulating mAbs would affect the function of HTR8 trophoblast cells expressing HtrA3 [33,34]. Initially, real-time cell proliferation was monitored to assess the effects of these mAbs on cell growth and survival. Neither 10H10 nor 6G6 (both at 5 μg/ml) significantly altered cell growth (data not shown). However, 10H10 significantly increased the migration (Figure 4A & B) as well as invasion (Figure 5A & B) of HTR8 cells, confirming previous reports that HtrA3 inhibition enhances cellular migration and invasion without affecting growth [33,34]. In contrast, when the experiment was repeated for mAb 6G6, no significant effects on cell migration or invasion were observed (data not shown).

The 3-dimensional homology model of HtrA3

The trimeric human HtrA3 homology model shown in Figure 6A is structurally analogous to the full length human HtrA1 solution structure determined by Eigenbrot and coworkers [27] using SAXS. The overlay of the trimeric HtrA3 homology model with the HtrA1 SAXS envelope is shown in Figure S1. The full sequence comparison of the four human HtrA subtypes is given in Figure 2. The N-terminal IGFB-Kazal domains are connected to the HtrA3 protease domains by a flexible linker of ~20 amino acids (designated Linker 1 in Figure 2) and are situated in between the protease domains of adjacent monomers

(Figure 6A). Although a total of eight disulfide bridges can be predicted in the N-terminal IGFB-Kazal domains, only seven are observed in the crystal structure, one is missing due to structural disorder [27]. These disulfide bridges appear to be conserved in both HtrA3 and HtrA4 (the eight pairs of Cys residues involved in the HtrA1 disulfide bridges are identified in Figure 2 by paired numbers 1-1', 2-2' and so on above the HtrA3 sequence). In the HtrA3 model, the sixteen Cys residues are able to form equivalent disulfide bridges. Interestingly, the other domains of HtrA1, HtrA3 and HtrA4 do not contain any Cys residues. The N-terminal domain of HtrA2 has a different architecture to that of the other three HtrA subtypes (Figure 1A) and it does not contain the same Cys residue pattern (Figure 2).

The protease domain of the HtrA family is structurally conserved and adopts a canonical trypsin fold consisting of two β-barrels with a couple of α-helices attached [27,28]. The catalytic Ser-His-Asp triad is located in between the two β-barrels. The HtrA3 protease catalytic Ser residue has been determined by mutagenesis to be Ser-305 and is contained within the mammalian protease GNSGGPL sequence motif [7,41,42]. Likewise, the catalytic HtrA3 His-191 residue has been identified by the mammalian HtrA protease TNAHV sequence motif. The putative catalytic Asp-227 residue was identified by analogy with HtrA1 and HtrA2 from sequence alignments (Figure 2) [7,41]. The location of the catalytic site in each of the three modeled HtrA3 protease domains is indicated by the red asterisks in Figure 6A and the catalytic triad residues His-191, Asp-227 and Ser-305 are highlighted in Figure 6B. In HtrA1 the sensor loop L3 (defined in Figure 2) is located on the catalytic face of the protease domain and interacts directly with the enzyme substrate. This is a key step in the activation of the HtrA1 protease [28]. The distance between the HtrA1 L3 sensor loop and the adjacent catalytic site is ~10 Å. Due to the structural conservation of the HtrA protease domain and the sequence similarity between HtrA1 and HtrA3, we would expect the HtrA3 L3 sensor loop (corresponding to residues 280–293, Figure 2) and catalytic triad to be a similar distance apart.

The C-terminal HtrA3 PDZ domain, is a protein-protein interaction domain, is comprised of five β-strands which form a β-sandwich structure and three α-helices. The residues involved in protein or peptide binding are Arg-360, Thr-363, Gln-389, Glu-390, Ala-392, Ser-419, Ser-420 and Gln-423 [colored light blue in Figure 7B & D, [44]]. The PDZ domains are connected to the HtrA3 protease domains by a flexible linker (~11 amino acids, designated Linker 2 in Figure 2), and are located ~28 Å distant from the protease domains (Figure 6A).

Although the trimeric HtrA3 homology model fits the HtrA1 SAXS envelope extremely well, the N-terminal region, the protease domain and the PDZ domain could adopt alternative packing arrangements in solution due to the flexible linkers connecting them. It has also been reported that the N-terminal and PDZ domains of HtrA1 are not required for protease activity [27–29]. Previous studies have confirmed that HtrA3-S lacking the PDZ domain is also proteolytically active [34,42]. Whether the N-terminal of HtrA3 is required for protease activity remains to be determined.

Mapping the five mAb epitopes onto the HtrA3 model

Antibody epitopes can be either continuous or discontinuous on the target protein surface. Mapping the location of antibody epitopes by screening linear peptide libraries is a common initial approach adopted by both academia and Big Pharma [47,48]. If there is sequence or shape similarity between peptides in the library then false positives can arise using this mapping. The linear epitopes, determined by peptide library screening for each of the

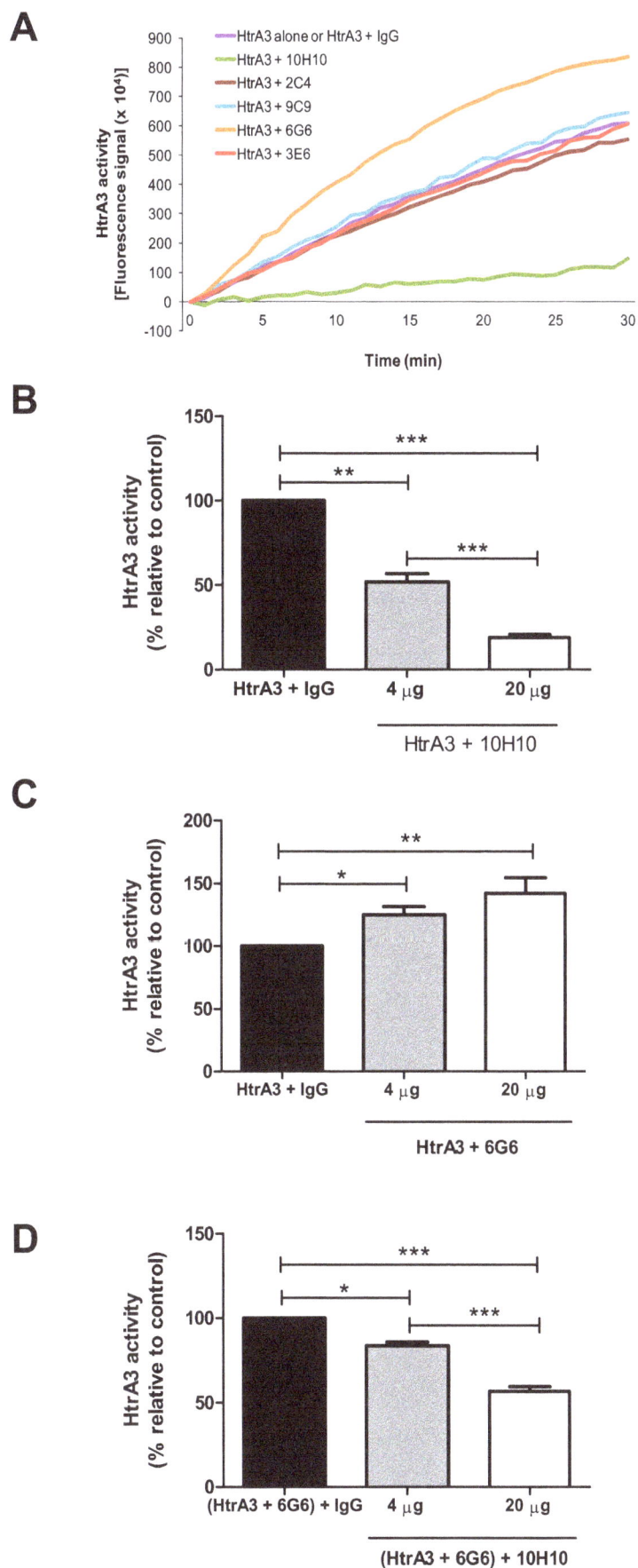

Figure 3. Effects of HtrA3 mAbs on substrate cleavage and dose-dependent modulation of HtrA3 activity by mAbs. (A) Representative real-time progressive curves of substrate cleavage by recombinant wild type human HtrA3 (HtrA3-L) in the presence of 20 μg/ml individual HtrA3 mAb (6G6, 9C9, 10H10, 3E6 and 2C4) or control mAb. (B) Inhibition of HtrA3 activity by mAb 10H10. (C) Enhancement of HtrA3 activity by mAb 6G6. (D) Inhibition of HtrA3 activity by mAb 10H10 subsequent to 6G6 stimulation. The data are expressed as changes in the rate of substrate cleavage relative to the control (B & C: control = HtrA3 with IgG control mAb, D: control = HtrA3 with mAb 6G6 at 20 μg/ml). Data are mean ± SEM from 3 independent experiments, *, P<0.05, **, P<0.01, ***, P<0.001.

five HtrA3-specific mAbs (Figure 1B & 2) [41], were mapped onto the 3D HtrA3 model (Figure 6–8) to visualize their location and to understand how they may modulate protease activity. Antibodies 10H10, 2C4 and 3E6 have continuous epitopes whereas mAbs 9C9 and 6G6 appear to have discontinuous epitopes.

The putative HtrA3 L3 sensor loop (residues 280–293, Figure 2) corresponds to the epitope of the inhibitory mAb 10H10 (residues 278–292; colored dark blue in Figure 2 & 6A-C). An antibody binding to residues 278–292 would not only prevent a substrate from interacting with the L3 sensor loop, but also block its access to the catalytic triad, thereby inhibiting the protease activity of HtrA3. Such a mode of action is consistent with 10H10 being a neutralizing mAb (Figure 3).

HtrA3 mAb 2C4 was raised against a single peptide corresponding to residues 230–239 (colored yellow in Figure 2, 6A, D & E). In the HtrA3 homology model, residues 230–239 lie on an outer surface loop of the protease domain on the opposite face to the catalytic site (Figure 6D & E) and would be accessible by an antibody without impacting on the catalytic residues. The

location of the 2C4 epitope is consistent with this mAb having no impact on the *in vitro* enzyme activity of HtrA3 (Figure 3A).

The mAb 3E6 was shown by Western blot to recognize HtrA3-L but not the short isoform (HtrA3-S) [41], hence the PDZ domain appears to be required for the binding of this mAb. The epitope for mAb 3E6, corresponding to residues 403–417 (colored red in Figure 2, 6A-E, 7A & B), lies solely in the PDZ domain and is largely surface exposed. It is located on the opposite surface to the PDZ domain protein/peptide binding groove (colored light blue in Figure 7B). An antibody would be able to bind to this epitope without blocking access to the PDZ domain protein/peptide binding groove and would also have no direct contact with the protease domain or its catalytic site when the PDZ domain is extended ~28 Å out from the central protease domain. This is highly consistent with the HtrA3 activity data (Figure 3A). The lack of any influence by mAb 3E6 on protease activity is also in accordance with the observation that the HtrA1 PDZ domain does not have any involvement in protease activity or enzyme regulation [27,28].

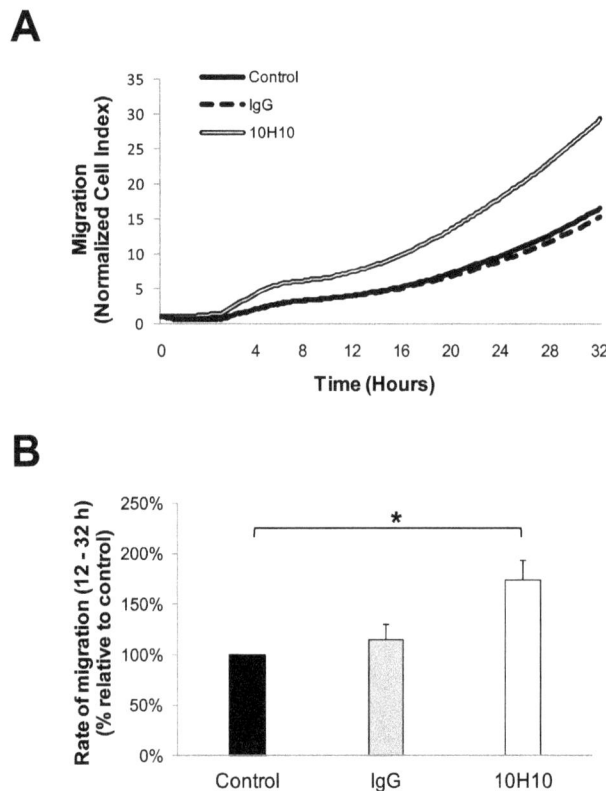

Figure 4. Enhancement of trophoblast migration by mAb 10H10. (A) Representative cell index curves of HTR8 cells for migration in the absence or presence of 10H10 or control mAb (5 μg/ml) measured with xCELLigence system. (B) Exogenous addition of mAb 10H10 (5 μg/ml) significantly increased HTR8 cell migration at 12–32 hours, compared to untreated controls. Data are mean ± SEM from 3 independent experiments, *, P<0.05.

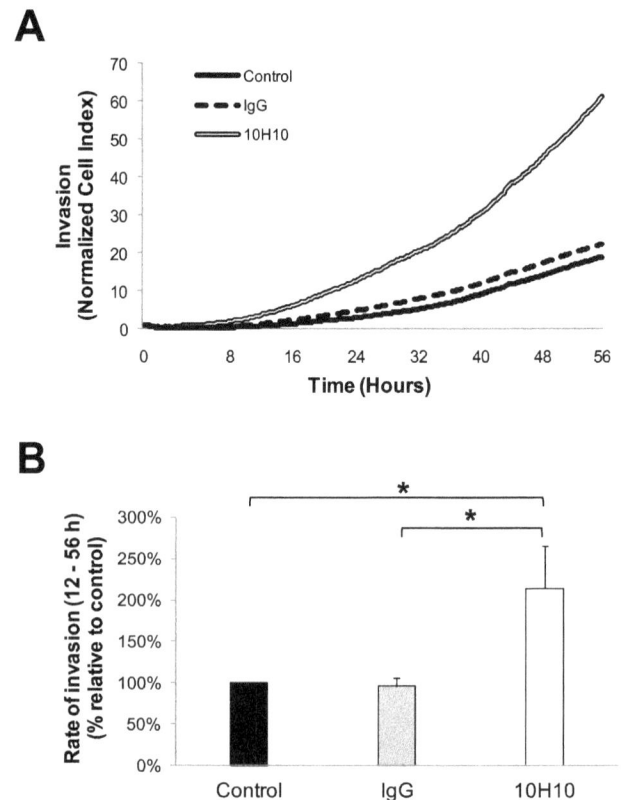

Figure 5. Enhancement of trophoblast invasion by mAb 10H10. (A) Representative cell index curves of HTR8 cells for invasion in the absence or presence of 10H10 or control mAb (5 μg/ml) measured with xCELLigence system. (B) Exogenous addition of mAb 10H10 (5 μg/ml) significantly increased HTR8 cell invasion at 12–56 hours, compared to untreated controls. Data are mean ± SEM from 3 independent experiments, *, P<0.05.

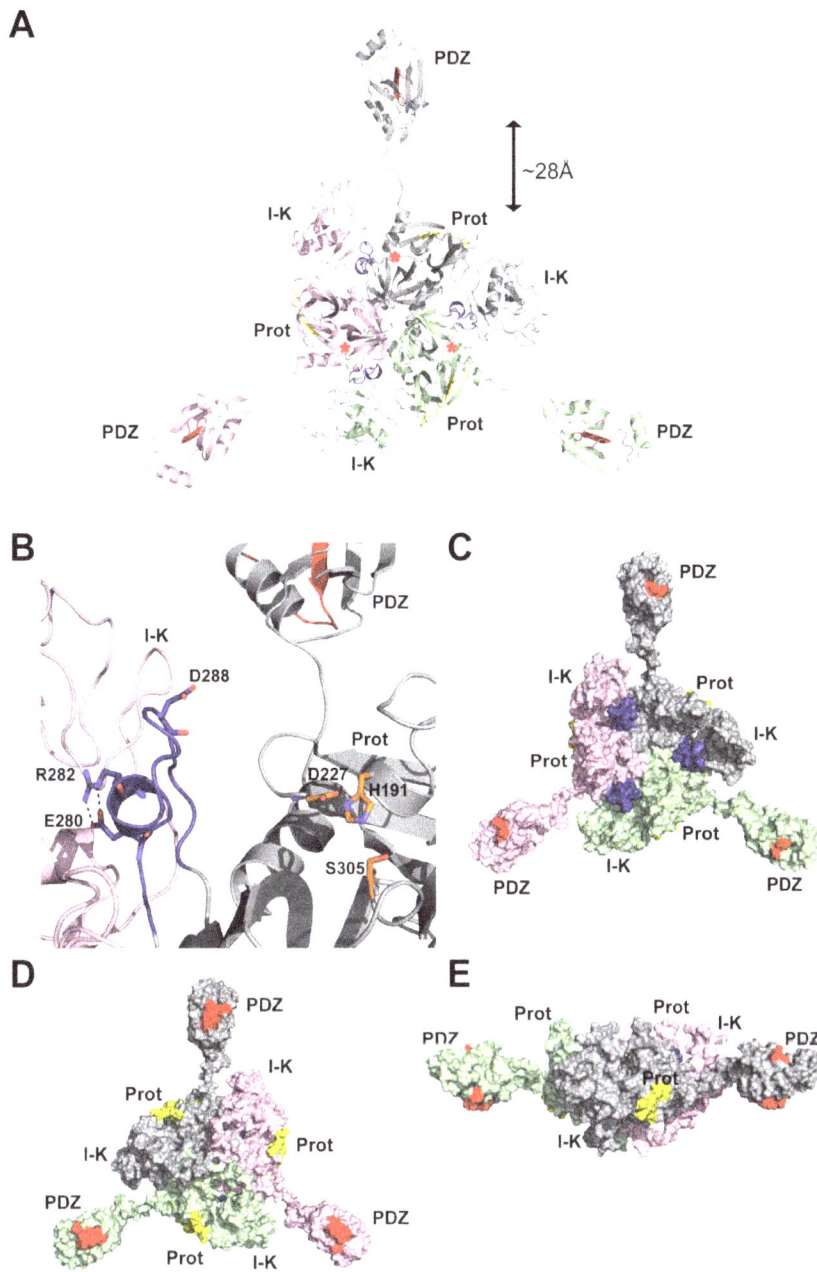

Figure 6. The trimeric HtrA3 homology model and location of the epitopes for mAbs 10H10, 2C4 and 3E6. (A) Cartoon representation of the trimeric HtrA3 (HtrA3-L) homology model, with each of the three HtrA3 monomers colored differently (grey, light green and light pink). Individual domains within each monomer are labeled: I-K; N-terminal combined IGFB-Kazal domains; Prot; central protease domain, and PDZ; the PDZ domain. Red asterix indicates the location of the catalytic site in each monomer. Epitope for the inhibitory mAb 10H10, residues 278–292 in the protease domain, is colored dark blue and corresponds to the putative HtrA3 sensor loop L3 [28]. Epitopes for mAbs 2C4 (residues 230–239) in each protease domain and 9C9 (residues 403–417) located in the PDZ domain are colored yellow and red, respectively. View directly above the protease catalytic sites. (B) Close up view of the 10H10 inhibitory epitope (dark blue) and modeled protease catalytic triad (orange sticks) for one HtrA3 monomer (grey cartoon). Catalytic triad residues His-191, Asp-227 and Ser-305 are displayed as orange sticks; Glu-280, Arg-282 and Asp-288 of the putative L3 sensor loop are displayed as dark blue sticks. Salt bridge/hydrogen bond interactions between Glu-280 and Arg-282 are shown as dashed black lines. The adjacent HtrA3 monomer is shown as a light pink cartoon. (C) Same view as in panel (A), protein is depicted as a molecular surface. (D) View of the non-catalytic face of HtrA3, i.e. a rotation of +180° about the Y-axis from the view shown in panels (A) and (C). (E) Side view of the grey monomer [−90° rotation about the X-axis, followed by a +120° rotation about the Y-axis from the view shown in panel (A)]. One letter amino acid codes have been used for labels.

Like 2C4 and 3E6, mAb 9C9 also has no effect upon HtrA3 protease activity (Figure 3A); but it appears to have a discontinuous epitope as three linear peptides in the library screen displayed affinity for this mAb. The three linear peptides correspond to HtrA3 protease domain residues 223–242, 283–297 and 308–322 (colored dark green, brown and magenta, respectively in Figure 2, 8A & B). Residues 283–297 (colored brown in Figure 2 & 8A) are located on the catalytic face of HtrA3 and encompass the majority

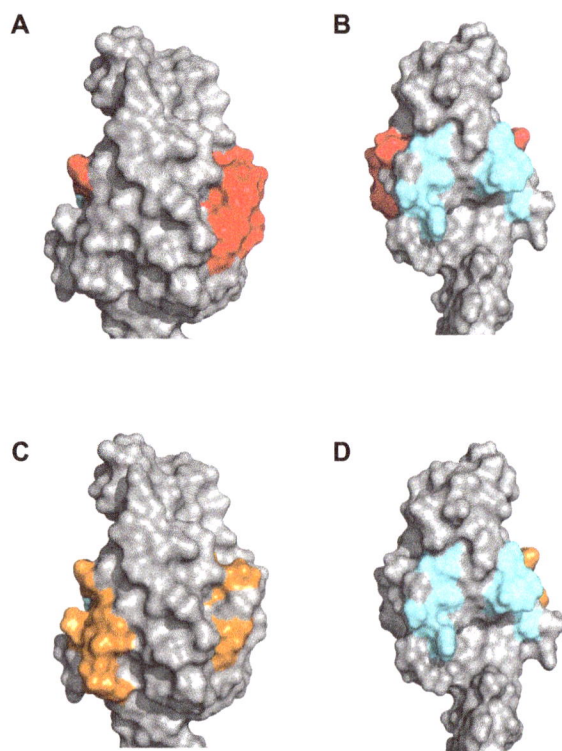

Figure 7. Comparison of the PDZ domain epitopes for mAbs 3E6 and 6G6. The PDZ domain for the grey monomer is shown as a molecular surface. (A) The 3E6 (residues 403-417) epitope is colored red. (B) View is a +180° rotation about the Y-axis from the view in panel (A). The light blue patches are the HtrA3 residues (Arg-360, Thr-363, Gln-389, Glu-390, Ala-392, Ser-419, Ser-420 and Gln-423) involved in protein/peptide interactions. (C) Same view as in panel (A), the 6G6 epitope corresponding to residues 398-412 is colored orange. (D) The 6G6 epitope, same view as in panel (B).

of the putative sensory loop L3 (residues 280–293). In addition, residues 224–227 (four residues in the epitope region 223–242) and 320–322 (three residues in the epitope region 308–322) are located on the protease catalytic face (dark green and magenta patches, respectively, located next to the brown patches in Figure 8A). If these residues (224–227, 283–297 and 320–322) were part of the 9C9 epitope, then this mAb would be expected to inhibit HtrA3 activity, in a similar manner to mAb 10H10, but this is not the case (Figure 3A). There is a 53% sequence similarity between the linear peptides corresponding to residues 283–297 and 308–322. Given that mAb 9C9 has no effect upon HtrA3 enzyme activity, we propose that residues 283–297 are not part of the 9C9 epitope but a false positive of the peptide library mapping assay. The region corresponding to residues 223–242 (dark green in Figure 2, 8A & B) encompasses the epitope of mAb 2C4 (residues 230–239, colored yellow in Figure 2, 6A, D & E). Residues 308–322 (colored magenta in Figure 2, 8A & B) are located on two β-strands and their connecting loop. Visual inspection of these two regions in the HtrA3 model showed that residues 228–242 from the 223–242 peptide and residues 308–319 from the 308–322 peptide are surface exposed and located on the non-catalytic face of the protease domain (colored dark green and magenta, respectively in Figure 8B). In each monomer these two regions would be accessible by a single antibody and their location on the HtrA3 non-catalytic face is consistent with the 9C9 mAb having no effect upon protease activity.

In contrast to mAb 10H10, 6G6 stimulates HtrA3 protease activity (Figure 3). The discontinuous 6G6 epitope identified by the linear peptide library screen corresponds to residues 73–92 (IGFB domain, colored dark grey in Figure 2 & 8C-E), 133–147 (Kazal domain and linker to the protease domain, colored black in Figure 2 and white in Figure 8D & E), 288–302 (protease domain, colored blue in Figure 2 & 8C), 313–327 (protease domain, colored purple in Figure 2 & 8C-E) and 398–412 (PDZ domain, colored orange in Figure 2, 7C, D & 8C-E). Residues 288–302 (colored blue in Figure 2 & 8C) cover the majority of the putative HtrA3 L3 sensor loop (residues 280–293) located on the protease domain catalytic face. Antibody interaction with residues 288–302 would be expected to block HtrA3 enzyme activity; however, mAb 6G6 acts in the opposite manner and stimulates HtrA3 enzyme activity (Figure 3). HtrA3 residues 288–302 has 53% sequence similarity to residues 313–327 (colored purple in Figure 2 & 8C-E) and we therefore propose that the linear peptide corresponding to HtrA3 residues 288–302 is a false positive in the epitope mapping assay. Although residues 73–78 (six residues in the epitope region 73–92) and 322–327 (six residues in the epitope region 313–327) are located on the protease catalytic face (dark grey and purple patches, respectively, located in the IGFB and protease domains near the blue patches in Figure 8C), the remainder of these two epitopes (i. e. residues 79–92 and 313–321) are located on the opposite face of HtrA3 (Figure 8D). Residues 79–92 (from epitope peptide 73–92), 313–321 (from epitope peptide 313–327) and 133–147 are located in close proximity to each other on the opposite face of HtrA3 to the protease catalytic sites (colored dark grey, purple and white, respectively, in Figure 8D & E). Distances between some of these epitope regions are shown in Figure 8E. Located>53 Å from these three regions in the HtrA3 model, the fifth linear peptide identified by mAb 6G6 corresponds to residues 398–412 in the PDZ domain (colored orange in Figure 7C, D & 8C-E) and overlaps with the linear peptide identified by mAb 3E6 (corresponding to residues 403–417, colored red in Figure 7A & B). How mAb 6G6 stimulates HtrA3 proteolytic activity is not immediately clear from the homology model. Under non-denaturing conditions, HtrA3-L was recognized by mAb 6G6 whereas HtrA3-S was not [41]; consistent with the Western blot data [41] and indicating that the PDZ domain is part of the mAb epitope. It is unlikely that all four regions corresponding to residues 73–92, 133–147, 288–302 and 313–327 are false positives of the linear peptide epitope mapping assay (although we do propose that 288–302 is a false positive). These data suggest two scenarios: (1) residues 398–412 in the PDZ domain alone are responsible for the stimulatory activity of mAb 6G6 or (2) all four regions (residues 73–92, 133–147, 313–327 and 398–412) are required for mAb 6G6 binding. The first scenario seems unlikely given the similarity to the mAb 3E6 epitope (compare Figure 7A-D). The second scenario is more probable, but a conformational change would be required to relocate the PDZ domain closer to the protease domain and the other epitope regions (i. e. movement in the direction of the arrow in Figure 8E).

Discontinuous antibody epitopes with extensive antigen interfaces have been previously reported, one prime example is the influenza virus N9 neuraminidase-NC41 Fab complex [49]. The interface between N9 neuraminidase and Fab NC41 buries 1815 Å² of surface area and the three extremities of the triangle-shaped interface are 27 Å, 30 Å and 27 Å apart. The discontinuous Fab NC41 epitope on the N9 neuraminidase surface consists of nineteen residues located in five separate segments. The tertiary structure places a number of short loops in close proximity on the N9 neuraminidase surface and these form the discontinuous epitope; a scenario not dissimilar to the one proposed here for

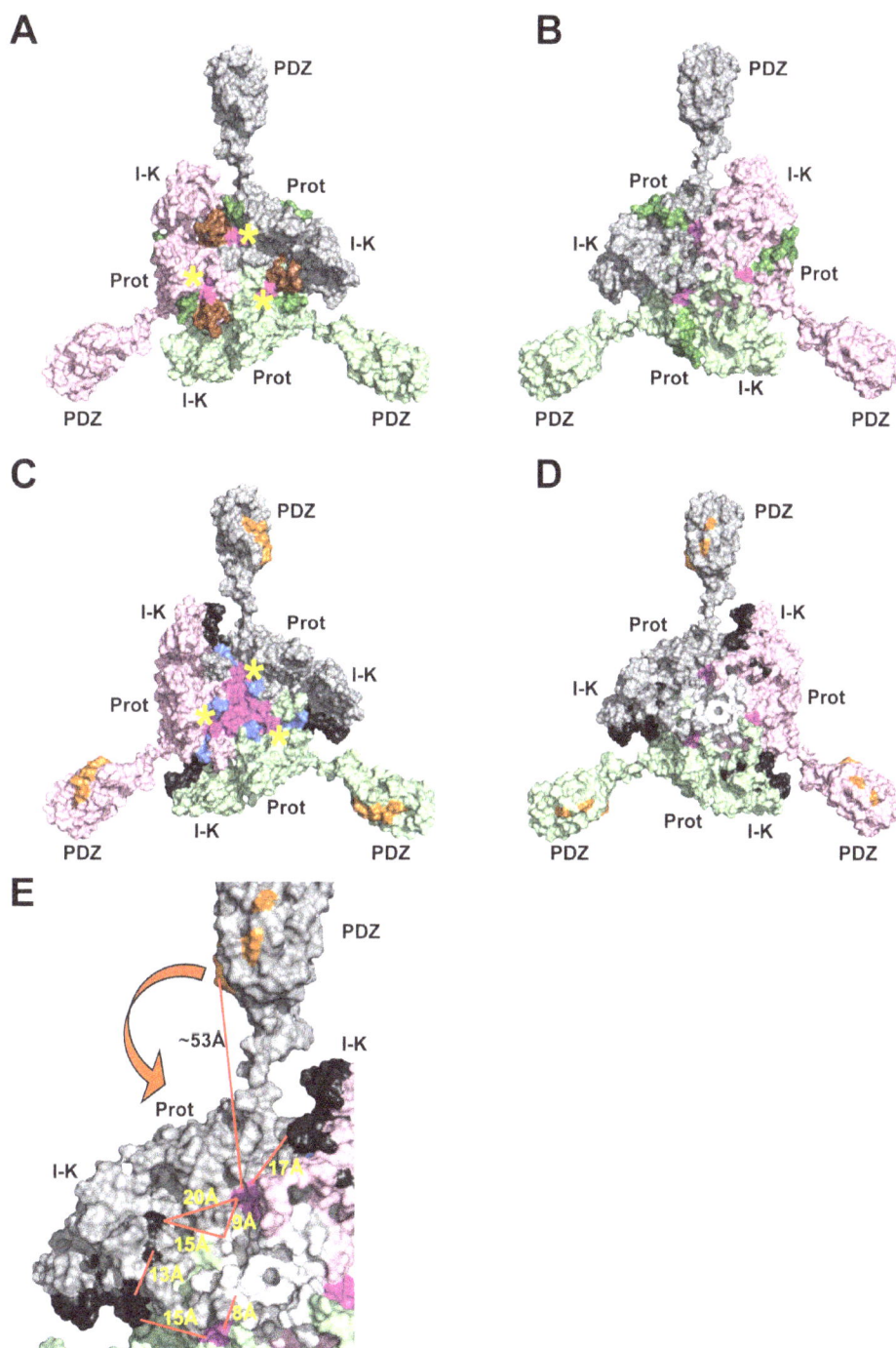

Figure 8. Location of the epitopes for mAbs 9C9 and 6G6. (A) Same view as in Figure 6A & C. The epitope regions for mAb 9C9 are shown: residues 283–297 (colored brown), residues 224–227 (from the 223–242 peptide, colored dark green) and residues 320–322 (from the 308–322 peptide, colored magenta). (B) A +180° rotation about the Y-axis from view shown in panel (A), showing the non-catalytic face of HtrA3. Residues 228–242 are colored dark green and 308–319 are colored magenta. (C) The epitope regions for mAb 6G6 are shown: residues 73–78 (from the 73–92 peptide, colored dark grey), residues 288–302 (colored blue), residues 322–327 (from the 313–327 peptide, colored purple), and residues 398–412 (colored orange). (D) A +180° rotation about the Y-axis from view shown in panel (C), showing the non-catalytic face of HtrA3. Residues 79–92 are colored dark grey and 313–321 are colored purple. The 6G6 epitope residues 133–147 (colored white) are now visible on the protease domain surface. (E) Close up view of panel (D) showing the 6G6 epitope regions on the non-catalytic face of HtrA3. Distances between some of the epitope regions are shown in Å. The arrow indicates the direction the PDZ domain would need to move to place the 6G6 epitope residues 398–412 (colored orange) in closer to the residues 313–321 (colored purple), 133–147 (colored white) and 79–92 (colored dark grey). In panels (A) and (C), the location of the catalytic site in each monomer is indicated by a yellow asterix.

mAb 6G6. The distances between the mAb 6G6 epitope regions on the non-catalytic face of HtrA3 (Figure 8E) are well within the interface dimensions reported for the N9 neuraminidase, although the PDZ domain would need to move closer to the protease domain to bring the PDZ epitope region within 15–25 Å of the other epitope regions.

All five mAbs are highly specific to HtrA3 and did not detect human HtrA1, HtrA2 [41] or HtrA4 (Figure 1C). Sequence comparison of the HtrA3 mAb epitope regions within the human HtrA family members are shown in Figure 2. The 10H10 epitope (HtrA3 residues 278–292) in HtrA3 encompasses almost the entire L3 sensor loop. Allowing for conservative amino acid substitution there are only two, four and one residues in the L3 sensor loop of HtrA1, HtrA2, and HtrA4 respectively that differ from HtrA3 (Figure 2). Of these, HtrA3 Glu-280 interacts with Arg-282 to maintain a short α-helical segment (Figure 6B). Glu-280 in HtrA3 is replaced by Gly-303, Pro-281 and Gly-301 in HtrA1, HtrA2 and HtrA4 respectively, thereby removing the possibility of an interaction with Lys-305 (HtrA1), Arg-283 (HtrA2) or Lys-303 (HtrA4) to maintain the α-helical conformation observed in the HtrA3 model. It is thus possible that subtle conformational differences in the L3 sensor loop are responsible for the specificity of mAb 10H10 to HtrA3.

The stimulating mAb 6G6 has a discontinuous epitope including residues from the HtrA3 IGFB, Kazal, protease and PDZ domains (Figures 1B & 2). As the N-terminal architecture of HtrA2 is very different to that of HtrA1, HtrA3 or HtrA4 [7], it is not surprising that mAb 6G6 does not recognize HtrA2. The mAb 6G6 epitope region 313–327 is conserved amongst HtrA family members, therefore specificity of 6G6 for HtrA3 over HtrA1 and HtrA4 is most likely imparted by sequence differences with HtrA3 residues 73–92, 133–147 and 398–412 (Figure 2). The HtrA3 specificity of mAb 9C9 is most likely imparted by residues 223–242 as the epitope region 308–322 is highly conserved amongst the human HtrA subtypes. For mAbs 2C4 (HtrA3 residues 230–239) and 3E6 (HtrA3 residues 403–417) the sequence differences among the four human HtrA family members are likely to be responsible for HtrA3 specificity (Figure 2).

Discussion

HtrA proteases have been implicated in the pathogenesis of several diseases such as cancers, neurodegenerative disorders and arthritis. However, there are no drugs available to treat diseases involving HtrA dysregulation. In this study, we characterized five highly specific HtrA3 mAbs and demonstrated that two of these specifically modulated (one inhibited and other stimulated) HtrA3 activity, providing unique research tools to investigate the molecular functions of HtrA3. The inhibitory HtrA3 mAb was further confirmed to increase trophoblast HTR8 cell migration and invasion in culture, demonstrating the potential therapeutic applications of this mAb to treat diseases associated with HtrA3 dysregulation. Furthermore, guided by published HtrA1 SAXS data [27], we constructed a 3D HtrA3 homology model to visualize the location of the mAb epitopes and to gain an understanding of their mechanism of action. Since mammalian HtrA1 and HtrA4 share similar domain organizations to that of HtrA3, the knowledge gained in this study about regulating HtrA3 activity may have boarder applications.

Western blot confirmed that all five HtrA3 mAbs were highly specific to HtrA3; none recognized any other human HtrA family members (HtrA1, HtrA2 or HtrA4). Among these five mAbs, 10H10 inhibited whereas 6G6 stimulated the proteolytic activity of HtrA3 in an *in vitro* protease activity assay. The differential effects of these two activity-modulating antibodies are striking. Together they provide us with highly specific tools to investigate HtrA3 cellular functions.

It was further confirmed by *in vitro* cell-based experiments that modulating the activity of HtrA3 by mAbs 10H10 or 6G6 had no effect on the growth of HTR8 trophoblast cells. The migration and invasion of HTR8 cells was significantly increased by the inhibitory mAb 10H10, consistent with our previous findings that HtrA3 is a negative regulator of invasion [33,34]. The HtrA3 homology model offers an explanation as to why mAb 10H10 inhibits HtrA3 activity. The mAb epitope is located near the HtrA3 protease catalytic site and binding of the mAb would be expected to block substrate access to the catalytic site. The homology model was also able to suggest why mAbs 2C4 and 3E6 had no effect upon the proteolytic activity of HtrA3; their epitopes do not impact upon the protease domain catalytic site. Given that mAbs 9C9 and 6G6 do not inhibit protease activity like mAb 10H10, we propose that the peptides corresponding to HtrA3 residues 293–297 (mAb 9C9) and 288–302 (mAb 6G6) are false positives arising from the *in vitro* linear peptide library screen.

The epitope of mAb 9C9 would appear to be discontinuous and consist of two separate regions, residues 228–242 from the 223–242 peptide and 308–319 from the 308–322 peptide. The HtrA3 homology model indicates that neither region would impact on the HtrA3 catalytic site, consistent with mAb 9C9 having no affect upon protease activity. The mAb 6G6 epitope also appears to be discontinuous and involve all four domains of HtrA3: IGFB, Kazal, protease and PDZ domains. As 6G6 recognizes only the HtrA3-L isoform, its epitope region in the PDZ domain (residues 398–412) would appear to be crucial for mAb binding. The mechanism by which mAb 6G6 stimulates HtrA3 enzyme activity is not clear from the homology model. Three regions of the discontinuous 6G6 epitope corresponding to residues 79–92 (from the 73–92 peptide, located in the IGFB domain), 133–147 (Kazal domain) and 313–321 (from the 313–327 peptide, located in the protease domain) are located within close proximity to each other on the opposite face of HtrA3 to the protease catalytic sites (Figure 8E). However, the epitope region in the PDZ domain is> 53 Å distant from these three regions. While it may be possible for an antibody to span such a large distance, it is more likely that there is a domain rearrangement bringing the PDZ domain closer to the other three epitope regions (indicated by the arrow in Figure 8E). It has been postulated by Truebestein and coworkers [28] that although the PDZ domain of human HtrA1 is not involved in enzyme activation, it may play a role in substrate processing by holding the protein/peptide substrate for C-terminal cleavage in the protease catalytic site. The mAb 6G6 may facilitate a more efficient substrate-holding process; consistent with our experimental observation that mAb 6G6 in a dose-dependent manner increased substrate cleavage. Future crystallographic and mutagenesis studies will establish how mAbs 6G6 and 9C9 interact with HtrA3 and confirm the blocking mechanism of mAb 10H10.

For HtrA1, it is thought that substrate-induced remodeling alters the conformation of the catalytic site loops, leading to enzyme activation. It has also been suggested that the catalytic site loops of HtrA1 undergo a disorder-to-order transition yielding a stably folded activation domain [28]. However, the structural rearrangement which occurs in the presence of the mAb 6G6 to possibly produce an activated form of HtrA3 is yet to be experimentally determined. HtrA protease substrate-induced activation may also coincide with conversion from lower to higher order oligomers [3]. Truebestein and coworkers [28] demonstrated that denatured citrate synthase as a substrate causes HtrA1 protease activation by stabilizing a higher-order multimer in a

PDZ-independent manner. The possibility that HtrA3 may form higher order oligomers requires further biochemical and structural investigation, but HtrA3 is predicted to exist at least in the trimeric form. N-terminal aromatic residues in the modeled HtrA3 protease domain (Phe-140, Phe-142 and Phe-255) are important for mediating an intermolecular oligomerization network that forms a ring of π-π interactions to stabilize the trimer, analogous to the HtrA1 crystal structures [28]. To date, no full length 3D structure of HtrA1 or HtrA3 has been determined crystallographically. Since the epitope of mAb 6G6 appears to involve all four domains, it may prove to be a useful tool in the structural determination of HtrA3-L and allow the characterization of rearrangements that may occur upon trimerization and substrate binding at the HtrA3 protease domain.

Since HtrA3 is downregulated in a number of cancers, it is proposed to be a tumor suppressor [10,37–39]. Previous studies have shown that HtrA1 is upregulated and activated during chemotherapy-induced cytotoxicity [50]. In addition, HtrA3 attenuates cell survival with either etoposide or cisplatin treatment in a manner dependent on serine protease function [10]. Thus, we tested whether stimulating HtrA3 activity with mAb 6G6 would modulate cell growth through a mechanism involving irreversible proteolysis of factors crucial to cell survival. However, our *in vitro* experiments with trophoblast HTR8 cells did not show a potential functional effect of mAb 6G6 on cell growth, migration or invasion, and the reasons for this require further investigation.

Various studies have demonstrated the utility of antibodies in treating human diseases such as transplant rejections, cancers, rheumatoid arthritis, Crohn's disease and antiviral prophylaxis [reviewed in [51]]. Modulating HtrA3 activity using 6G6 or 10H10 in pathological conditions such as preeclampsia, arthritis and tumor progression [1,10,11,15,16] would provide novel therapeutic applications. HtrA3 is expressed and secreted by a wide range of tissues although at different levels [7]. Placental HtrA3 is secreted into the maternal circulation and its serum levels are higher during early pregnancy in women who later develop preeclampsia [15,41]. Hence regulating HtrA3 activity with the inhibitory mAb 10H10 may provide a potential therapeutic opportunity in treating diseases where HtrA3 is abnormally high.

In summary, our study characterized the binding properties of five highly specific HtrA3 mAbs and identified that two of these modulated HtrA3 activity. These mAbs will provide useful tools to investigate the functional importance of HtrA3. They will also provide opportunities to characterize the effects of HtrA3 dysregulation in a number of diseases and cellular processes. Owing to the similarities in domain architecture, the knowledge gained in targeting HtrA3 may also have relevance to mammalian HtrA1 and HtrA4.

Supporting Information

Figure S1 Comparison of the trimeric HtrA3 homology model with the HtrA1 SAXS envelope. (A) Cartoon representation of the trimeric HtrA3 homology model. Each of the monomers has been colored differently. The locations of the IGFB-Kazal (I-K), protease (Prot) and PDZ domains are indicated. The three catalytic sites are identified by yellow asterisks. View is looking directly down onto the catalytic face of the protease domains. (B) Side view of the HtrA3 homology model, i.e. view is a -90^0 rotation about the X-axis from panel (A). (C) & (D) Overlay of the HtrA1 SAXS envelope (transparent light grey surface) onto the trimeric HtrA3 homology model and view as in panels (A) & (B), respectively.

Acknowledgments

We are grateful to Charles Eigenbrot and Mark Ultsch of Genentech for the use of their HtrA1 SAXS envelope in the construction of the HtrA3-L homology models.

Author Contributions

Conceived and designed the experiments: HS TLN GN. Performed the experiments: HS YW TLN. Analyzed the data: HS TLN MWP GN. Contributed reagents/materials/analysis tools: MWP GN. Wrote the paper: HS TLN MWP GN.

References

1. Clausen T, Kaiser M, Huber R, Ehrmann M (2011) HTRA proteases: regulated proteolysis in protein quality control. Nat Rev Mol Cell Biol 12: 152–162.

2. Clausen T, Southan C, Ehrmann M (2002) The HtrA family of proteases: implications for protein composition and cell fate. Mol Cell 10: 443–455.

3. Krojer T, Pangerl K, Kurt J, Sawa J, Stingl C, et al. (2008) Interplay of PDZ and protease domain of DegP ensures efficient elimination of misfolded proteins. Proc Natl Acad Sci U S A 105: 7702–7707.

4. Faccio L, Fusco C, Chen A, Martinotti S, Bonventre JV, et al. (2000) Characterization of a novel human serine protease that has extensive homology to bacterial heat shock endoprotease HtrA and is regulated by kidney ischemia. J Biol Chem 275: 2581–2588.

5. Gray CW, Ward RV, Karran E, Turconi S, Rowles A, et al. (2000) Characterization of human HtrA2, a novel serine protease involved in the mammalian cellular stress response. Eur J Biochem 267: 5699–5710.

6. Hu SI, Carozza M, Klein M, Nantermet P, Luk D, et al. (1998) Human HtrA, an evolutionarily conserved serine protease identified as a differentially expressed gene product in osteoarthritic cartilage. J Biol Chem 273: 34406–34412.

7. Nie GY, Hampton A, Li Y, Findlay JK, Salamonsen LA (2003) Identification and cloning of two isoforms of human high-temperature requirement factor A3 (HtrA3), characterization of its genomic structure and comparison of its tissue distribution with HtrA1 and HtrA2. Biochem J 371: 39–48.

8. Wang LJ, Cheong ML, Lee YS, Lee MT, Chen H (2012) High-temperature requirement protein A4 (HtrA4) suppresses the fusogenic activity of syncytin-1 and promotes trophoblast invasion. Mol Cell Biol 32: 3707–3717.

9. Zumbrunn J, Trueb B (1996) Primary structure of a putative serine protease specific for IGF-binding proteins. FEBS Lett 398: 187–192.

10. Beleford D, Rattan R, Chien J, Shridhar V (2010) High temperature requirement A3 (HtrA3) promotes etoposide- and cisplatin-induced cytotoxicity in lung cancer cell lines. J Biol Chem 285: 12011–12027.

11. Chien J, Campioni M, Shridhar V, Baldi A (2009) HtrA serine proteases as potential therapeutic targets in cancer. Curr Cancer Drug Targets 9: 451–468.

12. Coleman HR, Chan CC, Ferris FL 3rd, Chew EY (2008) Age-related macular degeneration. Lancet 372: 1835–1845.

13. Grau S, Baldi A, Bussani R, Tian X, Stefanescu R, et al. (2005) Implications of the serine protease HtrA1 in amyloid precursor protein processing. Proc Natl Acad Sci U S A 102: 6021–6026.

14. Hara K, Shiga A, Fukutake T, Nozaki H, Miyashita A, et al. (2009) Association of HTRA1 mutations and familial ischemic cerebral small-vessel disease. N Engl J Med 360: 1729–1739.

15. Li Y, Puryer M, Lin E, Hale K, Salamonsen LA, et al. (2011) Placental HtrA3 is regulated by oxygen tension and serum levels are altered during early pregnancy in women destined to develop preeclampsia. J Clin Endocrinol Metab 96: 403–411.

16. Milner JM, Patel A, Rowan AD (2008) Emerging roles of serine proteinases in tissue turnover in arthritis. Arthritis Rheum 58: 3644–3656.

17. Krojer T, Garrido-Franco M, Huber R, Ehrmann M, Clausen T (2002) Crystal structure of DegP (HtrA) reveals a new protease-chaperone machine. Nature 416: 155–159.

18. Krojer T, Sawa J, Schafer E, Saibil HR, Ehrmann M, et al. (2008) Structural basis for the regulated protease and chaperone function of DegP. Nature 453: 885–890.

19. Mohamedmohaideen NN, Palaninathan SK, Morin PM, Williams BJ, Braunstein M, et al. (2008) Structure and function of the virulence-associated high-temperature requirement A of Mycobacterium tuberculosis. Biochemistry 47: 6092–6102.

20. Shen QT, Bai XC, Chang LF, Wu Y, Wang HW, et al. (2009) Bowl-shaped oligomeric structures on membranes as DegP's new functional forms in protein quality control. Proc Natl Acad Sci U S A 106: 4858–4863.

21. Wilken C, Kitzing K, Kurzbauer R, Ehrmann M, Clausen T (2004) Crystal structure of the DegS stress sensor: How a PDZ domain recognizes misfolded protein and activates a protease. Cell 117: 483–494.

22. Kim DY, Kim KK (2005) Structure and function of HtrA family proteins, the key players in protein quality control. J Biochem Mol Biol 38: 266–274.

23. Meltzer M, Hasenbein S, Mamant N, Merdanovic M, Poepsel S, et al. (2009) Structure, function and regulation of the conserved serine proteases DegP and DegS of Escherichia coli. Res Microbiol 160: 660–666.

24. Polur I, Lee PL, Servais JM, Xu L, Li Y (2010) Role of HTRA1, a serine protease, in the progression of articular cartilage degeneration. Histol Histopathol 25: 599–608.

25. Suzuki Y, Takahashi-Niki K, Akagi T, Hashikawa T, Takahashi R (2004) Mitochondrial protease Omi/HtrA2 enhances caspase activation through multiple pathways. Cell Death Differ 11: 208–216.

26. Zurawa-Janicka D, Skorko-Glonek J, Lipinska B (2010) HtrA proteins as targets in therapy of cancer and other diseases. Expert Opin Ther Targets 14: 665–679.

27. Eigenbrot C, Ultsch M, Lipari MT, Moran P, Lin SJ, et al. (2012) Structural and functional analysis of HtrA1 and its subdomains. Structure 20: 1040–1050.

28. Truebestein L, Tennstaedt A, Monig T, Krojer T, Canellas F, et al. (2011) Substrate-induced remodeling of the active site regulates human HTRA1 activity. Nat Struct Mol Biol 18: 386–388.

29. Krem MM, Rose T, Di Cera E (1999) The C-terminal sequence encodes function in serine proteases. J Biol Chem 274: 28063–28066.

30. Nie G, Li Y, Hale K, Okada H, Manuelpillai U, et al. (2006) Serine peptidase HTRA3 is closely associated with human placental development and is elevated in pregnancy serum. Biol Reprod 74: 366–374.

31. Nie G, Li Y, He H, Findlay JK, Salamonsen LA (2006) HtrA3, a serine protease possessing an IGF-binding domain, is selectively expressed at the maternal-fetal interface during placentation in the mouse. Placenta 27: 491–501.

32. Nie GY, Li Y, Minoura H, Batten L, Ooi GT, et al. (2003) A novel serine protease of the mammalian HtrA family is up-regulated in mouse uterus coinciding with placentation. Mol Hum Reprod 9: 279–290.

33. Singh H, Endo Y, Nie G (2011) Decidual HtrA3 negatively regulates trophoblast invasion during human placentation. Hum Reprod 26: 748–757.

34. Singh H, Makino SI, Endo Y, Nie G (2010) Inhibition of HTRA3 stimulates trophoblast invasion during human placental development. Placenta 31: 1085–1092.

35. Than NG, Romero R, Hillermann R, Cozzi V, Nie G, et al. (2008) Prediction of preeclampsia - a workshop report. Placenta 29 Suppl A: S83–85.

36. Beleford D, Liu Z, Rattan R, Quagliuolo L, Boccellino M, et al. (2010) Methylation induced gene silencing of HtrA3 in smoking-related lung cancer. Clin Cancer Res 16: 398–409.

37. Bowden MA, Di Nezza-Cossens LA, Jobling T, Salamonsen LA, Nie G (2006) Serine proteases HTRA1 and HTRA3 are down-regulated with increasing grades of human endometrial cancer. Gynecol Oncol 103: 253–260.

38. Narkiewicz J, Klasa-Mazurkiewicz D, Zurawa-Janicka D, Skorko-Glonek J, Emerich J, et al. (2008) Changes in mRNA and protein levels of human HtrA1, HtrA2 and HtrA3 in ovarian cancer. Clin Biochem 41: 561–569.

39. Singh H, Li Y, Fuller PJ, Harrison C, Rao J, et al. (2013) HtrA3 Is Downregulated in Cancer Cell Lines and Significantly Reduced in Primary Serous and Granulosa Cell Ovarian Tumors. J Cancer 4: 152–164.

40. Tocharus J, Tsuchiya A, Kajikawa M, Ueta Y, Oka C, et al. (2004) Developmentally regulated expression of mouse HtrA3 and its role as an inhibitor of TGF-beta signaling. Dev Growth Differ 46: 257–274.

41. Dynon K, Heng S, Puryer M, Li Y, Walton K, et al. (2012) HtrA3 as an early marker for preeclampsia: specific monoclonal antibodies and sensitive high-throughput assays for serum screening. PLoS One 7: e45956.

42. Singh H, Makino S, Endo Y, Li Y, Stephens AN, et al. (2012) Application of the wheat-germ cell-free translation system to produce high temperature requirement A3 (HtrA3) proteases. Biotechniques 52: 23–28.

43. Graham CH, Hawley TS, Hawley RG, MacDougall JR, Kerbel RS, et al. (1993) Establishment and characterization of first trimester human trophoblast cells with extended lifespan. Exp Cell Res 206: 204–211.

44. Runyon ST, Zhang Y, Appleton BA, Sazinsky SL, Wu P, et al. (2007) Structural and functional analysis of the PDZ domains of human HtrA1 and HtrA3. Protein Sci 16: 2454–2471.

45. Sali A, Potterton L, Yuan F, van Vlijmen H, Karplus M (1995) Evaluation of comparative protein modeling by MODELLER. Proteins 23: 318–326.

46. Laskowski RA, MacArthur MW, Moss DS, Thornton JM (1993) PROCHECK: a program to check the stereochemical quality of protein structures. Journal of Applied Crystallography 26: 283–291.

47. Bohrmann B, Baumann K, Benz J, Gerber F, Huber W, et al. (2012) Gantenerumab: a novel human anti-Abeta antibody demonstrates sustained cerebral amyloid-beta binding and elicits cell-mediated removal of human amyloid-beta. J Alzheimers Dis 28: 49–69.

48. Krawczyk A, Krauss J, Eis-Hubinger AM, Daumer MP, Schwarzenbacher R, et al. (2011) Impact of valency of a glycoprotein B-specific monoclonal antibody on neutralization of herpes simplex virus. J Virol 85: 1793–1803.

49. Tulip WR, Varghese JN, Laver WG, Webster RG, Colman PM (1992) Refined crystal structure of the influenza virus N9 neuraminidase-NC41 Fab complex. J Mol Biol 227: 122–148.

50. Chien J, Aletti G, Baldi A, Catalano V, Muretto P, et al. (2006) Serine protease HtrA1 modulates chemotherapy-induced cytotoxicity. J Clin Invest 116: 1994–2004.

51. Brekke OH, Sandlie I (2003) Therapeutic antibodies for human diseases at the dawn of the twenty-first century. Nat Rev Drug Discov 2: 52–62.

5

Ionising Radiation Immediately Impairs Synaptic Plasticity-Associated Cytoskeletal Signalling Pathways in HT22 Cells and in Mouse Brain: An *In Vitro/In Vivo* Comparison Study

Stefan J. Kempf[1], Sonja Buratovic[2], Christine von Toerne[3], Simone Moertl[1], Bo Stenerlöw[4], Stefanie M. Hauck[3], Michael J. Atkinson[1,5], Per Eriksson[2], Soile Tapio[1]*

1 Institute of Radiation Biology, Helmholtz Zentrum München, German Research Center for Environmental Health GmbH, Neuherberg, Germany, 2 Department of Environmental Toxicology, Uppsala University, Uppsala, Sweden, 3 Research Unit Protein Science, Helmholtz Zentrum München, German Research Center for Environmental Health GmbH, Neuherberg, Germany, 4 Division of Biomedical Radiation Sciences, Rudbeck Laboratory, Uppsala University, Uppsala, Sweden, 5 Chair of Radiation Biology, Technical University Munich, Munich, Germany

Abstract

Patients suffering from brain malignancies are treated with high-dose ionising radiation. However, this may lead to severe learning and memory impairment. Preventive treatments to minimise these side effects have not been possible due to the lack of knowledge of the involved signalling pathways and molecular targets. Mouse hippocampal neuronal HT22 cells were irradiated with acute gamma doses of 0.5 Gy, 1.0 Gy and 4.0 Gy. Changes in the cellular proteome were investigated by isotope-coded protein label technology and tandem mass spectrometry after 4 and 24 hours. To compare the findings with the *in vivo* response, male NMRI mice were irradiated on postnatal day 10 with a gamma dose of 1.0 Gy, followed by evaluation of the cellular proteome of hippocampus and cortex 24 hours post-irradiation. Analysis of the *in vitro* proteome showed that signalling pathways related to synaptic actin-remodelling were significantly affected at 1.0 Gy and 4.0 Gy but not at 0.5 Gy after 4 and 24 hours. We observed radiation-induced reduction of the miR-132 and Rac1 levels; miR-132 is known to regulate Rac1 activity by blocking the GTPase-activating protein p250GAP. In the irradiated hippocampus and cortex we observed alterations in the signalling pathways similar to those *in vitro*. The decreased expression of miR-132 and Rac1 was associated with an increase in hippocampal cofilin and phospho-cofilin. The Rac1-Cofilin pathway is involved in the modulation of synaptic actin filament formation that is necessary for correct spine and synapse morphology to enable processes of learning and memory. We suggest that acute radiation exposure leads to rapid dendritic spine and synapse morphology alterations via aberrant cytoskeletal signalling and processing and that this is associated with the immediate neurocognitive side effects observed in patients treated with ionising radiation.

Editor: Xiangming Zha, University of South Alabama, United States of America

Funding: The research was supported by a grant from the European Community's Seventh Framework Programme (EURATOM) contract no 295552 (CEREBRAD - Cognitive and Cerebrovascular Effects Induced by Low Dose Ionizing Radiation). The funder had no role in study design, data collection, analysis and interpretation, decision to publish, or preparation of the manuscript.

Competing Interests: The authors have declared that no competing interests exist.

* Email: soile.tapio@helmholtz-muenchen.de

Introduction

Ionising radiation is frequently used during treatment of central nervous system (CNS) malignancies. Normally, the patient is exposed to a total radiation dose of 20–50 Gy that is given in fractions of 2–4 Gy to reduce the side-effects. Still, immediate detrimental decline in cognition and visual memory are widely observed [1,2]. Epidemiological studies indicate that even moderate radiation doses may lead to acute and permanent deficits in learning and memory [3–5], in particular if the exposure occurred during childhood [6,7].

Approximately 200,000 children worldwide were treated with X-rays for ringworm of the scalp (*Tinea capitis*), with head doses ranging from 0.7 to 1.7 Gy [8,9]: Long-term side-effects on cognition were evaluated 10 to 29 years later, showing that psychiatric disorders were more often diagnosed in exposed children than in not exposed ones [10]. A follow-up study with 11,000 irradiated Israeli *Tinea capitis* children showed similar long-term effects after radiation exposure including lower examination scores, intelligence quotients, and a small increase in the frequency of mental retardation [4].

The cognitive damage in people exposed early in life may be a consequence of the immature state of the brain when ionising radiation was applied. On its way to adolescence, the brain undergoes various remodelling processes on molecular and structural levels called the brain growth spurt [11]. It includes fundamental neuronal architecture changes such as growth of axons and dendrites to enable the formation and deletion of synaptic contacts [12]. The brain is especially susceptible to damage if exposed to ionising radiation during this developmental period. It has been shown that toxic agents given to mice within the susceptibility window around postnatal day ten lead to disruption of adult brain function [13,14]. Further, a synergistic effect between toxicants and ionising radiation given on postnatal day ten has been shown [15]. Interestingly, the brain growth spurt is species-dependent as in human beings it lasts until the age of three to four years whereas in rodents it corresponds to the second and fourth postnatal weeks [16].

Especially the hippocampus is a highly radiation-sensitive brain region involved in learning and memory consolidation. Irradiation may lead to changes in the neurogenic niche of the dentate gyrus of the hippocampus by depleting neural stem and progenitor cells [17–20]. Nevertheless, the low frequency of life-long newly generated neurons in this region may suggest that other brain regions and biological targets may also be of importance in the manifestation of long-lasting cognitive defects after radiation treatment. It has been shown recently that the mature neuronal networks of the hippocampus are highly-radiation sensitive [21]. Thus, ionising radiation may have adverse effects on the effective neurotransmission by altering the synaptic plasticity of the brain. Synaptic plasticity is a dynamic process involving rapid cytoskeletal organisation on the dendrite and spine morphology to modulate signal transmission. Defects in synaptic plasticity and dendrite or spine morphology have been observed in cognitive diseases such as Alzheimer's [22], Rett syndrome [23] and Down's syndrome [24], emphasising not only the role of the hippocampus but also that of the cortex in this process.

The aim of this study was (i) to determine the role of synaptic plasticity-associated cytoskeletal signalling pathways in the acute radiation response *in vitro* and *in vivo* and (ii) to compare these alterations. We show here that dendritic spine morphology-associated proteins and signalling pathways such as the Rac1-Cofilin pathway were rapidly altered after *in vitro* exposure to a dose of 1.0 Gy in primary immortalised neurons of the mouse hippocampal cells (HT22). Similar alterations were confirmed in the hippocampus and cortex of NMRI mice irradiated on postnatal day ten that represents a developmental stage within the brain growth spurt in mice.

Materials and Methods

Ethics statement, irradiation of animals and tissue collection

Experiments were carried out in accordance with the European Communities Council Directive of 24 November 1986 (86/609/EEC), after approval from the local ethical committees (Uppsala University and the Agricultural Research Council) and by the Swedish Committee for Ethical Experiments on Laboratory Animals. All animal experiments were performed under trained personnel, and all efforts were made to minimise animal suffering.

Male NMRI mice were total body irradiated on postnatal day 10 (PND 10) with a single exposure to gamma irradiation (^{137}Cs, 0.20 Gy/min) at doses of 0 (sham-irradiated control) and 1.0 Gy (Rudbeck Laboratory, Uppsala University). Dose verification was done with an ionisation chamber (Markus chamber type 23343, PTW-Freiburg) and was homogeneous within $\pm 3\%$ over the 10 cm dish area where mice were positioned during irradiation procedure. Neonates from each litter were irradiated together.

Animals were sacrificed via cervical dislocation. Brains were excised and transferred to ice-cold PBS, rinsed carefully, and dissected under stereomicroscopic inspection under cold conditions. Hippocampi and cortices without meninges from each hemisphere were separately sampled, gently rinsed in ice-cold PBS and snap-frozen in liquid nitrogen. Samples were stored at $-80°C$ until isolation of protein and RNA.

Irradiation and harvesting of cells

HT22 cells (immortalised primary neurons from the mouse hippocampus) were kindly provided from J. Lewerenz (Department of Neurology, University Hospital Hamburg-Eppendorf, Hamburg, Germany) [25]. The cells were grown in high glucose DMEM media (PAA Laboratories, E15-840) supplemented with 10% foetal bovine serum (PAA Laboratories, A15-101) without antibiotics in T75 tissue flask at $37°C$ with 5% CO_2 in air. They were irradiated in the exponential growth phase with doses of 0 Gy (sham), 0.5 Gy, 1.0 Gy or 4.0 Gy of γ-rays (^{137}Cs, 0.48 Gy/min) (HWM-D 2000, Waelischmiller Engineering, Germany). For each dose group and time point, four independent flasks were seeded and irradiated. At four and 24 hours post-irradiation cells were rinsed with ice-cold PBS and enzymatically detached with accutase (Invitrogen). After blocking of the accutase reaction with media containing 10% foetal bovine serum and splitting of cell volume in two equal parts for total RNA and protein isolation of each tissue flask, the cells were centrifuged and washed once with ice-cold PBS. This centrifugation and washing step was repeated, followed by cell pelleting via centrifugation. Pelleted cells were frozen at $-80°C$ until total protein and RNA content were isolated.

Isolation of total protein and RNA

a) Isolation of total protein. HT22 cell pellets or individual frozen hippocampi and cortices were homogenised with 6 M guanidine hydrochloride (SERVA Electrophoresis GmbH, Germany) on ice using a manual plastic mortar. Homogenates were briefly vortexed, sonicated, and cleared by centrifugation ($20,000 \times g$, 1 hour, $4°C$). The supernatants were collected and stored at $-20°C$ before further use. Total protein content was determined using Bradford assay (Thermo Fisher) following the manufacturer's instructions.

b) Total RNA isolation. Total RNA from HT22 cell pellets or individual frozen hippocampi and cortices was isolated and purified by mirVana^{TM} Isolation Kit (Ambion) according to the manufacturer's instructions. Total RNA was eluted with nuclease-free water. The optical density (OD) ratio of 260/280 was measured using a Nanodrop spectrophotometer (PeqLab Biotechnology; Germany); it ranged between 1.9 and 2.1. Eluates were stored at $-20°C$ until further analysis.

Mass spectrometry-based proteome analysis

a) Isotope coded protein label (ICPL) analysis of proteins, 1D PAGE separation and in-gel digest. In total, four individual replicates of HT22 cells were used for proteomic analysis at each radiation dose and time point. Total protein lysates were labelled with ICPL reagents (SERVA Electrophoresis GmbH, Germany) according to the manufacturer's instructions. Briefly, individual protein lysates (20 μg in 20 μl of 6 M guanidine hydrochloride from each biological sample) were reduced, alkylated and labelled with the respective ICPL-reagent as follows: control with ICPL-0, 0.5 Gy sample with ICPL-4, 1.0 Gy sample with ICPL-6 and 4.0 Gy sample with ICPL-10. All labelled samples representing each radiation dose at one time point (4 and 24 hours) were combined and overnight precipitated with 80% acetone at $-20°C$ to purify the labelled protein content.

Biological replicates from the *in vivo* mouse study included animals from at least three different litters. Four biological replicates from hippocampus and five from cortex were used for both control and irradiated groups. The samples were labelled with ICPL reagents as follows: control with ICPL-0 and 1.0 Gy sample with ICPL-6. These labelled samples were further treated as described for the HT22 cells.

Protein precipitates were separated by 12% SDS-polyacrylamide gel electrophoresis followed by Coomassie Blue staining. Gel lanes were cut into at least four equal slices, destained, and trypsinised overnight as described recently [26]. Peptides were extracted and acidified with 1% formic acid followed by analysis via mass spectrometry.

b) LC-MS/MS analysis. LC-MS/MS analysis was performed as described previously on a LTQ-Orbitrap XL (Thermo Fisher) [27]. Briefly, pre-fractionated samples were automatically injected and loaded onto the trap column and after 5 min, peptides were eluted and separated on the analytical column by reversed phase chromatography operated on a nano-HPLC (Ultimate 3000, Dionex) with a nonlinear 170 min gradient using 35% acetonitrile in 0.1% formic acid in water (A) and 0.1% formic acid in 98% acetonitrile (B) at a flow rate of 300 nl/min. The gradient settings were: 5–140 min: 14.5–90% A, 140–145 min: 90% A −95% B, 145–150 min: 95% B followed by equilibration for 15 min to starting conditions. From the MS pre-scan, the 10 most abundant peptide ions were selected for fragmentation in the linear ion trap if they exceeded an intensity of at least 200 counts and were at least doubly charged. During fragment analysis, a high-resolution (60,000 full-width half maximum) MS spectrum was acquired in the Orbitrap with a mass range from 200 to 1500 Da.

c) Identification and quantification of proteins. MS-MS spectra were searched against the ENSEMBL mouse database (Version: 2.4, 56416 sequences) via MASCOT (version 2.3.02; Matrix Science) with a mass tolerance of 10 ppm for peptide precursors and 0.6 Da for MS-MS peptide fragments, including not more than one missed cleavage. Fixed modifications included carbamidomethylation of cysteine and ICPL-0, ICPL-4, ICPL-6 and ICPL-10 for lysine. Proteins were identified and quantified based on the ICPL pairs using the Proteome Discoverer software (Version 1.3– Thermo Fisher). To ensure that only high-confident identified peptides were used for protein quantification, the MASCOT percolator algorithm was applied [28]. The percolator is an algorithm that improves the discrimination between correct and incorrect spectrum identifications and gives a q value sising the statistical confidence assigned to each peptide-spectra-match [29]. The q value was set to 0.01 representing strict peptide ranking. Only the best ranked peptides were used. Such peptides were filtered against Decoy database resulting into a false discovery rate (FDR) of each LC-MS-run; the significance threshold was set to 0.01 to ensure that only highly confident peptide identifications were used for protein quantification. Proteins from each LC-MS-run were normalised against the median of all quantifiable proteins. Proteins were considered significantly deregulated if they fulfilled the following criteria: (i) identification by at least two unique peptides in n-1 mass-spectrometry runs (n: number of biological replicates), (ii) quantification with an ICPL-variability of ≤30% and (iii) a fold-change of ≥1.3 or ≤ −1.3. The threshold of ±1.3 is based on our average experimental technical variance of the multiple analysis of hippocampal and cortical technical replicates (13.8%).

Data deposition of proteomics experiments

The raw-files of the obtained MS-MS spectra can be found under http://storedb.org/project_details.php?projectid=38 with the ProjectID 38.

Bioinformatics analysis

Deregulated proteins were assigned to functional classes using PANTHER classification system software (http://www.pantherdb.org) and the general annotation from UniProt (http://uniprot.org). To identify radiation-affected signalling pathways, a signalling pathway analysis was performed with all deregulated proteins for each dose group using INGENUITY Pathway Analysis (IPA) (http://www.ingenuity.com) applying databases of experimental and predictive origin.

Quantification of Rac1, cofilin and phospho-cofilin expression levels via immunoblotting

Protein extracts of cells and brain tissues (15 μg) were separated on 12% SDS polyacrylamide gels and transferred to nitrocellulose membranes (GE Healthcare) via BIO-RAD Criterion™ Blotter system at 100 V for 2 h. Membranes were blocked with Roti^R-Block solution (Roth), washed and incubated overnight at 4°C with primary antibody dilutions as recommended by the manufacturer (GAPDH – sc-47724 [murine monoclonal IgG1 raised against recombinant GAPDH of human origin; Santa Cruz], Rac1– ab33186 [murine monoclonal IgG2b raised against full-length recombinant Rac1 of human origin; Abcam], cofilin –3312 [rabbit polyclonal antibody produced by immunising rabbits with a synthetic peptide corresponding to residues surrounding Ser3 of human cofilin origin; Cell Signalling], p-Cofilin (Ser3) –3311 [rabbit polyclonal antibody by immunising animals with a synthetic phospho-peptide corresponding to residues surrounding Ser3 of human cofilin; Cell Signalling]). After a washing step, blots were incubated with appropriate horseradish peroxidase-conjugated secondary antibody in 8% milk for 1 h at room temperature and developed using ECL system (GE Healthcare) using standard protocol from the manufacturer. GAPDH was not significantly deregulated based on the global proteomics results in any sample and was therefore used as a loading control. Immunoblots were quantified with TotalLab TL100 software (www.totallab.com) using software-suggested background correction. Three or four biological replicates were used for statistical analysis (unpaired Student's t-test) with a significance threshold of 0.05.

Quantification of microRNA miR-132 via quantitative PCR

RNA isolates of cells and brain tissues (10 ng) were used to quantify microRNA miR-132 expression levels using the TaqMan Single MicroRNA Assay (Applied Biosystems) according to the manufacturer's protocol. Steps included a reverse transcription and real-time PCR (StepOnePlus) via Taqman-primers (mmu-miR-132 (ID000457), snoRNA135 (ID001239) – Life Technologies). Expression levels of miRNA were calculated based on the $2^{-\Delta\Delta Ct}$ method with normalisation against endogenous snoRNA135 [30]. Changes were considered significant if they reached a p-value of ≤0.05 (unpaired Student's t-test, n = 4 [*in vitro*] and n = 3 [*in vivo*] per dose group and time point).

Results

Acute effects of ionising radiation on synaptic plasticity-associated cytoskeletal signalling pathways
in vitro

We used HT22 cells as an *in vitro* model to detect radiation-induced alterations in synaptic plasticity-associated cytoskeletal signalling pathways. HT22 cells were irradiated with doses of 0.5 Gy, 1.0 Gy and 4.0 Gy, followed by a global quantitative proteome analysis 4 and 24 hours post-irradiation. The protein quantification showed a dose-dependency in the number of significantly deregulated proteins (4 h/24 h: 0.5 Gy −1/12, 1.0 Gy −31/34, 4.0 Gy −50/91) (Figure 1 A and B). Table S1 in File S2 shows the complete list of deregulated proteins. The Venn diagrams in Figure S1 A – C in File S1 show the overlapping proteins between the two time points (4 and 24 hours) at 0.5 Gy, 1.0 Gy and 4.0 Gy. Importantly, all overlapping proteins showed the same direction of deregulation at both time points (Figure S1 D – F in File S1).

Bioinformatics analysis of signalling pathways using the Ingenuity Pathway Analysis (IPA) software showed that synaptic-plasticity associated cytoskeletal remodelling pathways were affected by radiation exposure in HT22 cells. RhoGDI signalling (4 h: 4.0 Gy; 24 h: 1.0 Gy, 4.0 Gy – p<0.05), actin cytoskeleton signalling (4 h: 1.0 Gy, 4.0 Gy; 24 h: 1.0 Gy, 4.0 Gy – p<0.05), regulation of actin-based motility by Rho (4 h: 4.0 Gy; 24 h: 4.0 Gy – p<0.05) and signalling by Rho family GTPases (4 h:

1.0 Gy; 24 h: 1.0 Gy – p<0.05) were the most important pathways affected (Figure 1 C). Importantly, these pathways were not significantly altered at 0.5 Gy for any time point (Figure 1 C). The shared deregulated proteins from each pathway are shown in Figure S2 in File S1 consisting of the Rho family GTPase Rac, the kinases PAK and LIMK and the actin cytoskeleton-remodelling cofilin. All these proteins are involved in axonal maturation, spine- and synapse formation, maturation and -morphology via regulation of actin polymerisation (Rac1-Cofilin pathway) [31–33].

Ionising radiation impairs the RhoGTPase Rac1 *in vitro*

To validate the radiation-induced change in the Rac1-Cofilin pathway in HT22 cells, the Rho family GTPase Rac1 as the main upstream modulator of this pathway was quantified via immunoblotting. The analysis demonstrated that the expression levels of the Rac1 protein were significantly down-regulated at 1.0 Gy and

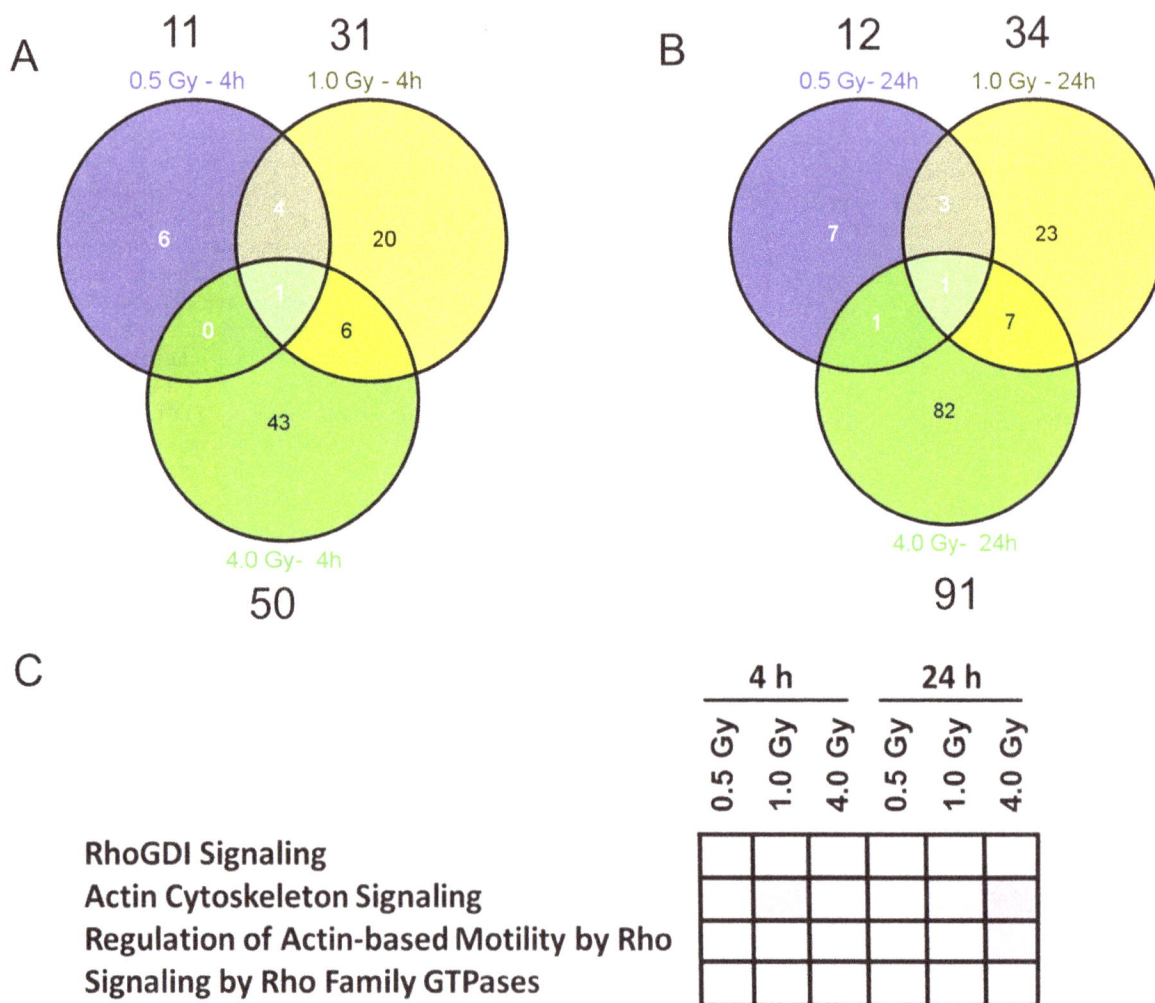

Figure 1. Mass spectrometry-based proteomics of *in vitro* irradiated HT22 cells. Venn diagrams showing the number of all and shared deregulated proteins from HT22 cells exposed to 0.5 Gy, 1.0 Gy and 4.0 Gy 4 hours (A) and 24 hours (B) post-irradiation using global proteomics approach; n = 4. The number above each dose shows the total number of deregulated proteins at this dose. Altered cytoskeletal signalling pathways at all doses using the Ingenuity Pathway Analysis (IPA) software are shown (C). Higher colour intensity represents higher significance (p-value) of the pathway. All coloured boxes have a p-value of ≤0.05; white boxes have a p-value of ≥0.05 and are not significantly altered. p-values at 4 hours: RhoGDI signalling (0.5 Gy: 0.3, 1.0 Gy: 0.289, 4.0 Gy: 0.0155), actin cytoskeleton signalling (0.5 Gy: 0.4, 1.0 Gy: 0.00819, 4.0 Gy: 0.0267), regulation of actin-based motility by Rho (0.5 Gy: 0.45, 1.0 Gy: 0.147, 4.0 Gy: 0.025), signalling by Rho Family GTPases (0.5 Gy: 0.4, 1.0 Gy: 0.049, 4.0 Gy: 0.43). p-values at 24 hours: RhoGDI signalling (0.5 Gy: 0.3, 1.0 Gy: 0.043, 4.0 Gy: 0.044), actin cytoskeleton signalling (0.5 Gy: 0.49, 1.0 Gy: 0.0106, 4.0 Gy: 0.00608), regulation of actin-based motility by Rho (0.5 Gy: 0.51, 1.0 Gy: 0.16, 4.0 Gy: 0.00971), signalling by Rho Family GTPases (0.5 Gy: 0.53, 1.0 Gy: 0.049, 4.0 Gy: 0.13).

4.0 Gy at both time points (Figure 2 A and B). The Rac1 protein levels at 0.5 Gy were significantly decreased at 4 hours but returned to control levels after 24 hours post-irradiation (Figure 2 A and B). These data are in good agreement with the changes in Rac1 expression levels obtained by global proteomics approach at 24 hours post-irradiation (1.0 Gy: $-1.43 \pm 20.3\%$; 4.0 Gy: $-1.40 \pm 15.9\%$) (Table S1 in File S2).

As miR-132 is involved in the regulation of the Rac1-Cofilin pathway via blocking of the GTP hydrolysis protein p250GAP [31], the levels of this microRNA were quantified. The miR-132 levels were significantly decreased at 0.5 Gy but not at higher doses 4 hours post-irradiation whereas 24 hours post-irradiation miR-132 levels were still significantly decreased at 0.5 Gy and were also decreased at the higher doses (Figure 2 C). Importantly, the changes observed for miR-132 expression originated from the alterations in the expression of miR-132 and not in the expression of endogenous standard snoRNA135 (Figure S5 in File S1). Figures S5 E and S5 F in File S1 show the variation of the n-fold changes of snoRNA135 used for miRNA normalisation after 4 hours and 24 hours post-irradiation and the ΔCt values between miR-132 and snoRNA135 in irradiated HT22 cells, respectively. Only small variances in the expression of snoRNA135 were detectable whereas the ΔCt values were more affected meaning that miRNA changes originated from radiation-induced differences in miR-132 expression profile. The error bars were in all conditions comparable.

Acute effects of ionising radiation on synaptic plasticity-associated cytoskeleton signalling pathways in vivo

To examine possible radiation-induced effects on spine- and synapse formation, maturation and morphology, we irradiated male NMRI mice on postnatal day ten – a developmental stage of maximal cytoskeleton remodelling in dendritic spines and synapses within the brain growth spurt period. We irradiated neonates using a dose of 1.0 Gy and performed experiments 24 hours post-irradiation as this dose was the lowest showing persistent effect on the Rac1 expression and signalling pathway alterations in the irradiated HT22 cells (Figure 1 and 2). The analysis was performed using both hippocampus and cortex as these two brain regions are involved in learning and memory formation [34,35].

The protein quantification via global proteomics showed that 66 and 60 proteins were deregulated in the hippocampus and cortex 24 hours after exposure to 1.0 Gy, respectively (Figure 3 A). Eight proteins were found to be shared and up-regulated in these two brain regions (Figure S3 in File S1). Table S2 in File S2 shows the complete list of the in vivo deregulated proteins.

Signalling pathway analysis using the IPA software tool demonstrated that, similar to the in vitro cellular study, synaptic-plasticity associated cytoskeletal pathways were affected in vivo by radiation exposure: RhoGDI signalling (hippocampus and cortex – $p < 0.05$), actin cytoskeleton signalling (hippocampus – $p < 0.05$), regulation of actin-based motility by Rho (hippocampus – $p < 0.05$) and signalling by Rho family GTPases (hippocampus and cortex – $p < 0.05$) (Figure 3 B).

Ionising radiation impairs the RhoGTPase Rac1 and its downstream target cofilin in vivo

Quantification of Rac1 protein levels by immunoblotting confirmed a significant decrease in hippocampus and cortex at 1.0 Gy (Figure 4 A and B) to a similar extent as in the HT22 cells (Figure 2 A and B). In accordance with decreasing miR-132 levels in HT22 cells 24 hours post-irradiation with 1.0 Gy (Figure 2 C), we observed significantly decreased miR-132 levels in both hippocampus and cortex at 1.0 Gy 24 hours post-irradiation (Figure 4 C). HT22 data highlighted that the observed deregulation

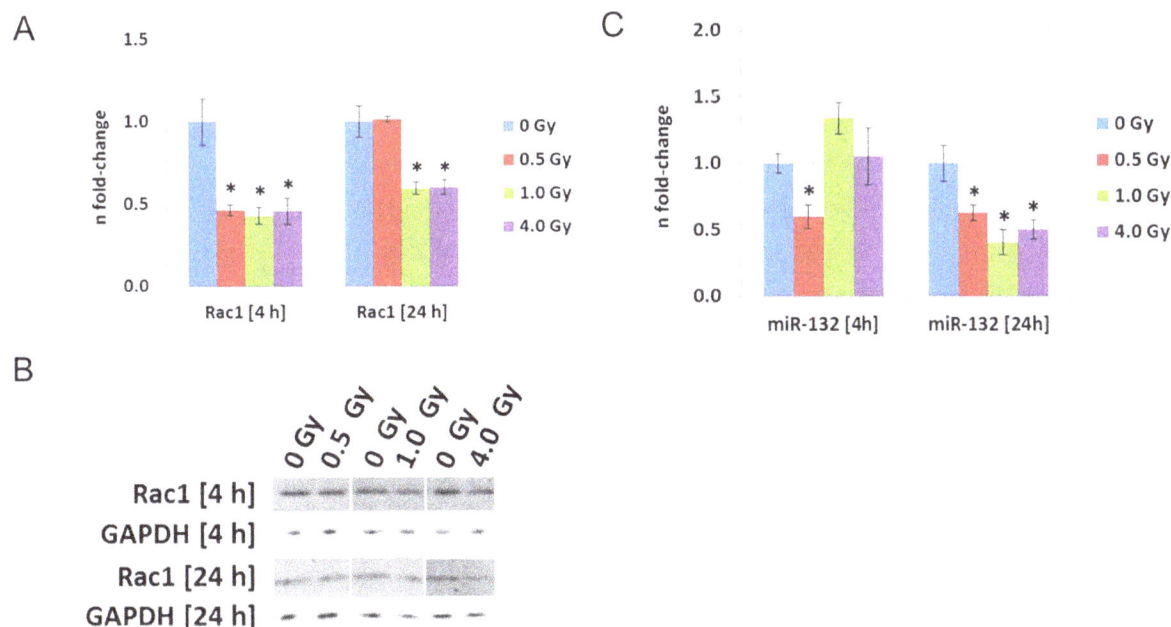

Figure 2. Immunoblotting and miRNA quantification of using HT22 cells. Data from immunoblotting (A–B) and miRNA quantification (C) associated to the Rac1-Cofilin pathway in HT22 cells irradiated with 0 Gy, 0.5 Gy, 1.0 Gy and 4.0 Gy 4 hours and 24 hours post-irradiation. The columns represent the fold-changes with standard errors of the mean (SEM); n = 3 for immunoblotting, n = 4 for miRNA quantification. The visualisation of protein bands shows the representative change from the biological replicates. *p<0.05; **p<0.01; ***p<0.001 (unpaired Student's t-test). Normalisation was performed against endogenous GAPDH and endogenous snoRNA135 for immunoblotting and miRNA quantification, respectively.

A

66   60

1.0 Gy - H   1.0 Gy - C

58   8   52

B

RhoGDI Signaling
Actin Cytoskeleton Signaling
Regulation of Actin-based Motility by Rho
Signaling by Rho Family GTPases

1.0 Gy [H]   1.0 Gy [C]

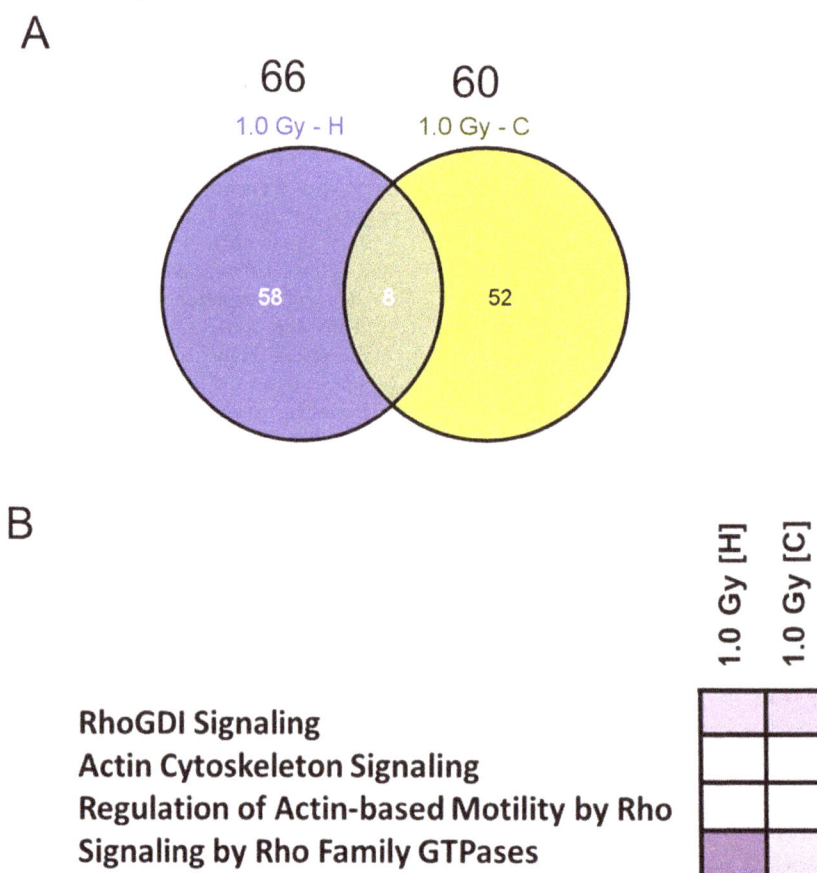

Figure 3. Mass spectrometry-based proteomics of the hippocampus and cortex of irradiated NMRI mice. Venn diagram of all deregulated and shared hippocampal [H] and cortical [C] proteins from global proteomics analysis using doses of 0 Gy and 1.0 Gy 24 hours post-irradiation (A). Hippocampus: n = 4 and cortex: n = 5. The number above each dose shows the total number of deregulated proteins at this dose. Associated cytoskeletal signalling pathways of all deregulated proteins using the Ingenuity Pathway Analysis (IPA) software in hippocampus [H] and cortex [C] are shown in (B). Higher colour intensity represents higher significance (p-value) whereas all coloured boxes have a p-value of ≤0.05; white boxes have a p-value of ≥0.05 and are not significantly altered. p-values: RhoGDI signalling (hippocampus: 0.000535, cortex: 0.000459), actin cytoskeleton signalling (hippocampus: 0.049, cortex: 0.189), regulation of actin-based motility by Rho (hippocampus: 0.0389, cortex: 0.261), signalling by Rho Family GTPases (hippocampus: 0.00000271, cortex: 0.00166).

of miR-132 is due to miR-132 deregulation itself (Figure S5 E and S5 F in File S1). Similar results were obtained in the evaluation of hippocampal and cortical ΔCt values of miR-132 and snoRNA135 (Figure S5 A – S5 D in File S1).

To further evaluate possible downstream effects of Rac1 in the Rac1-Cofilin pathway, we quantified the expression of cofilin as well as phosphorylated cofilin which is the inactive form. Immunoblotting showed that both cofilin and phospho-cofilin levels were significantly increased in the hippocampus by irradiation with 1.0 Gy but there was no significant expression alteration in the cortex (Figure 4 A and B).

Comparison of radiation-induced proteome changes
in vitro and *in vivo*

Deregulated proteins from *in vitro* cellular experiments (24 hours, 1.0 Gy) and from murine hippocampus and cortex (24 hours, 1.0 Gy) were grouped according to protein affiliations by the PANTHER software tool (Table S1 and S2 in File S2). The analysis showed that a high degree of proteins were categorised into protein classes involved in (i) cytoskeleton-associated processes, (ii) G-protein-associated processes and (iii) cell adhesion-associated processes (Table S1 and S2 in File S2– PANTHER

protein classes highlighted in orange). Proteins involved in these classes were then manually curated for the class "cytoskeleton-based synaptic plasticity" based on literature [36,37] and the UniProt database. In total, 9 out of 34 proteins, 25 out of 66 proteins and 28 out of 61 proteins from all deregulated proteins in HT22 cells, hippocampus and cortex, respectively, were found to belong to this class of proteins (Figure 5). The percentage of synaptic plasticity proteins of all deregulated proteins is shown in Figure 5. Thus, we observed an increase in deregulated proteins associated to synaptic-cytoskeletal processes after acute radiation exposure in HT22 cells (26.5%) as well as hippocampus (37.9%) and cortex (45.9%) (Figure 5). Proteins involved in other protein classes and signalling pathways act as transcription−/translation-factors, general metabolic enzymes or transfers/carriers (Table S1 and S2 in File S2). It is important to note that only one protein (Gmps - GMP synthase) was overall shared and up-regulated (Figure S4 A and S4 B in File S1). This protein is involved in the *de novo* synthesis of guanine nucleotides to provide GTP.

Overall, these results suggest that synaptic cytoskeleton-associated signalling pathways were to a high degree influenced after acute radiation both *in vitro* and *in vivo*. Moreover, a regulatory

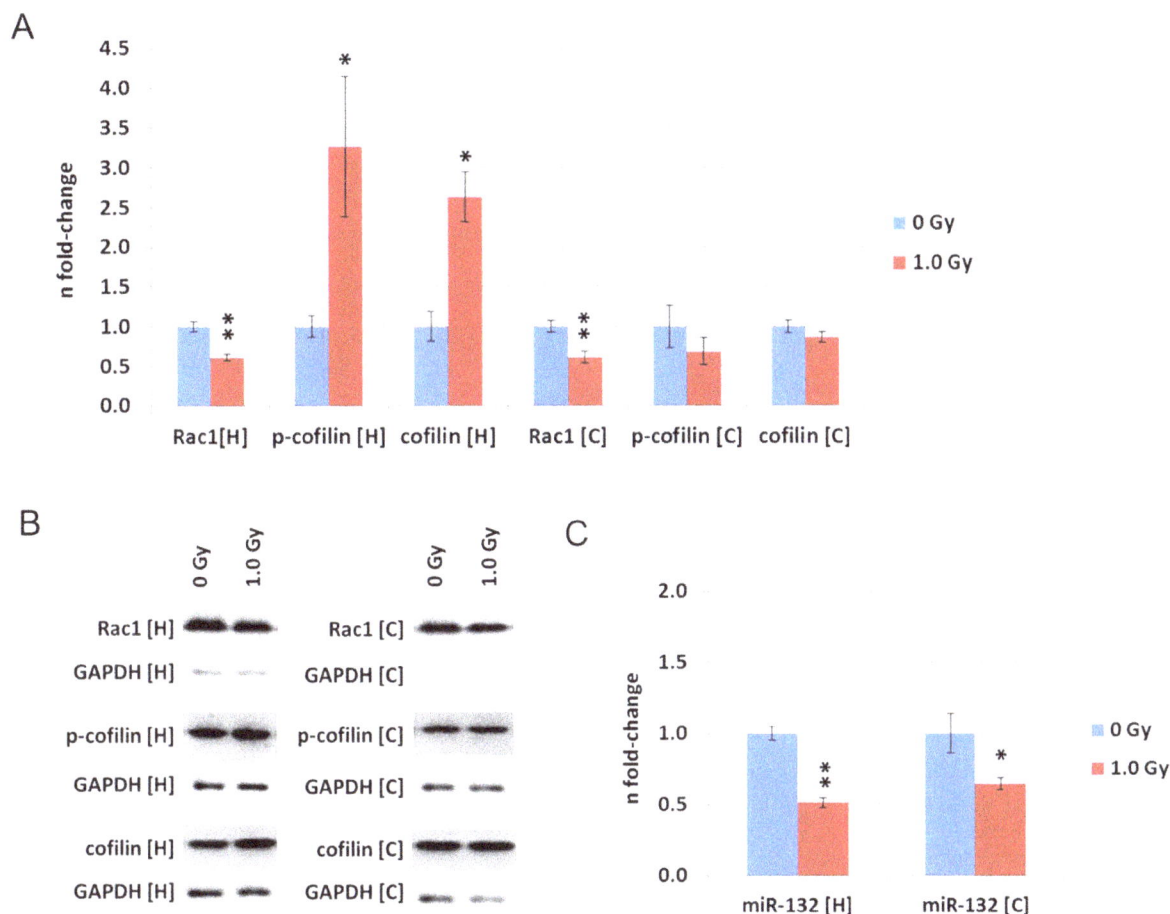

Figure 4. Immunoblotting and miRNA quantification of the *in vivo* data. Data from immunoblotting (A–B) and miRNA quantification (C) associated to the Rac1-Cofilin pathway in hippocampus [H] and cortex [C] from NMRI mice exposed on postnatal day 10 with doses of 0 Gy and 1.0 Gy. The measurement was performed 24 hours post-irradiation. The columns represent the fold-changes with standard errors of the mean (SEM); immunoblotting. n = 4 for Rac1 detection; n = 3 for p-cofilin and cofilin detection; n = 3 for miRNA quantification. The visualisation of protein bands shows the representative change from the biological replicates. *p<0.05; **p<0.01; ***p<0.001 (unpaired Student's t-test). Normalisation was performed against endogenous GAPDH and endogenous snoRNA135 for immunoblotting and miRNA quantification, respectively.

network involving miRNA and GTPases is indicated in this process.

Discussion

The aim of this study was to elucidate the biological mechanisms involved in the acute radiation-induced side-effects on learning and memory as seen in patients treated with radiotherapy. We first used immortalised primary neurons from mouse hippocampus (HT22 cells) to get information about affected signalling pathways. We then validated these data *in vivo* by using the lowest radiation dose inducing non-transient signalling pathway alterations in the cell culture system to irradiate male NMRI mice on postnatal day ten with subsequent analysis of the affected signalling pathways in the hippocampus and cortex.

Pathway analysis of the *in vitro* data showed that the doses of 1.0 Gy and 4.0 Gy, but not 0.5 Gy, significantly altered the expression of proteins functionally involved in RhoGDI signalling, actin cytoskeleton signalling, regulation of actin-based motility by Rho as well as signalling by Rho family GTPases.

Rho family GTPases are key regulators of actin cytoskeleton and are essential for orchestrating spine and synapse morphology

[37]. Their activity is controlled at least in part via RhoGDI proteins [38]. Overall, these pathways shared several proteins such as Rac1, PAK, LIMK and cofilin that all are constituents of the Rac1-Cofilin pathway. Quantification of Rac1 and miR-132 levels demonstrated a dose- and time-dependent reduction in both only at doses of 1.0 Gy and 4.0 Gy and 24 hours post irradiation. miR-132 is known to indirectly positively regulate Rac1 activity by blocking the GTPase-activating protein p250GAP [39,40]. Thus a decrease in miR-132 would lead to a decrease in Rac1 activity as we observed.

Similar signalling pathways were affected in murine hippocampus and cortex at 1.0 Gy 24 hours after the exposure as in the cell culture system. Also the Rac1 and miR-132 levels were similarly down-regulated as *in vitro* at this experimental set-up. Overall, the decrease in Rac1 expression and activity as observed *in vitro* and *in vivo* may lead to a presumptive aberrant actin remodelling in dendritic spines.

It has been shown that selective deletion of Rac1 in excitatory neurons *in vivo* affects spine structure, impairs synaptic plasticity and spatial learning [41]. Chemical inactivation of Rac1 impairs long-term plasticity in the mouse hippocampus [32]. In contrast, activation of the cerebral Rac1 leads to rearrangement of cerebral

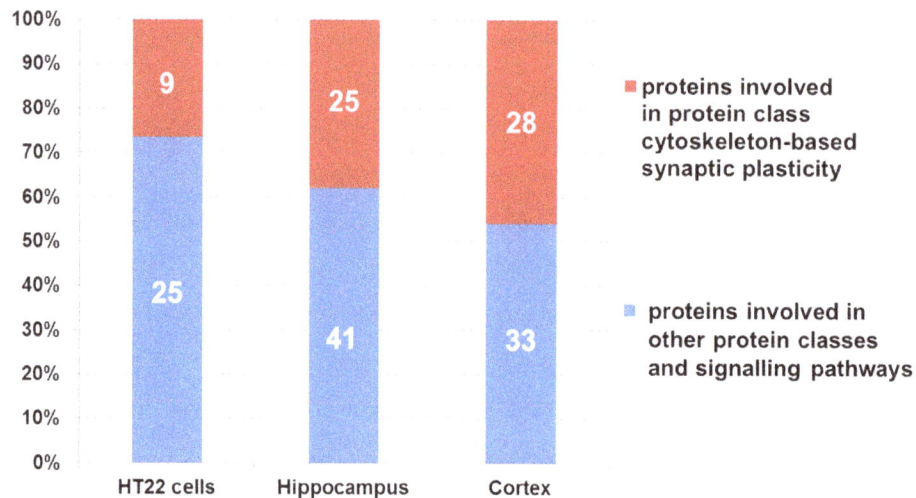

Figure 5. Mass spectrometry-based proteomics - comparison of the *in vitro* and *in vivo* data. The number of deregulated proteins from the *in vitro* and *in vivo* global proteomics analysis (1.0 Gy, 24 hours post-irradiation) belonging to the protein class "cytoskeleton-based synaptic plasticity" (Table S1 and S2) compared to all deregulated proteins in the respective cellular/organ system. The figure shows that 9 (26.5%), 25 (37.9%) and 28 (45.9%) proteins from all deregulated proteins of HT22 cells, hippocampus and cortex, respectively, can be grouped into this class. This class is based on the protein affiliations into sub-protein classes obtained from the PANTHER software (PANTHER protein class) as shown in Table S1 and S2 (PANTHER protein classes are highlighted in orange) involving cytoskeleton -associated processes, G-protein -associated processes and cell adhesion - associated processes.

actin cytoskeleton and improvement of learning and memory for several weeks in mice [42]. Similarly, it was demonstrated that inhibition of Rac1 leads to disruption of F-actin flow in hippocampal rat neuron cultures [43] and to a progressive elimination of dendritic spines in rats [44].

Recently, a miRNA profiling cohort study with Alzheimer's patients illustrated a strong decrease in miR-132 levels in the prefrontal cortex and hippocampus [45]. The deregulation of miR-132 seemed to occur predominately in neurons displaying Tau hyper-phosphorylation [45] emphasising the role of miR-132 in cognitive diseases. Additionally, it was shown that miR-132 is down-regulated in temporal cortical areas and in the CA1 region of hippocampal neurons of human Alzheimer's brain [46].

To get further insight into the downstream effects of Rac1 in the Rac1-Cofilin pathway, we quantified alterations in the end product of this pathway. Cofilin and phospho-cofilin levels were both increased in the hippocampus but not in the cortex after 1.0 Gy exposure (24 hours post-irradiation). It has been shown that also fascin plays an important role in the organisation of actin filament bundles whereas cofilin may play a cooperative role in the disassembly of filopodial actin filaments *in vitro* [47]. Proteomics data of the irradiated hippocampus showed an increase in fascin1 (Fscn1) levels whereas in the cortex we did not observe any significant alterations (Table S2 in File S2). Thus, it remains speculative whether an increase in cofilin and fascin protein levels is necessary for the acute radiation-induced actin remodelling only in the hippocampus but not in the cortex. Dephosphorylated cofilin binds to actin resulting in enhanced filament severing [48] and thus actin depolymerisation. In contrast, phosphorylated cofilin enhances actin filament turnover [49]. However, while phosphorylated cofilin is impaired in actin severing function, dephosphorylated cofilin is not necessarily active, since it may undergo inactivation by other means, including cellular sequestration [50,51]. Thus, it is important to note that cofilin regulation in the dendritic spine context is probably not a simple switched on-(dephosphorylated cofilin) and switched off- (phosphorylated

cofilin) mechanism but is also dependent on the relative concentration of cofilin to actin [51,52].

We suggest that radiation exposure alters the basal expression of cofilin and leads thus to aberrant actin signalling and processing in dendritic spines. Although we observe that both total cofilin and phospho-cofilin are increased in irradiated hippocampus potentially to equilibrate cofilin/phospho-cofilin ratio in a normal range and allow the recovery from synaptic damages, the relative concentration of actin to cofilin, whether phosphorylated or not, may be consequently changed. This hypothesis has to be confirmed by further experiments such as the evaluation of morphometric parameters relevant to actin/cofilin regulation after irradiation. Nevertheless, imbalances in the actin/cofilin ratio may lead to altered spine morphology. Dendritic spines and their ability to form synapses play an important role in modulating and storing of information [53]. Filamentous actin represents the major cytoskeletal component in dendritic spines to ensure morphological integrity [54,55]. Thus, it seems likely that morphological defects in spine shape, size and number are dependent on local actin dynamics and signalling. In fact, spines are able to induce rapid actin-based remodelling processes to change their morphology within seconds [56] to react efficiently to stressors such as ionising radiation.

However, it remains enigmatic if the acute alterations in spine architecture signalling pathways we observed here lead to persistent spine morphology changes. Chakraborti et al. showed that a high-dose (10 Gy) irradiation of the brain in young adult mice resulted in alterations in dendritic spine density and morphology in the hippocampus lasting up to one month [57]. Moreover, even doses as used in our study (1.0 Gy) are able to trigger persisting changes in dendritic complexity, synaptic protein levels, spine density and morphology in murine hippocampal neurons [21]. It has also been shown that immature filopodia in the hippocampus were more sensitive to irradiation compared to mature spines [21]. As filopodia are cytoplasmic projections containing actin filaments cross-linked into bundles by Rho family GTPases [43], these data are in good agreement with the acute alterations in the Rac1-Cofilin pathway found in our study. Importantly, the Rac1 protein regulates the synaptic maturation and integration of adult born neurons in the hippocampus [58,59]. Deregulation of Rac1 may even play an important role in defects of hippocampal adult neurogenesis that are observed in several radiation exposure studies [17–20].

Overall, we show that ionising radiation leads to acute changes in cytoskeletal signalling pathways associated with spine morphology *in vitro* and *in vivo*, by altering the molecular players within the Rac1-Cofilin-pathway. An understanding of the signalling pathway alterations after acute radiation exposure is essential in order to prevent immediate side-effects affecting learning and memory in accidentally exposed persons and patients treated with radiotherapy against brain tumours.

Supporting Information

File S1 Supporting figures. Figure S1, Mass spectrometry-based proteomics changes in HT22 cells as a function of time. Venn diagrams of all and shared deregulated proteins as a function of time (4 hours vs. 24 hours) from HT22 cells exposed to 0.5 Gy (A), 1.0 Gy (B) and 4.0 Gy (C) from global proteomics approach; n = 4 for each radiation dose. Detailed protein information (protein name, fold-changes, protein variability) of time-dependent overlapping deregulated proteins is given in D, E and F. Figure S2, Visualisation of cytoskeletal synaptic-plasticity signalling pathways and their overlapping proteins. Visualisation of pathways involved in RhoGDI signalling, regulation of actin-based motility by Rho, regulation by Rho Family GTPases and actin cytoskeleton signalling and their overlapping proteins as highlighted in red boxes. The images were downloaded from Ingenuity Signalling Pathway Analysis (IPA) software. Figure S3, Mass spectrometry-based proteomics of *in vivo* – comparison of brain regions. Overlapping deregulated proteins in hippocampus [H] and cortex [C] with protein names, fold-changes, and protein variability 24 hours post-irradiation using global proteomics approach; hippocampus: n = 4 and cortex: n = 5. Figure S4, Mass spectrometry-based proteomics of *in vivo and in vitro* – comparison of overlapping proteins in a similar experimental set-up. Overlapping deregulated proteins in HT22 cells, hippocampus [H] and cortex [C] at 1.0 Gy 24 hours post-irradiation are shown (A). The protein Gmps that was found deregulated in all analysis is shown with protein name, fold-changes, and protein variability using global proteomics approach (B); HT22 cells and hippocampus:

n = 4 and cortex: n = 5. Figure S5, Visualisation of the fold changes of snoRNA135 and ΔCt values between miRNA-135 and snoRNA135 in control and irradiated cells and tissues. The columns represent the fold changes of snoRNA135 in miRNA normalisation (A, C, E) and ΔCt values of the differences between the Ct values of miRNA-135 and snoRNA135 (B, D, and F) with standard errors of the mean (SEM) *in vitro* (HT22 cells: n = 4 for 4 hours and 24 hours post-irradiation) and *in vivo* (hippocampus [H]: n = 3 for 24 hours post-irradiation; cortex [C]: n = 3 for 24 hours post-irradiation) experimental set-ups.

File S2 Supporting tables. Table S1, Deregulated proteins found in mass spectrometry-based proteomics in *in vitro* irradiated HT22 cells. Complete detailed list of deregulated proteins (name, unique peptides, n-fold-change, variability and counts per biological replicate) obtained from HT22 cell experiment 4 hours and 24 hours post-irradiation at doses of 0.5 Gy, 1.0 Gy and 4.0 Gy from global proteomics analysis. Deregulated proteins at 1.0 Gy 24 hours after radiation exposure were categorised into protein classes using PANTHER classification system software and the general annotation from UniProt as indicated by an asterisk. Table S2, Deregulated proteins found in mass spectrometry-based proteomics of the hippocampus and cortex of irradiated NMRI mice. Complete detailed list of deregulated proteins in hippocampus and cortex (name, unique peptides, n-fold-change, variability and counts per biological replicate) obtained from NMRI mice experiment 24 hours post-irradiation at doses of 1.0 Gy from global proteomics analysis. Deregulated proteins at 1.0 Gy 24 hours after radiation exposure were categorised into protein classes using PANTHER classification system software and the general annotation from UniProt as indicated by an asterisk.

Acknowledgments

We thank Stefanie Winkler and Sandra Helm for their outstanding technical assistance. We thank J. Lewerenz for sharing the HT22 cells.

Author Contributions

Contributed reagents/materials/analysis tools: CT SM BS SH PE. Conceived and designed the experiments: SK PE ST. Performed the experiments: SK SB CT BS. Analyzed the data: SK. Contributed to the writing of the manuscript: SK ST MA PE SH SM.

References

1. Hoffman KE, Yock TI (2009) Radiation therapy for pediatric central nervous system tumors. J Child Neurol 24: 1387–1396.
2. Spiegler BJ, Bouffet E, Greenberg ML, Rutka JT, Mabbott DJ (2004) Change in neurocognitive functioning after treatment with cranial radiation in childhood. J Clin Oncol 22: 706–713.
3. Hall P, Adami HO, Trichopoulos D, Pedersen NL, Lagiou P, et al. (2004) Effect of low doses of ionising radiation in infancy on cognitive function in adulthood: Swedish population based cohort study. BMJ 328: 19.
4. Ron E, Modan B, Floro S, Harkedar I, Gurewitz R (1982) Mental function following scalp irradiation during childhood. Am J Epidemiol 116: 149–160.
5. Pearce MS, Salotti JA, Little MP, McHugh K, Lee C, et al. (2012) Radiation exposure from CT scans in childhood and subsequent risk of leukaemia and brain tumours: a retrospective cohort study. Lancet 380: 499–505.
6. Fouladi M, Gilger E, Kocak M, Wallace D, Buchanan G, et al. (2005) Intellectual and functional outcome of children 3 years old or younger who have CNS malignancies. J Clin Oncol 23: 7152–7160.
7. Kempf SJ, Azimzadeh O, Atkinson MJ, Tapio S (2013) Long-term effects of ionising radiation on the brain: cause for concern? Radiat Environ Biophys 52: 5–16.
8. Schulz RJ, Albert RE (1968) Follow-up study of patients treated by x-ray epilation for tinea capitis. 3. Dose to organs of the head from the x-ray treatment of tinea capitis. Arch Environ Health 17: 935–950.
9. Cipollaro AC, Kallos A, Ruppe JP, Jr. (1959) Measurement of gonadal radiations during treatment for tinea capitis. N Y State J Med 59: 3033–3040.
10. Omran AR, Shore RE, Markoff RA, Friedhoff A, Albert RE, et al. (1978) Follow-up study of patients treated by X-ray epilation for tinea capitis: psychiatric and psychometric evaluation. Am J Public Health 68: 561–567.
11. Dobbing J, Sands J (1979) Comparative aspects of the brain growth spurt. Early Hum Dev 3: 79–83.
12. Huttenlocher PR, Dabholkar AS (1997) Regional differences in synaptogenesis in human cerebral cortex. J Comp Neurol 387: 167–178.
13. Eriksson P, Ankarberg E, Fredriksson A (2000) Exposure to nicotine during a defined period in neonatal life induces permanent changes in brain nicotinic receptors and in behaviour of adult mice. Brain Res 853: 41–48.
14. Eriksson P (1997) Developmental neurotoxicity of environmental agents in the neonate. NeuroToxicology 18: 719–726.
15. Eriksson P, Fischer C, Stenerlow B, Fredriksson A, Sundell-Bergman S (2010) Interaction of gamma-radiation and methyl mercury during a critical phase of neonatal brain development in mice exacerbates developmental neurobehavioural effects. NeuroToxicology 31: 223–229.
16. Dobbing J, Sands J (1973) Quantitative growth and development of human brain. Arch Dis Child 48: 757–767.
17. Rola R, Raber J, Rizk A, Otsuka S, VandenBerg SR, et al. (2004) Radiation-induced impairment of hippocampal neurogenesis is associated with cognitive deficits in young mice. Experimental Neurology 188: 316–330.

18. Allen AR, Eilertson K, Sharma S, Schneider D, Baure J, et al. (2013) Effects of radiation combined injury on hippocampal function are modulated in mice deficient in chemokine receptor 2 (CCR2). Radiat Res 180: 78–88.

19. Raber J, Rola R, LeFevour A, Morhardt D, Curley J, et al. (2004) Radiation-induced cognitive impairments are associated with changes in indicators of hippocampal neurogenesis. Radiat Res 162: 39–47.

20. Mizumatsu S, Monje ML, Morhardt DR, Rola R, Palmer TD, et al. (2003) Extreme sensitivity of adult neurogenesis to low doses of X-irradiation. Cancer Res 63: 4021–4027.

21. Parihar VK, Limoli CL (2013) Cranial irradiation compromises neuronal architecture in the hippocampus. Proc Natl Acad Sci U S A 110: 12822–12827.

22. Tsamis IK, Mytilinaios GD, Njau NS, Fotiou FD, Glaftsi S, et al. (2010) Properties of CA3 dendritic excrescences in Alzheimer's disease. Curr Alzheimer Res 7: 84–90.

23. Armstrong DD, Dunn K, Antalffy B (1998) Decreased dendritic branching in frontal, motor and limbic cortex in Rett syndrome compared with trisomy 21. J Neuropathol Exp Neurol 57: 1013–1017.

24. Becker LE, Armstrong DL, Chan F (1986) Dendritic atrophy in children with Down's syndrome. Ann Neurol 20: 520–526.

25. Sahin M, Saxena A, Joost P, Lewerenz J, Methner A (2006) Induction of Bcl-2 by functional regulation of G-protein coupled receptors protects from oxidative glutamate toxicity by increasing glutathione. Free Radic Res 40: 1113–1123.

26. Merl J, Ueffling M, Hauck SM, von Toerne C (2012) Direct comparison of MS-based label-free and SILAC quantitative proteome profiling strategies in primary retinal Muller cells. Proteomics 12: 1902–1911.

27. von Toerne C, Kahle M, Schafer A, Ispiryan R, Blindert M, et al. (2013) Apoe, Mbl2, and Psp plasma protein levels correlate with diabetic phenotype in NZO mice-an optimized rapid workflow for SRM-based quantification. J Proteome Res 12: 1331–1343.

28. Yentrapalli R, Azimzadeh O, Sriharshan A, Malinowsky K, Merl J, et al. (2013) The PI3K/Akt/mTOR Pathway Is Implicated in the Premature Senescence of Primary Human Endothelial Cells Exposed to Chronic Radiation. PLoS One 8: e70024.

29. Brosch M, Yu L, Hubbard T, Choudhary J (2009) Accurate and sensitive peptide identification with Mascot Percolator. J Proteome Res 8: 3176–3181.

30. Shaltiel G, Hanan M, Wolf Y, Barbash S, Kovalev E, et al. (2013) Hippocampal microRNA-132 mediates stress-inducible cognitive deficits through its acetyl-cholinesterase target. Brain Struct Funct 218: 59–72.

31. Saneyoshi T, Fortin DA, Soderling TR (2010) Regulation of spine and synapse formation by activity-dependent intracellular signaling pathways. Curr Opin Neurobiol 20: 108–115.

32. Martinez LA, Tejada-Simon MV (2011) Pharmacological inactivation of the small GTPase Rac1 impairs long-term plasticity in the mouse hippocampus. Neuropharmacology 61: 305–312.

33. Kuhn TB, Meberg PJ, Brown MD, Bernstein BW, Minamide LS, et al. (2000) Regulating actin dynamics in neuronal growth cones by ADF/cofilin and rho family GTPases. J Neurobiol 44: 126–144.

34. Kirwan CB, Wixted JT, Squire LR (2008) Activity in the medial temporal lobe predicts memory strength, whereas activity in the prefrontal cortex predicts recollection. J Neurosci 28: 10541–10548.

35. Clopath C (2012) Synaptic consolidation: an approach to long-term learning. Cogn Neurodyn 6: 251–257.

36. Fortin DA, Srivastava T, Soderling TR (2012) Structural modulation of dendritic spines during synaptic plasticity. Neuroscientist 18: 326–341.

37. Tolias KF, Duman JG, Um K (2011) Control of synapse development and plasticity by Rho GTPase regulatory proteins. Prog Neurobiol 94: 133–148.

38. Dovas A, Couchman JR (2005) RhoGDI: multiple functions in the regulation of Rho family GTPase activities. Biochem J 390: 1–9.

39. Magill ST, Cambronne XA, Luikart BW, Lioy DT, Leighton BH, et al. (2010) microRNA-132 regulates dendritic growth and arborization of newborn neurons in the adult hippocampus. Proc Natl Acad Sci U S A 107: 20382–20387.

40. Impey S, Davare M, Lesiak A, Fortin D, Ando H, et al. (2010) An activity-induced microRNA controls dendritic spine formation by regulating Rac1-PAK signaling. Mol Cell Neurosci 43: 146–156.

41. Haditsch U, Leone DP, Farinelli M, Chrostek-Grashoff A, Brakebusch C, et al. (2009) A central role for the small GTPase Rac1 in hippocampal plasticity and spatial learning and memory. Mol Cell Neurosci 41: 409–419.

42. Diana G, Valentini G, Travaglione S, Falzano L, Pieri M, et al. (2007) Enhancement of learning and memory after activation of cerebral Rho GTPases. Proc Natl Acad Sci U S A 104: 636–641.

43. Tatavarty V, Das S, Yu J (2012) Polarization of actin cytoskeleton is reduced in dendritic protrusions during early spine development in hippocampal neuron. Mol Biol Cell 23: 3167–3177.

44. Nakayama AY, Harms MB, Luo L (2000) Small GTPases Rac and Rho in the maintenance of dendritic spines and branches in hippocampal pyramidal neurons. J Neurosci 20: 5329–5338.

45. Lau P, Bossers K, Janky R, Salta E, Frigerio CS, et al. (2013) Alteration of the microRNA network during the progression of Alzheimer's disease. EMBO Mol Med 5: 1613–1634.

46. Wong HK, Veremeyko T, Patel N, Lemere CA, Walsh DM, et al. (2013) De-repression of FOXO3a death axis by microRNA-132 and −212 causes neuronal apoptosis in Alzheimer's disease. Hum Mol Genet 22: 3077–3092.

47. Breitsprecher D, Koestler SA, Chizhov I, Nemethova M, Mueller J, et al. (2011) Cofilin cooperates with fascin to disassemble filopodial actin filaments. J Cell Sci 124: 3305–3318.

48. Pavlov D, Muhlrad A, Cooper J, Wear M, Reisler E (2007) Actin filament severing by cofilin. J Mol Biol 365: 1350–1358.

49. Maloney MT, Bamburg JR (2007) Cofilin-mediated neurodegeneration in Alzheimer's disease and other amyloidopathies. Mol Neurobiol 35: 21–44.

50. van Rheenen J, Condeelis J, Glogauer M (2009) A common cofilin activity cycle in invasive tumor cells and inflammatory cells. J Cell Sci 122: 305–311.

51. Bamburg JR, Bernstein BW (2010) Roles of ADF/cofilin in actin polymerization and beyond. F1000 Biol Rep 2: 62.

52. Andrianantoandro E, Pollard TD (2006) Mechanism of actin filament turnover by severing and nucleation at different concentrations of ADF/cofilin. Mol Cell 24: 13–23.

53. Kasai H, Matsuzaki M, Noguchi J, Yasumatsu N, Nakahara H (2003) Structure-stability-function relationships of dendritic spines. Trends Neurosci 26: 360–368.

54. Cohen RS, Chung SK, Pfaff DW (1985) Immunocytochemical localization of actin in dendritic spines of the cerebral cortex using colloidal gold as a probe. Cell Mol Neurobiol 5: 271–284.

55. Fifkova E, Delay RJ (1982) Cytoplasmic actin in neuronal processes as a possible mediator of synaptic plasticity. J Cell Biol 95: 345–350.

56. Fischer M, Kaech S, Wagner U, Brinkhaus H, Matus A (2000) Glutamate receptors regulate actin-based plasticity in dendritic spines. Nat Neurosci 3: 887–894.

57. Chakraborti A, Allen A, Allen B, Rosi S, Fike JR (2012) Cranial irradiation alters dendritic spine density and morphology in the hippocampus. PLoS One 7: e40844.

58. Vadodaria KC, Jessberger S (2013) Maturation and integration of adult born hippocampal neurons: signal convergence onto small Rho GTPases. Front Synaptic Neurosci 5: 4.

59. Luo L (2002) Actin cytoskeleton regulation in neuronal morphogenesis and structural plasticity. Annu Rev Cell Dev Biol 18: 601–635.

6

Chloroform-Assisted Phenol Extraction Improving Proteome Profiling of Maize Embryos through Selective Depletion of High-Abundance Storage Proteins

Erhui Xiong[꒐]**, Xiaolin Wu**[꒐]**, Le Yang, Fangping Gong, Fuju Tai, Wei Wang***

Collaborative Innovation Center of Henan Grain Crops, College of Life Science, Henan Agricultural University, Zhengzhou 450002, China

Abstract

The presence of abundant storage proteins in plant embryos greatly impedes seed proteomics analysis. Vicilin (or globulin-1) is the most abundant storage protein in maize embryo. There is a need to deplete the vicilins from maize embryo extracts for enhanced proteomics analysis. We here reported a chloroform-assisted phenol extraction (CAPE) method for vicilin depletion. By CAPE, maize embryo proteins were first extracted in an aqueous buffer, denatured by chloroform and then subjected to phenol extraction. We found that CAPE can effectively deplete the vicilins from maize embryo extract, allowing the detection of low-abundance proteins that were masked by vicilins in 2-DE gel. The novelty of CAPE is that it selectively depletes abundant storage proteins from embryo extracts of both monocot (maize) and dicot (soybean and pea) seeds, whereas other embryo proteins were not depleted. CAPE can significantly improve proteome profiling of embryos and extends the application of chloroform and phenol extraction in plant proteomics. In addition, the rationale behind CAPE depletion of abundant storage proteins was explored.

Editor: Silvia Mazzuca, Università della Calabria, Italy

Funding: This work was supported by National Natural Science Foundation of China (http://www.nsfc.gov.cn/; grant no. 31371543, to WW), Plan for Scientific Innovation Talent of Henan Province (http://www.hnkjt.gov.cn/; grant no. 144200510012, to WW) and State Key Laboratory of Crop Biology in Shandong Agricultural University (http://klab.sdau.edu.cn/; grant no. 2012KF01, to WW). The funders had no role in study design, data collection and analysis, decision to publish, or preparation of the manuscript.

* Email: wangwei@henau.edu.cn

[꒐] These authors contributed equally to this work.

Introduction

Maize (*Zea mays* L.) is one of the most important cereal crops worldwide [1]. For a long time, maize has been a staple food of the world's population and a primary nutrient source for animal feed. In recent years, maize has been used for biofuel production [2]. As one of the most frequently harvested organs in agriculture, maize seed contains about 10% proteins, in which 60–80% are storage proteins, mainly existing in embryo as nutrient reservoir for seed germination and early seedling establishment. Likewise, maize embryo is very important for human and livestock nutrition due to its high contents of protein and oil [3].

In maize embryo, vicilin (or globulin-1) is the most abundant, basic (arginine-rich) storage protein. Typically, vicilins are sparsely glycosylated trimeric clusters and each subunit contains two conserved cupin domains, characteristic of Cupin_2 globulin superfamily [4–6]. Maize vicilin is encoded by a single but polymorphic *Glb1* gene [7–8]. The expression of *Glb1* is embryo specific [7] and regulated by ABA [9]. Recently, maize vicilin was identified as a novel allergen [1]. Due to its high abundance and composition complexity, maize vicilins impede embryo proteomics analysis to a great extent.

Depleting high abundance proteins is often an essential step in enhanced proteomics of complex samples, e.g. depleting storage proteins from legume seeds [10], RuBisCO from leaf extract [11], agglutinin from *Pinellia ternata* tuber extract [12] and albumin and IgG from serum [13–14]. Currently, no methods have been reported to deplete vicilins or abundant storage proteins from maize embryo extracts. We reported here a chloroform-assisted phenol extraction (CAPE) method for vicilin depletion from maize embryo extracts. CAPE is also effective for depletion of abundant storage proteins in dicot (soybean and pea) seeds.

Materials and Methods

Plant material and sampling

Maize (*Zea mays* L. cv. Zhengdan 958), soybean (*Glycine max* L.) and pea (*Pisum sativum* L.) seeds were bought from Henan Qiule Seed Industry Science & Technology Co. Ltd (Zhengzhou, China). Mature maize seed consists of three genetically distinct components: embryo, endosperm and coat. The embryo is the young organism before it emerges from the seed. Dry maize seeds were soaked in water for 2 h to soften starchy endosperm. Then, the embryos were manually took out, rinsed and used for protein extraction. Likewise, soybean and pea seeds were soaked in water for 2 h to remove seed coat and whole embryos were used for protein extraction.

Reagents

All chemicals used were of analytical grade. High purity deionized water (18 MO.cm) was used throughout the experiment. Chloroform, buffered phenol (pH 8.0) and a cocktail of protease inhibitors were purchased from Sigma-Aldrich Co. LLC (St. Louis, MO, USA). Electrophoresis reagents and IPG strips were obtained from GE Healthcare Life Sciences (Pittsburgh, PA, USA).

CAPE

The CAPE protocol includes three parts (**Figure 1**). It is designed for 600 μl embryo extract to be processed in 2.0 ml Eppendorf tube and can be scaled up for bigger volumes.

a. Protein extraction. Embryo tissues (0.1 g fresh weight) were homogenized in a cold mortar in 1.0 ml of buffer containing 0.25 M Tris-HCl (pH 7.5), 1% SDS, 14 mM DTT and a cocktail of protease inhibitors (4°C). *Note: for embryo from developing, young seeds, less extract buffer (e.g. 0.1 g versus 0.5 ml) can be used.* The homogenate was centrifuged at 12,000 g for 10 min (4°C). The resulting supernatant was still murky with a white solid layer floating on top, indicating the presence of large amount of lipids. The supernatant (embryo extract) was centrifuged again as above and subjected to chloroform denaturing.

As comparisons, aliquots of the embryo extract were mixed with equal volume of chloroform or buffered phenol (pH 8.0). After phase separation by centrifugation, proteins in each parts of the mixture were analyzed using SDS-PAGE.

b. Chloroform denaturing. Transfer 600 μl of the embryo extract to each 2.0 ml Eppendorf tube and add chloroform at a volume ratio of 1:0 (extract/chloroform), 1:0.25, 1:0.33, 1:0.5 and 1:1, corresponding to 0, 150, 200, 300 and 600 μl chloroform, respectively. The mixture was thoroughly mixed by shaking for 10 min at room temperature (RT). *Note: Chloroform denaturing can be extended to 2 h as a pause point.* In the resulting white emulsion, embryo proteins were violently denatured by chloroform. The emulsion, without centrifugation, was directly subjected to phenol extraction.

c. Phenol extraction. Add 600 μl of buffered phenol (pH 8.0) to each tube containing the chloroform-treated extracts and thoroughly shook for 10 min at RT. *Note: In the presence of chloroform, organic phase was more hydrophobic and heavier, resulting in a quick and clear phase separation, with the organic phase (containing proteins) in the bottom.* After phase separation by centrifugation, the organic phase was transferred to new 2.0 ml Eppendorf tubes (300 μl each) and precipitated with 5 volumes of methanol containing 0.1 M ammonium acetate for 1 h (−20°C).

Proteins were precipitated by centrifugation at 15,000 g for 10 min (4°C) and washed with cold acetone and 80% cold acetone each once. Air-dried proteins were dissolved in a buffer of choice. Protein content was determined using Bio-Rad Protein Assay Kit and bovine serum albumin as a standard.

SDS-PAGE

SDS-PAGE was performed using 13.5% resolving and 5% stacking polyacrylamide gels in a Bio-Rad system (Bio-Rad, Hercules, CA, USA) using the Laemmli buffers (Laemmli 1976). The gels were stained with Coomassie brilliant blue (CBB) R overnight and destained in 7% acetic acid until a clear background.

Two dimensional gel electrophoresis (2-DE)

Isoelectric focusing (IEF) was performed using 11-cm linear pH 4–7 IPG strips with the Ettan III system (GE Healthcare, USA). About 600 μg proteins were loaded into the strip by passive rehydration overnight at RT. The IEF voltage was set at 250 V for 1 h, 1,000 V for 4 h, finally increasing to 8,000 V for 4 h, and holding for 10 h (20°C). Focused strips were equilibrated in Buffer I (0.1 M Tris–HCl, pH 8.8, 2% SDS, 6 M urea, 30% glycerol, 0.1 M DTT) and then in Buffer II (same as Buffer I, but with 0.25 M iodoacetamide instead of DTT) for 15 min each. SDS-PAGE was run on a 13.5% gel with 0.1% SDS in the gel and the running buffer. The gels were stained with 0.1% CBB G-250 overnight and destained in 7% acetic acid until a clear background. Digital images of the gels were processed using PDQUEST software (Bio-Rad).

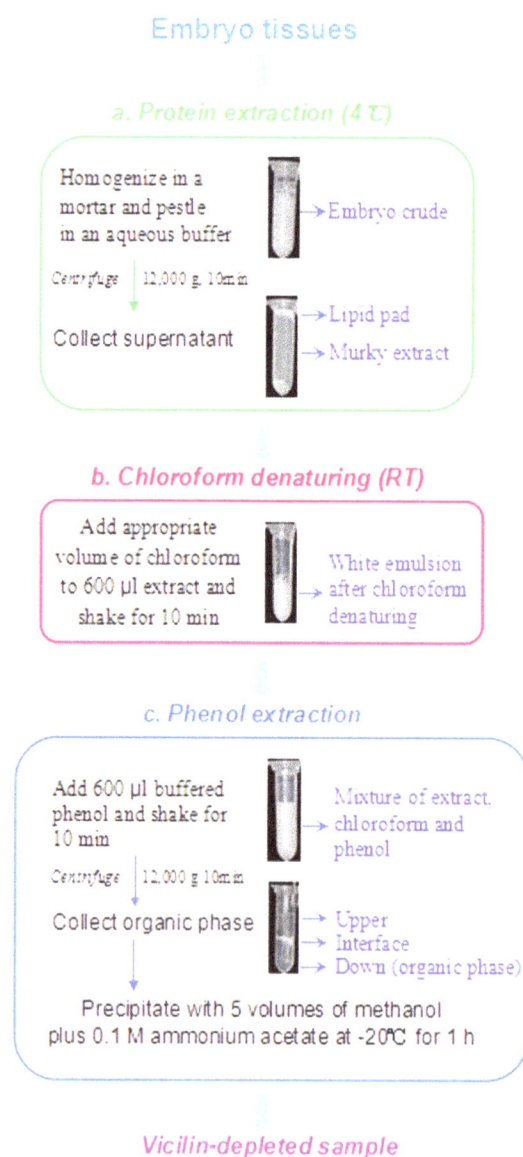

Figure 1. Schematic overview of the CAPE protocol. It was demonstrated to process 600 μl maize embryo extract in 2.0 ml Eppendorf tube.

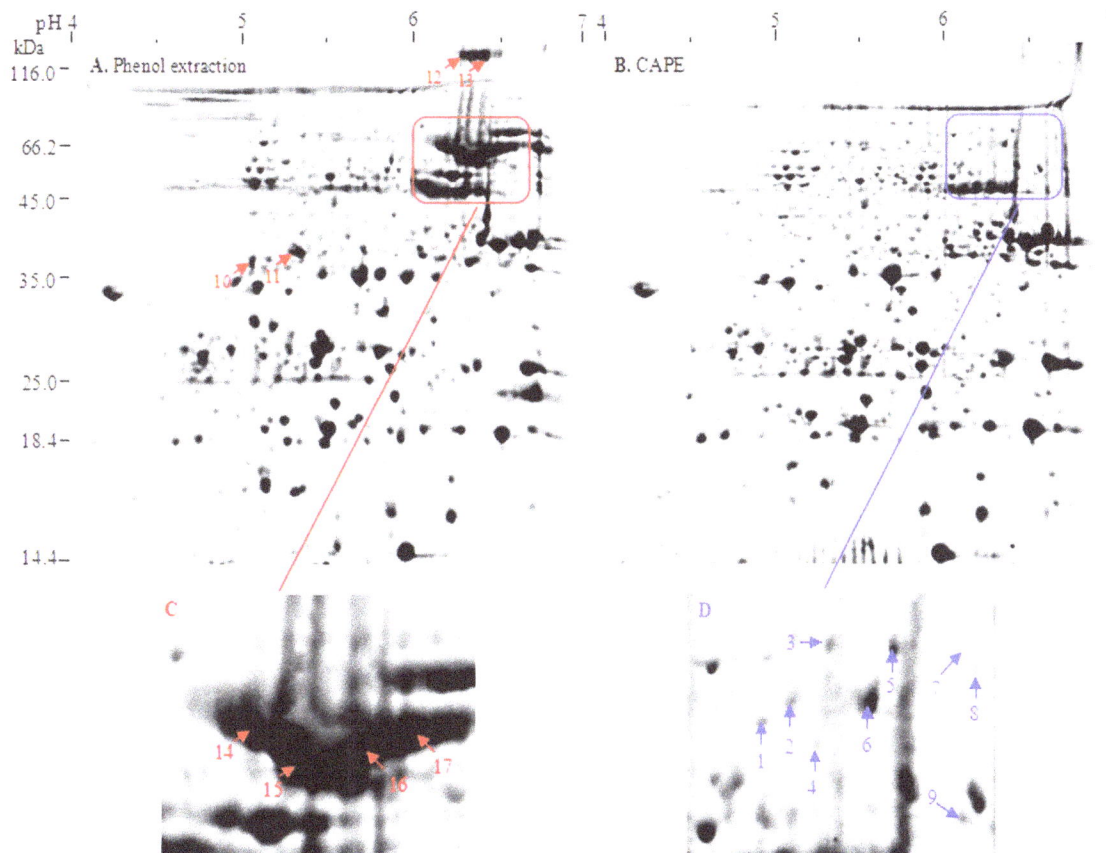

Figure 2. Depletion of maize vicilins by CAPE. A: Phenol extraction (control). **B**: CAPE. Maize embryo proteins (600 µg) were separated by IEF with 11-cm linear pH 4–7 IPG strips and then by SDS-PAGE. The magnified regions indicate before and after vicilin depletion.

MS/MS and protein identification

Protein spots of interest were reduced (10 mM DTT), alkylated (50 mM iodoacetic acid), and then digested with 20 ng/µl trypsin for 16 h at 37°C in 50 mM ammonium bicarbonate. The supernatants were vacuum-dried and dissolved in 10 µl of 0.1% trifluoroacetic acid, and 0.5 µl of the solution was added onto a matrix consisting of 0.5 µl of 5 mg/ml 2, 5-dihydroxybenzoic acid in water: acetonitrile (2:1). The digested fragments were analyzed using a MALDI-TOF/TOF analyzer (ultraflex III, Bruker, Germany). MS/MS spectra were acquired in the positive ion mode and automatically submitted to Mascot 2.2 (http://www.matrixscience.com) for peptide mass finger printing against the NCBInr 20140323 database (38032689 sequences). The taxonomy used was Viridiplantae (green plants) (1749470 sequences). The search parameters were as follows: type of search: MALDI-TOF ion search; enzyme: trypsin; fixed modifications: carbamidomethyl (C); variable modifications: acetyl (N-terminal) and oxidation (M); mass values: monoisotopic; protein mass: unrestricted; peptide mass tolerance: ±100 ppm; fragment mass tolerance: ±0.5 Da; max missed cleavages: 1; instrument type: MALDI-TOF-TOF. Only significant scores defined by Mascot probability analysis greater than "identity" were considered for assigning protein identity. All of the positive protein identification scores were significant (P<0.05, score>50).

Results and Discussion

Development of the CAPE protocol

Currently, methods based on trichloroacetic acid (TCA)/ acetone precipitation or phenol extraction are widely used in plant proteomics [15–16]. Most recently, we described in detail sample preparation method integrating TCA/acetone precipitation and phenol extraction for crop proteomics [15]. However, 2-DE image of maize embryo, based on these reported available protocols, was dominated by high-abundance vicilins (**Figure 2**), which inevitably masked low-abundance proteins with similar pIs and sizes and impeded to dig deep into embryo proteome. In preliminary experiments, we tried many protocols, but all failed to effectively deplete vicilins from maize embryo extracts. Fortunately, we found that chloroform denaturing prior to phenol extraction can effectively and selectively deplete vicilins from maize embryo extract.

We first optimized the CAPE method regarding the amount and treatment time of chloroform before phenol extraction. A ratio 1:1 (v:v) of extract/chloroform was found to be sufficient to deplete vicilins from maize embryo extract, as in the case of soybean and pea seeds (**Figure 3**). As demonstrated by SDS-PAGE, major vicilin (or abundant storage protein) bands (indicated by asterisk) decreased continually in abundance with the increase of chloroform amount. No significant difference of

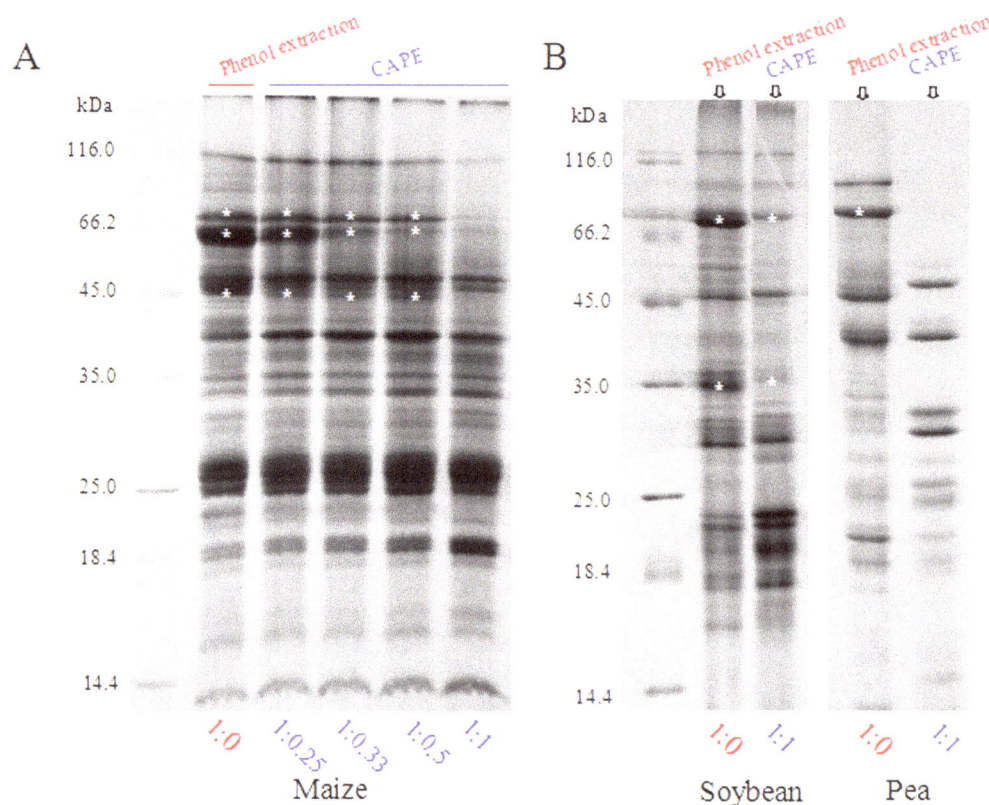

Figure 3. Optimization of chloroform ratio used in CAPE. Embryo extracts were mixed well with chloroform at different ratios. Proteins obtained from CAPE and phenol extraction (control) were compared by SDS-PAGE. The asterisks (*) indicated vicilins or abundant storage proteins.

incubation time of chloroform denaturing was observed between 10–30 min. Thus, a ratio of 1:1 (extract/chloroform) and 10 min incubation in CAPE were used in subsequent experiments. For dilute embryo extract, a 1:0.5 (extract/chloroform) was sufficient.

By phenol extraction, the smear region of vicilins was most striking in the top right edge of 2-DE image of maize embryo (**Figure 2**). These vicilins represented at least 20% of total embryo proteins, estimated on their band intensities in 1-D gel. MS/MS analysis (**Table 1**) and sequence alignment by ClustalW2 (http://www.ebi.ac.uk/Tools/msa/clustalw2/) (**Figure 4**) confirmed the vicilins identity of these spots. Maize vicilins contain two typical Cupin domains and share 79% homology with soybean vicilins. Obviously, maize vicilins consist of several major isoforms with similar sizes but different pIs (6.1–6.7), consistent with a previous study that vicilin from maize flour is composed of a string of six spots in 2-D gel, all of them allergenic [1]. Probably, this phenomenon resulted from three causes: (a) maize vicilin is encoded by a single but polymorphic *Glb1* gene [7–8], which may contribute the subtle difference of the gene products (protein isoforms); (b) vicilin is a glycoprotein [17] and exists as a heterogeneous mixture after glycosylation; and (c) proteolytic modification occurred during vicilin synthesis also contributes the heterogeneity in vicilin species [18–19].

The selectivity, efficiency and reproducibility of CAPE depletion of maize vicilins was evaluated using 2-DE (**Figure 2**). After vicilin depletion, 665 ($\pm$5) CBB-stained spots were detected in maize embryo extract, compared to 628 ($\pm$6) spots without the

depletion. In particular, spots 1–17 (**Figure 2**) were selectively removed or newly detected and other representative 12 spots (spots 18–29, **Figure S1**) were enriched in abundance to an extent after CAPE. Moreover, CAPE was highly reproducible based on the 2-DE images from three independent experiments (**Figure S1**).

The 17 spots mentioned above were chosen from the depletion and control gels for MS/MS identification (**Table 1**). The spots lacked in the depletion gel were almost vicilins, indicating a highly selective depletion of CAPE for maize vicilins. Two smaller (spot 10 and 11) and two bigger vicilins (spot 12 and 13) were also removed by CAPE. After vicilin depletion, several low-abundance proteins (**Figure 2**, the enlarged region C and D) were detectable in the places originally taken by vicilins, including globulin-1 S allele precursor (spots 1–5) and TPA: malate synthase (spot 6), which had 42% and 53% identity with maize vicilin, respectively, according to BLAST analysis (http://www.uniprot.org/blast/).

The utility of CAPE for dicot seeds was evaluated using soybean seeds. 2-DE analysis showed that CAPE was able to selectively deplete 19 abundant protein spots from soybean embryo extracts. These spots were concentrated in four regions of 2-DE image (**Figure 5, a–d**). Recently, a fractionation technique based on 10 mM Ca^+ precipitation was used to deplete abundant storage proteins from soybean seeds [10]. Estimated on protein sizes and pIs, the depleted protein spots between CAPE (this study) and Ca^+ fractionation precipitation [10] are almost the same, especially in the regions **a** and **b**. However, the quality of 2-DE image of soybean embryos after CAPE depletion was superior to that

```
1     -------MVSARIVVLLATLLCAAAAVASSWEDDNHHHHGGHKSGQCVRRCEDRPWHQRPR   Zea mays
      ------MGKGLTFLLVLVLVIS----------------------------VKGYEE---   Daucus carota
      ------MGNKTTLLLLLFVLCHGVATT-----------------------TMAFHD---   Glycine max
      MKQRAKMWKKEALVMLLIIAVLGNAIGIKEEAEA--------------AEEEEWWRE---   Cucurbita maxima
             *.   .:::  :                              :.:

61    CLEQCREEEREKRQERSRHEADDRSGEGSSEDEREQEKEKQKDRRPYVFDRRSFRRVVRS   Zea mays
      ---------------------------------DE------SGSEGSGRKLFILHISVEVVRS   Daucus carota
      ---------------------------------DE------GGDKKSPKSLFLMSNSTRVFKT   Glycine max
      ---------------------------------RE------EEREFRSKEQFLLEDSKRVIET   Cucurbita maxima
                                        *      :   *   .  .*..:

121   EQGSLRVLRPFDEVSRLLRGIRDYRVAVLEANPRSFVVPSHTDAHCICYVAEGEGVVTTI   Zea mays
      EAGGMKVVKGITG--KFV--DKPMHIGFIYMEPKSLFDPQYLDSNLILFIRRGEAKVGSI   Daucus carota
      DAGEMRVLKSHGG--RIF--YRHMHIGFISMEPKSLFVPQYLDSNLIIFIRRGEAKLGFI   Glycine max
      EAGEMRVIRSPAS--RIL--DRPMHIGFITMEPKSLFVPQYLDSSLILFVRRGEVKVGLI   Cucurbita maxima
      : * ::*::        ::.   :  :.:  :*:*:. *.: *:  * ::.**  :  *

181   ENGERRSYTIKQGHVFVAPAGAVTYLANTDGRKKLVITKILHT--ISVPGEFQFFFGPGG   Zea mays
      RNDKLVEQDLKTGDIYTIDAGSVFYIENTGEGQRLQIICSIDTSESLTWHAFQSFFIGGS   Daucus carota
      YDDELAERRLKTGDLYMIPSGGAFYLVNIGEGQRLHVICSIDPSTSLGLETFQSFYIGGG   Glycine max
      YKDELAERRMKGGDVYRIPAGSVFYMVNVGEGQRLQIICSIDKSESLSYGTFQSFFIGGG   Cucurbita maxima
        .  :.  .:* *:.::  :*:. *: *    ::* :  :.      ** *: **

241   RNPESFLSSFSKSIQRAAYKTSSDRLERLFGRHGQDKGIIVRATEEQ--------------   Zea mays
      RNPSSILAGFDKETLSTAFNVSVSELEEFLSPEPSGATVYISPESKSPNLWTHFINLEHH   Daucus carota
      ANSHSVLSGFEPAILETAFNESRTVVEEIFSKELDGPIMFVD-DSHAPALWTKFLQLKKD   Glycine max
      TYPVSVLAGFDQDTLATARNVSYTRLRRILSRQRQGPIVYVS-DTESPGVWSKFLQVKDG   Cucurbita maxima
      *.*:.*.    :*:: *  :..::.. .  :  :.

301   --TRELRRH---------ASEGGHGPHWPLPP--------------RGFSRGPYSL     Zea mays
      QKKAHLKKFVLFEGDVDVTESKEERPSWSLGKLVKSLFINENKENKDKVR-DSGDDVYNL   Daucus carota
      DKEQQLKKMMQDQ----EEDEEEKQTSRSWRKLLETVFGKVNEKIEN-KDTAGSPASYNL   Glycine max
      DKGNKIANINEDG----E--EAEKNKPWSWRNLVSLIFGNENRDKTKRTRTGKSPDSYNL   Cucurbita maxima
            .:.                                        *.*

361   LD-QRPSIANQHGQLYEADARSFHDLAFHDVSVSFANITAGSMSAPLFNTRSFKIAYVPN   Zea mays
      YD-RNPDFQNSYGWSLAVDDSQYKPLNHSGIGVYLVNLTAGSMMAPHINPTASEYGIVLR   Daucus carota
      YDDKKADFKNAYGWSKALHGGEYPPLSFPDIGVLLVKLSAGSMLAPHVNPISDEYTIVLS   Glycine max
      YD-KTPDFSNAYGWSVALDFHEYSPLGHSGIGVYLVNLTAGSMMAPHINPTAAEYGIVLR   Cucurbita maxima
       * :  .: *  :*    .: *  :.* :.::::**** **.* .:  *

421   GKGYAEIVCPHFQSQGGESERERDKGRESEEEEEESSEEQEEAGQGYHTIRARLSPGTAF   Zea mays
      GSGSIQIVFFNGT----------------------------------LAMNTKVNEGDVF   Daucus carota
      GYGELHIGYPNGS----------------------------------RAMKTKIKQGDVF   Glycine max
      GTGTIQIVYFNGT----------------------------------SAMDTEVTEGDVF   Cucurbita maxima
      * * .* *.                                       :: :.:. *.*

481   VVPAGHPRVAVASRDSNLQIVCFFVHADRNEKVFLAGADNVLQKLDRVAKALSFASKAEE   Zea mays
      WIPRYFHSVKFHQEQAPWSFFGFTTSSQRNHPQFLVGRGSLFQTMFGRELVVSFGSTEKK   Daucus carota
      VVPRYFPFCQVASRDGPLEFFGFSTSARKNKPQFLAGAASLLRTLMGPELSAAFGVSEDT   Glycine max
      WVPRYFPFCQIASRTGPFEFFGFTTSSRRNRPQFLACANSIFHTLRSPAVATAFDITEDD   Cucurbita maxima
          :*   .     .: *. :  :*.  **.   .:::.:        :* .  .

541   VDEVLGSRREKGFLPGPEESGGHEEREQEEEEEREEEHGGRGRERERHGREEREKEEEREGR   Zea mays
      FEKFIYAQNESTILSTASVAPPDDVNRVIL------KKGKREKMIPK---LAKKLSND-M   Daucus carota
      LRRAVDAQHEAVILPSAWAAPPENAGKLKM------EEEPNAI---------RSFANDVV   Glycine max
      LDRLLSAQYEVVILPSAEIAPPHKEEEEKRR------RREEGRRERERESERERREEWTRR   Cucurbita maxima
      .  .  :: *  :*       :   .. .                      :.

601   HGGREEREEEEERHGRGRRREEVAETLMRMVTARM   Zea mays          tr|Q03865|Q03865_MAIZE
      MMGFE-----------------------------   Daucus carota      gi|1276946|gb|AAC15238.1|
      MDVF------------------------------   Glycine max        gi|410067729|dbj|BAB21619.2|
      LEAF------------------------------   Cucurbita maxima   gi|691752|dbj|BAA06186.1|
```

Figure 4. ClustalW alignment of vicilin sequences from *Zea mays*, *Daucus carota*, *Glycine max* and *Cucurbita maxima*. Asterisks (*) under sequences indicate identically conserved residues while double dots (:) indicate conserved substitutions, and singe dots (·) indicate semi-conserved substitutions. The two Cupin1 domains in maize vicilin are indicated with dotted lines under the sequences. The amino acid sequences matched by MS/MS identification were indicated with solid lines above the sequences.

Table 1. MS/MS identification of maize embryo proteins of interest.

Spot	Protein/Organism	Uniprot accession	pI/Mr	Mascot Score	Matched peptides	Coverage (%)	Molecular function
1	Globulin-1 S allele precursor/Maize	B6UGJ0	6.16/50.3	204	K.LLAFGADEEQQVDR.V R.FTHELLEDAVGNYR.V	6	Nutrient reservoir activity
2	Globulin-1 S allele precursor/Maize	B6UGJ0	6.16/50.3	179	K.LLAFGADEEQQVDR.V R.FTHELLEDAVGNYR.V	6	Nutrient reservoir activity
3	Globulin-1 S allele precursor/Maize	B6UGJ0	6.16/50.3	301	R.LLDMDVGLANIAR.G K.LLAFGADEEQQVDR.V R.FTHELLEDAVGNYR.V R.FEEFFPIGGESPESFLSVFSDDVIQASFNTR.R	15	Nutrient reservoir activity
4	Globulin-1 S allele precursor/Maize	B6UGJ0	6.16/50.3	140	K.LLAFGADEEQQVDR.V R.FTHELLEDAVGNYR.V	6	Nutrient reservoir activity
5	Globulin-1 S allele precursor/Maize	B6UGJ0	6.16/50.3	284	K.LLAFGADEEQQVDR.V R.FTHELLEDAVGNYR.V R.FEEFFPIGGESPESFLSVFSDDVIQASFNTR.R	13	Nutrient reservoir activity
6	TPA: malate synthase, glyoxysomal/Maize	P49081	6.25/62.2	284	R.IWNGVFQR.A R.VQNWQWLR.H R.DALDFVAGLQR.E R.AGQGAGFGPFFYLPK.M R.AGHDGTWAAHPGLIPAIR.E R.ATVLVETLPAVFQMNEILHELR.E	14	Malate synthase activity
7-9	Unknown protein	-	-	-	Failed to be identified by MS/MS analysis	-	-
10	Vicilin/Maize	Q03865	6.23/66.6	137	R.GPYSLLDQRPSIANQHGQLYEADAR.S R.DSNLQIVCFEVHADRNEK.V	7	Nutrient reservoir activity
11	Vicilin/Maize	Q03865	6.23/66.6	117	R.LSPGTAFVVPAGHPFVAVASR.D R.HASEGGHGPHWPLPPFGESR.G R.GPYSLLDQRPSIANQHGQLYEADAR.S	11	Nutrient reservoir activity
12	Vicilin/Maize	Q03865	6.23/66.6	1014	R.RPYVFDR.R R.VLRPFDEVSR.L R.VAVLEANPR.S K.QGHVFVAPAGAVTYLANTDGR.K K.ILHTISVPGEFQFF FGPGGR.N K.GYAEIVCPHR.Q R.GPYSLLDQRPSIANQHGQLYEADAR.S R.RSEEEEESSEEQEEAGQGYHTIR.A R.LSPGTAFVVPAGHPFVAVASRDSNLQIVCFEVHADRNEK.V	28	Nutrient reservoir activity
13	Vicilin/Maize	Q03865	6.23/66.6	931	R.RPYVFDR.R R.VLRPFDEVSR.L R.VAVLEANPR.S K.ILHTISVPGEFQFFGPGGR.N R.HASEGGHGPHWPLPPFGESRGPYSLLDQRPSIANQHGQLYEADAR.S K.GYAEIVCPHR.Q R.RSEEEEESSEEQEEAGQGYHTIR.A R.LSPGTAFVVPAGHPFVAVASR.D	25	Nutrient reservoir activity
14	Vicilin/Maize	Q03865	6.23/66.6	759	R.RPYVFDR.R R.VLRPFDEVSR.L R.SFVVPSHTDAHCICVVAEGEGVVTTIENGER.R K.ILHTISVPGEFQFFFGPGGR.N R.HASEGGHGPHWPLPPFGESRGPYSLLDQRPSIANQHGQLYEADAR.S R.RSEEEEESSEEQEEAGQGYHTIRLSPGTAFVVPAGHPFVAVASRDSNLQIVCFEVHADR.N	29	Nutrient reservoir activity
15	Vicilin/Maize	Q03865	6.23/66.6	1068	R.RPYVFDR.R R.VLRPFDEVSR.L R.SFVVPSHTDAHCICVVAEGEGVVTTIENGER.R K.QGHVFVAPAGAVTYLAN TDGRKK K.ILHTISVPGEFQFFFGPGGR.N R.HASEGGHGPHWPLPPFGESRGPYSLLDQRPSIANQHGQLYEADAR.S R.RSEEEEESSEEQEEAGQGYHTIR.A R.LSPGTAFVVPAGHPFVAVASR.D R.DSNLQIVCFEVHADRNEK.V	34	Nutrient reservoir activity
16	Vicilin/Maize	Q03865	6.23/66.6	1164	R.RPYVFDR.R R.VLRPFDEVSR.L R.SFVVPSHTDAHCICVVAEGEGVVTTIENGER.R K.QGHVFVAPAGAVTYLANTD GRKK K.ILHTISVPGEFQFFFGPGGR.N R.HASEGGHGPHWPLPPFGESRGPYSLLDQRPSIANQHGQLYEADAR.S R.RSEEEEESSEEQEEAGQGYHTIR.A R.LSPGTAFVVPAGHPFVAVASR.D R.DSNLQIVCFEVHADRNEK.V	34	Nutrient reservoir activity
17	Vicilin/Maize	Q03865	6.23/66.6	1180	K.QGHVFVAPAGAVTYLANTDGR.K R.VLRPFDEVSR.L K.ILHTISVPGEFQFFFGPGGR.N R.RPYVFDR.R R.GPYSLLDQRPSIANQHGQLYEADAR.S R.RSEEEEESSEEQEEAGQGYHTIR.A R.LSPGTAFVVPAGHPFVAVASRDSNLQIVCFEVHADRNEK.V	31	Nutrient reservoir activity

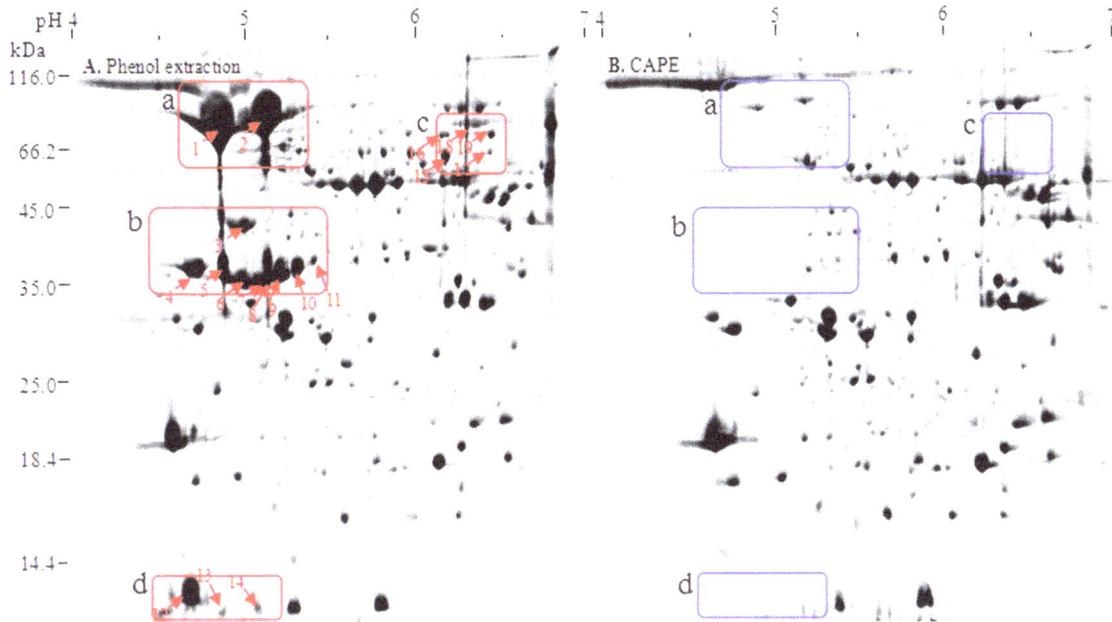

Figure 5. CAPE depletion of abundant storage proteins in soybean embryo extracts. A, phenol extraction (control); **B**, CAPE. Proteins (600 μg) were separated by IEF with 11-cm linear pH 4–7 IPG strips and then by SDS-PAGE (13.5% gel). The red and blue rectangle regions indicate before and after vicilin depletion.

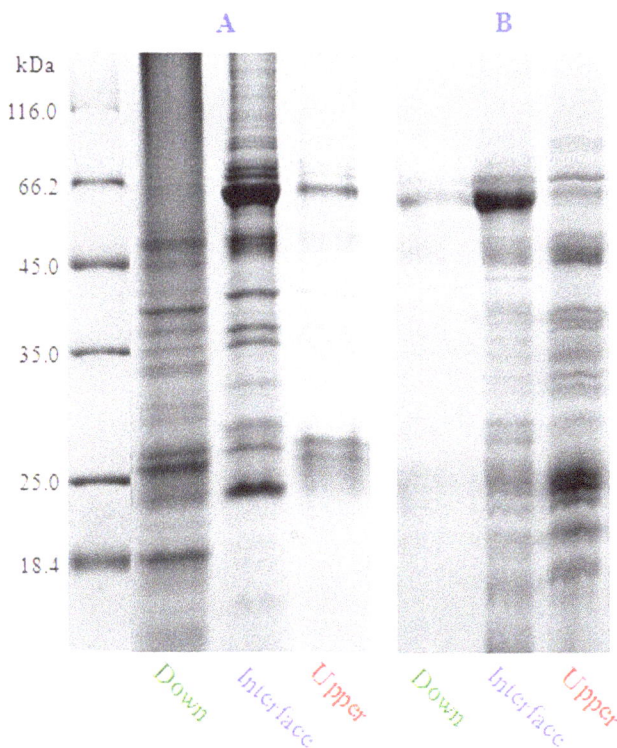

Figure 6. The effect of chloroform on the behavior of maize vicilins during phase separation. After phase separation, the proteins from upper phase, interface and down phase were analyzed using SDS-PAGE. **A**, CAPE. **B**, chloroform extraction.

obtained by Ca^+ fractionation precipitation. Therefore, CAPE was able to deplete abundant storage proteins from both monocot and dicot embryo extracts.

Obviously, CAPE improved proteome profiling of embryos in two ways: first, it selectively depletes vicilins (or abundant storage proteins) and allows the detection of low-abundance spots which were masked in 2-DE gels; second, it allows greater protein loads and meantime reduces spot 'tailing' in 2-DE, thus having potential to resolve more spots, especially using sensitive proteomic approaches (e.g. iTRAQ, DIGE).

Rationale of CAPE to selectively deplete abundant storage proteins

The novelty of CAPE is selectively depletion of vicilins or abundant storage proteins from embryo extracts. So, the rationale behind CAPE was explored.

We first examined the performance of maize vicilins during phase separation in CAPE. The proteins recovered from upper phase, interface and down phase of the chloroform/phenol/aqueous mixture were compared using SDS-PAGE (**Figure 6**). As a result, the overwhelming majority of maize vicilins were found to aggregate in the interface, while only a small amount remained in the upper phase and no vicilin band was detected by CBB staining in the down phase. Obviously, compared to other embryo proteins, maize vicilins were more susceptible to chloroform denaturing and prone to aggregate. These strong aggregates were not be resolubilized in the chloroform/phenol phase of CAPE. However, this was open to question: why were vicilins (or specific abundant storage proteins) depleted selectively by CAPE? Probably, the rationale behind CAPE was due to the specific interaction of chloroform and proteins, depending on the physicochemical property, spatial structure, especially amino acid sequence of proteins.

Likewise, a chloroform/aqueous buffer mixture, maize vicilins aggregated in the interface (**Figure 6**). As phenol extraction and other protein extraction protocols, CAPE inevitably resulted in the loss of a small amount of proteins in the interface and the upper phase.

Chloroform is a commonly used protein denaturing agent. Due to highly compatible with phenol and alcohol, chloroform is routinely used in molecular biology to separate proteins from DNA or RNA [20–21]. Moreover, chloroform is an effective organic solvent to extraction membrane protein [22–24]. The CAPE method reported here extends the application of chloroform, especially phenol extraction, in seed protein analysis. In addition, our results showed that vicilins are prone to aggregate in the interface during phase separation in CAPE; therefore, CAPE also provides a simple method to enrich and partially purify vicilins (or specific abundant storage proteins) from embryo extracts for downstream studies.

In conclusion, the present work reports a novel CAPE method for high-abundance storage proteins depletion from both monocot and dicot embryo extracts. The method is simple, low cost and compatible with proteomic analysis. It can significantly improve embryo proteome profiling and extends the application of chloroform and phenol extraction in proteomics.

Supporting Information

Figure S1   Depletion of maize vicilins by CAPE. A and B represented two groups of independent experiments. Maize embryo proteins were separated by IEF with 11-cm linear pH 4–7 IPG strips and then by SDS-PAGE. C, graphic column of relative spot volume. P, phenol extraction. C, CAPE.

Author Contributions

Conceived and designed the experiments: XLW WW. Performed the experiments: EHX XLW LY. Analyzed the data: WW EHX FPG. Contributed reagents/materials/analysis tools: FJT. Wrote the paper: WW EHX FPG.

References

1. Fasoli E, Pastorello EA, Farioli L, Scibilia J, Aldini G, et al. (2009) Searching for allergens in maize kernels via proteomic tools. J Proteomics 72: 501–510.

2. Colmsee C, Mascher M, Czauderna T, Hartmann A, Schlüter U et al. (2012) OPTIMAS-DW: A comprehensive transcriptomics, metabolomics, ionomics, proteomics and phenomics data resource for maize. BMC Plant Biol 12: 245.

3. Shewry PR, Halford NG (2002) Cereal seed storage proteins: structures, properties and role in grain utilization. J Exp Bot 53: 947–958.

4. Shewry PR, Napier JA, Tatham AS (1995) Seed storage proteins: structures and biosynthesis. Plant Cell 7: 945–956.

5. Gibbs PE, Strongin KB, McPherson A (1989) Evolution of legume seed storage proteins—a domain common to legumins and vicilins is duplicated in vicilins. Mol Biol 6: 614–623.

6. Astwood JD, Silvanovich A, Bannon GA (2002) Vicilins a case study in allergen pedigrees. J Allergy Clin Immunol 110: 26–27.

7. Belanger FC, Kriz AL (1989) Molecular characterization of the major maize embryo globulin encoded by the *Glb1* gene. Plant Physiol 91: 636–643.

8. Belanger FC, Kriz AL (1991) Molecular basis for allelic polymorphism of the maize *Globulin-1* gene. Genetics 129: 863–872.

9. Kriz AL, Wallace MS, Paiva R (1990) Globulin gene expression in embryos of maize viviparous mutants: Evidence for regulation of the *Glb1* gene by ABA. Plant Physiol 92: 538–542.

10. Krishnan HB, Oehrle NW, Natarajan SS (2009) A rapid and simple procedure for the depletion of abundant storage proteins from legume seeds to advance proteome analysis: A case study using *Glycine max*. Proteomics 9:3174–88.

11. Kim YJ, Lee HM, Wang Y, Wu J, Kim SG et al. (2013) Depletion of abundant plant RuBisCO protein using the protamine sulfate precipitation method. Proteomics 13: 2176–2179.

12. Wu XL, Xiong EH, An SF, Gong FP, Wang W (2012) Sequential extraction results in improved proteome profiling of medicinal plant *Pinellia ternata* tubers, which contain large amounts of high-abundance proteins. PLoS One 7: e50497.

13. Liu B, Qiu FH, Courtney V, Xu Y, Zhao MZ, et al. (2011) Evaluation of three high abundance protein depletion kits for umbilical cord serum proteomics. Proteome Science 9: 24.

14. Bellei E, Bergamini S, Monari E, Fantoni LI, Cuoghi A et al. (2011) High-abundance proteins depletion for serum proteomic analysis: concomitant removal of non-targeted proteins. Amino Acids 40: 145–156.

15. Wu XL, Xiong EH, Wang W, Scali M, Cresti M (2014) Universal sample preparation integrating trichloroacetic acid/acetone precipitation and phenol extraction for crop proteomics. Nat Protoc 9: 362–374.

16. Wu XL, Gong FP, Wang W (2014) Protein extraction from plant tissues for two-dimensional gel electrophoresis and its application in proteomic analysis. Proteomics 14: 645–658.

17. Gerlach JQ, Bhavanandan VP, Haynes PA, Joshi L (2009) Partial characterization of a vicilin-like glycoprotein from seeds of flowering tobacco (*Nicotiana sylvestris*). J Bot Doi:10.1155/2009/560394.

18. Higgins TJ, Newbigin EJ, Spencer D, Llewellyn DJ, Craig S (1988) The sequence of a pea vicilin gene and its expression in transgenic tobacco plants. Plant Mol Biol 11: 683–695.

19. Fukuda T, Prak K, Fujioka M, Maruyama N, Utsumi S (2007) Physicochemical properties of native adzuki bean (*Vigna angularis*) 7S globulin and the molecular cloning of its cDNA isoforms. J Agric Food Chem 55:3667–3674.

20. Sambrook J, Russell DW (2006) Purification of nucleic acids by extraction with phenol: chloroform. CSH Protoc doi:10.1101/pdb.prot4455.

21. Chomczynski P, Sacchi N (2006) Single-step method of RNA isolation by acid guanidinium thiocyanate-phenol-chloroform extraction: Twenty-something years on. Nat Protoc 1: 581–585.

22. Mirza SP, Halligan BD, Greene AS, Olivier M (2007) Improved method for the analysis of membrane proteins by mass spectrometry. Physiol Genomics 30: 89–94.

23. DI Girolamo F, Ponzi M, Crescenzi M, Alessandroni J, Guadagni F (2010) A simple and effective method to analyze membrane proteins by SDS-PAGE and MALDI mass spectrometry. Anticancer Res 30: 1121–1129.

24. Reigada R (2014) Electroporation of heterogeneous lipid membranes. Biochim Biophys Acta 1838:814–821.

7

A Low Dimensional Approach on Network Characterization

Benjamin Y. S. Li[1]*, Choujun Zhan[2], Lam F. Yeung[1], King T. Ko[1], Genke Yang[3]

1 Department of Electronic Engineering, City University of Hong Kong, Hong Kong, Hong Kong, **2** Department of Electronic and Information Engineering, The Hong Kong Polytechnic University, Hong Kong, Hong Kong, **3** Department of Automation, Shanghai Jiao Tong University, Shanghai, China

Abstract

In many applications, one may need to characterize a given network among a large set of base networks, and these networks are large in size and diverse in structure over the search space. In addition, the characterization algorithms are required to have low volatility and with a small circle of uncertainty. For large datasets, these algorithms are computationally intensive and inefficient. However, under the context of network mining, a major concern of some applications is speed. Hence, we are motivated to develop a fast characterization algorithm, which can be used to quickly construct a graph space for analysis purpose. Our approach is to transform a network characterization measure, commonly formulated based on similarity matrices, into simple vector form signatures. We shall show that the $N \times N$ similarity matrix can be represented by a dyadic product of two N-dimensional signature vectors; thus the network alignment process, which is usually solved as an assignment problem, can be reduced into a simple alignment problem based on separate signature vectors.

Editor: Vince Grolmusz, Mathematical Institute, Hungary

Funding: This project is supported by CityU Strategic Research Grant 7003016. http://www.cityu.edu.hk/ro/dlSRG.htm. The funders had no role in study design, data collection and analysis, decision to publish, or preparation of the manuscript.

Competing Interests: The authors have declared that no competing interests exist.

* Email: yeesli2-c@my.cityu.edu.hk

Introduction

In recent years, network mining has received a considerable amount of attentions. One important aspect of network mining is to measure the dissimilarities among networks since they can provide information to reproduce a graph space and allow analysis to be performed [1,2]. Yet due to the complex nature of networks, this is considered to be a challenging task [1–4].

Although it is challenging, many algorithms have been developed to solve the network comparison problem. Umeyama formulated the problem into a combinatorial optimization problem and was solved via eigendecomposition [5], Singh et al. proposed a Page rank like similarity matrix IsoRank and employed it on the search of optimal assignment [6], Li et al. proposed an integer quadratic programming approach and was solved using an ellipsoid trust region method with interior point technique [7], etc. These methods mainly consider a one-to-one comparison and may not be efficient to handle large data set. Signature extraction is one effective way to treat such large volume of data. While representing the data with a signature vector, the comparison among complex objects can be reduced to comparison between signature vectors. In addition, for pairwise comparisons, the aforementioned optimization problem is no longer required to be solved in every pairs of networks. A typical type of signature vector is the motif count vector, which summarizes the network structures by the occurrence frequencies of specific subgraphs [8–12].

Although many effective algorithms are being designed for motif counting, the computational demand is still high and the computation complexity increases as more motifs are considered.

In this paper, an alternative approach with a balance between precision and computational efficiency is proposed. Eigenvector signature distance (EVSD) is a dissimilarity measure for large-scale pairwise network comparison based on signature vector extraction techniques. The basic idea of EVSD is to represent a network by a signature vector, which is the Perron-Frobenius (PF) vector of a network's adjacency matrix. In this paper we shall show that the network comparison and alignment problem can be reduced from a matrix alignment problem into a vector alignment problem. Consequently, this vector alignment problem can be solved by simple sorting operations. Hence the complexity is reduced from $O(N^4)$ to $O(NlogN)$.

The optimal distance EVSD can be further reconstructed into an agreement measure, eigenvector signature agreement (EVSA), which can be used to quantify the similarity between two networks. The distribution of EVSA has been studied through pairwise comparisons of artificially generated networks. Results have shown that EVSAs of networks with similar structure are notably higher than EVSAs of networks with dissimilar structures. In addition, comparison between EVSA and another state-of-art signature induced similarity measures, Graphlet Degree Distribution Agreement (GDDA), will be given in Section 3.2.2. The comparison results show that classifications based on EVSA have a relatively stable distribution, which can provide a more convincing and consistent inference.

Methods

In this section we first formulate the network comparison problem, and then we show how the problem can be reduced and solved via the decomposition of Blondel's similarity matrix. Finally, based on the solution of this problem, we introduce Eigenvector Signature Distance and Eigenvector Signature Agreement to quantify networks' dissimilarity and similarity respectively

2.1 Preliminary

A graph $G(V,E)$ consists of the vertices set V and the edges set E. The edges set is a collection of all edges, each edge can be represented in the form of (u,v), where $u, v \in V$. A network can be quantified by a number of statistical measures. For instance, degree of node v_i, $deg(v_i)$, which is the total number of edges connected to node v_i. The average degree, $deg_{avg} = \frac{2e}{v}$, where $e = |E|$ and $v = |V|$ respectively. The graph density $\sigma = \frac{2e}{v(v-1)} = \frac{deg_{avg}}{v-1}$, is the ratio of the number of existing edges to the largest possible number of edges.

A graph G can also be represented by its adjacency matrix, $A = \{a_{ij}\} \in \mathbb{R}^{v \times v}$, entry $a_{ij} = 1$ shows that there is an edge connected from node i to node j, otherwise $a_{ij} = 0$. If the graph is undirected, the adjacency matrix will be symmetric. If the graph is connected, the adjacency matrix is irreducible.

2.2 Network Comparisons and Signature Vector

The network comparison problem can be considered as finding a graph distance metric d, which quantifies the difference between networks. There are many candidate measures that can be used [13–15]. Most of these measures need to deal with the problem caused by large amount of vertex mapping variations. For instance, if two networks of size 10 are being compared, then there exist 3628800 variations of vertex mapping. Computing the graph distance using the metric d on all the vertex mappings will be computationally demanding. Picking an arbitrary vertex mapping may yield a distance measure that is inappropriate for comparison.

An appropriate measure would be,

$$d(g,h) = \min_{Q \in \mathbb{P}} f(g,h,Q) \tag{1}$$

where g and h are graphs, $Q : V_g \to V_h$ is a mapping that maps nodes in graph g to h, $\mathbb{P}$ is the set of all permutation matrices. f represents a metric that quantifies dissimilarity between g and h under the mapping Q. A practical formulation of (1) could be designed with the aid of a similarity matrix. Similarity matrix is a node based similarity measure which stores all node-to-node pairwise similarity information between two networks. With such matrix, the problem can be transformed into searching a suitable mapping of nodes between two networks, which at the same time maximize the sum of all node pair similarities. That is

$$P_0 : \max_{Q \in \mathbb{P}} trace(S(G,QHQ^T)) \tag{2}$$

where G and H are the adjacency matrices of graph g and h respectively. $S(G,H)$ is the similarity matrix which s_{ij} is the similarity between node i in network g and node j in network h.

Note that, there are more than one way to interpret similarity between two networks. In this paper we employed the similarity matrix proposed by Blondel et al. [16]. According to Blondel et al., the similarity matrix for graphs g and h with adjacency matrices G and H can be computed by the following iterative process.

$$Z_{k+1}(G,H) = \frac{HZ_k G^T + H^T Z_k G}{\|HZ_k G^T + H^T Z_k G\|_F} \tag{3}$$

where $\|\cdot\|_F$ is the Frobenius norm. Z_k will finally converge to the similarity matrix.

While g and h are undirected graphs, the iterative process can be simplified into

$$Z_{k+1} = \frac{HZ_k G}{\|HZ_k G\|_F} \tag{4}$$

Thus P_0 becomes

$$P_1 : \arg\max_{Q \in \mathbb{P}} \ trace(\lim_{k \to \infty} Z_k(G,QHQ^T)) \tag{5}$$

On the other hand, by multiplying Q^T to both side of the iterative process, we have

$$Q^T Z_{k+1}(G,QHQ^T) = \frac{HQ^T Z_k G}{\|HQ^T Z_k G\|_F} \tag{6}$$

Note that $\|HQ^T Z_k G\|_F = \|QHQ^T Z_k G\|_F$ as Frobenius norm is unitarily invariant.

Let $S_Q = \lim_{k \to \infty} Z_k(G,QHQ^T)$ and $S = \lim_{k \to \infty} Z_k(G,H)$, from (6) we have the following relationship

$$Q^T S_Q = S \tag{7}$$

Thus P_1 becomes

$$P_2 : \arg\max_{Q \in \mathbb{P}} \ trace(S_Q) = \arg\max_{Q \in \mathbb{P}} \ trace(QS) \tag{8}$$

P_2 is an $N \times N$ assignment problem which can be solved by the Hungarian method in $O(N^4)$ time. It is computational demanding for large N. The efficiency can be dramatically reduced by the following decomposition on the S matrix.

According to the well know Von Mises iteration method [17], the following iterative processes converge and $x = \lim_{k \to \infty} x_k$, $y = \lim_{k \to \infty} y_k$ are the Perron-Frobenius (PF) vectors of G and H respectively.

$$x_{k+1} = \frac{Gx_k}{\|Gx_k\|_2} \ \text{and} \ y_{k+1} = \frac{Hy_k}{\|Hy_k\|_2} \tag{9}$$

Let y_k^Q be defined by $y_k = Q^T y_k^Q$. From (9) we have

$$y_{k+1}^Q = \frac{QHQ^T y_k^Q}{\|QHQ^T y_k^Q\|_2} \qquad (10)$$

And we can see that y_k^Q is the PF vector of QHQ^T. Note that $\|QHQ^T y_{k+1}\|_2 = \|HQ^T y_{k+1}\|_2$ as Euclidean norm is unitarily invariant.

Combining (9) and (10) we have

$$S_Q = Qyx^T \qquad (11)$$

Thus P_2 becomes

$$P_3 : \arg\max_{Q \in \mathbb{P}} trace(Qyx^T) = \arg\max_{Q \in \mathbb{P}} x^T Qy \qquad (12)$$

Since product of permutation matrices is still a permutation matrix, problem P_3 can be rewritten into the following form,

$$P_4 : \max_{Q_x, Q_y \in \mathbb{P}} x^T Q_x Q_y y \qquad (13)$$

and the problem becomes finding optimal permutation on x and y such that their inner product is maximized. According to the rearrangement inequality [18], the inner product is maximized when the two vectors are sorting in descending order. That is,

$$P_5 : \max_{Q_x, Q_y \in \mathbb{P}} x^T Q_x Q_y y = \hat{x}^T \hat{y} \qquad (14)$$

where $\hat{x}_i = x_{\Pi_x(i)}$ and $\hat{y}_i = y_{\Pi_y(i)}$, Π_x and Π_y are the optimal mapping such that

$$\hat{x}_1 \geq \cdots \geq \hat{x}_N \text{ and } \hat{y}_1 \geq \cdots \geq \hat{y}_N \qquad (15)$$

Here $\hat{x}$ and $\hat{y}$ are considered as the eigenvector signatures (EVS) of the two networks. By combining equations (14) and (15), we can see that the original problem P_1 can be solved via a simple sorting algorithm with the aid of our proposed decomposition. Hence, once the eigenvector is computed, only $O(N\log N)$ time is needed to complete the alignment. When comparing with the case without the decomposition, the computational time of P_3 is reduced from $O(N^4)$ to $O(N\log N)$ on the alignment process. In addition, in the case of pairwise comparison among M networks, as signatures are only required to be compute and sorted once, thus the time complexity of our proposed method is only $O(MN\log N)$. Yet without the decomposition, an assignment problem is required to be solved in each comparison, hence the time complexity of the entire task is $O(M^2 N^4)$.

2.3 Eigenvector Signature Distance and Eigenvector Signature Agreement

The optimal value of P_0 can be shown to be related to the Euclidean distance between EVSs of two networks. The relation can be summarized by the following property.

Property 2.1 $\bar{c}(A_1, A_2) < \bar{c}(A_3, A_4)$ *iff* $\min_{Q \in \mathbb{P}} \|x_1 - Qx_2\|_2 > \min_{Q \in \mathbb{P}} \|x_3 - Qx_4\|_2$, *where* $\mathbb{P}$ *is the set of permutation matrices,* $A_1, ..., A_4$ *are adjacency matrices and* $x_1, ..., x_4$ *are their respective EVSs.*

Proof of Property 2.1. Suppose $\bar{c}(A_1, A_2) < \bar{c}(A_3, A_4)$, we have

$$\max_{Q \in \mathbb{P}} x_1^T Qx_2 < \max_{Q \in \mathbb{P}} x_3^T Qx_4 \qquad (16)$$

Hence

$$\min_{Q \in \mathbb{P}} \sqrt{2 - 2x_1^T Qx_2} > \min_{Q \in \mathbb{P}} \sqrt{2 - 2x_3^T Qx_4} \qquad (17)$$

That is

$$\min_{Q \in \mathbb{P}} \|x_1 - Qx_2\|_2 > \min_{Q \in \mathbb{P}} \|x_3 - Qx_4\|_2 \qquad (18)$$

On the other hand, if $\min_{Q \in \mathbb{P}} \|x_1 - Qx_2\|_2 > \min_{Q \in \mathbb{P}} \|x_3 - Qx_4\|_2$, we have

$$\min_{Q \in \mathbb{P}} \sqrt{2 - 2x_1^T Qx_2} > \min_{Q \in \mathbb{P}} \sqrt{2 - 2x_3^T Qx_4} \qquad (19)$$

which is equivalent to

$$\min_{Q \in \mathbb{P}} 1 - x_1^T Qx_2 > \min_{Q \in \mathbb{P}} 1 - x_3^T Qx_4 \qquad (20)$$

According to Perron-Frobenius Theorem, all entries of $x_1, ..., x_4$ are nonnegatives, as Q is a permutation matrix so entries of Q should also be nonnegatives. Thus,

$$x_1 Qx_2 \geq 0 \text{ and } x_3^T Qx_4 \geq 0 \text{ fo all } Q \in \mathbb{P} \qquad (21)$$

Therefore

$$\max_{Q \in \mathbb{P}} x_1^T Qx_2 < \max_{Q \in \mathbb{P}} x_3^T Qx_4 \qquad (22)$$

Hence,

$$\bar{c}(A_1, A_2) < \bar{c}(A_3, A_4) \qquad (23)$$

According to the above property, for topological similar networks, the Euclidean distance between their corresponding EVSs will be smaller and vice versa. This indicates that the Euclidean distance between EVSs can be considered as the measure of dissimilarity between networks. With this, a network measure can be defined by:

$$d_{EVS}(g, h) \triangleq \frac{\|\hat{x} - \hat{y}\|_2}{\sqrt{2 - 2/\sqrt{N}}} \qquad (24)$$

where g and h are graphs, $\hat{x}$ and $\hat{y}$ are their EVS respectively. The denominator $\sqrt{2-2/\sqrt{N}}$ is to normalize the measure so that $d_{EVS}(g,h) \in [0,1]$. This choice of value can be explained by the following property.

Property 2.2 *The farthest pair of vectors in the set* $W = \{w|w \in \mathbb{R}^N, \|w\|_2 = 1, w_1 \geq ... \geq w_N \geq 0\}$ *are* $[1,0,...,0]^T$ *and* $[1/\sqrt{N},...,1/\sqrt{N}]^T$.

Proof of Property 2.2. Let $x = [1,0,...,0]^T$, $y = [1/\sqrt{N},...,1/\sqrt{N}]^T$ and $W = \{w|w \in \mathbb{R}^N, \|w\|_2 = 1, w_1 \geq ... \geq w_N \geq 0\}$.

We show x and y are the farthest pair of vectors in the set W by contradiction.

Suppose x is not the farthest vector to y in the set W. There exists a vector $a \in W$ where

$$\|a-y\|_2 > \|x-y\|_2 \tag{25}$$

$$a^T y < x^T y \tag{26}$$

$$\sum_{i=1}^{N} \frac{a_i}{\sqrt{N}} < \frac{1}{\sqrt{N}} \tag{27}$$

$$\sum_{i=1}^{N} a_i < 1 \tag{28}$$

Since $0 \leq a_i \leq 1$

$$\sum_{i=1}^{N} a_i > \sum_{i=1}^{N} a_i^2 = 1 \tag{29}$$

which contradicts $a \in W$.

Suppose y is not the farthest vector to x in the set W. There exist a vector $b \in W$ where

$$\|x-b\|_2 > \|x-y\|_2 \tag{30}$$

$$x^T b < x^T y \tag{31}$$

$$b_1 < \frac{1}{\sqrt{N}} \tag{32}$$

Since $b_1 > ... > b_N$

$$b_j < \frac{1}{\sqrt{N}}, \text{ for } j \, 2,...,N \tag{33}$$

Thus

$$\sum_{i=1}^{N} b_i < N \times \frac{1}{N} = 1 \tag{34}$$

which contradicts $b \in W$.

Combine the above proofs, and the fact that W is a closed and connected subset of a unit sphere, it can be concluded that x and y are the farthest vector pair in W.

In property 2.2, the set W is the set of EVS, hence $[1,0,...,0]^T$ and $[1/\sqrt{N},...,1/\sqrt{N}]^T$ are the farthest pair of EVS and the maximal value of d_{EVS} is $\|[1,0,...,0]^T - [1/\sqrt{N},..., 1/\sqrt{N}]^T\|_2 = \sqrt{2-2/\sqrt{N}}$.

As $d_{EVS}(g,h)$ measures the difference between networks, alternatively, an agreement measure can be defined as a complement of EVSD,

$$\alpha_{EVS} = 1 - d_{EVS}(g,h) \tag{35}$$

Eigenvector Signature Agreement (EVSA) is the network similarity measure induced by the EVS. It represents a measure of similarity between two networks; the higher the value, the more similar the networks. It is a normalized measure and lies within the range [0, 1]. An illustrative example of EVSA computation can be found in File S1.

Results and Discussions

To illustrate the effectiveness of the proposed EVSA similarity score, standard test networks were used and will be given in Section 3.1. Then in Section 3.2, an application of EVSA is demonstrated with the analysis of protein-protein interaction (PPI) networks for a family of herpesvirus.

3.1 Control Test on Standard Network Models

In this section, a test based on artificially generated networks is conducted to illustrate the use of EVSA. The network models will be given in Section 3.1.1 and the results of the test can be found in Section 3.1.2. These models are chosen to conduct the algorithm test as their structure and properties are well known.

3.1.1 Network models

Four network models were chosen for testing: the Erdős Rényi random graph (ER), Barabási Albert model (BA), Geometric random graph (GEO), and Stickiness model (STICKY).

Erdős Rényi random graph (ER) A network is generated randomly without considering any geometric or probability distribution constraints. It starts with N isolated nodes and $\frac{N(N-1)}{2}$ candidate edges. Candidate edges are all possible vertex pairs for edge being attached to which can be defined as the set $\{(i,j)|i \in V, j \in V, i \neq j\}$. For each candidate edge, there is a constant probability p for an edge to be attached [19].

Barabási Albert model (BA) A scale-free network is generated through the preferential attachment scheme. Unlike the ER model, the probability of edges attachment in this model are not constant. It is directly proportional to the degree of nodes. Thus the resulting networks will reflect "the rich get richer" phenomenon. The degree distribution of a scale free network follows a power law. That is $P(k) \sim ck^\gamma$, where $P(k)$ is the population of nodes having degree k, c and γ are constants [20].

Geometric random graph (GEO) A networks is generated randomly by the following procedures. Initially, nodes are randomly distributed in an N-dimensional Euclidean space. For any node pairs having geometric distance smaller than the threshold radius r, a link will be attached among them. In this paper the three-dimensional case is considered [21].

Stickiness model (STICKY) It is a network model designed for PPI networks. By providing the degree sequence of a network, the Stickiness model can be used to generate a set of networks having the same degree sequence. There are two main assumptions, i) the higher degree nodes have more reaction domain, i.e. these nodes can interact more frequently, ii) a stickiness index is defined; where a node pair both have a higher stickiness index, they are more willing to interact with each other. The stickiness index of nodes helps to control the expected degree sequence of the generated network [22]. The Stickiness model is designed to mimic a network based on the degree sequence, which is only used in the test among standard models; and in the study of herpesvirus PPI networks (Section 3.2), but not in the performance analysis (Section 3.2.2).

3.1.2 Control Test Results. The control test was performed by first generating a reference network from each model, parameters were adjusted such that the size and average degree were 500 and approximately 10 respectively. Then perturbed the reference network by different ways: (a) randomly attaching k edges on the reference network, where $k = \lfloor \phi E \rfloor$, E is the total number of edges in the reference network and $\phi \in [0,1]$ is an adjustable parameter for testing purpose (see Table 1); (b) randomly selecting ϕ of the nodes in the reference network, replacing the interconnection of the selected nodes by a ER network; (c), (d) and (e) are similar to (b), but the injected network are GEO3D type, BA type and STICKY type respectively. EVSA between the reference networks and perturbed networks were then computed. The control test was repeated 50 times for each case and the mean EVSA were summarized in Table 1.

While the reference network was being perturbed by random attachment, most of the topology remained the same, thus in most of the cases the mean EVSA values were high. Yet in the case of GEO3D, the topology of reference network follows a geometric constrains, random attachment of edges violated this constrain and caused a large difference in topology. According to the results, randomly attaching 50% extra edges caused the mean EVSA score changed from 1 to 0.5323. On the other hand, in tests (b) to (e), the EVSA scores were found to be close among different types of injection with the same reference network and ϕ value. For instance, injecting a GEO3D network and BA network on a ER network with $\phi = 0.5$ caused their mean EVSA values dropped from 1 to 0.9597 and 0.9609 respectively. This indicated that even the interconnection between part of the nodes were replaced by various types of network, the similarity between the original network and the perturbed network can still be reflected by their high EVSA.

3.2 Analysis on Protein-Protein Interaction Networks

In this section, five herpesviral PPI networks are analyzed using EVSA. The employed data set is described in Section 3.2.1, and the results can be found in Section 3.2.2.

3.2.1 Protein-Protein Interaction Networks. The PPI network is a network that consists of proteins and their interactions within a single organism, for example baker's yeast and human. In the PPI network a protein is represented in the form of nodes. For every pair of proteins having interaction, a link will be attached in between the corresponding nodes. Here, five herpesvirus PPI networks were chosen for the evaluation, namely Epstein-Barr virus (EBV), herpes simplex virus (HSV), Kaposi's sarcoma-associated herpesvirus (KSHV), murine cytomegalovirus (mCMV)

Table 1. This table summarized the EVSA values obtained from a series of control tests.

Type of Reference Network	ϕ	(a)	(b)	(c)	(d)	(e)
ER	0	1	1	1	1	1
	0.1	0.9814	0.9937	0.9935	0.9932	0.9934
	0.2	0.9716	0.9856	0.9869	0.9865	0.9873
	0.3	0.9607	0.9769	0.9778	0.9779	0.9770
	0.4	0.9513	0.9659	0.9683	0.9674	0.9669
	0.5	0.9440	0.9598	0.9597	0.9609	0.9582
GEO3D	0	1	1	1	1	1
	0.1	0.8061	0.9873	0.9683	0.9804	0.9693
	0.2	0.6914	0.9720	0.9410	0.9458	0.9460
	0.3	0.6213	0.9533	0.8760	0.9129	0.9156
	0.4	0.5705	0.9325	0.8477	0.8367	0.8841
	0.5	0.5323	0.8926	0.7994	0.8123	0.8262
BA	0	1	1	1	1	1
	0.1	0.9786	0.9898	0.9892	0.9900	0.9888
	0.2	0.9607	0.9735	0.9785	0.9732	0.9751
	0.3	0.9431	0.9616	0.9603	0.9520	0.9548
	0.4	0.9258	0.9466	0.9412	0.9279	0.9340
	0.5	0.9087	0.9304	0.9142	0.9120	0.9146

The first column indicates the network class of the reference network, the second column indicates the ϕ values in the control tests. Column (a) to (e) are the mean EVSA value obtained from 50 times of the corresponding test. (a) is the random edge attachment test, (b) is the ER injection test, (c) is the GEO3D injection test, (d) is the BA injection test and (e) is the STICKY injection test.

Table 2. Summary Statistics of Yeast and Human PPI Network obtained from Rito et al.

Name	# Nodes	# Edges	Graph Density	Average Degree	Experiment Type	Organism	Reference
YIC	796	841	0.0027	2.11	Yeast two-hybrid	S. Cerevisiae	Ito et al. (2000)
YHS	988	2455	0.0050	4.97	TAP-MS	S. Cerevisiae	Von Mering et al. (2002)
HSHS	1705	3186	0.0022	3.74	Yeast two-hybrid	Homo Sapiens	Stelzl et al. (2005)
BG MS	1923	3866	0.0021	4.02	Affinity Capture-MS	Homo Sapiens	BIOGRID (filtered)

and varicella-zoster virus (VZV). These five herpesviral PPI networks were collected by Peter Uetz et al. [23,24].

The following is an illustrative analysis on several PPI networks. The summary statistics of yeast and human PPI networks including graph density, are shown in Table 2. This is obtained from a study on GDDA [25]. According to that study, when the graph density is low, GDDA will suffer from a volatility issue.

Here a similar analysis is conducted on five herpesvirus PPI networks: Epstein Barr virus (EBV), Herpes simplex virus (HSV), Kaposi's sarcoma-associated herpesvirus (KSHV), murine cytomegalovirus (mCMV) and varicella-zoster virus (VZV). The summary statistics can be found in Table 3. In the herpesvirus case, the graph density is relatively higher (0.06 to 0.12) than the yeast-human case (0.0007 to 0.005), which indicates that the low graph density property is not consistence in all PPI networks. This inconsistency may be caused by the definition of graph density.

$$\rho = \frac{2e}{N(N-1)} \tag{36}$$

Graph density is sensitive to network size since $\rho \sim O(N^{-2})$, thus there exists an inconsistency between a small PPI network (e.g. herpesvirus) and a large PPI network (e.g. yeast). In this paper, instead of graph density, the average degree is being considered. The average degree is defined as

$$deg_{avg} = \frac{2e}{N} \tag{37}$$

As $deg_{dvg} = O(N^{-1})$, which is less sensitive to network size and hence is a more suitable candidate of changing variable in the performance evaluation for network similarity measures. This observation can be supported by the summary statistic as shown in Table 2 and 3; where average degree in yeast and human PPI network is around 2 to 5, in herpesvirus network it is about 4 to 7. It shows that average degree of node is more consistent than graph density among these three kinds of PPI networks. Note that by adjusting the average degree and network size, different ranges of graph density can also be covered.

3.2.2 Results on Protein-Protein Interaction Networks. The procedure can be summarized as follows: first pick one of the herpesvirus PPI networks and compute its total number of edges, for each of the four models (ER, BA, GEO3D, STICKY), generate 50 candidate networks where the parameters are adjusted so that they have approximately the same number of edges as the herpesvirus networks. Then the EVSA score between the selected herpesvirus network and each of its candidate networks are computed, the EVSA scores are then averaged for each model. This average EVSA score is considered as the similarity score between the query network (e.g. EBV) and the testing model (e.g. ER). The process is repeated until all herpesvirus networks have been tested.

Figure 1 shows the EVSA score of EBV, HSV, KSHV, mCMV and VZV matching with ER, GEO3D, SF and STICKY respectively. The EVSA scores between the STICKY model and the five herpesvirus networks are clearly higher than the other models. ER, GEO3D and BA have average EVSA scores around 0.8 and STICKY has average EVSA scores around 0.9. Since the STICKY model is distinct from the other three models in the matching test of herpesvirus PPI networks; we can observe that considering only ER, GEO3D, BA and STICKY models as candidates, STICKY is the closest model with respect to the

Table 3. Summary Statistics of HSV, VZV, KSPV, EBV, and mCMV PPI Network.

Name	# Nodes	# Edges	Graph Density	Average Degree	Experiment Type	Organism	Reference
HSV	48	100	0.0886	4.167	Yeast two-hybrid	Herpes simplex virus	Fossum et al. (2009)
VZV	55	159	0.1070	5.782	Yeast two-hybrid	Varicella zoster vrisu	Uetz et al. (2006)
KSPV	50	115	0.0938	4.6	Yeast two-hybrid	Kaposis sarcoma-associated herpesvirus	Uetz et al. (2006)
EBV	60	208	0.1175	6.933	Yeast two-hybrid	Epstein - Barr Virus	Fossum et al. (2009)
mCMV	111	393	0.0643	7.081	Yeast two-hybrid	Murine cytomegalovirus	Fossum et al. (2009)

herpesvirus family. This can also be interpreted that given only the degree sequence as information, one can reconstruct the PPI network of herpesvirus with similar topology using the stickiness model.

A similar analysis using GDDA was also conducted on the same five herpesviral PPI networks and yielded the same conclusion, STICKY is the best-fitting model of herpesvirus [26]. However, compared to the GDDA approach, the EVSA score is relatively stable among five herpesvirus PPI networks and the separation of EVSA scores between the STICKY model and the other network models is more apparent. This demonstrates that our results reinforce the observation of Kuchaiev et al. In Section 3.2.2, we will show that EVSA has higher stability and better classification performance.

Performance Evaluation

Finally, the performance of network classification using EVSA is given and compared with another signature based network classification techniques: GDDA. In Section 4.1, the design of the testing is introduced. Volatility and classification performance of these two similarity score systems are compared in Sections 4.2 and 4.3 respectively.

4.1 Designs of Experiment

In order to illustrate the performance of different similarity score systems, models introduced in Section 3.1.1 was employed to generate networks for testing purposes. The experiment was conducted as follows.

Given two network models, for example ER model and BA model: in a ER-BA comparison, networks randomly generated from ER model using parameter set $\{N_j, \theta_j^{ER}\}$ was considered as query network (G_Q^{ER}), where N_j is the network size and θ_j^{ER} is the rest of ER model parameters. Similarly, 50 candidate networks $\{G_1^{BA},...,G_{50}^{BA}\}$ with same size N_j was randomly generated from BA model. The parameters θ_j^{BA} were adjusted such that the resulting candidate networks had approximately the same edge number to the one in G_Q^{ER}.

The GDDA scores and EVSA scores between the query network and each of the candidate networks (i.e. $GDDA(G_Q^{ER},G_k^{BA})$ and $\alpha_{EVS}(G_Q^{ER},G_k^{BA})$ for $k=1,...,50$) can then be computed. The process was repeated for several parameter sets $\{N_i, \theta_i^{ER}\}$, until all the interested $\{N_i, \theta_i^{ER}\}$ were considered. In this paper, the interested networks were limited to those with size 50, 100, 500, 1000, 2000 nodes and average degree 2,3,...,10. According to Section 3.2.1, these values cover most of the PPI networks of human, yeast and herpesviruses.

Comparisons between models can be separated into two classes; the match and the mismatch set comparisons. Match set comparisons are comparisons between networks generated from the same model, for instance, ER-ER comparison. Mismatch set comparisons are comparisons of networks generated across different models, for instance, ER-BA comparison. The score distribution of these two classes of comparison were used to illustrate the classification performance in Section 4.3.

4.2 Volatility of EVSA and GDDA

The stability of similarity score is an important issue as it directly affects the confidence of classification and the reliability of the scores. It can be measured as the standard deviation of a score among similar comparisons. An ideal similarity score should be non-volatile. Here our proposed similarity score EVSA is

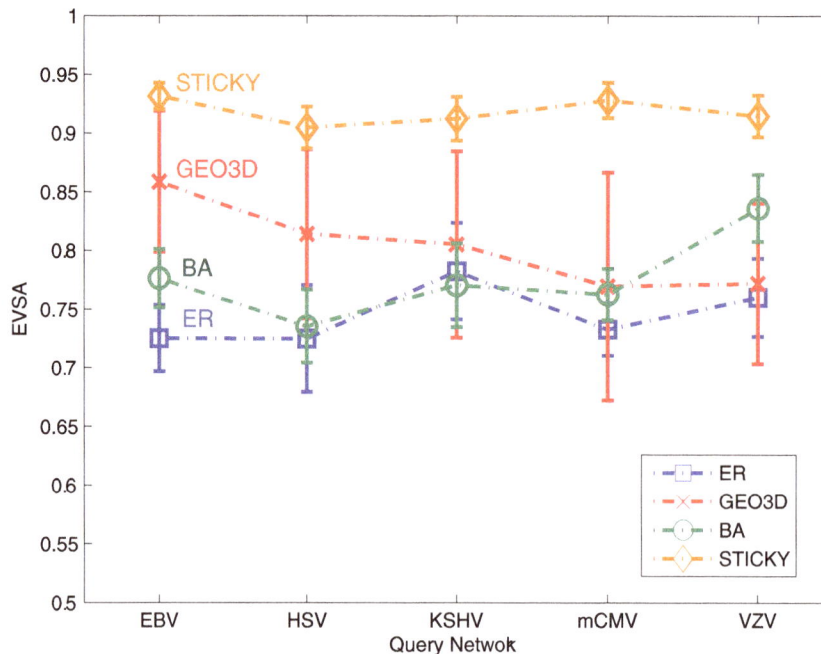

Figure 1. EVSA values of pairwise comparisons between PPI networks of EBV, HSV, KSHV, mCMV and VZV and network generated from ER,GEO3D,BA,STICKY models. The EVSA scores are averaged over 50 simulations and the error bar represents one standard deviation of the EVSA score.

compared with GDDA in terms of volatility. The EVSA scores were computed using the algorithm mentioned in previous sections. GDDA scores were computed using GraphCrunch 2.

Table 4 shows the standard deviation of EVSA scores and GDDA scores in ER-ER, BA-BA and GEO3D-GEO3D comparisons with different average degrees respectively. According to Table 4, in most of the cases EVSA has a lower deviation as compare to GDDA. However, exceptions are found in low density cases and these exceptions are caused by the volatility of the model itself. In a recent study of network comparisons [27], topologies of low density or low average degree graphs are considered as highly volatile. A sensitive similarity measures could reflect this fact in terms of high deviation. This explained the relatively higher deviation of EVSA on lower average degree networks. For instance, in ER-ER comparison among networks with an average degree of 2, the standard deviation of EVSA is 0.0703 which is higher than that of GDDA (0.0565). On the other hand, in a higher average degree graph, the EVSA yields a relatively lower deviation which reflects the stability of the measure. For instance, in BA-BA comparison among networks with an average degree of 10, the standard deviation of EVSA is 0.0326 which is much lower than GDDA's (0.0737). Thus from this experiment, the higher sensitivity and robustness of EVSA is demonstrated.

4.3 Classification Performance of EVSA and GDDA

Besides the volatility, the classification power of EVSA and GDDA is also an interesting aspect. The following is a test designed to reflect the classification power of a score. The basic idea of this test is to compare the distribution of scores in two cases, (i) networks generated from the same model and (ii) networks generated from different models.

In this experiment, networks were generated using ER, BA and GEO3D models. At the beginning of the test, a set of query networks was generated using different parameters through

different models. For each of the networks in the query set, a number of candidate networks were generated using all three models. The parameters of the network model were chosen to generate networks with the same vertex number and approximately the same edge number. So the networks have approximately equal average node degree and graph density. With this set of candidate networks a set of GDDA and EVSA scores can be computed between each query network and their candidate networks. Here the GDDA scores were computed using the network analysis application GraphCrunch 2 [26].

The GDDA and EVSA scores were further classified into agreement scores of matching and mismatching sets respectively. Matching and mismatching agreements are the set of agreement scores where query networks and candidate networks are generated from the same and different model types respectively.

The analysis results are summarized in Figures 2 and 3 in the form of a grey scale heat map. It shows the average GDDA and average EVSA in different parameter settings. The x-axis of the subplot represents the network size/vertex number; the y-axis represents the average degree of nodes. According to Figure 3, for GDDA, the difference between match-model comparisons (ER-ER, BA-BA, GEO3D-GEO3D) and mismatch-model comparisons (ER-BA, ER-GEO3D, BA-ER, etc) is not significant. While in the case of EVSA (Figure 2), notable difference is observed.

This reflects that the average EVSA scores among networks generated from the same model and those from different models have notable differences. It indicates that EVSA score can provide a clearer difference between the matching cases and the mismatching cases, or that the fuzziness and the mentioned difficulty on classification can be reduced.

To have a deeper understanding of the classification power of GDDA and EVSA, all the scores were being sampled and analyzed in terms of their distributions. The sampled data are illustrated in Figure 4 and 5 in histogram form. In EVSA, two sets of distributions can be found to be more divergent, which indicates

Table 4. Standard deviation of EVSA, GDDA in same model comparison over various average degrees.

| Average Degree | Standard Deviation | | | | | |
| | ER-ER | | BA-BA | | GEO3D-GEO3D | |
	EVSA	GDDA	EVSA	GDDA	EVSA	GDDA
2	0.0703	**0.0565**	0.0688	**0.0649**	**0.0309**	0.0460
3	**0.0481**	0.0726	**0.0608**	0.0731	**0.0218**	0.0332
4	**0.0266**	0.0467	0.0563	**0.0539**	0.0362	0.0502
5	0.0369	**0.0334**	0.0521	**0.0518**	**0.0318**	0.0332
6	**0.0282**	0.0485	**0.0523**	0.0567	**0.0182**	0.0428
7	**0.0206**	0.0526	**0.0392**	0.0499	**0.0288**	0.0766
8	**0.0181**	0.0484	**0.0332**	0.0605	**0.0362**	0.0464
9	**0.0190**	0.0505	**0.0501**	0.0671	**0.0218**	0.0401
10	**0.0173**	0.0558	**0.0326**	0.0737	**0.0309**	0.0618

that it provides a higher confidence for classification. In the GDDA case, the score distribution of matching models and mismatching models highly overlap when using GEO3D and BA networks as the query models. These overlapping regions are the "twilight zone" of the classification. The existence of such a region reduces the classification confidence and introduces fuzziness of classification.

To compare the divergence of matching and mismatching cases in GDDA and EVSA, the Jaccard distance is employed [28]. The Jaccard distance quantifies the divergence between two sets of distributions within a range from 0 to 1. A higher Jaccard distance

indicates that the two distributions are more divergent and vice versa. Definition of Jaccard distance can be found in File S1.

Table 5 summarizes the Jaccard distance between matching and mismatching sets of GDDA and EVSA in three cases respectively: ER, GEO3D and BA type query networks. In GDDA, using ER network as the query network can yield a good Jaccard distance in classification (0.9149). While in cases using GEO3D and BA network as the query network, the classification power varies over a large range in terms of Jaccard distance (0.3060 and 0.7477). A good performance in a single case but less desire in other cases may lead to fuzzy inference results in

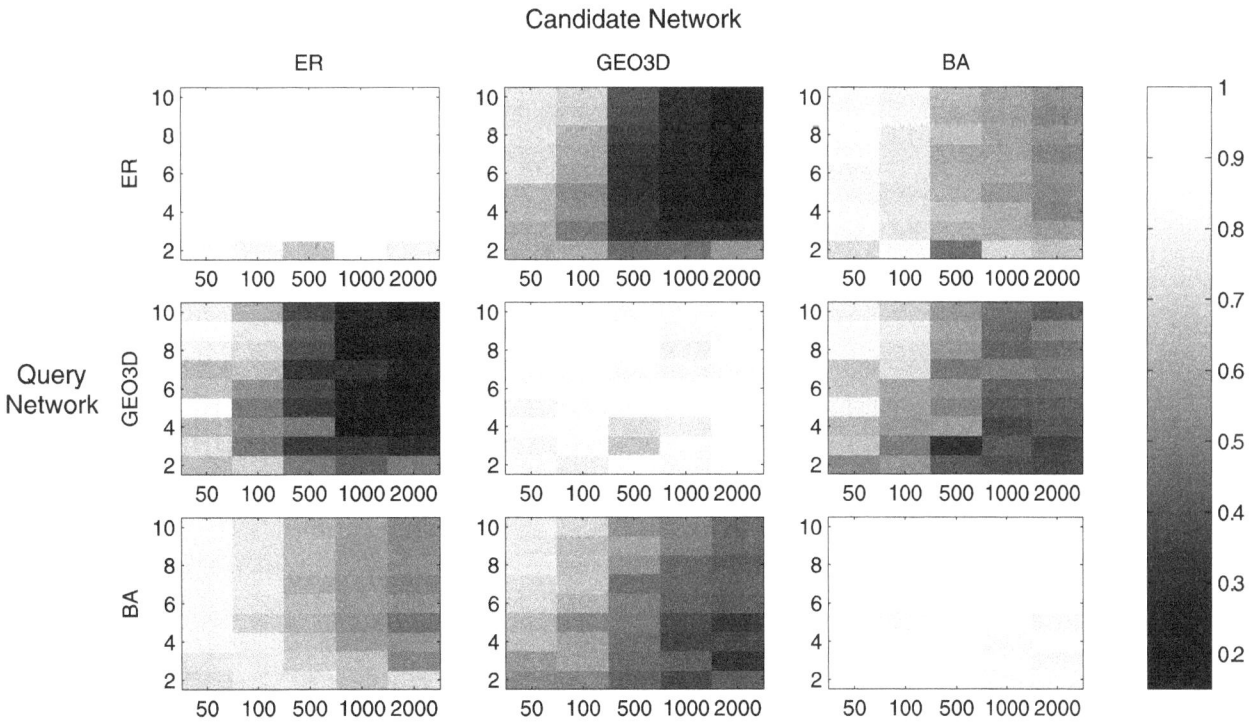

Figure 2. Heat map of average EVSA score in various comparisons. Each subplot represents the heat map of average EVSA score of a single type comparison. For instance, the bottom left subplot is the comparison of Scale-free (SF) network versus Erdos Renyi (ER) network, SF network is the Query Network and ER network is the candidate network. In each subplot, the y-axis represents the average degree, x-axis represents the network size (vertex number) the grey level represents the average EVSA score.

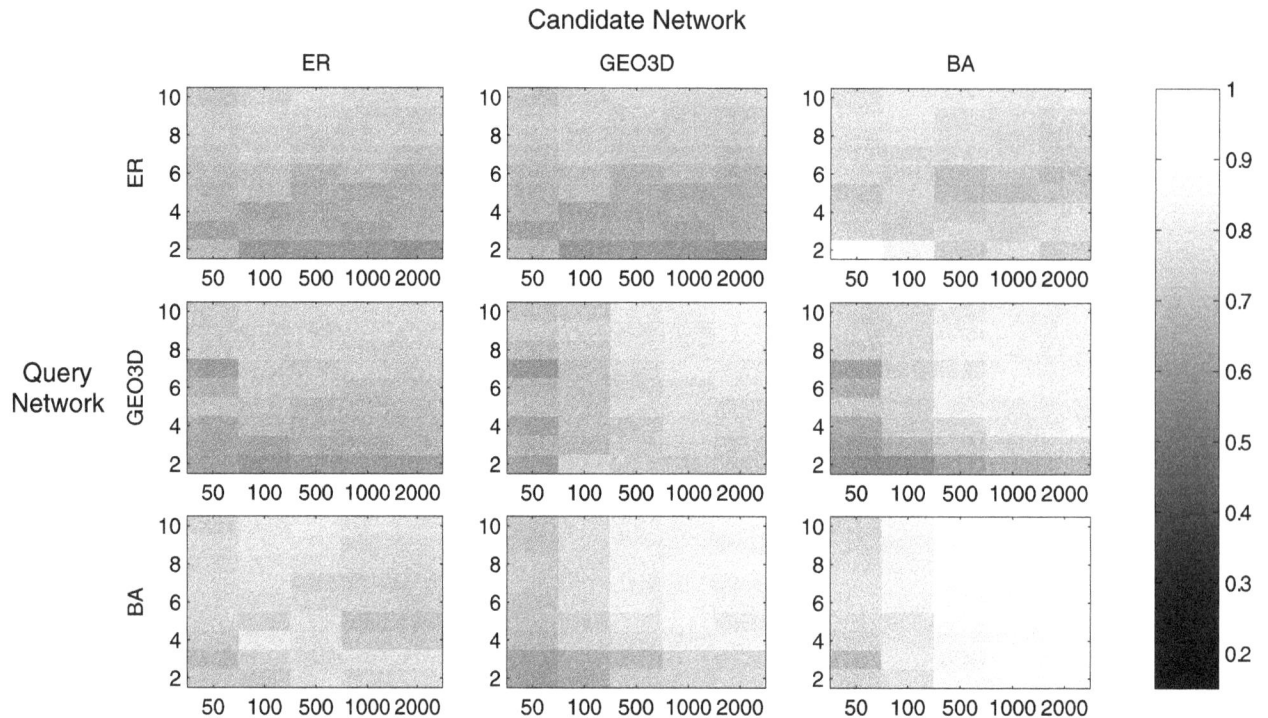

Figure 3. Heat map of average GDDA score in various comparisons. Each subplot represents the heat map of average GDDA score of a single type comparison. For instance, the bottom left subplot is the comparison of Scale-free (SF) network versus Erdos Renyi (ER) network, SF network is the Query Network and ER network is the candidate network. In each subplot, the y-axis represents the average degree, x-axis represents the network size (vertex number) the grey level represents the average.

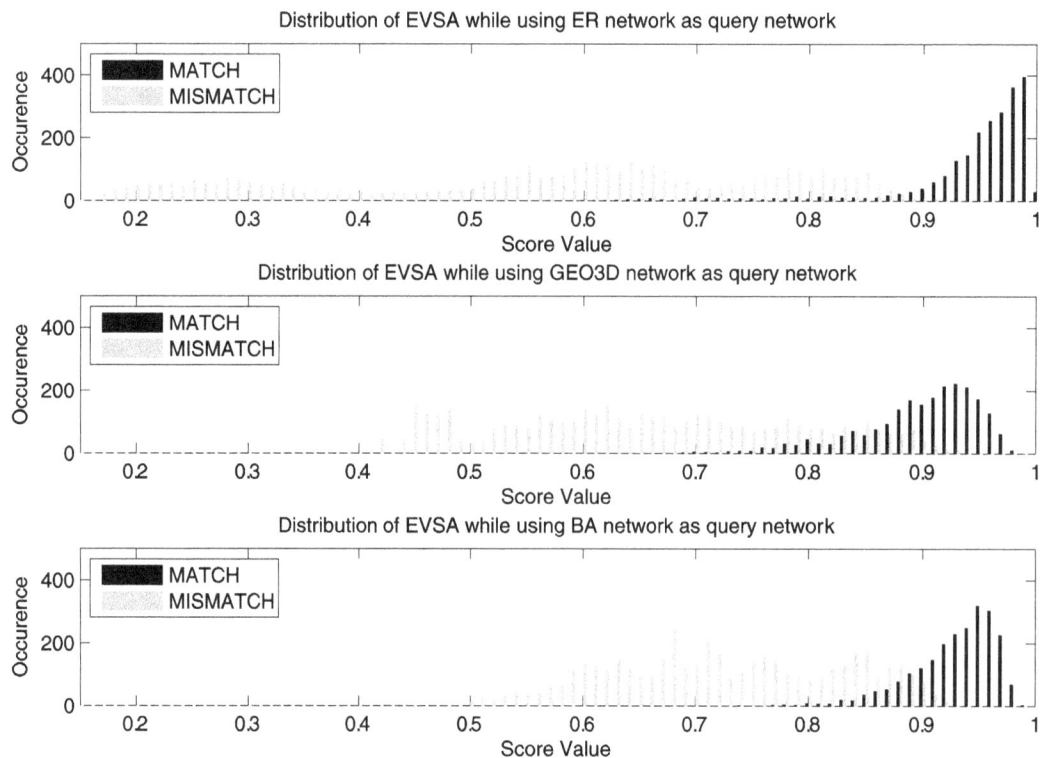

Figure 4. Sampled population of EVSA score. From top to bottom are a) ER network, b) GEO3D network, c) BA network as the query network used respectively.

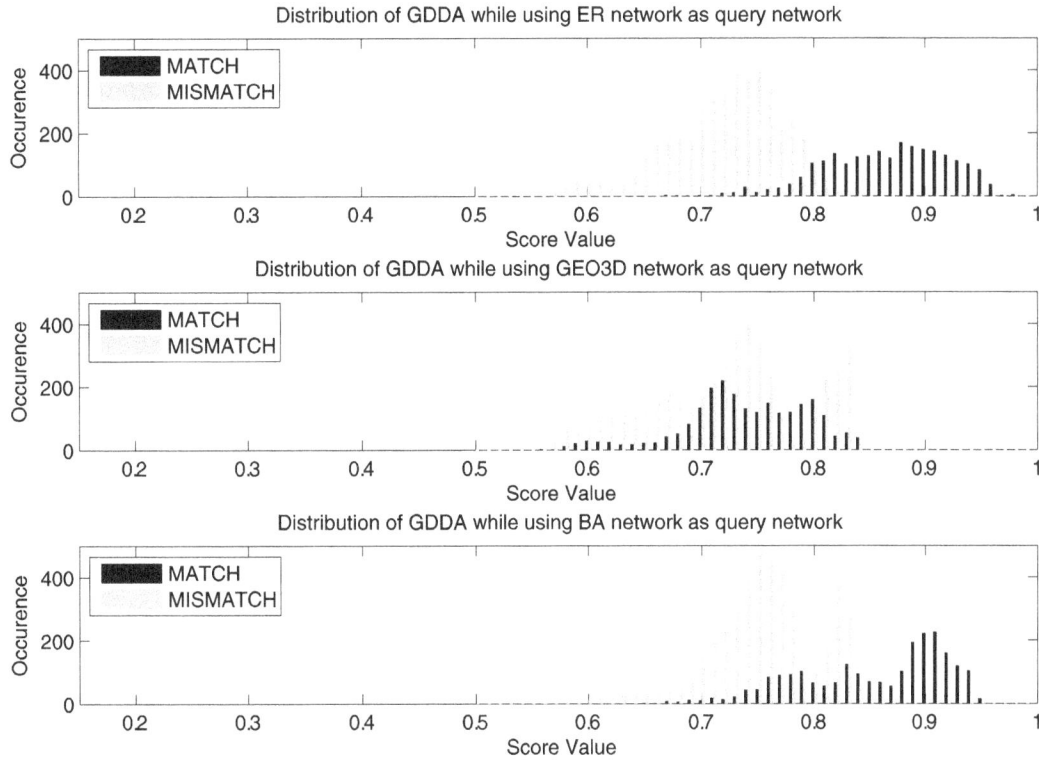

Figure 5. Sampled population of GDDA score. From top to bottom are a) ER network, b) GEO3D network, c) BA network as the query network used respectively.

classifying practical networks. In EVSA, all three types of query networks can yield a higher and more stable Jaccard distance around 0.9.

On the other hand we also evaluated the classification performance of GDDA and EVSA via applying them on a support vector machine (SVM) classifier as kernel values [29,30]. The SVM model classifies whether a network belongs or not to a specific class. The performances of the models are evaluated via a 5-fold cross validation and the results are summarized in Table 6. According to the results, model using EVSA as kernel value is relatively better that the one using GDDA in terms of accuracy, especially while classifying BA networks.

Conclusions

In this paper the Eigenvector Signature is proposed. With this signature, the original matrix alignment problem in network characterization can be transformed into a vector alignment problem. Consequently the computational complexity can be drastically reduced from $O(N^4)$ to $O(NlogN)$. In addition, an agreement measure - Eigenvector Signature Agreement (EVSA) is designed to quantify the similarity between networks.

Experimental results have shown that EVSA has a stable classification behaviour among various settings of different network models, including ER random graph, BA scale free network and geometric 3D random network. An application of EVSA on classifying herpesvirus PPI networks had also been conducted. The results are consistent with the previous studies, with much faster speed. Furthermore, performance analysis between EVSA and another signature based graph similarity measure GDDA had also been conducted. Results show that EVSA can achieve higher stability and better classification performance. Moreover, since EVSA quantifies network similarity, it can also be considered as a graph kernel. Hence kernel-based learning algorithms can also be applied.

Table 5. The Jaccard Distance of GDDA and EVSA between matching and mismatching network sets.

Query Network	Jaccard Distance	
	EVSA	GDDA
ER	0.9942	0.9149
GEO3D	0.9149	0.3060
BA	0.9455	0.7477

Table 6. The accuracy from 5-fold cross validation of SVM using EVSA and GDDA as kernel.

Class 1	Class 2	Accuracy	
		EVSA kernel	**GDDA kernel**
ER	Not ER	**81.48%**	77.78%
GEO3D	Not GEO3D	**92.59%**	81.48%
BA	Not BA	**96.30%**	66.67%

Supporting Information

File S1 Supporting information. Definition S1: Jaccard Index and Jaccard Distance. Computation S1: Example of EVSA Computation.

Acknowledgments

The authors would like to thank Peter Utez and Even Fossum who had kindly provided their herpesvirus PPI dataset and explained in details. The authors would also like to thank Robin Sarah Bradbeer and Anita Soley for their advices.

Author Contributions

Conceived and designed the experiments: BL LY KK GY. Performed the experiments: BL LY CZ. Analyzed the data: BL LY CZ KK GY. Contributed reagents/materials/analysis tools: BL LY CZ KK GY. Wrote the paper: BL LY CZ KK GY.

References

1. Vishwanathan S, Schraudolph NN, Kondor R, Borgwardt KM (2010) Graph kernels. The Journal of Machine Learning Research 11: 1201–1242.
2. Brandes U, Erlebach T (2005) Network analysis: methodological foundations, volume 3418. Springer.
3. Takigawa I, Mamitsuka H (2013) Graph mining: procedure, application to drug discovery and recent advances. Drug discovery today 18: 50–57.
4. Wegner JK, Sterling A, Guha R, Bender A, Faulon JL, et al. (2012) Cheminformatics. Communications of the ACM 55: 65–75.
5. Umeyama S (1988) An eigendecomposition approach to weighted graph matching problems. Pattern Analysis and Machine Intelligence, IEEE Transactions on 10: 695–703.
6. Singh R, Xu J, Berger B (2008) Global alignment of multiple protein interaction networks with application to functional orthology detection. Proceedings of the National Academy of Sciences 105: 12763–12768.
7. Li Z, Zhang S, Wang Y, Zhang XS, Chen L (2007) Alignment of molecular networks by integer quadratic programming. Bioinformatics 23: 1631–1639.
8. Milo R, Shen-Orr S, Itzkovitz S, Kashtan N, Chklovskii D, et al. (2002) Network motifs: simple building blocks of complex networks. Science Signaling 298: 824.
9. Pržulj N, Corneil DG, Jurisica I (2004) Modeling interactome: scale-free or geometric? Bioinformatics 20: 3508–3515.
10. Milenković T, Pržulj N (2008) Uncovering biological network function via graphlet degree signatures. Cancer informatics 6: 257.
11. Milenković T, Ng WL, Hayes W, Pržulj N (2010) Optimal network alignment with graphlet degree vectors. Cancer informatics 9: 121.
12. Pržulj N (2007) Biological network comparison using graphlet degree distribution. Bioinformatics 23: e177–e183.
13. Papadopoulos AN, Manolopoulos Y (1999) Structure-based similarity search with graph histograms. In: Database and Expert Systems Applications, 1999. Proceedings. Tenth International Workshop on. IEEE, pp. 174–178.
14. Chartrand G, Kubicki G, Schultz M (1998) Graph similarity and distance in graphs. Aequationes Mathematicae 55: 129–145.
15. Bunke H, Shearer K (1998) A graph distance metric based on the maximal common subgraph. Pattern recognition letters 19: 255–259.
16. Blondel VD, Gajardo A, Heymans M, Senellart P, Van Dooren P (2004) A measure of similarity between graph vertices: Applications to synonym extraction and web searching. SIAM review 46: 647–666.
17. Mises R, Pollaczek-Geiringer H (1929) Praktische verfahren der gleichungsauflösung. ZAMM-Journal of Applied Mathematics and Mechanics/Zeitschrift für Angewandte Mathematik und Mechanik 9: 58–77.
18. Hardy GH, Littlewood JE, Pólya G (1952) Inequalities. Cambridge university press.
19. Erdős P, Rényi A (1959) On random graphs. Publicationes Mathematicae Debrecen 6: 290–297.
20. Barabási AL, Albert R (1999) Emergence of scaling in random networks. science 286: 509–512.
21. Penrose M (2003) Random geometric graphs, volume 5. Oxford University Press Oxford.
22. Pržulj N, Higham DJ (2006) Modelling protein–protein interaction networks via a stickiness index. Journal of the Royal Society Interface 3: 711–716.
23. Uetz P, Dong YA, Zeretzke C, Atzler C, Baiker A, et al. (2006) Herpesviral protein networks and their interaction with the human proteome. Science 311: 239–242.
24. Fossum E, Friedel CC, Rajagopala SV, Titz B, Baiker A, et al. (2009) Evolutionarily conserved herpesviral protein interaction networks. PLoS pathogens 5: e1000570.
25. Rito T, Wang Z, Deane CM, Reinert G (2010) How threshold behaviour affects the use of subgraphs for network comparison. Bioinformatics 26: i611–i617.
26. Kuchaiev O, Stevanović A, Hayes W, Pržulj N (2011) Graphcrunch 2: Software tool for network modeling, alignment and clustering. BMC bioinformatics 12: 24.
27. Hayes W, Sun K, Pržulj N (2013) Graphlet-based measures are suitable for biological network comparison. Bioinformatics 29: 483–491.
28. Cha SH (2007) Comprehensive survey on distance/similarity measures between probability density functions. International Journal of Mathematical Models and Methods in Applied Science 1: 1.
29. Cortes C, Vapnik V (1995) Support-vector networks. Machine learning 20: 273–297.
30. Chang CC, Lin CJ (2011) LIBSVM: A library for support vector machines. ACM Transactions on Intelligent Systems and Technology 2: 27:1–27:27.

8

Lack of a 5.9 kDa Peptide C-Terminal Fragment of Fibrinogen α Chain Precedes Fibrosis Progression in Patients with Liver Disease

Santiago Marfà[1], Gonzalo Crespo[2], Vedrana Reichenbach[1], Xavier Forns[2], Gregori Casals[1], Manuel Morales-Ruiz[1], Miquel Navasa[2], Wladimiro Jiménez[1,3]*

1 Biochemistry and Molecular Genetics Service, Centro de Investigación Biomédica en Red de Enfermedades Hepáticas y Digestivas (CIBEREHD), Hospital Clínic, Institut d'Investigacions Biomèdiques August Pi i Sunyer (IDIBAPS), University of Barcelona, Barcelona, Spain, **2** Liver Unit, Centro de Investigación Biomédica en Red de Enfermedades Hepáticas y Digestivas (CIBEREHD), Hospital Clínic, Institut d'Investigacions Biomèdiques August Pi i Sunyer (IDIBAPS), University of Barcelona, Barcelona, Spain, **3** Departament de Ciencies Fisiologiques I, Centro de Investigación Biomédica en Red de Enfermedades Hepáticas y Digestivas (CIBEREHD), Hospital Clínic, Institut d'Investigacions Biomèdiques August Pi i Sunyer (IDIBAPS), University of Barcelona, Barcelona, Spain

Abstract

Early detection of fibrosis progression is of major relevance for the diagnosis and management of patients with liver disease. This study was designed to find non-invasive biomarkers for fibrosis in a clinical context where this process occurs rapidly, HCV-positive patients who underwent liver transplantation (LT). We analyzed 93 LT patients with HCV recurrence, 41 non-LT patients with liver disease showing a fibrosis stage F≥1 and 9 patients without HCV recurrence who received antiviral treatment before LT, as control group. Blood obtained from 16 healthy subjects was also analyzed. Serum samples were fractionated by ion exchange chromatography and their proteomic profile was analyzed by SELDI-TOF-MS. Characterization of the peptide of interest was performed by ion chromatography and electrophoresis, followed by tandem mass spectrometry identification. Marked differences were observed between the serum proteome profile of LT patients with early fibrosis recurrence and non-recurrent LT patients. A robust peak intensity located at 5905 m/z was the distinguishing feature of non-recurrent LT patients. However, the same peak was barely detected in recurrent LT patients. Similar results were found when comparing samples of healthy subjects with those of non-LT fibrotic patients, indicating that our findings were not related to either LT or HCV infection. Using tandem mass-spectrometry, we identified the protein peak as a C-terminal fragment of the fibrinogen α chain. Cell culture experiments demonstrated that TGF-β reduces α-fibrinogen mRNA expression and 5905 m/z peak intensity in HepG2 cells, suggesting that TGF-β activity regulates the circulating levels of this protein fragment. In conclusion, we identified a 5.9 kDa C-terminal fragment of the fibrinogen α chain as an early serum biomarker of fibrogenic processes in patients with liver disease.

Editor: Ratna B. Ray, Saint Louis University, United States of America

Funding: This work was supported by grants Dirección General de Investigación Científica y Técnica (SAF 2009-08839 and SAF 2012-35979 to W. Jiménez and SAF 2010-19025 to M. Morales-Ruiz) and from the Agència de Gestió d'Ajuts Universitaris i de Recerca (SGR 2009/1496). CIBEREHD is funded by the Instituto de Salud Carlos III. This work is co-financed by the European Union through the European Regional Development Fund (ERDF), "A way of making Europe". The funders had no role in study design, data collection and analysis, decision to publish, or preparation of the manuscript.

Competing Interests: The authors have declared that no competing interests exist.

* Email: wjimenez@clinic.ub.es

Introduction

Early detection of fibrosis progression and the development of portal hypertension is of major relevance in the prognosis and treatment of patients with chronic liver disease [1]. Indeed, early recognition of subjects prone to develop these alterations may allow prompt initiation of therapeutic interventions. Therefore, identification of noninvasive biomarkers related to the activation of the fibrogenic process is of major relevance, particularly in those subjects with sustained liver injury [2]. However, despite the numerous attempts to uncover such molecules, this objective has resulted elusive. This is likely related to the natural history of liver disease. With the exception of fulminant hepatic failure, liver

disease is an insidious process in which clinical detection and symptoms of hepatic decompensation may occur weeks, months or many years after the onset of injury, and healing may occur without clinical detection [3]. However, in particular clinical circumstances, i.e. patients infected with the hepatitis C virus (HCV), submitted to liver transplantation (LT), it is possible to expect recurrence of hepatic fibrosis and portal hypertension to occur within a short period of time [4]. Thus, these patients constitute a population particularly suitable to identify noninvasive markers of early fibrogenesis.

The current investigation took advantage of the faster development of hepatic fibrosis in HCV-positive LT patients. Serum samples were collected shortly after LT and high-throughput proteomic techniques were used to ascertain whether the proteomic pattern of these samples differs from the proteomic pattern expression obtained from serum samples of non-infected LT patients. Ultimately, the investigation was aimed to identify early circulating serum biomarkers of active fibrogenesis in patients with liver disease.

Materials and Methods

Patients

One hundred and nineteen patients admitted to the Liver Unit to undergo a liver biopsy from June 2001 to January 2006 were prospectively considered for this study. Exclusion criteria were presence of ascites, chronic kidney failure in hemodyalisis and moderate or severe acute graft rejection during the first three months, biliary complications or antiviral treatment during the first year after LT in the case of LT recipients. In addition 16 healthy volunteers were also included in the study.

The design of the study was two folded: first we assessed whether the serum proteomic profile of recurrent HCV-LT patients differs from that of non-recurrent HCV-LT patients. The serum proteomic profile and routine liver and renal function tests were initially analyzed in a training set of 10 HCV-RNA recurrent LT patients 6 months post LT that showed a fibrosis stage F$\geqq$1 at 1 year after LT. Paired hepatic venous pressure gradient (HVPG) determination was also available in 7 of these patients. The control group consisted in 9 patients without HCV-RNA recurrence, who underwent antiviral treatment before LT and achieve sustained virological response. In addition, serum samples were also collected from 41 non-LT patients with advanced liver disease. The HCV or hepatitis B virus (HBV) was present in 8 and 3 of these patients, respectively, whereas the etiology of liver disease was other than viral in the remaining (9 nonalcoholic steatohepatitis, NASH; 10 alcoholic liver disease, ALD; 4 autoimmune hepatitis, AH; and 7 cryptogenic). Thereafter, the results were validated in a test set of 83 HCV recurrent LT patients. Serum samples were also collected 6 months post-transplantation and the proteomic profile was evaluated along with liver and renal function tests. HVPG measurement in 53 of these patients was also performed.

Liver Biopsies and paired HVPG measurements

Percutaneous and transjugular liver biopsies and HVPG measurements were performed as we have previously described [5]. Fibrosis stage was scored using the Scheuer classification: no fibrosis (F0), minimal portal fibrosis (F1), periportal fibrosis (F2), fibrosis beyond the portal tract making septums (F3) and cirrhosis (F4) [6].

Serum fractionation

See (Data S1).

High-throughput proteomic processing of serum samples

Protein profiling was performed by surface-enhanced laser desorption/ionization time-of-flight mass spectrometry (SELDI-TOF-MS) using the eight-spot format CM10 (weak cationic exchange) ProteinChip arrays (Bio-Rad). In a preliminary study performed to set up the experimental conditions, 2 pooled serum samples from the 9 patients without HCV-RNA recurrence and the 10 patients included in the training set were loaded onto three different types of Protein Chip arrays: H50 (that binds proteins through reverse phase or hydrophobic interactions), CM10 (negatively charged surface that acts as a weak cation-exchanger)

and IMAC-30 (Immobilized Metal Affinity Capture surface preactivated with copper). The resulting spectra from each pool were compared and the CM10 array showed the highest number of peaks detected and the highest total signal intensity compared to H50 and IMAC-30; therefore only the CM10 array was used in the subsequent studies. Prior to sample loading, spots were equilibrated two times with 200 µl of CM binding/washing buffer (0.1 M sodium acetate, pH 4). Each sample was loaded in duplicate randomly in order to minimize any systematic error. Forty microliters of fractionated serum sample was incubated in 60 µL of CM binding buffer for 30 minutes on a shaker at room temperature. Afterwards, arrays were washed three times with 200 µL CM washing buffer for 5 minutes at room temperature. Unbound serum proteins were removed by washing twice with deionized water. Thereafter, arrays were air-dried and 1 µL of energy-absorbing matrix (saturated sinapinic acid in an aqueous solution containing 50% acetonitrile and 0.5% TFA) was added twice to each spot. The surface was allowed to air dry between each application. The array was read by using the ProteinChip PBS II reader (BioRad). Each spot was read at low (2500 nJ), medium (3000 nJ) and high (3500 nJ) energy laser intensities. The mass-to-charge ratio (m/z) was set from 1.000 to 25.000 m/z for the low-energy laser intensity, between 2.500 and 200.000 m/z for the medium-energy laser intensity and from 5.000 to 200.000 for the high-energy laser intensity. All spectra were calibrated using two external calibration standards (all-in-one peptide standard and all-in-one protein standard, BioRad). A peak resolution was optimized within 5.000 m/z, 12.000 m/z or 19.000 m/z according to low, medium or high energy laser intensity, respectively.

Data acquisition and analysis

All data were processed with the ProteinChip Data Manager Client 4.1 software (Bio-Rad). To minimize the possible random error and spectral outliers, all the raw data was normalized by the average total ion current across the group and all spectra differing by twice the standard deviation or more from the mean were deleted. Furthermore, the baseline was also corrected by adjusting the parameter to 30 times the expected peak width. For the peak selection, several parameters were selected for the identification of peak clusters. Thus, only peaks with a signal to noise equal or greater than 5; with a valley depth superior than three; found in a minimum of 20% of all spectra and with an m/z error below the 0.3% for the low-energy laser intensity spectra and below 2% for the medium- and high-energy laser intensity spectra, were considered. Subsequently, all peak clusters detected were verified manually. Relabeling, removal or addition of peaks was performed when necessary. To test the quality of the assay, pooled normal sera from two individuals was assessed. Five protein peaks randomly selected over the course of the study were used to calculate the coefficient of variance (CV) as described [7]. We then determined the reproducibility of the SELDI spectra, both within and between arrays (intra-assay and inter-assay, respectively). The intra-assay (spot to spot) CV was 11.95% for peak intensity and 0.02% for mass accuracy. The inter-assay (chip to chip) CV was 21.96% for peak intensity and 0.03% for mass accuracy.

Identification of candidate biomarker

See (Data S2).

Cell Culture

HepG2 cells were obtained from American Type Culture Collection (ATCC, Manassas, VA, USA). This immortalized, stable cell line can be repeatedly frozen, thawed and propagated. HepG2 cells were seeded (2×10^6 cells/well) in vented T-75 flasks

Table 1. Baseline characteristics of liver transplant recipients in non recurrent and recurrent hepatitis C patients.

	NON RECURRENT HCV SUBJECTS (n = 9)	RECURRENT HCV SUBJECTS (n = 10)	P*
Sex (M/F)	7/2	3/7	
Age (yr)	53.9±3.3	55.9±2.3	NS
Fibrosis score (1 yr after LT)			
F 1–2 (n)	-	4	
F 3–4 (n)	-	6	
HVPG (1 yr after LT)			
HVPG<6 mm Hg (n)	-	0	
HVPG 6–10 mm Hg (n)	-	2	
HVPG≥10 mm Hg (n)	-	5	
Bilirubin (mg/dl)	0.9±0.1	1.3±0.3	NS
Albumin (g/l)	43.1±1.1	37.8±2.1	<0.05
BUN (mg/dl)	27.4±3.5	34.0±3.4	NS
AST (U/l)	28.9±4.5	152.3±47.1	<0.001
ALT (U/l)	37.7±6.6	202.7±60.9	<0.001
Total proteins (g/l)	67.9±1.6	63.1±3.3	NS
PT(%)	93.2±2.4	86.3±3.4	NS
ELF score	9.9±0.3	11.7±0.5	<0.05

*in comparison to non recurrent HCV subjects (Mann-Whitney U test), NS: non significant. Results are given as mean±SEM.

and grown to confluence in Dulbecco's Modified Eagle Medium (DMEM), supplemented with 50 U/ml penicillin, 50 μg/ml streptomycin and 10% of fetal calf serum (FCS). Thereafter, cells were switched to 1% FCS and incubated (37°C) under normoxic (21% O_2, 5% CO_2) or hypoxic conditions (5% O_2, 5% CO_2) in a controlled O_2 water-jacketed CO_2 incubator (Forma Scientific Series II, 3131, Marietta, OH) or treated with TNF-α (10 ng/ml, Sigma, St Louis, MO), lipopolysaccharide (LPS, 10 ng/ml,

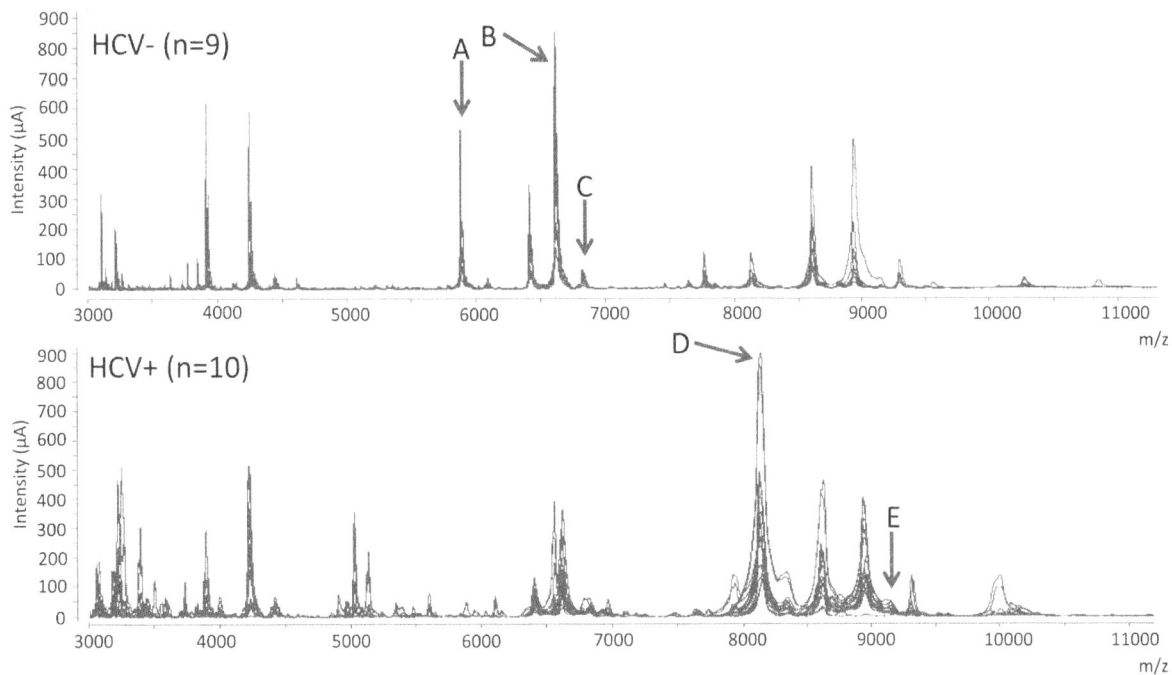

Figure 1. Differential proteomic profile between HCV− and HCV+ patients submitted to LT. Fragment ranging from 3000 to 11000 m/z of the differential SELDI-TOF-MS spectra of all serum samples obtained from patients 6 months after LT. The upper figure shows all overlapped spectra from HCV-negative patients (n = 9). The bottom figure shows the spectra from HCV-infected patients (n = 10). Arrows indicate peaks showing significantly different intensities between HCV− and HCV+ patients. Upper letters correspond to the identification peak noted in table 2.

Table 2. Intensity values (μA) of peaks showing different patterns of expression in the two groups of patients.

PEPTIDE	m/z (Da)	NON RECURRENT HCV SUBJECTS (n = 9)	RECURRENT HCV SUBJECTS (n = 10)
A	5905	257.1±43	5.99±3.49[d]
B	6639	687.4±96.3	176.8±32.5[c]
C	6845	71.9±8.9	27.13±5.42[b]
D	8144	76.9±22.5	392.8±64.8[d]
E	9172	15.1±6.4	26.03±3.37[a]
F	12986	6.6±1.2	1.99±0.41[a]

[a]$p<0.05$, [b]$p<0.01$, [c]$p<0.001$, [d]$p<0.0001$; in comparison to non recurrent HCV subjects (Mann-Whitney U test). Results are given as mean±SEM.

Sigma), AII (80 pM, Sigma), Endothelin-1 (2 nM, Sigma), Apelin (100 nM, Phoenix Pharmaceuticals, Burlingane, Ca), Fibronectin (10 ng/ml, Sigma), Interleukin-1β (20 ng/ml, Sigma) and TGF-β (10 ng/ml, R&D Systems, Minneapolis, Mn). All experiments carried out in cell lines were reproduced three times in at least 2 independent assays. Conditioned media were harvested, concentrated (80:1) using 3000 MW Amicon Ultra centrifugal filters (Millipore Corp) and the presence of the fibrinogen α C-chain was assessed by SELDI-TOF-MS as described above.

Messenger RNA expression of human fibrinogen α, β, and γ chains in HepG2 cells

See (Data S3).

Measurements and statistical analysis

The same day of the liver biopsy, 20 ml of blood were obtained in a fasting status. Serum was stored at −80°C, and serum albumin, aspartate aminotransferase (AST), alanine transaminase (ALT), bilirrubin and blood urea nitrogen (BUN) were measured with the ADVIA 2400 Instrument (Siemens Healthcare Diagnostics, Tarrytown, NY, USA). Amino-terminal propeptide of type III procollagen (PIIINP), hyaluronic acid (HA), and tissue inhibitor of matrix metalloproteinase type-1 (TIMP-1) were measured in all patients by a CE-marked random-access automated clinical immunochemistry analyzer that performs magnetic separation enzyme immunoassay tests (ADVIA Centaur, Siemens Healthcare Diagnostics, Tarrytown, NY, USA). The enhanced liver fibrosis (ELF) score was calculated using the algorithm recommended in the CE-marked assay [ELF = 2.278 + 0.851 ln(HA) + 0.751 ln(PIIINP) + 0.394 ln(TIMP-1)] blood tests.

Statistical analysis of the results was performed by the non parametric Mann-Whitney U test and the Kruskal-Wallis test with the Dunn post hoc test as appropriated. Quantitative data were analyzed using GraphPad Prism 5 (GraphPad Software, Inc. San Diego, CA).

Ethics Statement

We obtained written informed consent from all patients included in the study and the investigation was approved by the Investigation and Ethics Committee of the Hospital Clinic of Barcelona following the ethical guidelines of the 1975 Declaration of Helsinki.

Results

Mass Spectrogram of LT patients with early fibrosis recurrence significantly differs from that of non recurrent LT patients

The principal demographic values of patients included in the definition group are shown in Table 1. As per the selection criteria, most recurrent HCV patients showed higher fibrosis and ELF scores, elevated HVPG measures and greater AST and ALT values than non recurrent HCV subjects. Figure 1 depicts a portion of the spectra of all the samples investigated in this training group, ranging between 3000 and 11000 Daltons; the mass to charge ratio analyzed. The expression pattern of the spectrograms obtained from non recurrent HCV patients clearly differed from those of recurrent HCV subjects. Six statistically different peaks, identified in the figure as peptides A, B, C, D, and E, were detected. As shown in Table 2 the signal intensity of four of these peaks (A, B, C, and F) was markedly higher in non recurrent than in recurrent patients whereas an inverse situation was observed in the remaining two peaks (D and E). The most remarkable difference was detected on analyzing peptide A (5905 m/z), since it was fully evident in all samples obtained from non recurrent HCV patients but almost suppressed in the serum of recurrent HCV individuals.

Neither LT nor HCV infection account for Mass Spectrogram differences in patients with active fibrogenesis

An intriguing question arising from the above described results was to elucidate whether these findings result from the particular characteristics of HCV transplanted patients rather than to a differential feature characterizing early fibrogenic processes. Thus, the serum proteomic spectrum was next analyzed in healthy individuals, in non-LT fibrotic HCV infected patients and in non HCV fibrotic patients. The clinical and demographic characteristics of these subjects are shown in table 3. Fibrosis scores and liver function tests of the two groups of fibrotic patients were similar to those obtained in HCV LT patients. The mass spectrograms of all individuals included in this portion of the investigation are shown in figure 2. The upper, middle and lower parts correspond to healthy subjects, fibrotic non HCV subjects and fibrotic HCV patients, respectively. Since in the previous experiments the most striking differences were observed with peptide A, in this and the subsequent experiments we focused on the peak with a mass/charge ratio of 5905 Da. All serum samples analyzed from healthy subjects showed a spectrogram compatible with the presence of this peptide whereas pathological serum

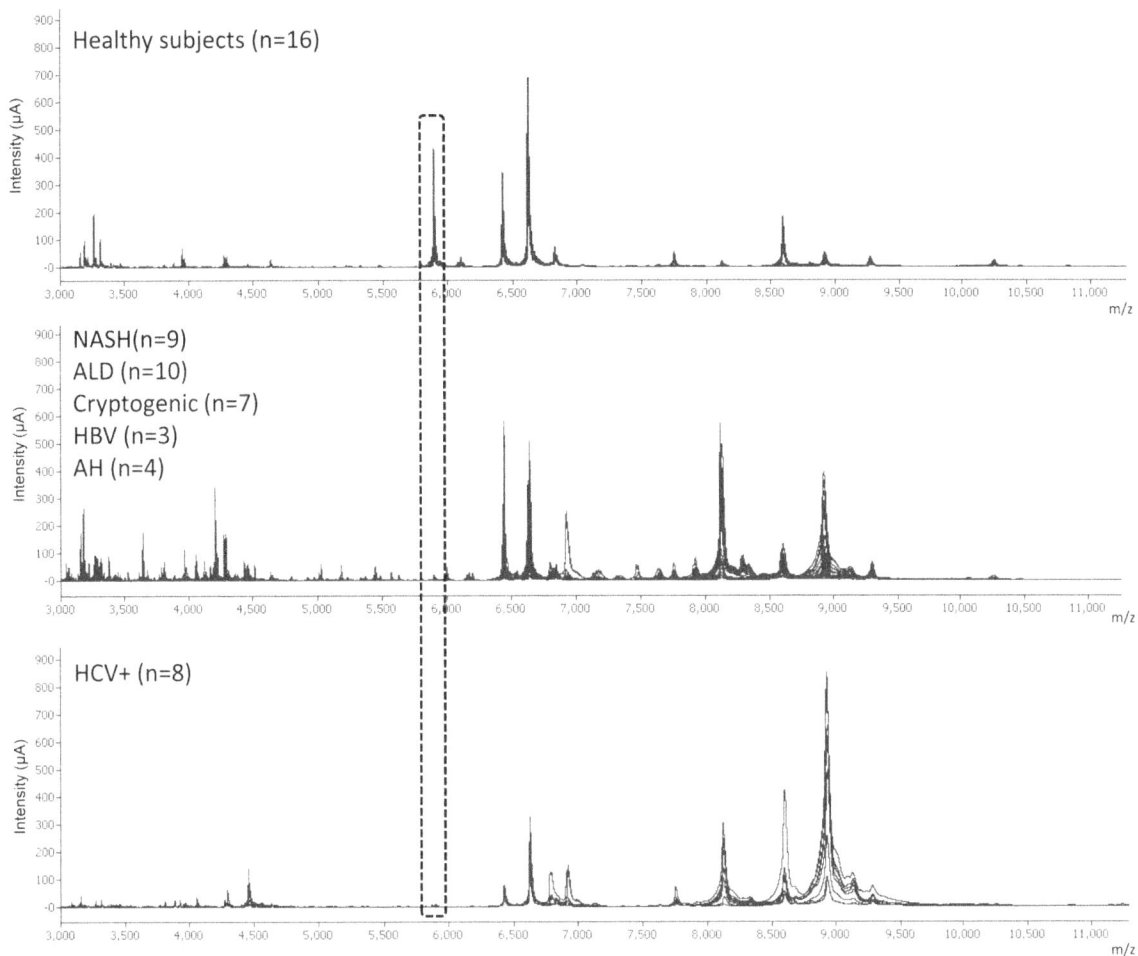

Figure 2. Lack of the 5.9 kDa protein peak in non-LT fibrotic patients. Portion of the SELDI-TOF-MS spectra comprised between 3000 and 11000 m/z of serum samples obtained from healthy subjects (n = 16) and patients with liver fibrosis of several etiologies (NASH, ALD, cryptogenic, AH, HBV and HCV). The lack of the protein peak at m/z 5905 is clearly associated with fibrogenesis regardless of its etiology.

samples, either from fibrotic non LT HCV patients or fibrogenic non HCV subjects showed the absence of this peptide or very low intensity peaks. These results indicate that neither HCV nor LT account for the suppressed expression of the A peptide in serum samples of fibrotic patients.

The Mass Spectrogram of most HCV infected patients lacks a 5.9 kDa fragment

To further confirm that peptide A behaves as an early serum biomarker of fibrosis, mass spectrometric analysis was performed in a test group of serum samples obtained from 83 HCV recurrent patients 6 months after LT. HVPG was assessed in 53 of these patients and the average value was of 5.5±0.8 mm Hg. All the serum samples showed a quite similar expression pattern and coincidences included both the different peptide fragments detected and the signal intensity of these fragments (Data S4). The most relevant finding was, however, that the mass spectrum corresponding to peptide A showed a very low peak intensity in all samples obtained from HCV recurrent patients. In fact, in 53 out of the 83 patients the peak intensity at m/z 5905 was under the background levels of 10 µA and in the remaining samples, intensities ranged between 11.5 and 81.4 µA (figure 3). Thus, these results confirm the findings obtained in the training group in a larger group of subjects.

The 5.9 kDa protein is a C-terminal fragment of the fibrinogen α chain

To isolate the protein of interest and to determine candidate protein identity, serum samples from two healthy subjects containing high SELDI intensity were pooled and separated by tricine-SDS-PAGE (figure 4B). The band at 5.9 kDa was excised trypsinized and analyzed by LC-MS/MS. As shown in figure 4C, two peptide sequences were identified, which matched with the human fibrinogen α C-chain at 3.23% coverage, suggesting that suppression of the fibrinogen α C-chain 5.9 kDa fragment is an early surrogate indicative of active fibrogenesis in patients with liver disease.

TGF-β reduces the expression of the fibrinogen α chain but not of the β and γ chain in HepG2 cells assays

To unveil potential mechanisms governing the release of the 5.9 kDa peptide C-terminal fragment of the fibrinogen α chain in the serum of patients under an early fibrogenic process, HepG2 cells were treated with well known proinflammatory stimuli (TNF-α, LPS and Il-1β) or profibrogenic substances (AII, ET-1, Apelin, Fibronectin and TGF-β) for 6 hours. With the exception of TGF-β, none of these substances induced significant changes in the expression of human fibrinogen chains messenger. However,

Table 3. Baseline characteristics of healthy subjects and HCV-infected and non-infected fibrotic patients.

	HEALTHY SUBJECTS (n = 16)	HCV INFECTED PATIENTS (n = 8)	NON INFECTED FIBROTIC PATIENTS (n = 33)
Sex (M/F)	(9/7)	(5/3)	(19/14)
Age (yr)	35.6±2.5	48.7±3.6	52.3±2.4***
Fibrosis score			
F 1–2 (n)	-	2	11
F 3–4 (n)	-	6	22
Etiology			
HCV (n)	-	8	-
HBV (n)	-	-	3
NASH (n)	-	-	9
ALD (n)	-	-	10
AH (n)	-	-	4
Cryptogenic (n)	-	-	7
Bilirubin (mg/dl)	0.6±0.1	0.5±0.06	1.2±0.2*,b
Albumin (g/l)	41.0±1.5	42.9±0.6	40.9±1.2
BUN (mg/dl)	15.2±2.1	14.4±1.6	18.1±2.2
AST (U/l)	20.9±1.6	64.0±7.9***	56.3±6.6***
ALT (U/l)	15.9±2.3	82.1±8.8***	79.1±15.4***
Total proteins (g/l)	71.3±1.2	78.6±2.3	75.4±1.7
PT(%)	96.5±0.9	89.8±3.1	86.4±2.4*
ELF score	8.6±0.4	9.6±0.6	9.7±0.2
INR	1.00±0.01	1.04±0.03	1.10±0.03

*$p < 0.05$, ***$p < 0.001$ in comparison to healthy subjects and ^{b}$p < 0.01$ in comparison to HCV infected patients (Kruskal-Wallis test with the Dunn pos hoc test). Results are given as mean±SEM.

TGF-β markedly reduced the expression of the α chain in cultured human hepatocytes. As shown in figure 5, this phenomenon was specific for the α chain since no significant changes were observed on analyzing the β and γ chain messengers. In addition, culture media of cells treated with TGF-β displayed diminished abundance of the 5.9 kDa fragment of the fibrinogen α C-chain in comparison to untreated cells.

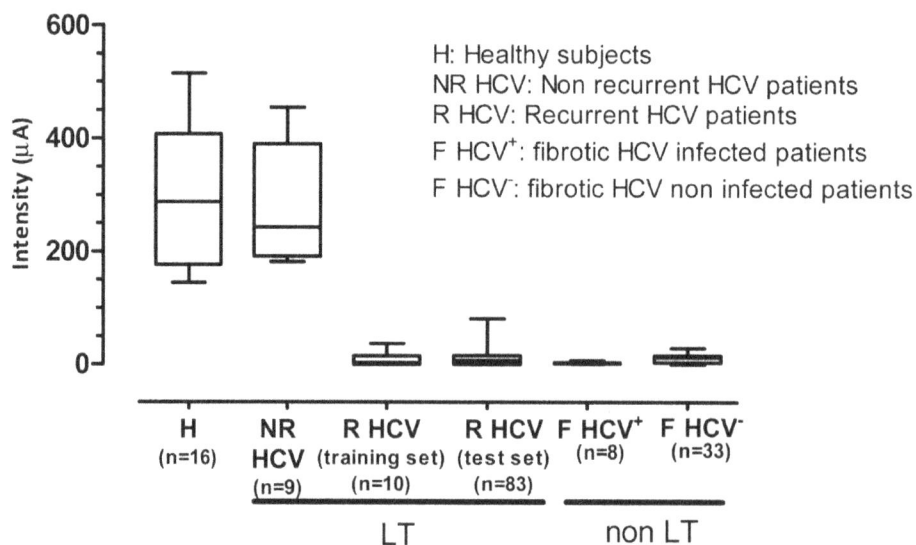

Figure 3. Comparison of the 5.9 kDa protein peak intensity between all the groups analyzed. SELDI-TOF-MS intensity values of the 5.9 kDa peak of the different groups of patients studied. Intensity of the peak was markedly suppressed in all patients under an active fibrogenic process.

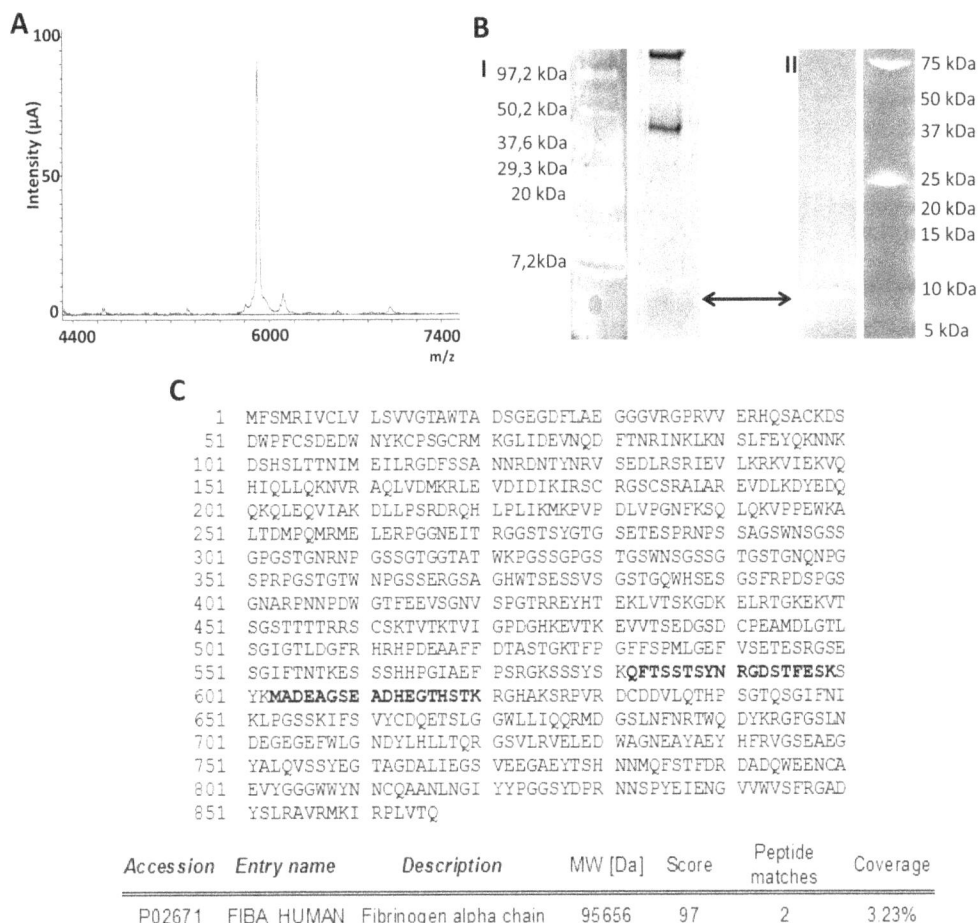

Figure 4. Isolation, separation and identification of the 5.9 kDa protein peak. A/ Spectra obtained by SELDI-TOF-MS showing the isolated protein peak after the purification process. B/ The isolated band indicated with arrows, after running two tris-tricine gels, one stained with sypro (I) and the other with oriole (II) staining. C/ Fragment of the Fibrinogen α-Chain identified as the differential protein peak by the amino acid sequencing. The two sequences identified are shown in bold.

Discussion

Evaluation of the extension and aggressiveness of the fibrogenic process in the injured liver is of major relevance for the diagnosis, prognosis and treatment of patients with hepatic disease [8]. The methods currently available to assess liver fibrosis include the serological determination of several parameters related to liver function and hepatic remodeling, imaging techniques, such as Fibroscan or ARFI and the use of invasive procedures such as HVPG measurement or liver biopsy, the latter still being the most widely accepted gold standard method for assessing liver fibrosis [9]. The specific limitations of each of these methods have been extensively discussed previously [10]. The risk of complications and low sensitivity for mild or moderate fibrosis are among the most remarkable limitation for invasive and non invasive methods, respectively [11]. Recently, a liver fibrosis score, namely ELF, which combines the serum concentrations of substances related to collagen metabolism (PIIINP) and tissue remodeling (TIMP-1 and HA), has progressively been incorporated among the most common diagnostic tools to evaluate liver fibrosis. However, whereas this technique was found to be highly accurate in patients with advanced fibrosis (F3–F4 stage) [12–14] it appeared to be less efficacious in the diagnosis of mild or moderate fibrosis (F1–F2

stage) [15,16]. Early detection of active fibrogenic activity, therefore, still remains an open challenge in liver disease.

Fibrosis progression evolves over long periods of time, with this representing one of the most relevant difficulties to identify specific early biomarkers of fibrosis. In the current investigation this issue was overcome by assessing the proteomic profile of HCV-positive LT recipients in a training set of serum samples. Blood samples were obtained at 6 months after LT and a liver biopsy was performed 1 year after surgery to define fibrosis stage. It is well known that fibrosis progression is accelerated in recurrent hepatitis C, with 15% to 47% of LT recipients developing fibrosis/cirrhosis within the first 3 years post transplantation [17]. Therefore, rapid fibrosis progression is a major characteristic of this group of patients and for this reason they are particularly suitable to uncover serum tags of hepatic fibrosis.

SELDI-TOF-MS technology or protein chip profiling combines mass spectrometry with a surface enhanced biochip which allows uniform and reproducible binding and desorption of biomarkers [18]. SELDI-TOF-MS also incorporates sample prefractionation. This markedly decreases the complexity of protein rich fluids, such as serum, allowing comparison of peak intensity between samples using large sample sets [19]. In the current study, serum proteins were fractionated by anion exchange chromatography based on their isoelectric points using a pH gradient. The resulting fraction

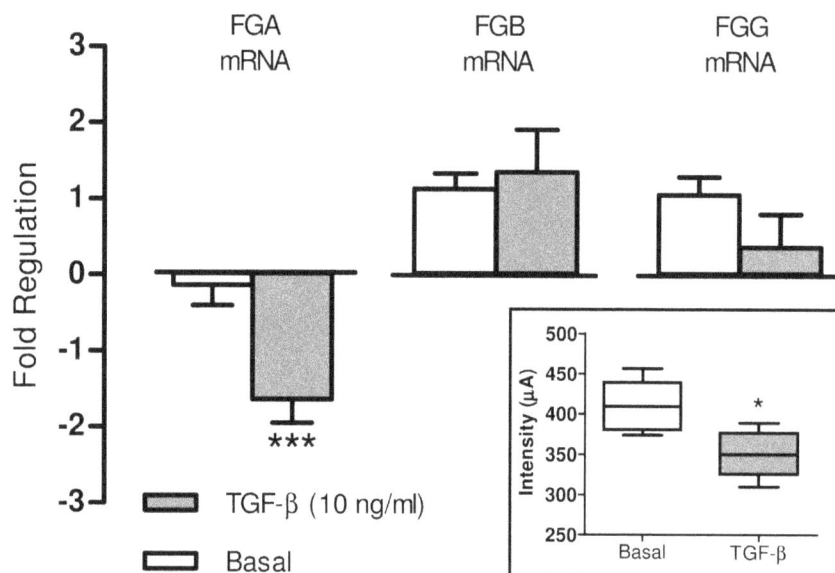

Figure 5. TGF-β reduces the fibrinogen α-chain expression in HepG2 cells. Fold regulation in Fibrinogen Alpha Chain (FGA), Beta Chain (FGB) and Gamma Chain (FGG) genes regulation in HepG2 cells after 6 hours of treatment with TGF-β (10 ng/ml). Results are given as mean ± SE; ***p<0.001 vs. basal. Statistical analysis was calculated by unpaired Student's t test. The insert shows the intensity values of the 5.9 kDa peak detected in the cellular supernatant of HepG2 cells after 48 hours of treatment with TGF-β (10 ng/ml). Results are given as mean ± SE; *p<0.05 vs. basal. Statistical analysis was calculated by Mann-Whitney U test.

was bound to a weak cation exchange surface to create an array of Protein Chip spots. This surface was selected according to its higher accuracy and reproducibility yields. Using this technology we were able to simultaneously detect relative protein expression levels over a range of molecular masses of 2 to 180 kDa, although the 2–20 kDa range appeared to be the most sensitive. By means of this profiling system, we found at least 6 serum biomarkers that were differentially increased or decreased in recurrent HCV patients. Among them, a protein of 5.9 kDa (protein A) was fully suppressed in the serum of all the HCV patients included in the training set. In contrast, readily detectable levels of this protein were detected in all non-recurrent HCV patients. We assessed whether LT and/or HCV infection account for the different expression patterns of peak A in serum samples of non-transplanted HCV positive and HCV negative subjects with fibrosis. The demographic and biochemical characteristics of patients with fibrosis included in the training set of samples were quite similar to those displayed by the fibrotic patients of this protocol with the exception of hepatic enzymes which, as expected, were higher in LT recurrent HCV patients than in fibrotic non-transplant patients. Both, the proteomic profile of the HCV positive samples and the proteomic profile of fibrotic patients non-infected sera, showed no or very low intensity peaks at the 5.9 kDa spectra. This markedly differed from the proteomic analysis of the serum samples of healthy subjects included in this set of experiments because all displayed consistent amounts of protein A. Interestingly, different etiologies (NASH, ALD, HBV, AH, cryptogenic) accounted for liver fibrosis in negative HCV patients, further emphasizing the close relationship between the lack of the 5.9 kDa protein and the fibrogenic process.

Next, the spectral data obtained in the test set were applied for validation purposes. All serum samples included in the test set showed an intensity m/z 5905 peak well below the values found in both healthy subjects and non recurrent HCV patients. Indeed, in most of these samples the A peak was not detected (figure 3). Overall, our results showing markedly decreased expression of the

m/z 5905 in the spectral profile of all samples from patients with fibrosis further strengthen the highly sensitive diagnostic performance of this peak.

A major limitation of SELDI-TOF-MS technology is related to the unfeasibility of directly identifying the protein of interest. In fact, for the majority of protein identifications it is necessary to achieve the enrichment of the specific peak by chromatography procedures and purification by SDS gel electrophoresis with subsequent triptic digest. In our investigation, amino acid sequencing of the trypsin digest of the 5.9 kDa protein revealed it to be a fragment of the fibrinogen α C-chain. Human fibrinogen is a circulating 340 kDa glycoprotein which has been shown to be of hepatic origin *in vivo*. Moreover, inflammatory stimuli may induce *in vitro* secretion of this glycoprotein in non hepatic cells including epithelial cells, granulosa cells, cervical carcinoma cells and trophoblasts [20]. However, current evidence strongly suggests that the largest site of human plasma fibrinogen is the hepatocytes [21]. It is comprised of two symmetric half molecules bound by a disulphide knot, each consisting in one set of three different polypeptide termed Aα, Bβ and γ. Each of these polypeptides is encoded by a separate gene located on chromosome four. The predominant Aα of circulating fibrinogen contains 610 aa (63.5 kDa), the Bβ chain 461 aa (56 kDa) and the γ chain is heterogeneous, but the most abundant form consists of 411 aa (48 kDa). The protein shows extensive post translational modification including phosphorylation, sulphation, glycosylation and hydroxylation. The fibrinogen α C-domain of the human fibrinogen is the C-terminal two-thirds of the Aα chain that extends from the coiled oil portion of each half of the dimeric fibrinogen molecule [22,23]. The α C-fragments are released into circulation as natural by-products of fibrinolytic systemic activation and are therefore, found in the systemic circulation in healthy individuals [24]. Our results showing almost suppressed expression of the 5.9 kDa fragment of the α C-chain of fibrinogen in patients undergoing a fibrogenic process are in agreement with those previously reported by Nomura F et al in heavy drinkers [25].

Furthermore, these authors showed that serum levels of this fragment were recovered when alcohol intake has ceased for more than 3 months and they also extended their findings to HCV infected patients [26]. Later, this fragment was described as having diagnostic value in patients with acute respiratory syndrome [27], breast cancer [28] and pancreatic adenocarcinoma [29].

The regulation of total human fibrinogen by a number of proinflammatory agents has been previously investigated using the HepG2 hepatocellular carcinoma cell line [30]. This *in vitro* model faithfully recapitulates fibrinogen expression including α, β and γ fibrinogen [31] and has been used to study fibrinogen production and regulation *in vitro* [32]. Accordingly we subsequently assessed the potential regulatory role of several candidate mediators on α-fibrinogen expression in HepG2 cells. A number of proinflammatory/profibrogenic agents that have previously been involved in the pathogenesis of the fibroproliferative processes [33–35] were tested. Among them, only TGF-β showed significant regulatory activity on α-fibrinogen mRNA expression and decreased 5.9 kDa fibrinogen αC-fragment intensity. Of note was, however, that the fold change in the fibrinogen αC-fragment induced by TGF-β in HepG2 cells was makedly lower than that observed in samples from fibrotic patients. The marked differences between the *in vivo* and *in vitro* experimental conditions likely account for this discordance. For instance, HepG2 is a human derived carcinoma cell line that shows altered abundance of TGF-β receptors [36,37] which in turn could result in some sort of resistance to this cytokine. On the other hand it is well known that regulation of acute-phase proteins is mediated by a combination of cytokines thus raising the possibility that additional factors involved in inflammatory processes also regulate the expression of the 5.9 kDa fragment of fibrinogen [38]. Our results are in line with past studies in which TGF-β inhibited the induction of fibrinogen produced by IL-6 and decreased the synthesis of fibrinogen in HepG2 and HepB cells [38], respectively. These latter experiments also showed a parallel diminution in α-fibrinogen mRNA levels. This phenomenon seems to be mediated by post-transcriptional mechanisms since TGF-β did not modify fibrinogen gene transcription, suggesting that the effect of this cytokine in liver cells is regulated at the level of mRNA stability [39]. Overall, all these results indicate that TGF-β may regulate the synthesis of α-fibrinogen at the postranscriptional level.

In summary, the current investigation took advantage of the faster development of hepatic fibrosis in HCV-positive LT patients to identify early circulating serum biomarkers of active fibrogenesis in patients with liver disease. Using high throughput SELDI-TOF-MS technology we unveiled a differential protein pattern profile between early fibrosis recurrence and non recurrent LT patients. Six protein peaks displaying statistically significant different intensities were observed within a range of 1000 to 25000 m/z.

The peak located at 5905 m/z showed the most remarkable difference, since it was fully detected in non-recurrent LT patients but was almost suppressed in recurrent LT patients. Similar results were found when comparing samples of healthy subjects with those of non LT fibrotic patients both HCV positive and negative, indicating that our findings were not related to either LT or HCV infection. Identification of this protein peak showed more than a 99% coincidence with a C-terminal fragment of the fibrinogen α chain. Moreover, cell culture experiments demonstrated that TGF-β downregulates α-fibrinogen mRNA expression and decreases the peak intensity of the m/z 5.9 KDa protein in HepG2 cells. In conclusion, we identified a 5.9 kDa C-terminal fragment of the fibrinogen α chain as a serum biomarker of early fibrogenic processes in patients with liver disease. Since TGF-β inhibited α-fibrinogen mRNA expression in HepG2 cells it is temptative to speculate that the activation of this cytokine in the early phases of liver injury could be responsible for the impairment in the circulating levels of the fibrinogen α C-chain fragment in patients with active hepatic fibrogenesis.

Supporting Information

Data S1 Materials and Methods corresponding to the serum fractionation procedure.

Data S2 Materials and Methods corresponding to the identification of the candidate biomarker.

Data S3 Materials and Methods corresponding to the analysis of the messenger RNA expression of human fibrinogen α, β, and γ chains.

Data S4 Spreadsheet containing all protein peaks detected in all the samples included in the study.

Acknowledgments

The authors are indebted to Drs. F. Elortza and I. Iloro for their collaboration in the identification of the 5.9 kDa protein peak.

Author Contributions

Conceived and designed the experiments: MN WJ. Performed the experiments: SM VR. Analyzed the data: SM VR G. Crespo. Contributed reagents/materials/analysis tools: SM VR G. Casals WJ. Contributed to the writing of the manuscript: SM WJ. Sample recruitment: XF. Revising the article critically: G. Crespo XF MMR MN. Final approval of the version to be published: SM G. Crespo VR XF G. Casals MMR MN WJ.

References

1. Pinzani M, Vizzutti F (2008) Fibrosis and cirrhosis reversibility: clinical features and implications. Clin Liver Dis 12(4): 901–913,

2. Gressner OA, Weiskirchen R, Gressner AM (2007) Biomarkers of hepatic fibrosis, fibrogenesis and genetic pre-disposition pending between fiction and reality. J Cell Mol Med 11: 1031–1051.

3. Hernandez-Gea V, Friedman SL (2011) Pathogenesis of liver fibrosis. Annu Rev Pathol 6: 425–456.

4. Bataller R, Brenner DA (2005) Liver fibrosis. J Clin Invest 115: 209–218.

5. Blasco A, Forns X, Carrión JA, García-Pagán JC, Gilabert R, et al. (2006) Hepatic venous pressure gradient identifies patients at risk of severe hepatitis C recurrence after liver transplantation. Hepatology 43: 492–499.

6. Scheuer PJ (1995) The nomenclature of chronic hepatitis: time for a change. J Hepatol 22: 112–114.

7. Scarlett CJ, Saxby AJ, Nielsen A, Bell C, Samra JS, et al. (2006) Proteomic profiling of cholangiocarcinoma: diagnostic potential of SELDI-TOF MS in malignant bile duct stricture. Hepatology 44: 658–666.

8. Afdhal NH, Nunes D (2004) Evaluation of liver fibrosis: a concise review. Am J Gastroenterol 99(6): 1160–1174.

9. Castera L, Pinzani M (2010) Biopsy and non-invasive methods for the diagnosis of liver fibrosis: does it take two to tango? Gut 59(7): 861–866.

10. Fernandez-Varo G, Jimenez W (2011) Non invasive markers of liver fibrosis. Europ Gastr & Hepatol Rev 7(2): 93–96.

11. Martinez SM, Crespo G, Navasa M, Forns X (2011) Noninvasive assessment of liver fibrosis. Hepatology 53: 325–335.

12. Rosenberg WM, Voelker M, Thiel R, Becka M, Burt A, et al. (2004) Serum markers detect the presence of liver fibrosis: a cohort study. Gastroenterology 127(6): 1704–1713.

13. Lichitinghagen R, Pietsch D, Bantel H, Manns MP, Brand K, et al. (2013) The enhanced Liver Fibrosis (ELF) score: Normal values, influence factors and proposed cut-off values. J Hepatol 59: 236–242.

14. Martinez SM, Fernández-Varo G, González P, Sampson E, Bruguera M, et al. (2011) Assessment of liver fibrosis before and after antiviral therapy by different

serum markers panels in patients with chronic hepatitis C. Aliment Pharmacol Therap 33: 138–148.

15. Nguyen D, Talwakar JA (2011) Noninvasive assessment of liver fibrosis. Hepatology 53: 2107–2110.

16. Crespo G, Fernández-Varo G, Mariño Z, Casals G, Miquel R, et al. (2012) ARFI, Fibroscan©, ELF and their combinations in the assessment of liver fibrosis: a prospective study. J Hepatology 57: 281–287.

17. Berenguer M, Schuppan D (2013) Progression of liver fibrosis in post-transplant hepatitis C: mechanisms, assessment and treatment. J Hepatology 58: 1028–1041.

18. Semmes OJ, Feng Z, Adam BL, Banez LL, Bigbee WL, et al. (2005) Evaluation of serum protein profiling by surface-enhanced laser desorption/ionization time-of-flight mass spectrometry for the detection of prostate cancer: I. Assessment of platform reproducibility. Clin Chem 51(1): 102–112.

19. Engwegen JY, Gast MC, Schellens JH, Beijnen JH (2006) Clinical proteomics: searching for better tumour markers with SELDI-TOF mass spectrometry. Trends Pharmacol Sci 27: 251–259.

20. Weisel JW (2005) Fibrinogen and fibrin. Adv Protein Chem 70: 247–299.

21. Tennent GA, Brennan SO, Stangou AJ, O'Grady J, Hawkins PN, et al. (2007) Human plasma fibrinogen is synthesized in the liver. Blood 109: 1971–1974.

22. Herrick S, Blanc-Brude O, Gray A, Laurent G (1999) Fibrinogen. Int J Biochem Cell Biol 31: 41–46.

23. Mosesson MW, Siebenlist KR, Meh DA (2001) The structure and biological features of fibrinogen and fibrin. Ann N Y Acad Sci 936: 11–30.

24. Rudchenko S, Trakht I, Sobel JH (1996) Comparative structural and functional features of the human fibrinogen alpha C-domain and the isolated alpha C fragment. Characterization using monoclonal antibodies to defined COOH-terminal A alpha chain regions. J Biol Chem 271(5): 2523–2530.

25. Nomura F, Tomonaga T, Sogawa K, Ohashi T, Nezu M, et al. (2004) Identification of novel and downregulated biomarkers for alcoholism by surface enhanced laser desorption/ionization-mass spectrometry. Proteomics 4(4): 1187–1194.

26. Sogawa K, Noda K, Umemura H, Seimiya JM, Kuga T, et al. (2013) Serum fibrinogen α C-chain 5.9 kDa fragment as a biomarker for early detection in hepatic fibrosis related to hepatitis C virus. Proteomic Clin Appl 7: 424–431.

27. Pang RT, Poon TC, Chan KC, Lee NL, Chiu RW, et al. (2006) Serum proteomic fingerprints of adult patients with severe acute respiratory syndrome. Clin Chem 52(3): 421–429.

28. Belluco C, Petricoin EF, Mammano E, Facchiano F, Ross-Rucker S, et al. (2007) Serum proteomic analysis identifies a highly sensitive and specific discriminatory pattern in stage 1 breast cancer. Ann Surg Oncol 14(9): 2470–2476.

29. Koopmann J, Zhang Z, White N, Rosenzweig J, Fedarko N, et al. (2004) Serum diagnosis of pancreatic adenocarcinoma using surface-enhanced laser desorption and ionization mass spectrometry. Clin Cancer Res 10(3): 860–868.

30. Knowles BB, Howe CC, Aden DP (1980) Human hepatocellular carcinoma cell lines secrete the major plasma proteins and hepatitis B surface antigen. Science 209(4455): 497–499.

31. Farrell DH, Mulvihill ER, Huang SM, Chung DW, Davie EW (1991) Recombinant human fibrinogen and sulfation of the gamma' chain. Biochemistry 30(39): 9414–9420.

32. Mackiewicz A, Speroff T, Ganapathi MK, Kushner I (1991) Effects of cytokine combinations on acute phase protein production in two human hepatoma cell lines. J Immunol 146(9): 3032–3037.

33. Melgar-Lesmes P, Pauta M, Reichenbach V, Casals G, Ros J, et al. (2011) Hypoxia and proinflammatory factors upregulate apelin receptor expression in human stellate cells and hepatocytes. Gut 60(10): 1404–1411.

34. Reichenbach V, Muñoz-Luque J, Ros J, Casals G, Navasa M, et al. (2013) Bacterial lipopolysaccaride inhibits CB2 receptor expression in human monocytic cells. Gut 62(7): 1089–1091.

35. Melgar-Lesmes P, Casals G, Pauta M, Ros J, Reichenbach V, et al. (2010) Apelin mediates the induction of profibrogenic genes in human hepatic stellate cells. Endocrinology 151(11): 5306–5314.

36. Liu P, Menon K, Alvarez E, Lu K, Teicher BA (2000) Transforming growth factor-β and response to anticancer therapies in human liver and gastric tumors in vitro and in vivo. Int J Oncol 16: 599–610.

37. Dituri F, Mazzocca A, Peidrò FJ, Papappicco P, Fabregat I, et al. (2013) Differential inhibition of the TGF-β signaling pathway in HCC cells using the small molecule inhibitor LY2157299 and the D10 monoclonal antibody against TGF-β receptor type II. PLoS One 8(6): e67109.

38. Mackiewicz A, Ganapathi MK, Schultz D, Brabenec A, Weinstein J, et al. (1990) Transforming growth factor beta 1 regulates production of acute-phase proteins. Proc Natl Acad Sci U S A 87(4): 1491–1495.

39. Buenemann CL, Willy C, Buchmann A, Schmiechen A, Schwarz M (2001) Transformin growth factor-β1 induced Smad signaling, cell cycle arrest and apoptosis in hepatoma cells. Carcinogenesis 22: 447–452.

9

AdaBoost Based Multi-Instance Transfer Learning for Predicting Proteome-Wide Interactions between *Salmonella* and Human Proteins

Suyu Mei[1,2]*, Hao Zhu[2]*

1 Software College, Shenyang Normal University, Shenyang, China, **2** Bioinformatics Section, School of Basic Medical Sciences, Southern Medical University, Guangzhou, China

Abstract

Pathogen-host protein-protein interaction (PPI) plays an important role in revealing the underlying pathogenesis of viruses and bacteria. The need of rapidly mapping proteome-wide pathogen-host interactome opens avenues for and imposes burdens on computational modeling. For *Salmonella typhimurium*, only 62 interactions with human proteins are reported to date, and the computational modeling based on such a small training data is prone to yield model overfitting. In this work, we propose a multi-instance transfer learning method to reconstruct the proteome-wide *Salmonella*-human PPI networks, wherein the training data is augmented by homolog knowledge transfer in the form of independent *homolog instances*. We use AdaBoost instance reweighting to counteract the noise from *homolog instances*, and deliberately design three experimental settings to validate the assumption that the *homolog instances* are effective to address the problems of *data scarcity* and *data unavailability*. The experimental results show that the proposed method outperforms the existing models and some predictions are validated by the findings from recent literature. Lastly, we conduct gene ontology based clustering analysis of the predicted networks to provide insights into the pathogenesis of *Salmonella*.

Editor: Michael Hensel, University of Osnabrueck, Germany

Funding: The work is partly supported by China Postdoctoral Science Foundation Funded Projects (No. 2013M531869, No. 2014T70821). The funders had no role in study design, data collection and analysis, decision to publish, or preparation of the manuscript.

Competing Interests: The authors have declared that no competing interests exist.

* Email: meisygle@gmail.com (SM); zhuhao@smu.edu.cn (HZ)

Introduction

Pathogen-host protein-protein interaction (PPI) plays an important role in revealing the molecular-level dynamic mechanism of microbial pathogenesis. Fast and accurate reconstruction of proteome-wide pathogen-host PPI networks is essential to reveal the host cellular processes that pathogen proteins may interfere with. In recent years, high throughput experimental techniques have drastically accumulated much knowledge about *intra-species* PPI networks, though noisy and far incomplete [1,2]. Accordingly the majority of computational methods have been developed as the complement of labor-intensive biological experiments for *intra-species* PPI network reconstruction, e.g. yeast PPI network [3], *Arabidopsis thaliana* PPI network [4], human PPI network [5], etc. However, the current host-pathogen PPI networks are comparatively much smaller. The latest HIV-human PPI database [6] contains about 3,638 interactions, the *P.falciparum-H.sapiens* PPI dataset [7] contains about 1,112 interactions, and the small *Salmonella*-human PPI data [8] contains only 62 interactions. At present there are very few computational methods developed for pathogen-host PPI networks reconstruction, e.g. HIV-human PPI prediction [9–13], *P.falciparum-H.sapiens* PPI prediction [14] and *Salmonella*-human PPI prediction [15–17]. To improve the predictive performance, most of the reported methods

simultaneously leverage a catalog of biological feature information (see Table 1).

Salmonella typhimurium is a facultative intracellular pathogen that causes a variety of diseases from acute gastroenteritis to systemic infection. After invasion into the lumen of host small intestine, *Salmonella* secretes effectors that interact with the host cellular proteins to ensure its survival in the host cellular environment and gain control of the host immune response [18]. To gain more insight into the inflammation/immune signaling pathways that *Salmonella* induced or interfered with, we need to fast and accurately reconstruct the complete *Salmonella*-human PPI networks. Unfortunately, the current experimentally derived *Salmonella*-human PPI network contains only 62 interactions [8], much smaller than the HIV-human PPI network [6] and the *P.falciparum-H.sapiens* PPI network [7]. As a fast complement to experimental techniques, computational modeling can accelerate the reconstruction of *Salmonella*-human PPI networks at low cost. To our knowledge, only a few computational methods have been developed to date for *Salmonella*-human PPI prediction [15–17]. Schlekera et al. [15] used *protein sequence similarity* and *protein domain similarity* to predict *Salmonella*-human PPIs. Kshirsagar et al. proposed two machine learning methods to predict *Salmonella*-human PPIs [16,17], wherein the random forest method [16] imputed the missing feature information of *gene ontology* and *gene*

Table 1. Summary of feature information extracted from literature.

Integration of feature information	Literature
sequence k-mer, interlog, gene ontology, metabolic pathways	[7]
binding motif, gene expression profile, gene ontology, sequence similarity, post-translational modification, tissue distribution, PPI network topology	[9,10]
protein domain profile, sequence k-mer	[11]
structural similarity	[12]
protein domain profile, gene expression, gene ontology, gene co-expression	[14]

expression, and the multi-task learning method [17] proposed DC optimization to integrate the feature information of *gene ontology*, *gene expression* and *pathways*. Except the similarity based method [15], the other two methods both adopt data integration to improve the model performance. Data integration is a popular method to enrich the abundance of feature information, but it has two major disadvantages: (1) aggregating more features without augmenting the size of training data is prone to increase the risk of model overfitting on small data; (2) integration of multiple aspects of feature information poses more demanding data constraints on the computational modeling. If the required feature information is not available for the proteins to be predicted, the data integration methods [8–11,14,16–17] will fail to work. Even for those methods that exploit only one type of feature information, e.g. *protein structural similarity* [12], *gene ontology* [14], etc., the problem of *data unavailability* should also be properly addressed. Most of the types of effective feature information listed in Table 1 are derived from costly experiments and are likely to be not available for some proteins. Thus we need to deliberately consider effective substitution of missing feature information and design proper experimental setting to validate the feature substitution. Kshirsagar et al. [16] conducted explicit substitution for the missing feature information of *gene ontology* and *gene expression*, but did not explicitly estimate the model performance in the case of feature information substitution. As compared to non-sequence feature information, protein sequence is cheap to obtain and imposes the least demanding data constraints on computational modeling. However, it has been reported that protein sequence alone is not sufficient to train a satisfactory model for PPI prediction [19].

In this work, we propose a multi-instance transfer learning method to reconstruct the proteome-wide *Salmonella*-human PPI networks. In the method, *gene ontology* (*GO*) is used as discriminative features to represent proteins. Due to the incompleteness and scarcity of *gene ontology* knowledge, we treat the homolog *GO* information (the aggregated *GO* information from the homologs) as independent *homolog instance* to augment or substitute for the *target instances* (the *GO* information from the protein itself). The potential noise from *homologs* is counteracted by AdaBoost instances reweighting algorithm [20,21]. To validate the effectiveness of the method, we design the following three experimental settings: (1) *Single Instance Learning* as the baseline model that conducts no homolog knowledge transfer; (2) *Multi-instance Learning Novel* where the training data are represented with *target instances* and *homolog instances*, while the test data are represented with *homolog instances* only; (3) *Multi-instance Learning* where both the training data and the test data are represented with *target instances* and *homolog instances*. Last, we use the proposed method to reconstruct the proteome-wide *Salmonella*-human PPI networks, based on which we further conduct gene ontology based clustering analysis to provide valuable cues for further biomedical research.

Materials and Methods

Data and materials

Schleker et al. [8] experimentally derived 62 interactions between 25 *Salmonella* proteins and 51 human proteins, based on which Kshirsagar et al. [16,17] developed two machine learning methods for *Salmonella*-human PPI prediction. PPI prediction is generally treated as a problem of two-class classification where the PPIs are treated as *positive* data and a *negative* data is needed for computational modeling. At present the experimental *negative* data is hardly available and the common practice to generate *negative* data is random sampling. Random sampling is based on the assumption that the expected number of negatives (non-interacting protein pairs) is several orders of magnitude higher than the number of positives (interacting protein pairs) [22], such that the *negative* space is randomly sampled with larger probability than the *positive* space. The human proteins for negative data sampling are taken from the latest SwissProt database [23]. Besides the way of negative data sampling, the second problem is to determine the ratio of *positive* data to *negative* data. Here we adopt 1:1 ratio instead of highly skewed ratio like 1:100 [16,17] based on the two points: (1) it is hard to simulate the true ratio of *positive* data to *negative* data and simulation of the ratio makes little sense to computational modeling; (2) simply pooling so large a *negative* data is prone to yield an extremely unbalanced training data and thus yields a highly biased model.

To validate how well the proposed model generalizes to unseen data, we further need to construct a validation set from recent literature. We find 18 *Salmonella*-human PPIs in [18] and two novel interactions (SspH2, SGT1) & (SspH2, Nod1) in [24]. After excluding the non-protein interactions (e.g. cholesterol, inositol phospates) and the interactions that have been collected in [8], we obtain 7 novel interactions as validation set.

Transfer learning

As compared to traditional supervised learning, transfer learning focuses on useful knowledge/information transfer across related domains that are heterogeneously subjected to distinct statistical distributions [25]. One major merit of transfer learning is that there is no need to make the assumption of *independent and identical distribution* (*iid*) between target domain and auxiliary domain. Such the relaxation opens up wide avenues for transfer learning in the field of biological data analysis. In recent years, many sophisticated machine learning methods have been developed to exploit the auxiliary data for useful biological information transfer [26,27,28].

In this work, the homolog knowledge is exploited to make up for *data scarcity* as well as to address the concern of *data unavailability*. Unlike computing individual kernel matrices in [26,27] and training individual classifiers in [28], the homolog knowledge transfer is conducted here by means of independent *homolog instances* under AdaBoost learning framework [20,21]. The merit is that the independent *homolog instance* is used to augment and enhance the *target instance*, and especially substitute for the *target instance* when the required feature information is not available. Meanwhile the potential of negative knowledge transfer by *homolog instances* can be attenuated by AdaBoost instance reweighting algorithm[20,21].

GO feature construction

The homologs for each protein are extracted from SwissProt 57.3 database [23] using *PSI-BLast* [29]. We choose *default PSI-BLast* parameters setting (*E-value* = 10) to enlarge the coverage of homologs and *GO* terms. The *GO* terms are extracted from *GOA* database [30] (114 Release, as of 28 November, 2012). For each protein i, there are two sets of *GO* terms, one set contains the *GO* terms from homologs denoted as *homolog set* S_H^i, and the other set contains the *GO* terms from the protein itself denoted as *target set* S_T^i. Here the term *target* is used to denote the protein itself (comparative to *homolog*) instead of the pathogen *targeted* protein. Based on the denotations, two feature vectors for each PPI pair (i_1, i_2) are formally defined as follows:

$$B_T^{(i_1,i_2)}[g] = \begin{cases} 0, g \notin S_T^{i_1} \wedge g \notin S_T^{i_2} \\ 2, g \in S_T^{i_1} \wedge g \in S_T^{i_2} \\ 1, otherwise \end{cases} \quad (1)$$

$$B_H^{(i_1,i_2)}[g] = \begin{cases} 0, g \notin S_H^{i_1} \wedge g \notin S_H^{i_2} \\ 2, g \in S_H^{i_1} \wedge g \in S_H^{i_2} \\ 1, otherwise \end{cases} \quad (2)$$

where $B_T^{(i_1,i_2)}[g]$ denotes component g of the *target instance* $B_T^{(i_1,i_2)}$ and $B_H^{(i_1,i_2)}[g]$ denotes component g of the *homolog instance* $B_H^{(i_1,i_2)}$. Formula (1) and formula (2) mean that if the interacting proteins pair shares the same *GO* term g, the corresponding component in the feature vector $B_T^{(i_1,i_2)}$ or $B_H^{(i_1,i_2)}$ is set 2; if neither protein in the protein pair possesses the *GO* term g, the value is set 0; otherwise the value is set 1. The definition is symmetrical, i.e., the protein pair (i_1, i_2) and the protein pair (i_2, i_1) have identical feature representation.

Multi-instance transfer learning

In the scenario of traditional supervised learning, each data point is represented by only one instance, but only one instance may not be sufficient to depict a complex object in most cases. For example, as biological macromolecules protein molecule constantly changes spatial conformations and DNA molecule temporally changes expression levels. Full depiction of the temporal and spatial information needs more than one instance. For another example, evolutionary information is usually used for us to understand the molecular functions of novel proteins or the interactions between orthologs (*interlog*) [7]. In the case that we can not mingle the information of the protein itself with the information of the homologs, multi-instance representation is a good choice.

In this work, we depict each protein with two instances, the *target instance* and the *homolog instance*, The *target instance* is used to represent the *GO* information of the protein itself and the *homolog instance* is used to capture the evolutionary information of

the target protein. Besides enriching the feature information of *target instance*, the *homolog instance* serves the second purpose of substituting the *target instance* when the target *GO* information is not available. We can see that the homolog knowledge transfer by means of independent *homolog instances* is a novel way to simultaneously solve the problems of *data scarcity* and *data unavailability*. Despite the merits, *homolog instances* may carry a certain level of noise from evolutionary divergence and the noise probably does harm to the model performance. For the reason, we need to choose noise-resistant machine learning methods to counteract the noise contained in the *homolog instances*. To our knowledge, AdaBoost is an empirically established and theoretically proven machine learning method that boosts an ensemble of weak learners by instances reweighting [20,21,31]. It has been theoretically proven that by means of regularization technique, multiple rounds of instances reweighting help AdaBoost to achieve maximum margin between two-class hyper-planes [31]. The regularization technique penalizes the noise/outlier at the cost of high training error to achieve low generalization error. In this work, we adopt the latest variant Modest AdaBoost [21] that softens the weight distributions between easy-to-classify instances and the hard-to-classify instances. For completeness, Modest AdaBoost [21] is briefly described as follows:

1. Given training instances$(x_1, y_1), ..., (x_N, y_N)$, initialize instance weights $D_0(i) = 1/N, i = 1, ..., N$;
2. For m $= 1, ..., M$ and while $f_m \neq 0$

a. Use distribution $D_m(i)$ and weighted least squares to train weak classifier:

a.

$$h_m(x) = \arg\min_h \left(\sum_{i=1}^{N} D_m(i) \bullet (y_i - h(x_i))^2 \right) \quad (3)$$

b. Compute the *inverted* distribution

b.

$$\overline{D_m(i)} = 1 - D_m(i) / Z_m \quad (4)$$

c. Compute

c.

$$P_m^{+1}(x) = P_{D_m}(y = +1 \wedge h_m(x)); \overline{P}_m^{+1}(x) = P_{\overline{D_m}}(y = +1 \wedge h_m(x)) \quad (5)$$

$$P_m^{-1}(x) = P_{D_m}(y = -1 \wedge h_m(x)); \overline{P}_m^{-1}(x) = P_{\overline{D_m}}(y = -1 \wedge h_m(x)) \quad (6)$$

d. Set

d.

$$f_m(x) = (P_m^{+1}(1 - \overline{P}_m^{+1}) - P_m^{-1}(1 - \overline{P}_m^{-1}))(x) \quad (7)$$

Update the weight distributions

$$D_{m+1}(i) = D_m(i)\exp(-y_i f_m(x_i))\big/Z_m \qquad (8)$$

3. Construct the final classifier

3.

$$F(x) = \sum_{m=1}^{M} f_m(x) \qquad (9)$$

where M denotes the number of iterations in the training, $\overline{Z}_m, Z_m$ are normalizing coefficients that are chosen to satisfy $\sum_i^N \overline{D}_m(i) = \sum_i^N D_m(i) = 1$, P_m^{+1}, P_m^{-1} denote the probability that the weak classifier $h_m(x)$ assigns label $\{+1, -1\}$ to the instance x under the weight distribution D_m, and similarly $\overline{P}_m^{+1}, \overline{P}_m^{-1}$ denote the probability under the inverted weight distribution $\overline{D}_m$. Formula (8) suggests that the weight increases $(D_{m+1}(i) > D_m(i))$ for those misclassified instances $(y_i f_m(x_i) < 0)$. The weight distribution $\overline{D}_m$ inverted from D_m conversely assigns higher weights to those correctly-classified instances. We can see that D_m pays more attention to those easy-to-classify instances while $\overline{D}_m$ pays more attention to those hard-to-classify instances. Modest AdaBoost made a compromise between the two weight distributions to make soft the decision function $f_m(x)$ (see formula (7)).

In the test phase, each test protein pair (i_1, i_2) is represented by two instances, the *target instance* $B_T^{(i_1, i_2)}$ and the *homolog instance* $B_H^{(i_1, i_2)}$. The decision committee $\mathbb{F}(x)$ as defined in Formula (9) yields two outputs $\mathbb{F}(B_T^{(i_1, i_2)}), \mathbb{F}(B_H^{(i_1, i_2)})$ for the two instances $(B_T^{(i_1, i_2)}, B_H^{(i_1, i_2)})$. In multi-instance AdaBoost, the final decision value for the protein pair (i_1, i_2) is defined as follows:

$$V(i_1, i_2) = \begin{cases} \mathbb{F}(B_T^{(i_1, i_2)}), if\ |\mathbb{F}(B_T^{(i_1, i_2)})| > |\mathbb{F}(B_H^{(i_1, i_2)})| \\ \mathbb{F}(B_H^{(i_1, i_2)}), otherwise \end{cases} \qquad (10)$$

where $|\bullet|$ denotes the absolute value. The final class label for the protein pair (i_1, i_2) is defined as follows:

$$L(i_1, i_2) = \begin{cases} 1, if\ V(i_1, i_2) > 0 \\ 0, otherwise \end{cases} \qquad (11)$$

For those positive predictions, the decision values are further normalized to measure the confidence level of prediction:

$$\bar{V}(i_1, i_2) = V(i_1, i_2) - V_{min}\big/V_{max} - V_{min} \qquad (12)$$

where V_{min} denotes the minimum decision value and V_{min} denotes the maximum decision value.

Model evaluation and model selection

To validate the effectiveness of homolog knowledge transfer by means of independent *homolog instances*, we design three experimental settings: (1) *Single Instance Learning* that does not

consider homolog knowledge transfer; (2) *Multi-instance Learning Novel* where the training data point is represented with *target instance* and *homolog instance*, while the test data point is represented with the *homolog instance* only.; (3) *Multi-instance Learning* where both the training data point and the test data point is represented with *target instance* and *homolog instance*. The experimental setting (2) is explicitly designed to estimate the model robustness against *data unavailability*, and the experimental setting (3) is designed to validate the assumption that the *homolog instance* is effective to augment the training data and thus to solve the problem of *data scarcity* for *Salmonella*-human PPI prediction.

As regards with Modest AdaBoost [21], there are two hyper-parameters to be empirically determined, one parameter M is the rounds of training, and the other parameter is the base learner. Here M is chosen within $\{50, 100, 150, 200, 250, 300, 350\}$ and the base learner is a decision tree with the number of tree splits chosen within $\{1, 2, 3, 4, 5, 6, 7, 8, 9, 10\}$. The model performance is estimated by 10-fold cross validation using the following performance metrics: *ROC-AUC* (AUC of *Receiver Operating Characteristic*), *PR-AUC* (AUC of *Precision recall curve*), *SP* (*Specificity*), *SE* (*Sensitivity*) and *MCC* (*Matthews correlation coefficient*). We first derive several intermediate variables from confusion matrix M as defined in formula (13), and then we calculate SP, SE and MCC for each label (SP_l, SE_l and MCC_l) as defined in formula (14), based on which to further calculate the *overall accuracy* (*Acc*) and the *overall MCC* (*MCC*) as defined in formula (15).

$$p_l = M_{l,l}, q_l = \sum_{i=1, i \neq l}^{L} \sum_{j=1, j \neq l}^{L} M_{i,j}, r_l = \sum_{i=1, i \neq l}^{L} M_{i,l}, s_l = \sum_{j=1, j \neq l}^{L} M_{l,j}$$
$$p = \sum_{l=1}^{L} p_l, q = \sum_{l=1}^{L} q_l, r = \sum_{l=1}^{L} r_l, s = \sum_{l=1}^{L} s_l \qquad (13)$$

$$SP_l = P_l\big/p_l + r_l, l = 1, 2..., L$$
$$SE_l = P_l\big/p_l + s_l, l = 1, 2..., L \qquad (14)$$
$$MCC_l = (p_l q_l - r_l s_l)\big/\sqrt{(p_l + r_l)(p_l + s_l)(q_l + r_l)(q_l + s_l)}, l = 1, 2..., L$$

$$Acc = \sum_{l=1}^{L} M_{l,l}\big/\sum_{i=1}^{L} \sum_{j=1}^{L} M_{i,j}$$
$$MCC = (pq - rs)\big/\sqrt{(p+r)(p+s)(q+r)(q+s)} \qquad (15)$$

where the confusion matrix $M_{i,j}$ records the counts that class i are classified to class j and L denotes the number of class labels. The *AUC* metric is calculated on the basis of the decision values defined by formula (10). For comparison with the existing methods, we also report the *F1* score defined as follows:

$$F1\,score = 2 \times SP_l \times SE_l\big/SP_l + SE_l,$$
$$l = 1\,denotes\,the\,positive\,class \qquad (16)$$

Results

Model performance

Cross validation performance evaluation on benchmark data. We conduct 10-fold cross validation to validate the effectiveness of homolog knowledge transfer by means of independent *homolog instances*. The *ROC curves* for the three experimental settings are illustrated in Figure 1. From Figure 1, we can see that both the experimental setting *Multi-instance Learning* and the experimental setting *Multi-instance Learning Novel* outperform the baseline experimental setting *Single Instance Learning*, with *ROC-AUC* scores equal to 0.8335, 0.8176 and 0.8003, respectively. Besides *ROC curve*, *precision-recall curve* (*PR curve*) is another performance metric that is often used to measure the performance of two-class classification, especially in the scenario of highly skewed (extremely unbalanced) training data [32]. For comprehensive study, the *PR curves* for the three experimental settings are plot in Figure 2, with *PR-AUC* scores equal to 0.8678, 0.8369 and 0.8325, respectively. From Figure 1 and Figure 2, we can see that the homolog knowledge transfer by means of independent *homolog instances* does improve the model performance (*Multi-instance Learning* versus *Single Instance Learning*), and the *homolog instances* can be treated as effective substitute when the required feature information is not available (*Multi-instance Learning Novel* versus *Single Instance Learning*).

The *ROC Curve* and the *PR Curve* focus on the *positive* class (interaction), paying little attention to the *negative* class (non-interaction). In this work, we comprehensively survey the performance on both he *positive* class and the *negative* class, and use *MCC* to measure the predictive bias. The performance metrics in terms of *SP*, *SE* and *MCC* are given in Table 2. From Table 2, we can see that the *SE*, *SP* and *MCC* values show no significant variance between the *positive* class and the *negative* class in the three experimental settings, except the *SE* values (*positive* class: 0.8065, *negative* class: 0.6935) in the setting *Multi-instance Learning Novel*. The *MCC* values on the two classes suggest that there is little predictive bias. As compared to *ROC-AUC* scores and *PR-AUC* scores, the performance metrics of *overall accuracy* and *overall MCC* shows obvious performance

increase between *Multi-instance Learning* and *Single Instance Learning* (*overall accuracy* 78.23% versus 70.97%; *overall MCC* 0.6306 versus 0.5260), so does it between *Multi-instance Learning Novel* and *Single Instance Learning* (*overall accuracy* 75% versus 70.97%; *overall MCC* 0.5833 versus 0.5260). The results once again suggest that the *homolog instances* are effective to augment the training data and substitute for the missing target *GO* information. The very small performance difference between *Multi-instance Learning* and *Multi-instance Learning Novel* further verifies the effectiveness of *homolog instances*.

The *F1* score defined in formula (16) takes into account *SP* and *SE* on the *positive* class. As shown in Table 2, the three experimental settings *Multi-instance Learning*, *Multi-instance Learning Novel* and *Single Instance Learning* achieve *F1* scores 0.80, 0.76 and 0.71, respectively. The relatively significant increase of *F1* scores also demonstrates the effectiveness of homolog knowledge transfer by means of independent *homolog instances*.

Performance comparison with existing models. To further validate the model merits, we compare our proposed multi-instance transfer learning method with two reported machine learning methods, one is Random Forest [16], and the other is multi-task learning method [17]. As shown in Table 3, Random Forest [16] achieved rather low *SE* value (0.407) on the *positive* class, much lower than our method (*SE* = 0.8065), though a little higher *SP* values than our method (*SP* 0.817 versus 0.7692). Such a low *SE* value (0.407) suggests that Random Forest [16] is likely to be highly biased towards the *negative* class. In addition, Random Forest [16] also achieved much lower *F1* score than our method (*F1* score 0.52 versus 0.80).

The multi-task learning method [17] simultaneously exploited multiple PPI networks from the bacteria *S.typhi*, *B.anthracis*, *F.tularensis* and *Y.pestis*. The method reported the *F1* score only (0.758). Model evaluation with inadequate performance metric is one of major weaknesses to both the methods [16,17]. As

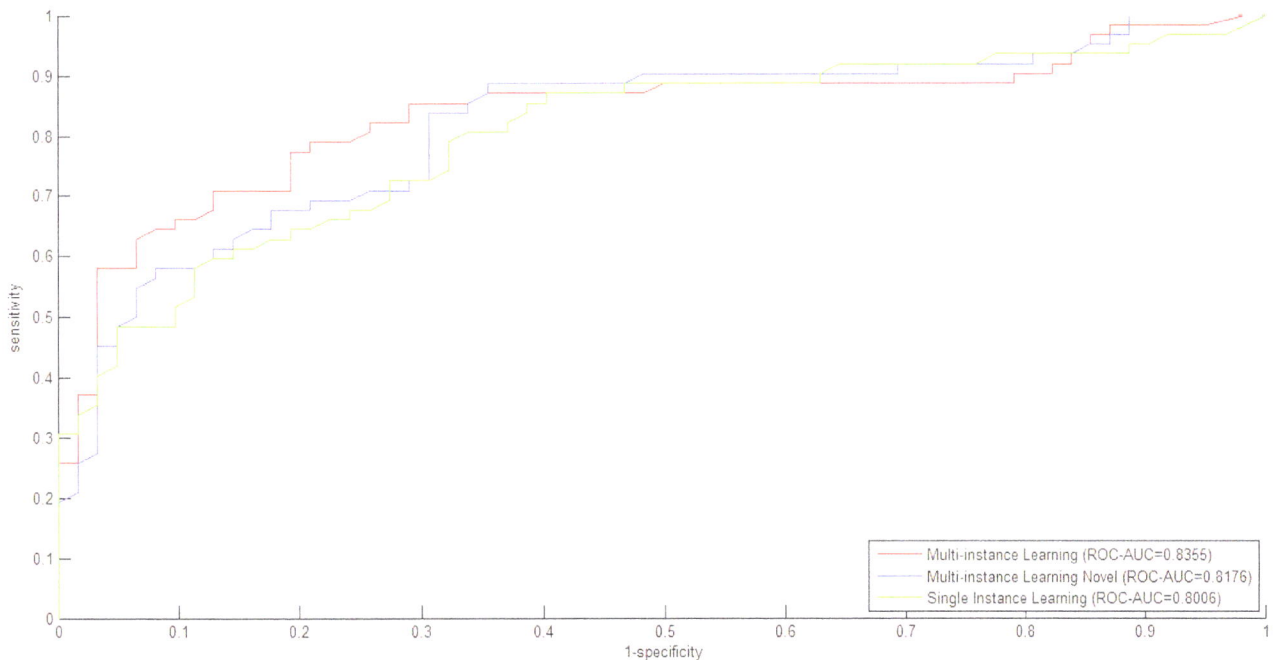

Figure 1. ROC curves for the three experimental settings (Multi-instance Learning, Multi-instance Learning Novel, Single Instance Learning).

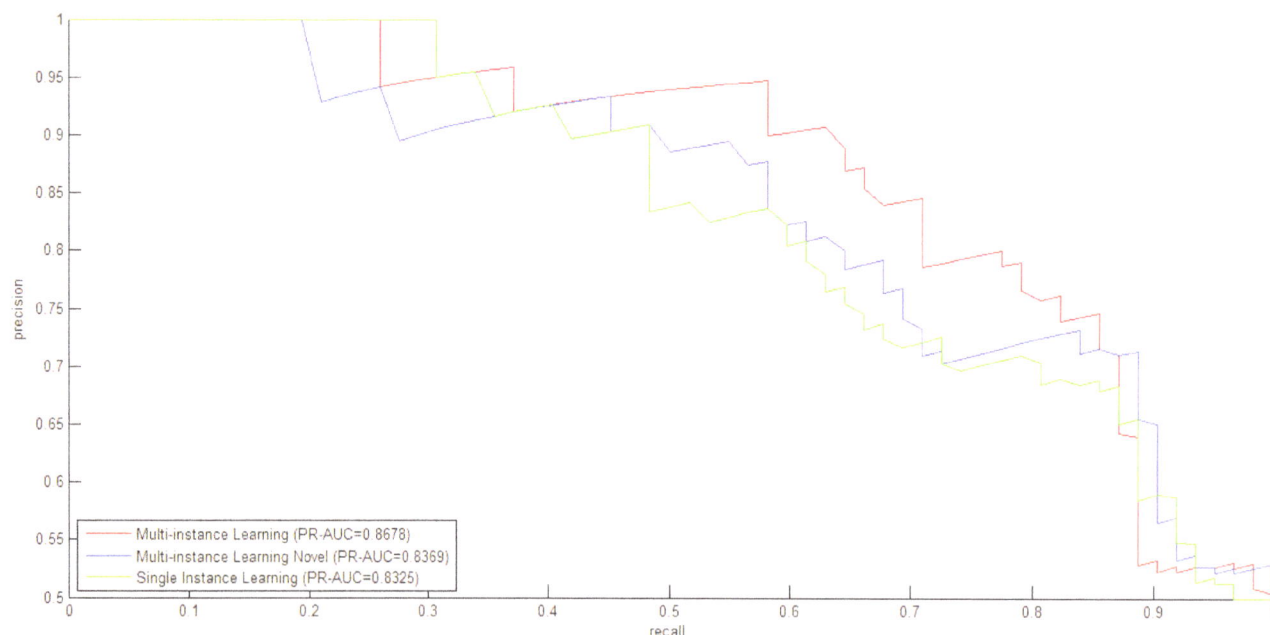

Figure 2. Precision-Recall curves for the three experimental settings (Multi-instance Learning, Multi-instance Learning Novel, Single Instance Learning).

compared to the Random Forest method [16] and the multi-task learning method [17], our proposed multi-instance transfer learning method has the following merits (1) we design the experimental setting *Single Instance Learning* as the baseline to verify the performance gain and robustness of information substitution by the *homolog instances*. Unfortunately, the Random Forest method [16] should have designed the baseline experimental setting of no imputing missing values, and the multi-task learning method [17] should have designed the single-task learning of only *S.typhi* without information from *B.anthracis*, *F.tularensis* and *Y.pestis*, as the baseline experimental setting. Without the baseline performance, we can not gain knowledge about the performance increase by imputing missing values [16] and optimizing multiple sub-tasks learning [17]; (2) Besides the baseline experimental setting, we also design an extreme experimental setting *Multi-instance Learning Novel* where the feature information of the test proteins is completely not available (e.g. novel proteins). Both the methods [16,17] did not test the model robustness against *data unavailability*; (3) Besides *gene ontology*, the two methods [16,17] also need other feature information such as *gene expression*, *RNAi expression*, *conserved pathways*, *Pfam interactions*, etc., which may also be not available. In our methods, only *gene ontology* is needed and thus the data constraint is much less demanding. Of course, the above-mentioned comparison is *rough* in a sense. The *randomness* introduced by data partition of cross validation and negative data sampling makes it hard to conduct strictly fair comparison between different computational methods. Furthermore, the performance increase of novel method against the existing methods benefits to a certain degree from the update of data (e.g. GOA update). Nevertheless, a relatively rough comparison helps us to gain some knowledge about the reliability of novel models. For the existing models, it is necessary to frequently update the model training using up-to-date data.

Schlekera et al. [15] used protein sequence similarity (sequence identity was 21%) to predict *interlogs*, achieving 49% recognition rate (i.e. $SE = 0.49$) on the dataset [8]. The SE value is equivalent to the Random Forest [16] ($SE = 0.407$) but much lower than the proposed multi-instance transfer learning method (*Multi-instance Learning* $SE = 0.8065$; *Multi-instance Learning Novel* $SE = 0.8065$; *Single Instance Learning* $SE = 0.7258$).

Model validation on data from recent literature. To further validate the model performance, we conduct independent test using the data from recent literature. The recent experimental findings are generally scarce and scattered in massive biomedical literature, and it is hard to manually collect adequate data for model validation. By manual search, we find 18 *Salmonella*-human PPIs in [18]. Excluding the non-protein interactions (e.g. cholesterol, inositol phospates) and the interactions that have been collected in [8], we obtain 5 novel interactions, together with two novel interactions (SspH2, SGT1) & (SspH2, Nod1) [24], we obtain 7 novel interactions as validation set. The predictions on the validation set are shown in Table 3, where the normalized decision values in the third column measure the confidence level of the predictions. All the 7 novel PPIs are correctly recognized by the proposed method. Interestingly, the two proteins O43765 and O95905 both are related to the gene *SGT1*, but only O43765 is predicted to interact with SspH2. According to *gene ontology* annotations, O95905 is annotated as "*Novel regulator of p53 stability and function*" (http://www.uniprot.org/uniprot/O95905) and O43765 is annotated as "*Co-chaperone that binds directly to HSC70 and HSP70 and regulates their ATPase activity*" (http://www.uniprot.org/uniprot/O43765). It has been reported in [24] that SGT1 interacting with SspH2 was *restricted* to those SGT1 proteins that have NLR *co-chaperone* function. O43765 fulfils *co-chaperone* function while O95905 does not, O43765 (*SGT1*) is predicted to interact with SspH2 while O95905 is not. The details of the predicted interactions between SspH2 and SGT1 are consistent with the experimental findings in [24].

Table 2. Model performance estimation by 10-fold cross validation.

Multi-instance transfer learning		Multi-instance Learning			Multi-instance Learning Novel			Single Instance Learning		
		SP	SE	MCC	SP	SE	MCC	SP	SE	MCC
Positive		0.7692	0.8065	0.6338	0.7246	0.8065	0.5936	0.7031	0.7258	0.5290
Negative		0.7966	0.7581	0.6284	0.7818	0.6935	0.5780	0.7167	0.6935	0.5234
Accuracy		78.23%			75%			70.97%		
MCC		0.6306			0.5833			0.5260		
ROC-AUC		0.8335			0.8176			0.8003		
PR-AUC		0.8678			0.8369			0.8325		
F1 Score		0.80			0.76			0.71		
*Random forest [16]	Positive	SP 0.817				SE 0.407				
	F1 Score	0.52								
*Multi-task learning [17]	F1 Score	0.758								

Note: * denotes the existing models.

Reconstruction of proteome-wide *Salmonella*-human PPI networks

Proteome-wide PPI predictions. To reconstruct the proteome-wide *Salmonella*-human PPI networks, the 25 *Salmonella* proteins in the 62 interactions [8] are used as pathogen proteins and the human proteins are taken from the file *uniprot_sprot_human. dat.gz* available at ftp://ftp.uniprot.org/pub/databases/uniprot/current_release/knowledgebase/taxonomic_divisions/. Excluding those uncurated proteins and those proteins that already appear in the 62 interactions [8], we totally obtain 20,334 reviewed human proteins. Hence the prediction space contains 508,350 (25×20,334) protein pairs. The predictions are given in File S1. Among the 20,334 human proteins, 6271 human proteins are predicted to interact with the 25 *Salmonella* proteins and 75,381 novel interactions are detected by our method. Comparatively, the Random Forest method [16] predicted 190,868 novel interactions between 22,653 human proteins and 486 *Salmonella* proteins, of which 461 *Salmonella* proteins were not included in the 62 PPIs [8]. After excluding the 461 *Salmonella* proteins and the related predictions, the Random Forest method [16] predicted 134,339 interactions, much larger than 75,381 interactions predicted by our method, suggesting a larger coverage of true interactions and meanwhile a higher risk of false positive predictions.

Overlap with existing models. The overlap of predictions between different computational methods is generally small. For instance, the overlap between the predictions by the three methods [9,12,13] contains only 4 HIV-human PPIs. Such a low overlap largely results from the two points: (1) the overlap analysis is not based on proteome-wide predictions; (2) the threshold of the final predictions is generally chosen to be much higher than the threshold for model estimation [10,16,17], and thus narrows down the scale of the predicted PPI networks. As regards with *Salmonella*-human PPI prediction, the Random Forest method [16] yielded 134,339 interactions, our method yields 75,381 interactions and there are 23,159 overlapped predicted PPIs (see File S2). The overlap rate between Random Forest [16] and our method is 12.41%. Here the overlap rate among K methods is formally defined as $|\bigcap_{i=1}^{K} O_i| \Big/ |\bigcup_{i=1}^{K} O_i|$, where O_i denoted the predictions yielded by the i-th method and $|A|$ denotes the cardinality of set A. The 23,159 overlapped PPIs are supposed to be more reliable. As compared to the low overlap rate among the three methods [9,12,13], the large overlap rate between Random Forest [16] and our method is largely due to the proteome-wide prediction space. Besides, as compared to large training data, the small training data is prone to yield a relatively larger portion of false positive predictions, which may also contribute to the large overlap rate. It is worth noting that if we further define as positives those predictions with $\bar{V}(i_1,i_2) > \delta(\bar{V}(i_1,i_2)$ is defined in formula (12)), the predicted interactions are supposed to be more reliable.

Schlekera et al. [15] used the 62 *Salmonella*-human interactions to derive *interlogs*, i.e. the interacting pairs between *Salmonella* protein *orthologs* and human protein *orthologs*. Unlike the Random Forest method [16] and our method, Schlekera et al. [15] did not predict the interactions between the 25 *Salmonella* proteins contained in the benchmark data with other human proteins, so we do not conduct analysis of network overlap with the method.

Discussions

Reconstruction of pathogen-host PPI networks is of importance to reveal the underlying mechanism of pathogen infection and host defence. At present, it is still a challenging task for labor-intensive

Table 3. Model validation on data from recent literatures.

Salmonella effector proteins	Targeted human proteins	Confidence level
AvrA	MTOR	0.3271
SipA	F-actin caspase-3	0.5157
SipC	F actin	0.5185
PibB2	Kinesin-1	0.2943
SseI	F-actin	0.5185
SspH2	SGT1	0.1059
SspH2	Nod1	0.5410

experimental techniques to accurately map the proteome-wide pathogen-host protein interactome. Computational modeling is a good complement to experimental methods for fast proteome-wide interactome mapping at low cost. Unfortunately, the current experimentally derived pathogen-host PPI networks are rather small, for instance, there are only 62 known PPIs between *Salmonella typhimurium* and human, such a small network would yield a computational model that does not generalize well. On the other hand, computational modeling need deliberate dwelling on the problem that the knowledge in biological databases (e.g. *gene ontology*) is currently not complete.

In this work, we conduct homolog knowledge transfer by means of independent *homolog instances* to serve the purposes: (1) augmenting the training data to reduce the risk of model overfitting; (2) enriching the feature information to improve the model performance; (3) substituting for the missing feature information to enhance the model robustness against *data unavailability*. The multi-instance transfer learning method is implemented under the framework of AdaBoost, where the *gene ontology* knowledge from the homologs is treated as independent *homolog instance* to augment the training data. The noise from the *homolog instances* could be counteracted by AdaBoost instance reweighting algorithm.

To validate the assumptions that the independent *homolog instances* are effective to solve the problems of *data scarcity* and *data unavailability* for *Salmonella*-human PPI prediction, we deliberately design three experimental settings: *Multi-instance Learning, Multi-instance Learning Novel* and *Single Instance Learning* and comprehensively survey the model performance by multiple performance metrics: *SP, SE, Accuracy, MCC, ROC-AUC* and *PR-AUC*. The results of 10-fold cross validation experiments show that the homolog knowledge transfer by means of independent *homolog instances* does improve the model performance and helps the model work properly in the extreme case that the *gene ontology* knowledge is completely not available. The experimental results also show that the proposed multi-instance transfer learning method significantly outperforms the existing machine learning methods with less demanding data constraints. To further validate how well the proposed model generalizes to unseen data, we collect 7 experimentally derived interactions from the recent literatures. Al78 interactions can be correctly recognized by our method. Interestingly, for the two proteins {O43765, O95905} that are related to gene *SGT1*, only the protein O43765 that fulfils the molecular function of *co-chaperone* is predicted to interact with SspH2, which is consistent with the recent findings.

Noteworthily, there are two concerns that need to be addressed, one concern is about the feasibility of training model on the small *Salmonella*-human PPI data, and the other concern is about false positive rate. Although the proposed model demonstrates sound performance from the point of view of cross validation and independent test experiments, the *smallness* of data may not convince us of the validity of the predictions as the large HIV data does [6,10,28]. The worry comes from whether or not it is feasible to use so small a data to train a satisfactory model. Actually, some machine learning methods like SVM (*support vector machine*) are derived from small-example statistical learning theory [35], where only a small number of *support vectors* (*SV*) are needed to define the two-class decision hyper-planes and a large number of non-*SVs* are discarded. The gracefulness of *sparseness* is often used to reduce the computational complexity [28,36,37]. It has been theoretically proven that the regularization technique helps AdaBoost to achieve large margin between two-class hyperplanes and *sparseness* like SVM [31], implying that it is still feasible to train a satisfactory model on such a small data for *Salmonella*-human PPI prediction. We note that the multi-instance transfer learning method is specifically proposed to augment very small data. For large data like HIV, the proposed method will increase the computational complexity and thus seems not to be a proper

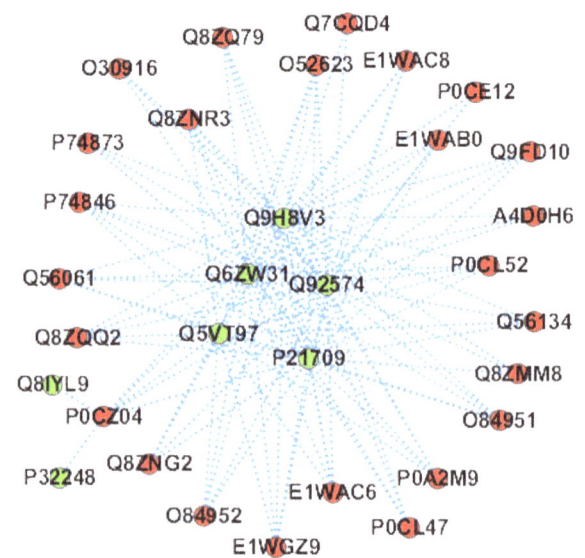

Figure 3. The predicted *Salmonella*-human PPI sub-network GO: 0032862 (*biological process: activation of Rho GTPase activity*). The red node denotes *Salmonella* protein and the green node denotes human protein.

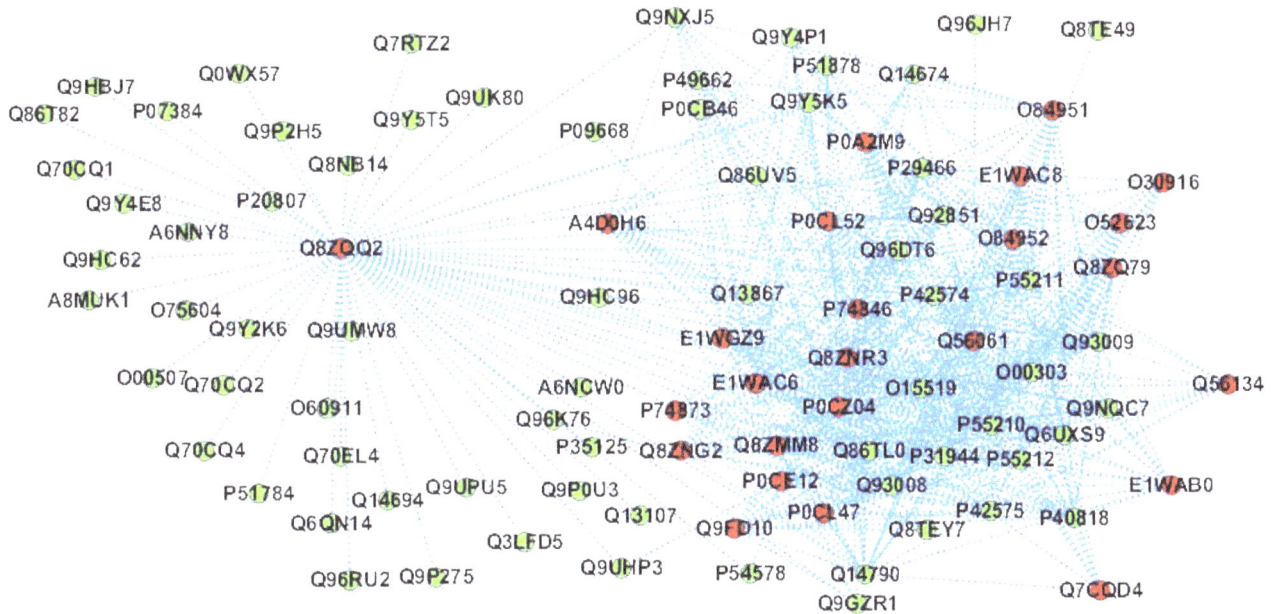

Figure 4. The predicted *Salmonella*-human PPI sub-network GO: 0008234 (*molecular function: cysteine-type peptidase activity*). The red node denotes *Salmonella* protein and the green node denotes human protein.

solution. As regards with the second concern, both biological experiments and computational predictions are supposed to yield a certain level of false positives. For biological experiments, different experimental techniques are generally specific to particular types of interactions, for instance, Y2H as an *in vivo* technique is highly effective at detecting transient interactions and can be readily applied to screen large genome-wide libraries, but is limited by its biases toward non-specific interactions [38]. Likewise computational method is also prone to yield false positive predictions, especially on small training data, and the 7 correctly recognized interactions (see Table 3) may be worried to result from high false positive rate. Actually, our proposed method greatly reduces the risk of false positive predictions as compared to the Random Forest [16] (predicted 75,381 interactions versus predicted 134,339 interactions). To ensure the validity of the predictions, we can increase the decision threshold as $\bar{V}(i_1,i_2) > \delta(\bar{V}(i_1,i_2)$ is defined in formula (12)).

The ultimate goal of model development and model estimation is to reconstruct reliable proteome-wide protein interaction networks between *Salmonella typhimurium* and Homo sapiens. To gain insights into the patterns of *Salmonella* infection and host response, the predicted PPI networks are further subjected to *gene ontology* based clustering analysis. We simply cluster together the *Salmonella* targeted human proteins on the basis of *gene ontology* terms. To provide valuable cues about the host protein complexes, molecular functions and signaling pathways that *Salmonella* proteins may interfere with, we assign to the same cluster the human partners that possess the same *GO* term, with each cluster corresponding to one PPI sub-network. All the *GO* terms of human proteins are classified into thee major classes, i.e., biological processes (P), molecular functions (F) and cellular compartments (C). For each major class, we further discuss the two cases: (1) all the 25 *Salmonella* proteins are involved in the PPI sun-network, denoted as P1, F1 and C1, respectively; (2) NOT all the 25 *Salmonella* proteins are involved in the PPI sun-network, denoted as P2, F2 and C2, respectively. P1, F1 and C1 are given in

File S3, File S4 and File S5, respectively. P2, F2 and C2 are given in File S6, File S7 and File S8, respectively. For the sake of quite a large number of predicted PPI sub-networks, we only demonstrate several PPI sun-networks as examples, interested readers are referred to File S3~File S8 for insightful biological cues.

PPI sub-network GO: 0032862 - activation of Rho GTPase activity

It has been reported in [33] that *Salmonella typhimurium* injects toxins SopE, SopE2 and SptP to change the GTP/GDP loading

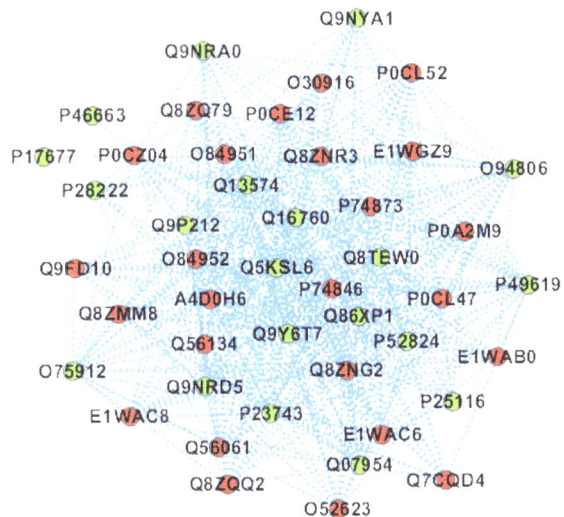

Figure 5. The predicted Salmonella-human PPI sub-network GO: 0007205 (biological process: activation of protein kinase C activity by G-protein coupled receptor protein signaling pathway). The red node denotes Salmonella protein and the green node denotes human protein.

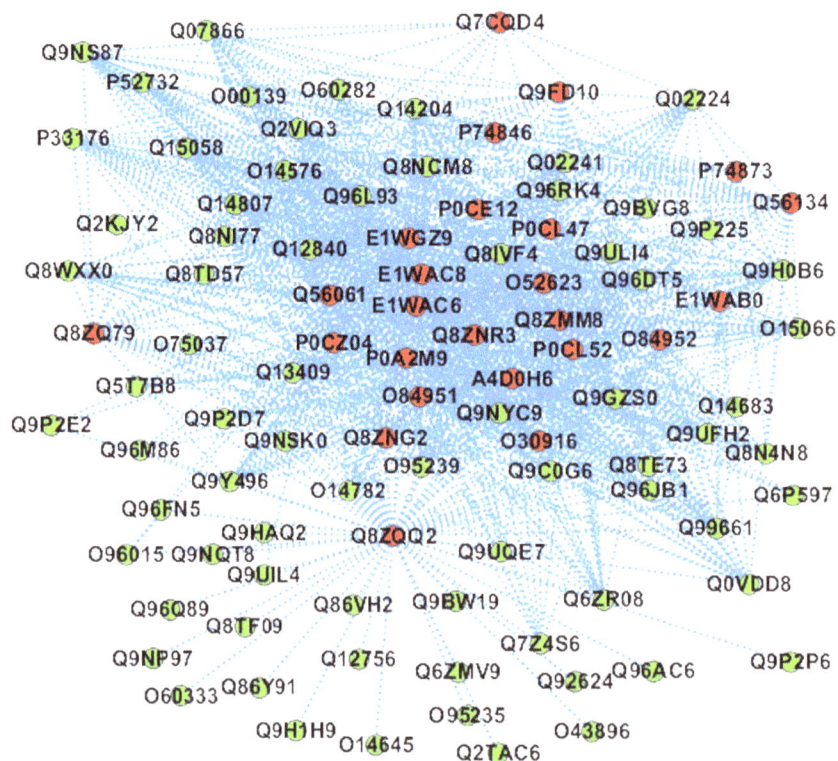

Figure 6. The predicted *Salmonella*-**human PPI sub-network GO: 0003777** (*molecular function: microtubule motor activity*)**.** The red node denotes *Salmonella* protein and the green node denotes human protein.

state of Rho GTPases by transient interactions, wherein SopE and SopE2 mimic eukaryotic G-nucleotide exchange factors (GEF) and thereby activate Rho GTPase signaling pathways in the infected host cells. In this work, the *Salmonella* proteins are predicted with 7 human proteins that are annotated with *GO* term GO: 0032862 (activation of Rho GTPase activity). The predicted PPI sub-network GO: 0032862 is illustrated in Figure 3. The two *Salmonella* SopE proteins (O52623, Q7CQD4) are predicted to interact with 5 human proteins {Q9H8V3, Q6ZW31, Q5VT97, Q92574, P21709}, suggesting that the SopE activates Rho GTPase signaling pathways by directly mimicking the molecular functions of the host proteins or indirectly interacting with the host proteins. Among the 7 human proteins, the two human proteins {Q9H8V3, Q92574} are predicted to be targeted by all the 25 *Salmonella* proteins while the two proteins {Q8IYL9, P32248} are predicted to be targeted by only one *Salmonella* protein.

PPI sub-network GO: 0008234 - cysteine-type peptidase activity

The predicted PPI sub-network is illustrated in Figure 4. In Figure 4, the *Salmonella* proteins are predicted to interact with 70 human proteins that fulfil the molecular function "cysteine-type peptidase activity" (GO: 0008234), wherein the *Salmonella* protein Q8ZQQ2 is a hub protein that is predicted to target 63 human proteins. The *Salmonella* protein slrP (Q8ZQQ2) is an effector protein that functions to alter the host cell physiology and promote bacterial survival in the host tissues. This protein is an E3 ubiquitin ligase that interferes with the host ubiquitination pathway and leads to significant decrease of thioredoxin activity and increase of the host cell death (http://www.uniprot.org/uniprot/Q8ZQQ2). The six human proteins {P5521, Q14790,

P42575, O00303, P55210, P55211} are predicted to be targeted by all the 25 *Salmonella* proteins, wherein P5521 is involved in the activation cascade of caspases responsible for apoptosis execution, whose over-expression promotes programmed cell death. The predictions suggest the *Salmonella* proteins may also interfere with the host apoptotic signaling pathways.

PPI sub-network GO: 0007205 - activation of protein kinase C activity by G-protein coupled receptor protein signaling pathway

The predicted PPI sub-network is illustrated in Figure 5. As shown in Figure 5, all the 20 predicted human partners except {P46663, P17677} are densely connected with the 25 *Salmonella* proteins. *Salmonella* protein spiC (P0CZ04) is predicted to target all the 20 human proteins and the human proteins {Q07954, P49619, Q86XP1, Q9NRD5, Q16760, P52824, Q9Y6T7, Q9P212, Q8TEW0, Q5KSL6, P23743} are predicted to be targeted by all the 25 *Salmonella* proteins. *Salmonella* protein spiC (P0CZ04) is a virulence protein that plays a central role in mammalian macrophage infection by inhibiting phagosome-lysosome fusion and cellular trafficking (http://www.uniprot.org/uniprot/P0CZ04). The human protein Q07954 is an endocytic receptor involved in endocytosis and in phagocytosis of apoptotic cells, and may modulate the cellular events such as APP metabolism, kinase-dependent intracellular signaling, neuronal calcium signaling as well as neurotransmission (http://www.uniprot.org/uniprot/Q07954). The molecular function annotations of the two proteins suggest that *Salmonella* protein spiC (P0CZ04) is highly likely to interact with the human protein Q07954.

PPI sub-network GO: 0003777 - microtubule motor activity

In [34], it is reported that *Salmonella typhimurium* grows within the host cells in a permissive compartment termed *Salmonella*-containing vacuole (SCV), and some SPI-2 effectors modulate microtubule motor activity on the SCV. In this work, the *Salmonella* proteins are predicted with 74 human proteins that are annotated with the *GO* term GO: 0003777 (microtubule motor activity) (see Figure 6), and the SPI-2 effector sseJ (Q9FD10) is predicted to interact with 26 human proteins that are involved in the molecular functions "microtubule motor activity". From Figure 6, we can see the predicted PPI sub-network GO: 0003777 is densely connected and the 10 human proteins {Q9H0B6, Q8NCM8, Q8NCM8, Q9GZS0, Q12840, P33176, Q02224, Q9NS87, O95239, O14576} are predicted to be targeted by all the 25 *Salmonella* proteins. The human protein Q9H0B6 (Kinesin light chain 2) is a light chain of Kinesin (a microtubule-associated force-producing protein that may play a role in organelle transport), and the light chain may function in coupling of cargo to the heavy chain or in the modulation of its ATPase activity (http://www.uniprot.org/uniprot/Q9H0B6). The prediction suggests that the *Salmonella* proteins may interfere with the host organelle transport and ATPase activity.

Supporting Information

File S1 XML file contains the proteome-wide predicted *Salmonella*-human PPIs.

File S2 XML file contains the overlapped interactions between our method and Random Forest [16].

File S3 XML file contains the clusters of interacting human partners that participate in specific *biological processes*. All the 25 *Salmonella* proteins are involved in the predicted PPI sub-networks.

File S4 XML file contains the clusters of interacting human partners that fulfil specific *molecular functions*. All the 25 *Salmonella* proteins are involved in the predicted PPI sub-networks.

File S5 Text file contains the clusters of interacting human partners that are localized in specific *cellular compartments*. All the 25 *Salmonella* proteins are involved in the predicted PPI sub-networks.

File S6 XML file contains the clusters of interacting human partners that participate in specific *biological processes*. NOT all the 25 *Salmonella* proteins are involved in the predicted PPI sub-networks.

File S7 XML file contains the clusters of interacting human partners that fulfil specific *molecular functions*. NOT all the 25 *Salmonella* proteins are involved in the predicted PPI sub-networks.

File S8 XML file contains the clusters of interacting human partners that are localized in specific *cellular compartments*. NOT all the 25 *Salmonella* proteins are involved in the predicted PPI sub-networks.

Acknowledgments

Thanks for the helpful comments from the anonymous reviewers.

Author Contributions

Conceived and designed the experiments: SM. Performed the experiments: SM. Analyzed the data: SM. Contributed reagents/materials/analysis tools: SM. Wrote the paper: SM. Revised the manuscript: HZ.

References

1. von Mering C, Krause R, Snel B, Cornell M, Oliver SG, et al. (2002) Comparative assessment of large-scale datasets of protein-protein interactions. Nature 417: 399–403.
2. Edwards A, Kus B, Jansen R, Greenbaum D, Greenblatt J, et al. (2002) Bridging structural biology and genomics: assessing protein interaction data with known complexes. Trends Genet 18: 529–536.
3. Wu X, Zhu L, Guo J, Zhang D, Lin K (2006) Prediction of yeast protein-protein interaction network: insights from the Gene Ontology and annotations. Nucleic Acids Res 34: 2137–2150.
4. DeBodt S, Proost S, Vandepoele K, Rouzé P, Peer Y, et al. (2009) Predicting protein-protein interactions in Arabidopsis thaliana through integration of orthology, gene ontology and co-expression. BMC Genomics 10: 288.
5. Shen J, Zhang J, Luo X, Zhu W, Yu K, et al. (2007) Predicting protein-protein interactions based only on sequences information. Proc Natl Acad Sci USA 104: 4337–41.
6. Fu W, Sanders-Beer BE, Katz KS, Maglott DR, Pruitt KD, et al. (2009) Human immunodeficiency virus type 1, human protein interaction database at NCBI. Nucleic Acids Res (Database Issue) 37: D417–22.
7. Wuchty S (2011) Computational prediction of host-parasite protein interactions between P. falciparum and H. sapiens. PLoS ONE 6: e26960.
8. Schleker S, Sun J, Raghavan B, Srnec M, Müller N, et al. (2012) The current Salmonella-host interactome. Proteomics Clin Appl 6: 117–133.
9. Tastan O, Qi Y, Carbonell J, Klein-Seetharaman J (2009) Prediction of interactions between HIV-1 and human proteins by information integration. In: Proceedings of the Pacific Symposium on Biocomputing (PSB-2009). pp 516–527.
10. Qi Y, Tastan O, Carbone J, Klein-Seetharaman, Weston J, et al. (2010) Semi-supervised multi-task learning for predicting interactions between HIV-1 and human proteins. Bioinformatics 26: i645–i652.

11. Dyer M, Muralib T, Sobrala B (2011) Supervised learning and prediction of physical interactions between human and HIV proteins. Infect Genet Evol 11: 917–923.
12. Doolittle J, Gomez S (2010) Structural similarity-based predictions of protein interactions between HIV-1 and Homo sapiens. Virology J 7: 82.
13. Mukhopadhyay A, Maulik U, Bandyopadhyay S (2012) A novel biclustering approach to association rule mining for predicting HIV-1-human protein interactions. PLoS One 7: e32289.
14. Dyer M, Muralib T, Sobrala B (2007) Computational prediction of host-pathogen protein-protein interactions. Bioinformatics 23: i159–i166.
15. Schleker S, Garcia-Garcia J, Klein-Seetharaman J, Oliva B (2012) Prediction and comparison of Salmonella-human and Salmonella-Arabidopsis interactomes. Chem Biodivers 9: 991–1018.
16. Kshirsagar M, Carbonell J, Judith K (2012) Techniques to cope with missing data in host–pathogen protein interaction prediction. Bioinformatics 28: i466–i472.
17. Kshirsagar M, Carbonell J, Judith K (2013) Multitask learning for host-pathogen protein interactions. Bioinformatics 29: i217–i226.
18. Srikanth CV, Regino ML, Hallstrom K, McCormick B (2011) Salmonella effector proteins and host-cell responses. Cell Mol Life Sci 68: 3687–3697.
19. Yu J, Guo M, Needham C, Huang Y, Cai L, et al. (2010) Simple sequence-based kernels do not predict protein-protein interactions. Bioinformatics 26: 2610–2614.
20. Freund Y, Schapire RE (1997) A decision-theoretic generalization of on-line learning and an application to boosting. J Comput Syst Sci 55: 119–139.
21. Vezhnevets A, Vezhnevets V (2005) Modest AdaBoost – teaching AdaBoost to generalize better. Graphicon 12: 987–997.
22. Jansen R, Gerstein M (2004) Analyzing protein function on a genomic scale: the importance of gold-standard positives and negatives for network prediction. Curr Opin Microbiol 7: 535–545.

23. Boeckmann B, Bairoch A, Apweiler R, Blatter MC, Estreicher A, et al. (2003). The SWISS-PROT protein knowledgebase and its supplement TrEMBL in 2003. Nucleic Acids Res 31: 365–370.

24. Bhavsar A, Brown N, Stoepe J, Wiermer M, Martin D, et al. (2013) The Salmonella type III effector SspH2 specifically exploits the NLR co-chaperone activity of SGT1 to subvert immunity. PLoS Pathog 9: e1003518.

25. Pan S, Yang Q (2010) A Survey on Transfer Learning. IEEE Trans Knowl Data Eng 22: 1345–1359.

26. Mei S, Wang F, Zhou S (2011) Gene ontology based transfer learning for protein subcellular localization. BMC Bioinformatics 12: 44.

27. Mei S (2012) Multi-label multi-kernel transfer learning for human protein subcellular localization. PLoS ONE 7: e37716.

28. Mei S (2013) Probability weighted ensemble transfer learning for predicting interactions between HIV-1 and human proteins. PLoS ONE 8: e79606.

29. Altschul S, Madden T, Schaffer A, Zhang J, Zhang Z, et al. (1997) Gapped BLAST and PSI-BLAST: a new generation of protein database search programs. Nucleic Acids Res 25: 3389–3402.

30. Barrell D, Dimmer E, Huntley R, Binns D, O'Donovan C, et al. (2009) The GOA database in 2009–an integrated Gene Ontology Annotation resource. Nucleic Acids Res (Database Issue) 37: D396–D403.

31. Meir R, Ratsch G (2003) An introduction to boosting and leveraging. Lect Notes Artif Int 2600: 118–183.

32. Davis J, Goadrich M (2006) The relationship between precision-recall and ROC curves. In: Proceedings of the 23rd International Conference on Machine Learning.

33. Schlumberger MC, Hardt WD (2005) Triggered phagocytosis by Salmonella: bacterial molecular mimicry of RhoGTPase activation/deactivation. Curr Top Microbiol Immunol 291: 29–42.

34. Wasylnka JA, Bakowski MA, Szeto J, Ohlson MB, Trimble WS, et al. (2008) Role for myosin II in regulating positioning of Salmonella-containing vacuoles and intracellular replication. Infect Immun 76: 2722–2735.

35. Vapnik V (1999) An Overview of Statistical Learning Theory. IEEE Trans Neural Netw 10: 988–999.

36. Mei S (2014) SVM ensemble based transfer learning for large-scale membrane proteins discrimination. J Theor Biol 340: 105–110.

37. Dong J, Adam K, Ching Y (2005) Fast SVM Training Algorithm with Decomposition on Very Large Data Sets. IEEE Trans Pattern Anal Mach Intell 27: 603–18.

38. Gonzalez MW, Kann MG (2012) Chapter 4: Protein Interactions and Disease. PLoS Comput Biol 8: e1002819.

10

Identification of Host-Immune Response Protein Candidates in the Sera of Human Oral Squamous Cell Carcinoma Patients

Yeng Chen[1,2]*, Siti Nuraishah Azman[3], Jesinda P. Kerishnan[1], Rosnah Binti Zain[2,4], Yu Nieng Chen[5], Yin-Ling Wong[1], Subash C. B. Gopinath[1]

1 Department of Oral Biology & Biomedical Sciences, Faculty of Dentistry, University of Malaya, Kuala Lumpur, Malaysia, 2 Oral Cancer Research and Coordinating Center, Faculty of Dentistry, University of Malaya, Kuala Lumpur, Malaysia, 3 Institute for Research in Molecular Medicine, Universiti Sains Malaysia, Georgetown, Penang, Malaysia, 4 Department of Oro-Maxillofacial and Medical Science, Faculty of Dentistry, University of Malaya, Kuala Lumpur, Malaysia, 5 Chen Dental Specialist Clinic, Kueh Hock Kui Commercial Centre, Jalan Tun Ahmad Zaidi Adruce, Kuching, Sarawak, Malaysia

Abstract

One of the most common cancers worldwide is oral squamous cell carcinoma (OSCC), which is associated with a significant death rate and has been linked to several risk factors. Notably, failure to detect these neoplasms at an early stage represents a fundamental barrier to improving the survival and quality of life of OSCC patients. In the present study, serum samples from OSCC patients (n = 25) and healthy controls (n = 25) were subjected to two-dimensional gel electrophoresis (2-DE) and silver staining in order to identify biomarkers that might allow early diagnosis. In this regard, 2-DE spots corresponding to various up- and down-regulated proteins were sequenced via high-resolution MALDI-TOF mass spectrometry and analyzed using the MASCOT database. We identified the following differentially expressed host-specific proteins within sera from OSCC patients: leucine-rich α2-glycoprotein (LRG), alpha-1-B-glycoprotein (ABG), clusterin (CLU), PRO2044, haptoglobin (HAP), complement C3c (C3), proapolipoprotein A1 (proapo-A1), and retinol-binding protein 4 precursor (RBP4). Moreover, five non-host factors were detected, including bacterial antigens from *Acinetobacter lwoffii*, *Burkholderia multivorans*, *Myxococcus xanthus*, *Laribacter hongkongensis*, and *Streptococcus salivarius*. Subsequently, we analyzed the immunogenicity of these proteins using pooled sera from OSCC patients. In this regard, five of these candidate biomarkers were found to be immunoreactive: CLU, HAP, C3, proapo-A1 and RBP4. Taken together, our immunoproteomics approach has identified various serum biomarkers that could facilitate the development of early diagnostic tools for OSCC.

Editor: Qing-Yi Wei, Duke Cancer Institute, United States of America

Funding: This study was supported by University of Malaya (UM)-High Impact Research(HIR)-the Ministry of Education (MoE) Grant UM/C/625/1/HIR/MOE/DENT/9 and UM.C/625/1/HIR/MOE/DENT/20. The funders had no role in study design, data collection and analysis, decision to publish, or preparation of the manuscript.

Competing Interests: The authors have declared that no competing interests exist.

* Email: chenyeng@um.edu.my

Introduction

Oral cancer represents the sixth most prevalent cancer in the world. Among the different types of oral cancer, oral squamous cell carcinoma (OSCC) arising from the oral mucosa accounts for more than 90% of these malignancies. Thus, OSCC is the most common malignancy affecting the head and neck region. Notably, there are nearly 300,000 new cases of oral cancer reported annually [1,2], and it was estimated that approximately 128,000 oral cancer patients died worldwide in 2008 [3]. Therefore, despite recent advances in the diagnosis and treatment of oral cancer (e.g., chemotherapy, radiotherapy, and surgical therapy) the survival rate of OSCC patients has remained less than 60% [4,5]. A fundamental barrier in improving the survival of OSCC patients is the fact that these malignancies often remain undetected until the later stages. In this regard, it was reported that a several-month delay in diagnosis could reduce the chance of survival from 80% to 40% [6]. Thus, in order to prevent the high OSCC-related mortality rate, recent attention has been focused on identifying

potential diagnostic molecular markers (e.g., cell cycle regulators) that might represent biological predictors of oral cancer [7]. Moreover, OSCC has been linked to several risk factors, including various bacterial pathogens [8], which might be useful for detecting OSCC.

Recent studies have indicated that early diagnosis, lifestyle modification, and effective treatment can prevent more than two-thirds of OSCC-related mortalities. However, currently available diagnostic methods do not allow for the detection of oral cancer in the early stages. To visualize malignant lesions in the oral cavity, different microscopic methods are available, which make use of various techniques, including autofluorescence, chemiluminescence, or dye-based tissue staining. However, due to the low sensitivity and specificity of these diagnostic strategies, clinicians generally use biopsies to detect OSCC [9]. Nevertheless, successful diagnosis through tissue biopsy is highly dependent on acquiring whole and complete tissue samples from patients for examination. In this regard, biopsies harvested from oral cancer patients are often associated with the soft tissues that surround the cancer tissue. In addition, oral cancer frequently involves the development of multiple primary tumors. Indeed, the occurrence of a second primary tumor is 3–7% higher per year in oral cancer when compared to other malignancies [10]. Therefore, the identification of suitable and reliable OSCC biomarkers is essential for achieving early detection and treatment, which can reduce mortality rates in OSCC patients. In this respect, antibody-based diagnostic tests that recognize specific tumor-associated antigens in cancer sera might represent a valid methodology [11].

Proteomic analysis allows the identification and quantification of proteins and peptides in biological samples [12]. However, through this approach, numerous post-translational forms of protein regulation, including regulating enzymes and low abundance proteins may remain undetected. Thus, in the present investigation, we employed an immunoproteomic approach and pooled human antibodies to detect host-specific response proteins in OSCC patients. Specifically, we used a well-characterized analytical platform combining two-dimensional gel electrophoresis (2-DE) and mass spectrometry (MS) to identify biomarkers in unfractionated sera from OSCC patients and normal controls. This immunoproteomics approach can be used to identify antigens targeted by the immune system in sera during disease progression. In addition, the immune responses are known to be involved in the mechanism of carcinogenesis [13]. Therefore, our comparative analyses revealed distinct OSCC biomarkers that might promote the development of specific diagnostic tests for early detection of oral cancer.

Materials and Methods

Serum samples

Twenty-five serum samples from OSCC patients were obtained from the Oral Cancer Research and Coordinating Center (OCRCC) at the University of Malaya (Kuala Lumpur). Additionally, 25 control serum samples were acquired from healthy individuals. All samples were collected with the verbal consent of patients, and the Dental Faculty at the University of Malaya and the Universiti Sains Malaysia Medical Ethics Committee (Ref: USMKK/PPP/JEPeM [213.3(09)]) approve this consent procedure. We obtained the permission from the above committee and they have cleared all the approval and having the record. This study was also conducted in accordance with International Conference on Harmonisation–Good Clinical Practice (ICH–GCP) guidelines and the Declaration of Helsinki.

Two-dimensional gel electrophoresis (2-DE)

We performed 2-DE as previously described [14]. Briefly, unfractionated human serum samples (10 µl) were lysed, rehydrated in lysis buffer (2M thiourea, 8M urea, 4% CHAPS, 1% dithreitol, and 2% pharmalyte), and subjected to isoelectric focusing in 13-cm rehydrated precast immobilized dry strips (pH 4–7; GE Healthcare, Sweden). Sodium dodecyl sulfate–polyacrylamide gel electrophoresis (SDS–PAGE) was performed using 8–18% gradient polyacrylamide gels in the presence of SDS for the second dimension separation. Gel silver staining was performed as previously described [15]. Silver staining and Coomassie Brilliant Blue staining for MS were conducted using slightly modified published methods [16].

Differential image acquisition and statistical analysis

The ImageScanner III (GE Healthcare, Sweden) was used to capture and store 2-DE gel images. PD-Quest 2-D gel analysis software (version 8.0.1, Bio-Rad) was employed to evaluate the differentially expressed protein spots. Identical spots were matched in the serial gels and normalized by correcting for spot quantification values and gel-to-gel variation unrelated to expression changes. For the normalization method, we used total densities from the gel images (i.e., raw quantity of each gel spot was divided by the total quantity of all spots within the gel). All protein concentration values are presented as means of percentage volume (% volume) $\pm$ standard deviations (SD). The student's t-test and one-way analysis of variance (ANOVA) were used to analyze differences between patients and controls. P-values less than 0.05 ($p<0.05$) were considered as statistically significant.

Mass spectrometry analysis and database search

Spots of interest were excised and subjected to in-gel tryptic digestion using a commercially available kit (Calbiochem, Germany). MS analysis and database searches were performed at the Proteomic Center within the Faculty of Biological Sciences at the National University of Singapore. Digested peptides were mixed with 1.2 µl of CHCA matrix solution (5 mg/ml of cyano-4-hydroxy-cinnamic acid in 0.1% trifluoroacetic acid [TFA] and 50% acetonitrile [ACN]) and spotted onto MALDI target plates. An ABI 4800 Proteomics Analyzer MALDI-TOF/TOF Mass Spectrometer was used for spectra analysis (Applied Biosystems, USA), and the MASCOT search engine (version 2.1; Matrix Science, UK) was employed for database searches. In addition, GPS Explorer software (version 3.6; Applied Biosystems, USA) was utilized along with MASCOT to identify peptides and proteins. Search parameters allowed for N-terminal acetylation, C-terminal carbamidomethylation of cysteine (fixed modification), and methionine oxidation (variable modification). Peptide and fragment mass tolerance were set to 100 ppm and ± 0.2 Da, respectively. The peptide mass fingerprinting (PMF) parameters were set as follows: one missed cleavage allowed in trypsin digest; monoisotropic mass value; ± 0.1 Da peptide mass tolerance; and 1+ peptide charge state.

Peptides were initially identified using the ProteinPilot proteomics software on the Mass Spectrometer (Applied Biosystems, USA). A score reflecting the relationship between theoretically and experimentally determined masses was calculated and assigned. Analyses were conducted using the International Protein Index (http://www.ebi.ac.uk/IPI) and NCBI Unigene human databases (version 3.38). A total of 100,907 entries were searched, and a score of >82 was considered as significant in the MASCOT NCBI database.

Immunoblotting

Our 2-DE gel immunoblotting protocol was organized into four categories: (1) normal sera probed with normal sera, (2) normal sera probed with OSCC sera, (3) OSCC sera probed with normal sera, and (4) OSCC sera probed with OSCC sera. After running the 2-DE gels, they were transferred onto nitrocellulose membranes using the Multiphor II Novablot semi-dry system (GE Healthcare, Sweden). Membranes were blocked with SuperBlock (Pierce, USA) and washed three times with Tris-buffered saline (TBS)–Tween-20. The membranes were subsequently incubated overnight (4°C) with pooled sera from patients or healthy subjects that contains the primary antibodies against various targets (1:200 dilution). Following washing, membranes were incubated for 1 hour at room temperature with horseradish peroxidase (HRP)-linked monoclonal anti-human immunoglobulin M (IgM) (1:5000; Invitrogen, USA). The membranes were again washed and then visualized using chemiluminescence substrate (Pierce, USA) and 18 cm × 24 cm films (Kodak, USA).

Functional annotation and protein interaction analyses

Functional annotation analysis was performed using DAVID v6.7 (Database for Annotation, Visualization, and Integrated Discovery), which provides a comprehensive set of functional annotation tools to understand the biological significance associated with large lists of genes or proteins [17]. This functional categorization is considered significant when a p-value of less than 0.05 (p<0.05) is obtained. STRING v9.1 (Search Tool for the Retrieval of Interacting Genes) was used to examine protein–protein interaction networks [18].

Results and Discussion

Image analysis of 2-DE serum protein profiles from healthy subjects and OSCC patients

Unfractionated serum samples were separated via 2-DE to generate high-resolution proteome profiles for healthy subjects (Fig. 1a) and OSCC patients (Fig. 1b). Protein spots in the 2-DE gels from OSCC patients (n = 25) and normal controls (n = 25) were analyzed using PDQuest 2-D gel analysis software. This comparative analysis revealed several up- and down-regulated proteins in the serum of OSCC patients (Table 1).

Identification of possible biomarkers using MS

Following digestion of the differentially expressed spots, MS analysis allowed us to identify the following seven host-specific proteins: leucine-rich α2-glycoprotein (LRG), alpha-1-B-glycoprotein (ABG), clusterin (CLU), PRO2044, haptoglobin (HAP), proapolipoprotein A1 (proapo-A1) and retinol-binding protein 4 precursor (RBP4) (Table 1). Furthermore, MS analysis revealed complement C3c (C3) as additional host-specific protein, which found to be immunoreactive and also five non-host factors (A1, A2, A3, A4, and A5), which corresponded to antigens from *Acinetobacter lwoffii, Burkholderia multivorans, Myxococcus* *xanthus, Laribacter hongkongensis*, and *Streptococcus salivarius*. In total, we identified 13 host and non-host specific protein spots, which were subsequently subjected to MS analysis (MALDI-TOF/TOF). Data regarding the identification (ID), MASCOT accession number, isoelectric point (pI), and molecular mass (Mr) for each protein are presented in Table 2 and 3.

Down-regulated host-specific proteins in OSCC patients

Comparing our candidate OSCC biomarkers to control samples, we found that LRG, ABG, CLU, and PRO2044 displayed 0.21-, 0.45-, 0.6-, and 0.63-fold down regulation, respectively (p<0.05). This indicated that CLU and PRO2044 represented the most significantly decreased protein spots. Notably, all of these down-regulated factors have been previously studied and could be important with regard to OSCC.

LRG is a protein that has been observed in patients with bacterial infections [19], severe acute respiratory syndrome [20], as well as various malignancies, including pancreatic [21], liver [22], and lung [23] cancers. Moreover, LRG has been suggested to play an anti-apoptotic role during stress. Nevertheless, Weivoda et al. [24] has reported low levels of LRG in patients with inflammatory arthritis, in spite of the fact that LRG can be produced in response to inflammation. Overall, the relevance of LRG in distinct patient groups remains ill defined.

ABG is one of the eight host-specific proteins observed in this study. Interestingly, decreased ABG expression has also been observed in pancreatic cancer [25]. On the other hand, up-regulated ABG was previously reported in bladder cancer [26], non-small cell lung cancer [27], and squamous cell carcinoma of the uterine cervix [28].

With regard to CLU, differential expression has been linked to oncogenesis and tumor growth in bladder [29], breast [30], colorectal [31], ovarian [14], pancreatic [32], and prostate [33] cancers. However, although many researchers have reported

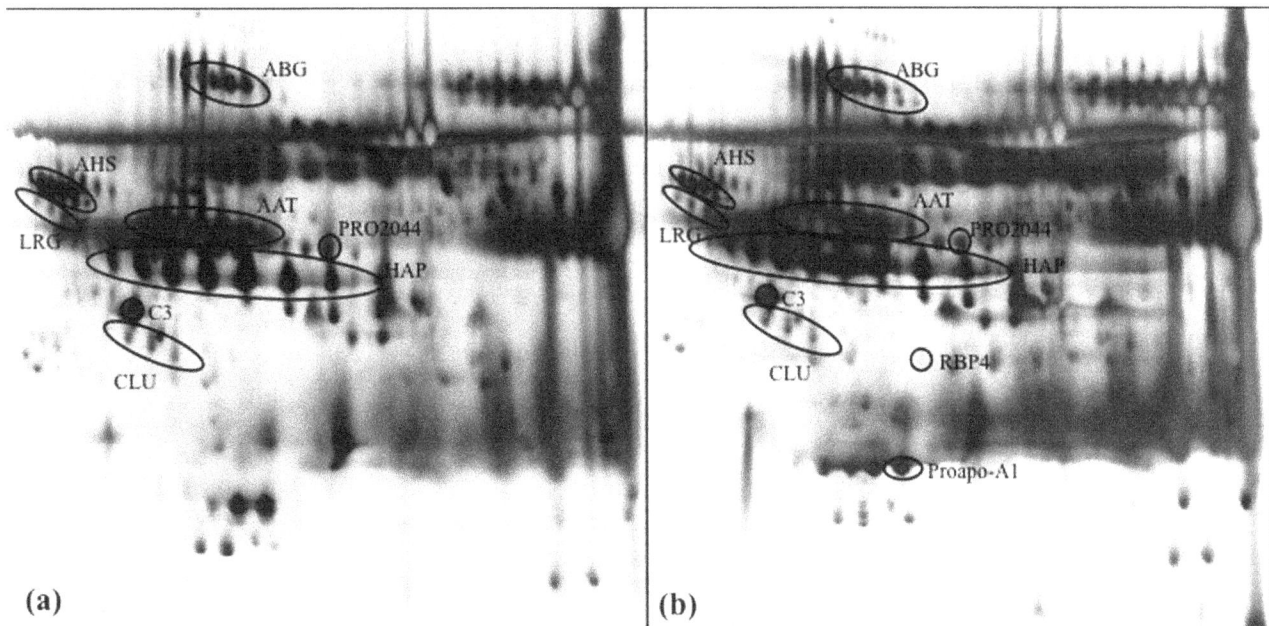

Figure 1. Representative 2-DE serum protein profiles of normal controls and OSCC patients. Unfractionated serum samples of (a) normal controls and (b) OSCC patients were subjected to 2-DE and silver staining. The labeled spot clusters are proteins which are consistently identified in profiles of normal controls and OSCC patients. α2-HS-glycoprotein (AHS) and α1-antitrysin (AAT) are high abundance proteins that typically appeared in protein profiles.

Table 1. The relative expression of host specific proteins among the sera of patients.

Protein Entry Name	Protein Name	Fold Change
LRG	Leucine-rich alpha 2-glycoprotein precursor	0.21 (down)
ABG	Alpha-1-B-glycoprotein	0.45 (down)
CLU	Clusterin	0.60 (down)
PRO2044	PRO2044	0.63 (down)
HAP	Haptoglobin	1.47 (up)
Proapo-A1	Proapolipoprotein	1.82 (up)
RBP4	Retinol binding protein 4	2.66 (up)

Fold change measures the degree of change in the protein of the OSCC patients (n = 25) when compared to normal controls (n = 25). This is measured by dividing the average spot intensity in the patients by the average spot intensity in the controls. (up) represents up-regulated expression whilst (down) represents down-regulated expression of protein spot.

increased CLU expression during tumorigenesis, we have observed significantly down-regulated CLU levels in sera from OSCC patients. Nevertheless, other reports have also described a similar loss of CLU expression in various tumors, including prostate cancer [34], pancreatic cancer [32], esophageal squamous cell carcinoma [35] and neuroblastoma [36]. Notably, these discrepancies may stem from the fact that there are differentially expressed CLU isoforms in human tissues and fluids that may exhibit distinct functions in tumors [14,37]. Even though it has been suggested that CLU down regulation could be associated with disease progression [32,38], this may depend on the type of cancer [39,40]. Notably, the true function of CLU has remained elusive despite extensive investigation. So far, CLU has been proposed to participate in the immediate cellular response to stress, which regulates cell growth and survival [41]. In this regard, its function appears isoform dependent, with both proapoptotic and anti-apoptotic forms [34].

Finally, our analysis revealed that PRO2044 (the C-terminal fragment of albumin, ALB) was the most down-regulated protein in the sera of OSCC patients. Interestingly, Kawakami et al. [22] observed that PRO2044 was also down regulated in hepatocellular carcinoma patients following curative radiofrequency ablation. In contrast, Jin et al. [42] reported that PRO2044 levels were increased in the cerebrospinal fluid of patients with Guillain–Barré syndrome, an acute inflammatory autoimmune disorder of the peripheral nervous system.

Up-regulated host-specific proteins in OSCC patients

In addition to down-regulated proteins, up-regulated host-specific antigens were also identified in OSCC patients, including HAP, proapo-A1, and RBP4. Compared to the control samples, these proteins displayed 1.47-, 1.82-, and 2.66-fold increases, respectively (p<0.05). RBP4 was found to be the most up-regulated.

Notably, a correlation between HAP expression and malignancies has been reported [43,44]. Indeed, similar to a previous study by Lai et al. [45], we found that HAP levels were significantly increased in sera from OSCC patients. Furthermore, since HAP is primarily hepatocyte-produced, a rise in its levels may indicate the occurrence of an acute phase response in OSCC [44]. Moreover, HAP has also been reported to participate in cell migration, extracellular matrix degradation, and arterial restructuring, suggesting its possible role in cancer [46]. In addition, HAP might act as an angiogenic agent that contributes to endothelial cell growth and differentiation [47].

With regard to proapo-A1 (or apo-A1), our findings are consistent with studies that have found elevated expression in various malignancies, including breast [48], colorectal [49], non-small cell lung [50] pancreatic [51], and hepatocellular [52] cancers. It is possible that increases in proapo-A1 might stem from reduced activity of proapo-A1 cleaving enzyme or higher turnover of apo-A1 [50,53].

We also observed an elevation of RBP4 levels in OSCC. It has been suggested that RBP4 over expression in cancer cells could result from an inhibition of phosphatidylinositol-3 kinase (PI3K) activity [54,55]. Moreover, RBP4 expression might be related to retinoid depletion, which is a common feature in cancer patients [56–58]. In addition, RBP4 levels can be influenced by transthyretin, which could reduce renal clearance of RBP4 [59].

Non-host specific proteins detected in OSCC patients

In addition to the above down- and up-regulated self-antigens, we also identified five non-host markers in the proteomic profiles of OSCC patients. These proteins were derived from various bacteria, including *A. lwoffii*, *B. multivorans*, *M. xanthus*, *L. hongkogenisis*, and *S. salivarius*. Although our detection of these factors could have resulted from nosocomial infections [60,61], heightened risk of malignancy has been linked to viruses, bacteria, and schistosomes [62,63]. In fact, the relationship between bacterial infections and cancer has been discussed for decades [64,65]. In this regard, there are several possible mechanisms by which bacteria could be oncogenic. For example, altered host responses during bacterial infection (e.g., chronic inflammation, antigen-driven lymphoproliferation, and hormone induction that promotes epithelial cell proliferation) have been suggested to influence oncogenesis [66]. Also, bacterial infection can lead to the production of toxin and/or carcinogenic metabolites that enhance oncogenesis [67]. In contrast, bacterial infections have also been suggested to play a protective role by altering host physiology and reducing cancer risk [68].

Oral cancer is considered to be a multi-factorial disease, as it can stem from exposure to several types of carcinogens, including microbial factors [69]. Indeed, several pieces of evidence have supported the association of microbial infection with oncogenesis. For instance, *Helicobacter pylori* has been linked to gastric cancer [70] and categorized by the World Health Organization International Agency for Research on Cancer (IARC) as a carcinogenic factor in humans [71–73]. Similarly, *Chlamydophila pneumoniae* has been associated with malignant lymphoma and lung cancer in males [74,75], whereas *Candida albicans* and *Streptococcus anginosus* have been linked to oral carcinoma [70–

Table 2. Mass spectrometric identification of host-specific protein spots from serum protein profiles using MASCOT search engine and the NCBI database.

Protein Name	MASCOT accession number	pI	Theorectical mass	Sequence coverage	Search score	Queries match	Expected value
Alpha-1-B-glycoprotein – human [Homo sapiens]	gi|69990	5.65	52479	40%	824	24	9.1e-078
Leucine-rich alpha-2-glycoprotein precursor [Homo sapiens]	gi|1641846	6.45	38382	40%	601	18	1.8e-055
Clusterin [Homo sapiens]	gi|2666585	5.60	16267	11%	39	4	29
Retinol binding protein 4 [Homo sapiens]	gi|1808832	5.76	23371	48%	276	14	2.9e-021
PRO2044 [Homo sapiens]	gi|6650826	6.97	39984	45%	355	17	3.7e-029
Haptoglobin [Homo sapiens]	gi|3337390	6.14	38722	32%	264	11	4.6e-020
Proapolipoprotein [Homo sapiens]	gi|178775	5.45	28944	53%	503	20	1.1e-045
Chain B, Human Complement Component C3 [Homo sapines]	gi|7810126	5.55	114238	20%	469	24	1.5e-040

Table 3. Mass spectrometric identification of non-host specific protein spots from serum protein profiles using MASCOT search engine and the NCBI database.

Protein Name	MASCOT accession number	pI	Theorectical mass	Sequence coverage	Search score	Queries match	Expected value
(A1) Predicted protein [Acinetobacter lwoffii SH145]	gi|262375905	8.89	9185	83%	56	12	30
(A2) Hypothetical protein BURMUCGD2M_4365 [Burkholderia multivorans CGD2M]	gi|221195969	8.31	4546	92%	51	7	1e+002
(A3) Hypothetical protein MXAN_1050 [Myxococcus xanthus DK 1622]	gi|108761930	5.05	10984	36%	52	8	82
(A4) Hypothetical protien LHK_003399 [Laribacter hongkongensis HLHK9]	gi|226939330	9.50	6340	68%	44	6	4.2e+002
(A5) Hemolysin A [Streptococcus salivarius SK126]	gi|228476878	5.47	30333	20%	50	10	1.2e+002

76]. It has also been demonstrated that OSCC patients possess significantly elevated concentrations of certain bacteria in their saliva. Thus, changes in salivary microflora could represent a noninvasive diagnostic tool for predicting oral cancer [70].

Confirmation of host-specific proteins by western blotting

Immunoglobulin M (IgM) antibodies are present in the circulation of normal humans and other mammalian species. IgM is initially secreted by B cells upon primary antigen stimulation [77,78] and participates in natural defenses against foreign pathogens as well as neoplastic cells and tumors [79]. In fact, autoantibodies against specific cancer antigens have been identified for several types of tumors, including colon, breast, lung, ovary, prostate, and head and neck. These antibodies have been found to recognize several overexpressed (e.g., Her2), mutated (e.g., p53), or tissue-restricted (e.g., testis-cancer antigens) proteins, which are produced by cancer cells and elicit immune responses [6]. Therefore, detection of such autoantibodies in patient sera could be exploited as a means of cancer diagnosis. Indeed, the specificity and sensitivity of the antibody response to low antigen levels make it an ideal screening/diagnostic tool for early identification of cancer biomarkers in serum-based assays.

In order to extend the results obtained from our immunoproteomics analyses, we performed 2-DE immunoblots using OSCC patient and control sera (Table 4). In this regard, we performed immunoblots based on the following four conditions (i.e., categories 1–4): (1) normal pooled sera probed with normal pooled sera; (2) normal pooled sera probed with OSCC pooled sera; (3) OSCC pooled sera probed with normal pooled sera; and (4) OSCC pooled sera probed with OSCC pooled sera.

The use of normal pooled sera against normal and OSCC pooled sera was to prove that the reaction was restricted to the tumor specificity. Therefore, only few host-specific proteins could be detected in the normal controls. Based on our results, only proapo-A1 could be detected in category 1 (Fig. 2a), while HAP showed immunogenicity in category 2 blots (Fig. 2b). The detection of HAP in category 2 blot reveals the supportive role

of the natural immunity against cancer cells apart from its involvement in the defense system against pathogens [79].

Based on the analysis in category 3 (Fig. 2C), our results show that the healthy control serums have the autoantibodies against OSCC serum antigens, CLU, C3, proapo-A1, and RBP4, which are the host-specific proteins that showed the most immunoreactivity in the immunoblot. These aberrant host-specific proteins were found to be more immunoreactive in the OSCC sera compared to the normal sera. In addition, other antigenic protein spots (A1, A2, A3, A4, and A5) were also detected in the category 3.

Finally, four reactive protein spots were observed in category 4, corresponding to C3, HAP, proapo-A1, and RBP4 (Fig. 2d). Therefore, the identified proteins are uniquely produced by the innate immune response of cancer cells. Altogether, five host-specific proteins were found to be immunoreactive in OSCC patients: CLU, C3, HAP, proapo-A1 and RBP4. Based on these analyses, our data support the rationale of using host-specific proteins as cancer biomarkers panel for early detection and diagnosis of OSCC [11].

Functional annotation and protein interaction analysis

Functional annotation analysis was performed for our eight candidate host-specific biomarkers using DAVID v6.7 (http://david.abcc.ncifcrf.gov/) (Table 5). An analysis of biological processes revealed that HAP, apo-A1, and RBP4 are involved in homeostasis. Indeed, cancer-related inflammation is considered to be an essential hallmark of cancer due to the tumor-promoting consequences of inflammatory responses [80]. Thus, the study of aberrant homeostatic mechanisms has indicated that there exists an interaction between cancer cells and host immune cells during carcinogenesis.

In addition, functional annotation analyses revealed that all of the host-specific proteins were located within the extracellular region and indicating that these proteins mediate immunogenic reactions against cancer cells. Therefore, this finding shows that these proteins could play an important role in the immunogenicity of carcinogenesis. Moreover, CLU, HAP, and apo-A1 were found to be associated with protein–lipid complexes that are responsible

Table 4. Host- and non-host specific proteins on the 2-DE immunoblots.

Antigenic Proteins	Category			
	(a)	(b)	(c)	(d)
1) Host specific proteins:				
CLU	–	–	/	–
HAP	–	/	–	/
C3	–	–	/	/
Proapo-A1	/	–	/	/
RBP4	–	–	/	/
2) Non-host specific proteins:				
Predicted protein [Acinetobacter lwoffii SH145]	–	–	/	–
Hypothetical protein BURMUCGD2M_4365 [Burkholderia multivorans CGD2M]	–	–	/	–
Hypothetical protein MXAN_1060 [Myxococcus xanthus DK 1622]	–	–	/	–
Hypothetical protein LHK_0039 [Laribacter hongkongensis HLHK9]	–	–	/	–
Hemolysin A [Streptococcus salivarius SK126]	–	–	/	–

/Proteins of the patients or normal pooled serum recognized by the primary antibody.
–Proteins of the patients or normal pooled serum not recognized by the primary antibody.

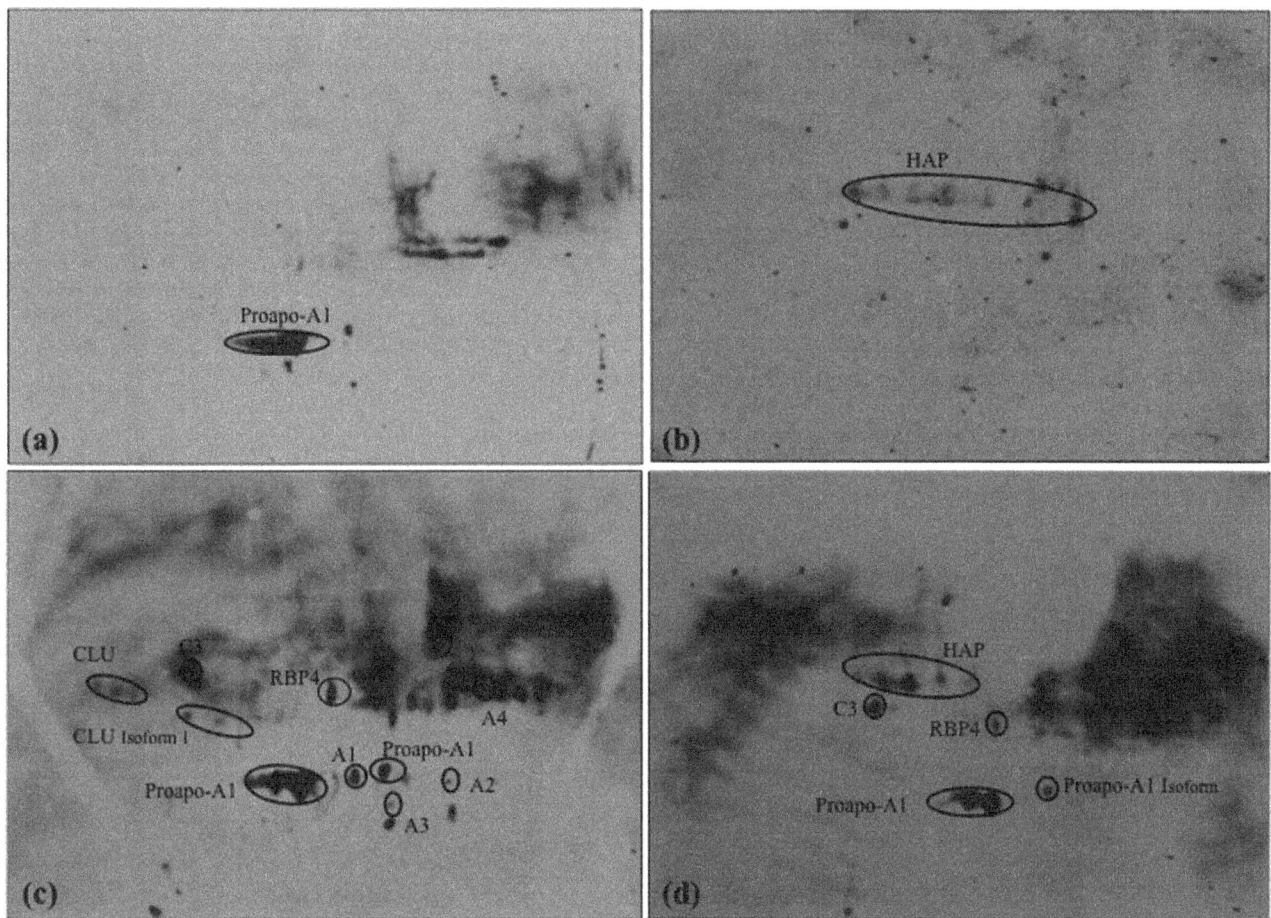

Figure 2. Results from 2-DE immunoblots for (a) normal pooled sera probed with normal pooled sera, (b) normal pooled sera probed with OSCC pooled sera, (c) OSCC pooled sera probed with normal pooled sera, (d) OSCC pooled sera probed with OSCC pooled sera. Unfractionated, pooled serum samples from control and OSCC patients were subjected to 2-DE and blotted onto nitrocellulose membranes, which were then probed with pooled sera and monoclonal anti-human IgM-HRP.

for cholesterol and lipid transportation. In this regard, cholesterol acts as a key contributor to carcinogenesis by promoting cell migration and mediating inflammatory processes. Indeed, lipoproteins may be the major suppliers of cholesterol to cancer cells via receptor-mediated mechanisms [81]. Therefore, alterations in

these proteins could alter lipid metabolism, increasing the risk of cancer development and progression.

A protein interaction network was generated using the STRING v9.1 (http://string-db.org/) database to identify potential binding partners for our host-specific biomarkers (Fig. 3). Interestingly, most of the host-specific proteins could be linked to

Table 5. Functional annotation analysis of identified host-specific proteins using DAVID v6.7.

	Term Name	Enrichment Score[+]	Protein Count (%)	Protein Entry Name	p value
Biological Process	Chemical process	2.28	3 (37.5)	APOA1, HAP, RBP4	1.9e-2
	Homeostasis/Hemostasisprocess	2.28	3 (37.5)	APOA1, HAP, RBP4	4.0e-2
Cellular Component	Extracellular region	5.05	8 (100)	A1BG, ALB, APOA1, C3, CLU, HAP, LRG1, RBP4	2.4e-6
	Spherical high-density lipoprotein particle	4.22	3 (37.5)	APOA1, CLU, HAP	7.2e-6
	High-density lipoprotein particle	4.22	3 (37.5)	APOA1, CLU, HAP	7.7e-5
	Plasma lipoprotein particle	4.22	3 (37.5)	APOA1, CLU, HAP	1.5e-4
	Protein-lipid complex	3.92	3 (37.5)	APOA1, CLU, HAP	1.5e-4

[+]The classification stringency was set to high.

ten predicted functional partners (ABCA1, apo-A2, apo-B, apo-C3, CFH, CFI, CR2, FN1, LCAT, and TTR). Moreover, when KEGG pathway analysis was performed using DAVID v6.7, complement and coagulation cascades as well as the peroxisome proliferator-activated receptor (PPAR) signaling pathway could be linked to our identified host-specific proteins. Notably, these two pathways have been predicted to play important roles in cancer biology [82].

There were a total of four proteins (C3, CFH, CFI, and CR2) found to be involved in complement and coagulation cascades ($p = 1.9 \times 10^{-4}$). These pathways may participate in homeostatic processes that influence the host defense response. Indeed, the complement cascade is activated early in the immune response and might be important in cancer immunotherapy, as evidence has suggested that activation of complement regulators can promote tumor growth [83]. Activation of the coagulation cascade during cancer highlights how altered homeostasis can contribute to tumorigenesis. Indeed, it was reported that coagulation pathways could induce angiogenesis to facilitate tumor growth and metastasis [84]. As stated in the table 5 based on DAVID functional annotation analysis, "homeostasis" could be correlated with the proteins ApoA1, HAP and RBP4. However, it was

reported that ApoA1, HAP have the possibility to relate with the process "hemostasis" [85,86].

Furthermore, apo-A1, apo-A2, and apo-C3 were found to be associated with the PPAR signaling pathway ($p = 6.1 \times 10^{-4}$). PPARs are members of the nuclear hormone superfamily, which can be activated by fatty acids and their derivatives. The PPAR signaling pathway regulates lipid metabolism, cell proliferation, and cell differentiation. It has also been reported to be involved in the regulation of cancer cell apoptosis, proliferation, and differentiation. PPAR signaling is proposed to modify tumor growth by affecting angiogenesis, inflammation, and immune cell functions in the tumor cell environment [87]. Thus, targeting the PPAR signaling pathway could represent a potential strategy for cancer therapy.

Conclusions

Using proteomic profiling, we were able to identify several differentially expressed host-specific proteins in the sera of OSCC patients, including LRG, ABG, CLU, PRO2044, HAP, proapo-A1, and RBP4. The immunogenicity of five of these proteins was further confirmed by western blot analyses (CLU, C3, HAP, proapo-A1 and RBP4). In addition, five non-host factors were

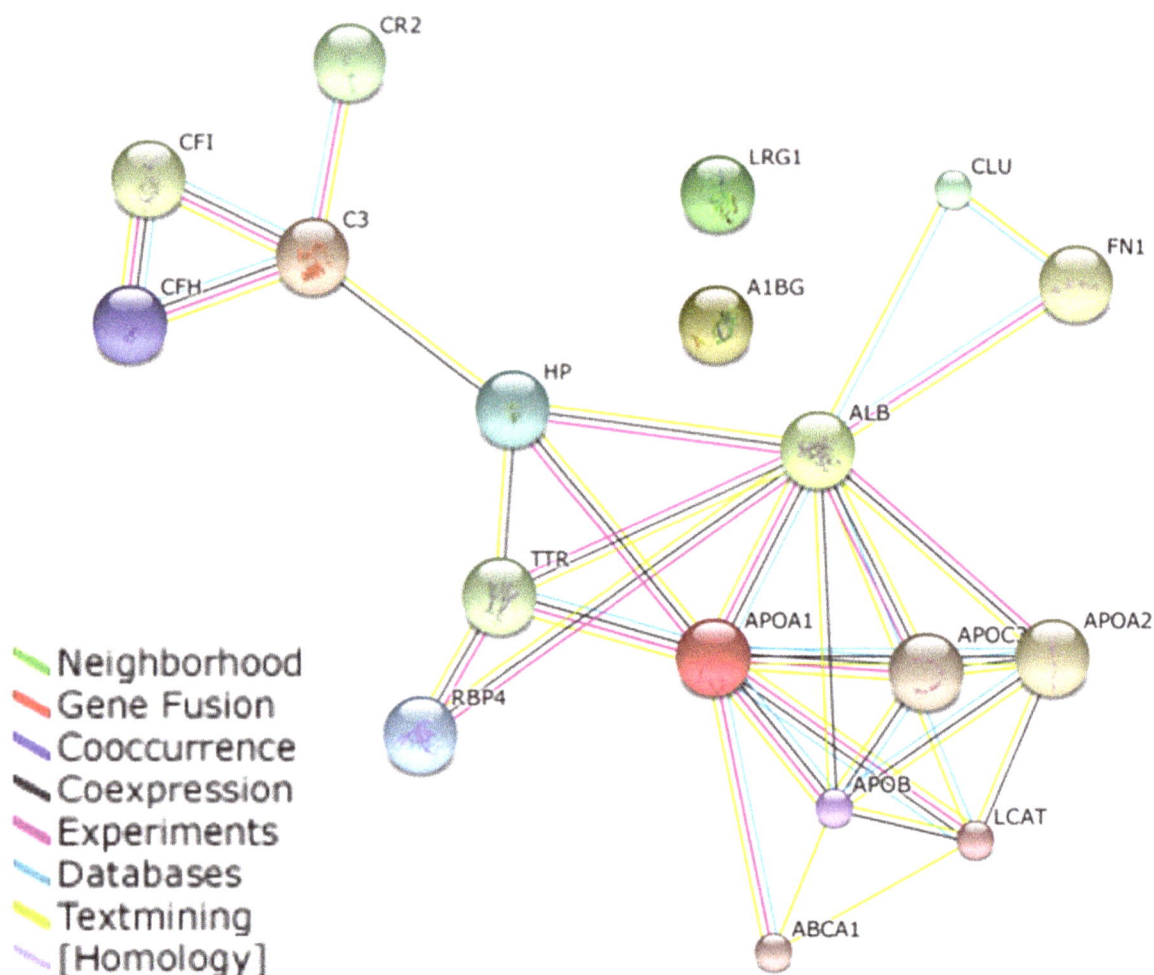

Figure 3. Interaction networks of identified host specific proteins using STRING v9.1. STRING database is a curated knowledge database of known and predicted protein-protein interactions. Most of the identified host-specific proteins have an established link with each other in the interaction network.

detected, including proteins from *A. lwoffii*, *B. multivorans*, *M. xanthus*, *L. hongkogenisis*, and *S. salivarius*. As previously suggested [88], combined proteomic and serological approaches, such as the one used in the present study, can reflect numerous events occurring *in vivo* simultaneously due to the fact that patient serum is complex and consists of many proteins. Our data indicate that immunoproteomics approach could be a promising application for biomarker discovery and disease progression. Our study could be used as a landmark in a more comprehensive study and can be applied to individual patient serums or to a larger sample size. Using these methods, we have identified distinct serum biomarkers that might facilitate the development of early diagnostic tools for OSCC and promote further understanding the 'host responses' that occur in OSCC patients.

References

1. Parkin DM, Bray F, Ferlay J, Pisani P (2005) Global cancer statistics, 2002. CA Cancer J Clin 55: 74–108.
2. Arellano-Garcia ME, Li R, Liu X, Xie Y, Yan X, et al. (2010) Identification of tetranectin as a potential biomarker for metastatic oral cancer. Int J Mol Sci 11: 3106–3121.
3. Jemal A, Bray F, Center MM, Ferlay J, Ward E, et al. (2011) Global cancer statistics. CA Cancer J Clin 61: 69–90.
4. Neville BW, Day TA (2002) Oral cancer and precancerous lesions. CA Cancer J Clin 52: 195–215.
5. Forastiere AA, Goepfert H, Maor M, Pajak TF, Weber R, et al. (2003) Concurrent chemotherapy and radiotherapy for organ preservation in advanced laryngeal cancer. N Engl J Med 349: 2091–2098.
6. Lin HS, Talwar HS, Tarca AL, Ionan A, Chatterjee M, et al. (2007) Autoantibody approach for serum-based detection of head and neck cancer. Cancer Epidemiol Biomarkers Prev 16: 2396–2405.
7. da Silva SD, Ferlito A, Takes RP, Brakenhoff RH, Valentin MD, et al. (2011) Advances and applications of oral cancer basic research. Oral Oncol 47: 783–791.
8. Whitmore SE, Lamont RJ (2014) Oral bacteria and cancer. PLoS Pathog 10: e1003933.
9. Lingen MW, Kalmar JR, Karrison T, Speight PM (2008) Critical evaluation of diagnostic aids for the detection of oral cancer. Oral Oncol 44: 10–22.
10. Day GL, Blot WJ (1992) Second primary tumors in patients with oral cancer. Cancer 70: 14–19.
11. Mou Z, He Y, Wu Y (2009) Immunoproteomics to identify tumor-associated antigens eliciting humoral response. Cancer Lett 278: 123–129.
12. Arnott D, Emmert-Buck MR (2010) Proteomic profiling of cancer-opportunities, challenges, and context. J Pathol 222: 16–20.
13. Goldszmid RS, Dzutsev A, Trinchieri G (2014) Host immune response to infection and cancer: unexpected commonalities. Cell Host Microbe 15: 295–305.
14. Chen Y, Lim BK, Peh SC, Abdul-Rahman PS, Hashim OH (2008) Profiling of serum and tissue high abundance acute-phase proteins of patients with epithelial and germ line ovarian carcinoma. Proteome Sci 6: 20.
15. Heukeshoven J, Dernick R (1988) Improved silver staining procedure for fast staining in PhastSystem Development Unit. I. Staining of sodium dodecyl sulfate gels. Electrophoresis 9: 28–32.
16. Shevchenko A, Wilm M, Vorm O, Mann M (1996) Mass spectrometric sequencing of proteins silver-stained polyacrylamide gels. Anal Chem 68: 850–858.
17. Huang da W, Sherman BT, Lempicki RA (2009) Systematic and integrative analysis of large gene lists using DAVID bioinformatics resources. Nat Protoc 4: 44–57.
18. Franceschini A, Szklarczyk D, Frankild S, Kuhn M, Simonovic M, et al. (2013) STRING v9.1: protein-protein interaction networks, with increased coverage and integration. Nucleic Acids Res 41: D808–815.
19. Bini L, Magi B, Marzocchi B, Cellesi C, Berti B, et al. (1996) Two-dimensional electrophoretic patterns of acute-phase human serum proteins in the course of bacterial and viral diseases. Electrophoresis 17: 612–616.
20. Chen JH, Chang YW, Yao CW, Chiueh TS, Huang SC, et al. (2004) Plasma proteome of severe acute respiratory syndrome analyzed by two-dimensional gel electrophoresis and mass spectrometry. Proc Natl Acad Sci U S A 101: 17039–17044.
21. Kakisaka T, Kondo T, Okano T, Fujii K, Honda K, et al. (2007) Plasma proteomics of pancreatic cancer patients by multi-dimensional liquid chromatography and two-dimensional difference gel electrophoresis (2D-DIGE): up-regulation of leucine-rich alpha-2-glycoprotein in pancreatic cancer. J Chromatogr B Analyt Technol Biomed Life Sci 852: 257–267.
22. Kawakami T, Hoshida Y, Kanai F, Tanaka Y, Tateishi K, et al. (2005) Proteomic analysis of sera from hepatocellular carcinoma patients after radiofrequency ablation treatment. Proteomics 5: 4287–4295.
23. Okano T, Kondo T, Kakisaka T, Fujii K, Yamada M, et al. (2006) Plasma proteomics of lung cancer by a linkage of multi-dimensional liquid chromatography and two-dimensional difference gel electrophoresis. Proteomics 6: 3938–3948.
24. Weivoda S, Andersen JD, Skogen A, Schlievert PM, Fontana D, et al. (2008) ELISA for human serum leucine-rich alpha-2-glycoprotein-1 employing cytochrome c as the capturing ligand. J Immunol Methods 336: 22–29.
25. Li C, Zolotarevsky E, Thompson I, Anderson MA, Simeone DM, et al. (2011) A multiplexed bead assay for profiling glycosylation patterns on serum protein biomarkers of pancreatic cancer. Electrophoresis 32: 2028–2035.
26. Kreunin P, Zhao J, Rosser C, Urquidi V, Lubman DM, et al. (2007) Bladder cancer associated glycoprotein signatures revealed by urinary proteomic profiling. J Proteome Res 6: 2631–2639.
27. Liu Y, Luo X, Hu H, Wang R, Sun Y, et al. (2012) Integrative proteomics and tissue microarray profiling indicate the association between overexpressed serum proteins and non-small cell lung cancer. PLoS One 7: e51748.
28. Jeong DH, Kim HK, Prince AE, Lee DS, Kim YN, et al. (2008) Plasma proteomic analysis of patients with squamous cell carcinoma of the uterine cervix. J Gynecol Oncol 19: 173–180.
29. Stejskal D, Fiala RR (2006) Evaluation of serum and urine clusterin as a potential tumor marker for urinary bladder cancer. Neoplasma 53: 343–346.
30. Redondo M, Villar E, Torres-Munoz J, Tellez T, Morell M, et al. (2000) Overexpression of clusterin in human breast carcinoma. Am J Pathol 157: 393–399.
31. Rodriguez-Pineiro AM, de la Cadena MP, Lopez-Saco A, Rodriguez-Berrocal FJ (2006) Differential expression of serum clusterin isoforms in colorectal cancer. Mol Cell Proteomics 5: 1647–1657.
32. Xie MJ, Motoo Y, Su SB, Mouri H, Ohtsubo K, et al. (2002) Expression of clusterin in human pancreatic cancer. Pancreas 25: 234–238.
33. Zellweger T, Chi K, Miyake H, Adomat H, Kiyama S, et al. (2002) Enhanced radiation sensitivity in prostate cancer by inhibition of the cell survival protein clusterin. Clin Cancer Res 8: 3276–3284.
34. Scaltriti M, Brausi M, Amorosi A, Caporali A, D'Arca D, et al. (2004) Clusterin (SGP-2, ApoJ) expression is downregulated in low- and high-grade human prostate cancer. Int J Cancer 108: 23–30.
35. Zhang LY, Ying WT, Mao YS, He HZ, Liu Y, et al. (2003) Loss of clusterin both in serum and tissue correlates with the tumorigenesis of esophageal squamous cell carcinoma via proteomics approaches. World J Gastroenterol 9: 650–654.
36. Santilli G, Aronow BJ, Sala A (2003) Essential requirement of apolipoprotein J (clusterin) signaling for IkappaB expression and regulation of NF-kappaB activity. J Biol Chem 278: 38214–38219.
37. Wei L, Xue T, Wang J, Chen B, Lei Y, et al. (2009) Roles of clusterin in progression, chemoresistance and metastasis of human ovarian cancer. Int J Cancer 125: 791–806.
38. Wu J, Xie X, Nie S, Buckanovich RJ, Lubman DM (2013) Altered expression of sialylated glycoproteins in ovarian cancer sera using lectin-based ELISA assay and quantitative glycoproteomics analysis. J Proteome Res 12: 3342–3352.
39. Lourda M, Trougakos IP, Gonos ES (2007) Development of resistance to chemotherapeutic drugs in human osteosarcoma cell lines largely depends on up-regulation of Clusterin/Apolipoprotein J. Int J Cancer 120: 611–622.
40. Redondo M, Rodrigo I, Alcaide J, Tellez T, Roldan MJ, et al. (2010) Clusterin expression is associated with decreased disease-free survival of patients with colorectal carcinomas. Histopathology 56: 932–936.
41. Trougakos IP, Lourda M, Agiostratidou G, Kletsas D, Gonos ES (2005) Differential effects of clusterin/apolipoprotein J on cellular growth and survival. Free Radic Biol Med 38: 436–449.
42. Jin T, Hu LS, Chang M, Wu J, Winblad B, et al. (2007) Proteomic identification of potential protein markers in cerebrospinal fluid of GBS patients. Eur J Neurol 14: 563–568.
43. Kuhajda FP, Katumuluwa AI, Pasternack GR (1989) Expression of haptoglobin-related protein and its potential role as a tumor antigen. Proc Natl Acad Sci U S A 86: 1188–1192.

Acknowledgments

The authors acknowledged the Oral Cancer Research and Coordinating Centre (OCRCC), University of Malaya (UM) for providing serum samples.

Author Contributions

Conceived and designed the experiments: YC SNA. Performed the experiments: YC SNA JPK. Analyzed the data: YC RBZ YNC YLW SCBG. Contributed reagents/materials/analysis tools: YC. Contributed to the writing of the manuscript: YC SNA JPK YLW SCBG.

44. Ahmed N, Barker G, Oliva KT, Hoffmann P, Riley C, et al. (2004) Proteomic-based identification of haptoglobin-1 precursor as a novel circulating biomarker of ovarian cancer. Br J Cancer 91: 129–140.

45. Lai CH, Chang NW, Lin CF, Lin CD, Lin YJ, et al. (2010) Proteomics-based identification of haptoglobin as a novel plasma biomarker in oral squamous cell carcinoma. Clin Chim Acta 411: 984–991.

46. Zhao C, Annamalai L, Guo C, Kothandaraman N, Koh SC, et al. (2007) Circulating haptoglobin is an independent prognostic factor in the sera of patients with epithelial ovarian cancer. Neoplasia 9: 1–7.

47. Ye B, Cramer DW, Skates SJ, Gygi SP, Pratomo V, et al. (2003) Haptoglobin-alpha subunit as potential serum biomarker in ovarian cancer: identification and characterization using proteomic profiling and mass spectrometry. Clin Cancer Res 9: 2904–2911.

48. Huang HL, Stasyk T, Morandell S, Dieplinger H, Falkensammer G, et al. (2006) Biomarker discovery in breast cancer serum using 2-D differential gel electrophoresis/MALDI-TOF/TOF and data validation by routine clinical assays. Electrophoresis 27: 1641–1650.

49. Yu B, Li SY, An P, Zhang YN, Liang ZJ, et al. (2004) Comparative study of proteome between primary cancer and hepatic metastatic tumor in colorectal cancer. World J Gastroenterol 10: 2652–2656.

50. Huang LJ, Chen SX, Huang Y, Luo WJ, Jiang HH, et al. (2006) Proteomics-based identification of secreted protein dihydrodiol dehydrogenase as a novel serum markers of non-small cell lung cancer. Lung Cancer 54: 87–94.

51. Mikuriya K, Kuramitsu Y, Ryozawa S, Fujimoto M, Mori S, et al. (2007) Expression of glycolytic enzymes is increased in pancreatic cancerous tissues as evidenced by proteomic profiling by two-dimensional electrophoresis and liquid chromatography-mass spectrometry/mass spectrometry. Int J Oncol 30: 849–855.

52. Wang HY (2007) Laser capture microdissection in comparative proteomic analysis of hepatocellular carcinoma. Methods Cell Biol 82: 689–707.

53. Harn HJ, Chen YL, Lin PC, Cheng YL, Lee SC, et al. (2010) Exploration of Potential Tumor Markers for Lung Adenocarcinomas by Two-Dimensional Gel Electrophoresis Coupled with Nano-LC/MS/MS. Journal of the Chinese Chemical Society 57: 180–188.

54. Kuppumbatti YS, Rexer B, Nakajo S, Nakaya K, Mira-y-Lopez R (2001) CRBP suppresses breast cancer cell survival and anchorage-independent growth. Oncogene 20: 7413–7419.

55. Farias EF, Marzan C, Mira-y-Lopez R (2005) Cellular retinol-binding protein-I inhibits PI3K/Akt signaling through a retinoic acid receptor-dependent mechanism that regulates p85-p110 heterodimerization. Oncogene 24: 1598–1606.

56. Kuppumbatti YS, Bleiweiss IJ, Mandeli JP, Waxman S, Mira YLR (2000) Cellular retinol-binding protein expression and breast cancer. J Natl Cancer Inst 92: 475–480.

57. Reynolds CP, Matthay KK, Villablanca JG, Maurer BJ (2003) Retinoid therapy of high-risk neuroblastoma. Cancer Lett 197: 185–192.

58. Lorkova L, Pospisilova J, Lacheta J, Leahomschi S, Zivny J, et al. (2012) Decreased concentrations of retinol-binding protein 4 in sera of epithelial ovarian cancer patients: a potential biomarker identified by proteomics. Oncol Rep 27: 318–324.

59. Kotnik P, Fischer-Posovszky P, Wabitsch M (2011) RBP4: a controversial adipokine. Eur J Endocrinol 165: 703–711.

60. Forster DH, Daschner FD (1998) Acinetobacter species as nosocomial pathogens. Eur J Clin Microbiol Infect Dis 17: 73–77.

61. Velasco E, Byington R, Martins CS, Schirmer M, Dias LC, et al. (2004) Bloodstream infection surveillance in a cancer centre: a prospective look at clinical microbiology aspects. Clin Microbiol Infect 10: 542–549.

62. Pisani P, Parkin DM, Munoz N, Ferlay J (1997) Cancer and infection: estimates of the attributable fraction in 1990. Cancer Epidemiol Biomarkers Prev 6: 387–400.

63. de Martel C, Ferlay J, Franceschi S, Vignat J, Bray F, et al. (2012) Global burden of cancers attributable to infections in 2008: a review and synthetic analysis. Lancet Oncol 13: 607–615.

64. Parsonnet J (1995) Bacterial infection as a cause of cancer. Environ Health Perspect 103 Suppl 8: 263–268.

65. Beebe JL, Koneman EW (1995) Recovery of uncommon bacteria from blood: association with neoplastic disease. Clin Microbiol Rev 8: 336–356.

66. Chang AH, Parsonnet J (2010) Role of bacteria in oncogenesis. Clin Microbiol Rev 23: 837–857.

67. Mager DL (2006) Bacteria and cancer: cause, coincidence or cure? A review. J Transl Med 4: 14.

68. Francois F, Roper J, Goodman AJ, Pei Z, Ghumman M, et al. (2008) The association of gastric leptin with oesophageal inflammation and metaplasia. Gut 57: 16–24.

69. Ogbureke KU, Bingham C (2012) Overview of Oral Cancer. In: Ogbureke KU, editor. Oral Cancer. Croatia: Intech. 3–20.

70. Hooper SJ, Wilson MJ, Crean SJ (2009) Exploring the link between microorganisms and oral cancer: a systematic review of the literature. Head Neck 31: 1228–1239.

71. Bjorkholm B, Falk P, Engstrand L, Nyren O (2003) Helicobacter pylori: resurrection of the cancer link. J Intern Med 253: 102–119.

72. Correa P, Houghton J (2007) Carcinogenesis of Helicobacter pylori. Gastroenterology 133: 659–672.

73. Peek RM Jr, Blaser MJ (2002) Helicobacter pylori and gastrointestinal tract adenocarcinomas. Nat Rev Cancer 2: 28–37.

74. Anttila TI, Lehtinen T, Leinonen M, Bloigu A, Koskela P, et al. (1998) Serological evidence of an association between chlamydial infections and malignant lymphomas. Br J Haematol 103: 150–156.

75. Kocazeybek B (2003) Chronic Chlamydophila pneumoniae infection in lung cancer, a risk factor: a case-control study. J Med Microbiol 52: 721–726.

76. Sasaki M, Yamaura C, Ohara-Nemoto Y, Tajika S, Kodama Y, et al. (2005) Streptococcus anginosus infection in oral cancer and its infection route. Oral Dis 11: 151–156.

77. Tchoudakova A, Hensel F, Murillo A, Eng B, Foley M, et al. (2009) High level expression of functional human IgMs in human PER.C6 cells. MAbs 1: 163–171.

78. Zouali M (2002) Antibodies. eLS.

79. Brandlein S, Pohle T, Ruoff N, Wozniak E, Muller-Hermelink HK, et al. (2003) Natural IgM antibodies and immunosurveillance mechanisms against epithelial cancer cells in humans. Cancer Res 63: 7995–8005.

80. Hanahan D, Weinberg RA (2011) Hallmarks of cancer: the next generation. Cell 144: 646–674.

81. Cruz PM, Mo H, McConathy WJ, Sabnis N, Lacko AG (2013) The role of cholesterol metabolism and cholesterol transport in carcinogenesis: a review of scientific findings, relevant to future cancer therapeutics. Front Pharmacol 4: 119.

82. Krupp M, Maass T, Marquardt JU, Staib F, Bauer T, et al. (2011) The functional cancer map: a systems-level synopsis of genetic deregulation in cancer. BMC Med Genomics 4: 53.

83. Kolev M, Towner L, Donev R (2011) Complement in cancer and cancer immunotherapy. Arch Immunol Ther Exp (Warsz) 59: 407–419.

84. Rickles FR, Patierno S, Fernandez PM (2003) Tissue factor, thrombin, and cancer. Chest 124: 58S–68S.

85. Talens S, Leebeek FW, Demmers JA, Rijken DC (2012) Identification of fibrin clot-bound plasma proteins. PLoS One 7: e41966.

86. Li D, Weng S, Yang B, Zander DS, Saldeen T, et al. (1999) Inhibition of arterial thrombus formation by ApoA1 Milano. Arterioscler Thromb Vasc Biol 19: 378–383.

87. Michalik L, Wahli W (2008) PPARs Mediate Lipid Signaling in Inflammation and Cancer. PPAR Res 2008: 134059.

88. Yeng C, Osman E, Mohamed Z, Noordin R (2010) Detection of immunogenic parasite and host-specific proteins in the sera of active and chronic individuals infected with Toxoplasma gondii. Electrophoresis 31: 3843–3849.

11

Altered Proteomic Polymorphisms in the Caterpillar Body and Stroma of Natural *Cordyceps sinensis* during Maturation

Yun-Zi Dong[1]❥, **Li-Juan Zhang**[1]❥, **Zi-Mei Wu**[1], **Ling Gao**[1], **Yi-Sang Yao**[1], **Ning-Zhi Tan**[1], **Jian-Yong Wu**[2,3], **Luqun Ni**[4], **Jia-Shi Zhu**[2,3,5]*

1 Pharmanex Beijing Clinical Pharmacology Center, Beijing, China, 2 Department of Applied Biology and Chemistry Technology, Hong Kong Polytechnic University, Hung Hom, Kowloon, Hong Kong, 3 Shenzhen TCM Pharmacy and Molecular Pharmacology Kay Laboratory, Hong Kong Polytechnic University, Shenzhen, Guangdong, China, 4 Department of Mechanical and Aerospace Engineering, University of California San Diego, La Jolla, CA, United States of America, 5 NS Center for Anti-Aging Research, Provo, UT, United States of America

Abstract

Objective: To examine the maturational changes in proteomic polymorphisms resulting from differential expression by multiple intrinsic fungi in the caterpillar body and stroma of natural *Cordyceps sinensis* (Cs), an integrated micro-ecosystem.

Methods: The surface-enhanced laser desorption/ionization time-of-flight mass spectrometry (SELDI-TOF MS) biochip technique was used to profile the altered protein compositions in the caterpillar body and stroma of Cs during its maturation. The MS chromatograms were analyzed using density-weighted algorithms to examine the similarities and cluster relationships among the proteomic polymorphisms of the Cs compartments and the mycelial products *Hirsutella sinensis* (Hs) and *Paecilomyces hepiali* (Ph). **Results:** SELDI-TOF MS chromatograms displayed dynamic proteomic polymorphism alterations among samples from the different Cs compartments during maturation. More than 1,900 protein bands were analyzed using density-weighted ZUNIX similarity equations and clustering methods, revealing integral polymorphism similarities of 57.4% between the premature and mature stromata and 42.8% between the premature and mature caterpillar bodies. The across-compartment similarity was low, ranging from 10.0% to 18.4%. Consequently, each Cs compartment (i.e., the stroma and caterpillar body) formed a clustering clade, and the 2 clades formed a Cs cluster. The polymorphic similarities ranged from 0.51% to 1.04% between Hs and the Cs compartments and were 2.8- to 4.8-fold higher (1.92%–4.34%) between Ph and the Cs compartments. The Hs and Ph mycelial samples formed isolated clades outside of the Cs cluster.

Conclusion: Proteomic polymorphisms in the caterpillar body and stroma of Cs change dynamically during maturation. The proteomic polymorphisms in Hs and Ph differ from those in Cs, suggesting the presence of multiple Cs-associated fungi and multiple *Ophiocordyceps sinensis* genotypes with altered differential protein expression in the Cs compartments during maturation. In conjunction with prior mycological and molecular observations, the findings from this proteomic study support the integrated micro-ecosystem hypothesis for natural Cs

Editor: Raffaella Balestrini, Institute for Sustainable Plant Protection, C.N.R., Italy

Funding: This study was supported by the Pharmanex Cordyceps sinensis Research Fund. The funder provided support in the form of full or partial salaries for the authors [YZD, LJZ, ZMW, LG, YSY, NZT, JSZ], the purchase of research materials and the payment of SELDI-TOF MS service fees. The funder did not have any additional role in the study design, data collection and analyses, decision to publish or preparation of the manuscript. The specific roles of these authors are articulated in the 'author contributions' section.

Competing Interests: The Pharmanex Cordyceps sinensis research fund was provided as a special supply for Cordyceps sinensis-related research by Nu Skin Enterprises USA. The Pharmanex Beijing Clinical Pharmacology Center was one of the Nu Skin Research Labs and was closed in Dec. 2013. The co-authors [YZD, LJZ, ZMW, LG, YSY, NZT] were staff scientists of Nu Skin Research Labs prior to its closure. Co-author JSZ was an employee of Nu Skin USA and left the company on Oct. 1, 2013 (currently an Adjunct Professor at Hong Kong Polytechnic Univ.). Co-authors LJZ and ZMW are currently employed by other institutes, and their jobs are completely unrelated to C. sinensis research. The remaining co-authors [YZD, LG, YSY, NZT] are currently unemployed, and JSZ remains in an adjunct position at the university. The majority of the work related to this paper was performed before the lab closure. The authors continued working on the project (e.g., data analyses, manuscript writing/revising, submitting/resubmitting) without salary support from Nu Skin after leaving the company in Oct. or Dec. 2013.

* Email: zhujosh@gmail.com

❥ These authors contributed equally to this work.

Introduction

For centuries, *Cordyceps sinensis* has been used as a precious medicinal product in China and other Asian countries and features a broad spectrum of health benefits, including anti-aging and lifespan-extension effects [1–3]. (Note: The Latin name *Cordyceps sinensis* (Berk.) Sacc. is used for both the teleomorph/holomorph of *C. sinensis* fungus and the wild product indiscriminately [4],[5]. The fungus was re-named *Ophiocordyceps sinensis* (Berk.) Sung et al. [6]; however, the Latin name for the wild product has remained unchanged. Because a consensus Latin name for the wild product has not been reached by mycological and TCM botanical taxonomists, we have temporarily used the term *O. sinensis* to refer to the fungus/fungi and continued to use the name *C. sinensis* to refer to the wild product.) Mycological and molecular approaches have demonstrated that *C. sinensis* comprises more than 90 intrinsic fungi from more than 37 genera and at least 6 *O. sinensis* genotypes [7–25]. Although an anamorph-teleomorph connection between *Hirsutella sinensis* and *O. sinensis* has been proposed based on the aggregation of indirect evidence [4–5],[24], integrated analyses have demonstrated large dissimilarities between the random amplified polymorphic DNA (RAPD) polymorphisms of *H. sinensis* and of the *C. sinensis* ascocarp and no studies to date have truly satisfied Koch's postulates by describing the successful artificial induction of *C. sinensis* sexual fruiting bodies and ascospores [9],[17–18],[25–29]. However, there has been no direct evidence to either approve or reject the *Paecilomyces hepiali* hypothesis for the *O. sinensis* anamorph [30]. *P. hepiali*, *H. sinensis* and several mutant genotypes of *O. sinensis* have been found to naturally coexist in the ascocarps and ascospores of natural *C. sinensis*, and the fungal complex showed a 39-fold enhancement of its infection potency over that of pure *H. sinensis* [31–32]. Other researchers have thus hypothesized that *C. sinensis* is an integrated micro-ecosystem with differential expressions by multiple intrinsic fungi in its compartments and have identified a culture-dependent microbial community or mycobiota in natural *C. sinensis* along with evidence of possible symbiotic interactions among the component fungi [13–23],[27–32]. We have previously reported dynamic changes in the differential fungal expression of at least 6 *O. sinensis* genotypes during *C. sinensis* maturation [19–23],[29]. However, no previous studies have compared the proteomes of *C. sinensis* and *H. sinensis* (the proposed anamorphic fungus of *O. sinensis*) or reported global changes in the macrocosmic proteomic polymorphisms in *C. sinensis* compartments during maturation. In contrast to the microcosmic studies that have focused specifically on individual protein species, we used the surface-enhanced laser desorption/ionization time-of-flight mass spectrometry (SELDI-TOF MS) protein chip technique in this study to macrocosmically profile the changes in proteomic polymorphisms in the *C. sinensis* caterpillar body and stroma during maturation [33–34]. We also examined the similarities and cluster relationships between the proteomic polymorphisms of *C. sinensis* and those of the mycelial fermentation products *H. sinensis* Bailing and *P. hepiali* Cs-4.

Materials and Methods

Collection of *C. sinensis*

Fresh *C. sinensis* specimens were purchased in a local market (Latitude 30°04′N, Longitude 101°95′E) in the Kangding County of Sichuan Province, China. Governmental permission was not required for *C. sinensis* purchases in local markets, and the collections of *C. sinensis* specimen sales by local farmers fall under the governmental regulations for traditional Chinese herbal products. Premature *C. sinensis* features a plump caterpillar body (sclerotium) and a short stroma ranging from 1.0 to 2.0 cm in length (Figure 1). Mature *C. sinensis* features a less plump caterpillar body and a long stroma with a length of>5.0 cm and an expanded portion densely covered with ascocarps close to the stroma tip. All fresh *C. sinensis* specimens were washed thoroughly on site in running water with gentle brushing, soaked in 0.1% mercuric chloride for 10 min for surface sterilization and washed again 3 times with sterile water. The specimens were immediately

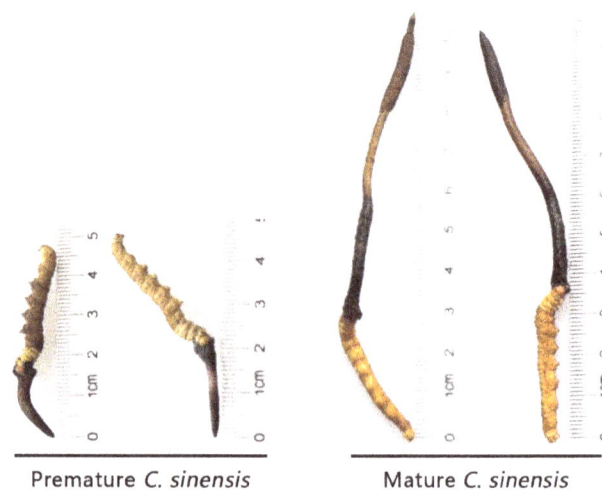

Premature *C. sinensis*   Mature *C. sinensis*

Figure 1. Freshly collected natural *C. sinensis* at 2 maturation stages with stromata of various heights.

frozen in liquid nitrogen for transportation and storage prior to further processing in the lab in Beijing [13].

Sample preparations for proteomic profiling

Ten *C. sinensis* specimens at each maturation stage were used in this study. The caterpillar bodies and stromata from the premature and mature *C. sinensis* specimens and the mycelial fermentation products *H. sinensis* Bailing (Bailing capsule, Lot #040811, #050403 and #051103, Zhejiang American-Sina Pharmaceutical Company, Hangzhou, Zhejiang, China) and *P. hepiali* Cs-4 (Jinshuibao capsule, Lot #JX12931, #20040608 and #20051020, Jiangxi TCM Pharmaceutical Company, Nanchang, Jiangxi, China) were individually ground into powder in liquid nitrogen. To evaluate the proteomic polymorphisms of the samples as group-averages at each maturation stage to minimize the influence of individual variations due to sampling and the lack of a more accurate method to measure the sample's maturation status, the powders (0.5 g each) of the *C. sinensis* compartment samples were pooled according to their compartment origins and maturational stages to form the following testing samples: premature stroma, premature caterpillar body, mature stroma and mature caterpillar body. Based on the pre-test results with insignificant variations in overall proteomic similarity, 0.95 for the 3 *H. sinensis* Bailing samples and 0.96 for the 3 *P. hepiali* Cs-4 samples, the powders of *H. sinensis* Bailing (Lot #051103) and *P. hepiali* Cs-4 (Lot #20051020) were selected for the formal study. The powder samples were dissolved in 600 μl of tris-glycine buffer (pH 8.3) and centrifuged at 14,000 rpm for 5 min at 4°C. The sample supernatants were used for proteomic profiling.

SELDI-TOF mass spectrometry

The supernatants prepared above were diluted in PBS to a concentration of 200–300 nM before application to a normal-phase biochip and analysis on a PBS-II protein chip reader (SELDI-TOF MS; BioSpace Ciphergen Biosystems, Fremont, CA, USA) [33–34]. The SELDI-TOF MS experiments were performed at the Universities' Confederated Institute for Proteomics at the School of Life Sciences, Beijing Normal University, Beijing, China. In brief, different proteins captured on the surface of protein chips were collected through SELDI-TOP mass spectrometry using a laser power of 215 (sensitivity 9; molecule size range: 0–60,000 Da). Following mass calibration, total ion current normalization and baseline subtraction, the molecular size ranges of proteins were manually selected for analyses, and the intensities (peak heights) were extracted using ProteinChip software (Ciphergen proteinchip 3.0.2).

Across-chromatogram normalization of densities of protein species

The SELDI-TOF MS chromatograms were scanned with Quantity One software (Bio-Rad Laboratories, Inc., Hercules, CA, USA). To conduct integrated proteomic profiling on the basis of chromatographic tracing at the molecular weight segments, the band trace quantities (OD*mm) of all protein bands in all chromatograms were normalized using the maximal chromatographic tracing scales for each molecular weight segmented tracing panel as the reference factor (Figure 2). The relative intensity/density was defined as the scanned band trace quantity (OD*mm) multiplied by the difference between the maximum scale "n" on the vertical axis of each chromatographic tracing panel and the baseline scale if the trace baseline was not exactly zero.

Similarity computation for proteomic polymorphisms

ZUNIX equations (http://www.ebioland.com/ZUNIX.htm; Beijing Bioland Technology, 2013) were used for similarity computations with band intensities/density weighting [28–29] while considering (i) mismatched protein bands and (ii) matched protein bands with dissimilar intensities/densities. The following ZUNIX equation (1) was used to compare the polymorphisms of 2 mass spectrometry chromatograms: $d_{ik} \geq 0$, $i = 1,2$, $k = 1,2, …, m$. We defined the measure of similarity as follows:

$$S = \frac{\sum\limits_{k=1}^{m} [2\,\mathrm{Min}\{d_{1k}, d_{2k}\}]}{\sum\limits_{k=1}^{m} (d_{1k} + d_{2k})} \quad (1)$$

where the similarity of the 2 densities d_{1k} and d_{2k} is defined as the common portion of their values.

The second ZUNIX equation (2) is suitable for comparing the proteomic polymorphisms in more mass spectrometry chromatograms, where $d_{ik} \geq 0$, $i = 1,2,…,n$, $k = 1,2,…,m$ and the description is as follows:

$$S = \frac{\sum\limits_{k=1}^{m} [n\,\mathrm{Min}\{d_{1k}, d_{2k}, …, d_{nk}\}]}{\sum\limits_{r=1}^{n} \sum\limits_{s=1}^{m} d_{rs}} \quad (2)$$

Density-weighted cluster analysis for the polymorphic protein fingerprints

For the mismatched protein species, a missing band at the given molecular weight location in a MS chromatogram was assigned a score of 0. The digital density data of all matched and unmatched protein bands in the compared chromatograms were ranked and arbitrarily assigned scores of 1–9 according to the ranks of their densities in 2 or more compared chromatograms [28–29]. The digital data scores were entered into PAUP 4.0B (Swofford, 2002; Sinauer Asso. Inc, Sunderland, MA, USA) to construct cluster trees (semi-quantitative density-weighted neighbor-joining distance method; bootstrap = 1000). In addition to the semi-quantitative algorithm provided by PAUP 4.0B, a fully quantitative cluster analysis was also performed with a parametric hierarchical clustering analysis (density-weighted furthest neighbor Pearson correlation average linkage distance method) in SPSS 10.1 (SPSS Inc., Chicago, IL, USA; Note: no bootstrap strategy was provided in the software package).

Results

Comparison of the protein fingerprint chromatograms of premature and mature *C. sinensis*

Figure 3 displays the SELDI-TOF MS chromatograms for the *C. sinensis* protein species at the 2 maturation stages in a molecular weight range of 0 to>60,000 Daltons. Using the density-weighted ZUNIX equation (1), a percentage similarity of 57.9% was observed between the protein fingerprint polymorphisms of pooled premature and mature *C. sinensis* samples, thus indicating altered protein expression during *C. sinensis* maturation.

Comparison of the polymorphic protein chromatograms of the caterpillar bodies and stromata of premature and mature *C. sinensis*

Figure 4 displays the SELDI-TOF MS chromatograms for the *C. sinensis* caterpillar body and stroma protein moieties at the 2 maturational stages in a molecular weight range of 0 to>60,000 Daltons. Using ZUNIX equation (2), an overall percentage similarity of 3.1% was observed between the proteomic polymorphisms for all *C. sinensis* caterpillar body and stroma samples at both maturation stages. Using ZUNIX equation (1) for pairwise comparisons, the calculated similarities from Figure 4 were 57.4% and 42.8% between the proteomic polymorphisms of the 2 maturation stages in the stromata or caterpillar bodies, respectively, but were much lower (10.0%–18.4%) for the across-compartment pair comparisons (Table 1). These similarities indicate major differences in the proteomic profile within the *C. sinensis* caterpillar body and stroma resulting from large differences in the compositions of the multiple intrinsic fungi from more than 37 genera and at least 6 mutant *O. sinensis* genotypes together with the transcription and translation of their fungal

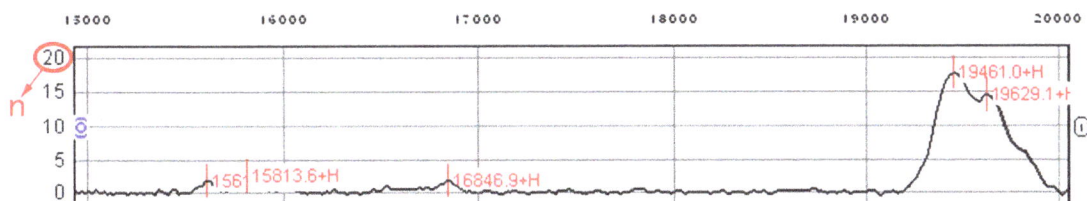

Figure 2. A representative SELDI-TOF MS chromatogram tracing. The tracing scale maximum, n, on the vertical axis of each chromatographic panel was used as the across-chromatogram normalization reference.

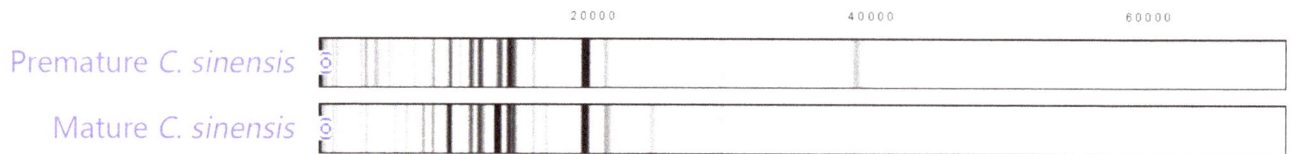

Figure 3. SELDI-TOF MS protein chromatograms to examine the protein fingerprints (molecular weight: 0 to>60,000 Daltons) and proteomic polymorphisms of premature and mature *C. sinensis*.

genes. In contrast to the large between-compartment differences in the proteomic profile, the within-compartment differences were moderate across the *C. sinensis* maturation stages.

Comparison of the *C. sinensis*, *H. sinensis* and *P. hepiali* sample protein fingerprint chromatograms in multiple molecular weight segments

The above-described results for the *C. sinensis* proteins are displayed integrally from molecular weights of 0 to>60,000 Daltons, as shown above in Figures 3 and 4. To increase the chromatographic resolution, Figure 5 displays 7 panels of the segmented SELDI-TOF MS chromatograms of all protein species in the *C. sinensis* caterpillar bodies and stromata at the 2 maturation stages as well as of the commercial mycelial fermentation products *H. sinensis* Bailing and *P. hepiali* Cs-4. Figure 5-A presents protein species in the molecular weight range from 0 to 5,000 Daltons, Figure 5-B ranges from 5,000 to 10,000 Daltons, Figure 5-C ranges from 10,000 to 15,000 Daltons, Figure 5-D ranges from 15,000 to 20,000 Daltons, Figure 5-E ranges from 20,000 to 30,000 Daltons, Figure 5-F ranges from 30,000 to 40,000 Daltons and Figure 5-G ranges from 40,000 to>60,000 Daltons.

The segmented mass spectrometry chromatograms shown in Figure 5 indicate large polymorphic differences between the protein profiles of the *C. sinensis* compartments at each of the 2 maturation stages, and the complex protein expression patterns resulting from multiple intrinsic fungi across the *C. sinensis* compartments underwent differential maturational fungal expression changes. The mass spectrometry chromatograms also indicate large overall differences between the proteomic polymorphisms of *C. sinensis* and those of the fermentation products *H. sinensis* Bailing and *P. hepiali* Cs-4.

Polymorphic similarities in the protein fingerprints of the *C. sinensis* compartments at 2 maturational stages and the mycelial fermentation products *H. sinensis* Bailing and *P. hepiali* Cs-4

The densities of all protein species in all 7 mass spectrometry chromatogram panels shown in Figure 5 were normalized using

the mass spectrometry tracing scales described in the Methods section and illustrated in Figure 2. The normalized densities were subjected to polymorphic similarity calculations with ZUNIX similarity equation (1) [28–29], to examine the similarities between the protein profiles of the *C. sinensis* compartment and those of the mycelial products *H. sinensis* Bailing and *P. hepiali* Cs-4. As shown in Table 2, the proteomic polymorphism similarities were low (0.51%–1.04%) between the *H. sinensis* Bailing and *C. sinensis* protein profiles and were 2.8- to 4.8-fold higher (1.92%–4.34%) between the *P. hepiali* Cs-4 and *C. sinensis* compartments than between the *H. sinensis* Bailing and *C. sinensis* compartments.

Density distributions of all protein bands and determination of the scoring cutoff value for the semi-quantitative analysis

After normalization, the scanned band trace quantities (OD*mm) of approximately 1,900 protein bands were sorted for arbitrary scoring in preparation for the cluster construction using the semi-quantitative density-weighted algorithm (Figure 6).

Density-weighted cluster analysis of the protein fingerprints of *C. sinensis* and the mycelial products *H. sinensis* Bailing and *P. hepiali* Cs-4

The highest density value in Figure 6 was divided by 9 to obtain the critical cut-off values for a semi-quantitative density grouping. Each density was assigned a score from 1 to 9 according to the above-mentioned cutoff values, and all arbitrarily assigned scores were used for the cluster construction according to the density-weighted algorithm provided by PAUP 4.0B software [28–29]. Figure 7 displays a cluster tree that was constructed with the density-weighted neighbor-joining algorithm (bootstrap = 1000). Similar to the percentage similarity results shown in Table 1, the caterpillar body samples formed 1 clade and the stroma samples formed another clade; these 2 clades then formed a *C. sinensis* cluster. The mycelial fermentation product *H. sinensis* Bailing formed an isolated clade that was separated from the *C. sinensis* cluster by the clade formed by *P. hepiali* Cs-4.

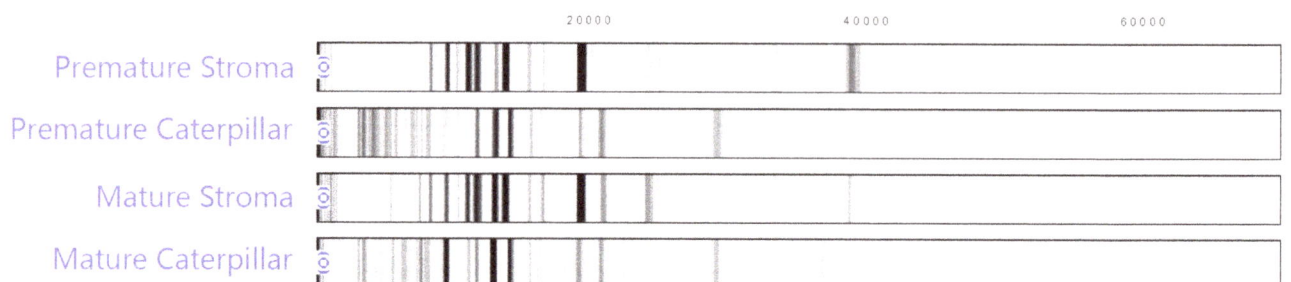

Figure 4. SELDI-TOF MS protein chromatograms to examine the protein fingerprints (molecular weight: 0 to>60,000 Daltons) and proteomic polymorphisms of the caterpillar bodies and stromata of premature and mature *C. sinensis*.

Table 1. Percentage similarities in the total protein profiles of the caterpillar bodies and stromata of the premature and mature *C. sinensis*, computed with the density-weighted ZUNIX equation (1).

	Stroma		Caterpillar Body	
	Premature	Mature	Premature	Mature
Premature stroma	—	—	—	—
Mature stroma	57.4%	—	—	—
Premature caterpillar body	10.0%	13.2%	—	—
Mature caterpillar body	18.4%	17.8%	42.8%	—

Although the PAUP 4.0B software offered the advantage of constructing cluster trees according to the bootstrap value (bootstrap = 1000), the program used only semi-quantitative algorithms. A fully quantitative, density-weighted algorithm included in the SPSS 10.1 software package was also employed to construct a cluster tree [28]. As shown in Figure 8, the *C. sinensis* sample clade formation pattern for *C. sinensis* samples generated using the fully quantitative algorithm was similar to that generated via semi-quantitation, as shown in Figure 7. The caterpillar body and stroma clades joined to form a *C. sinensis* cluster with a greater rescaled distance in the cluster tree that was indicative of large differences in polymorphic protein expression between the *C. sinensis* caterpillar bodies and stromata and reflective of the low similarity observed between the *C. sinensis*

compartments in Table 1. The mycelial fermentation products *H. sinensis* Bailing and *P. hepiali* Cs-4 formed a clade with a greater rescaled distance and were thus situated outside of the *C. sinensis* cluster.

Discussion

C. sinensis is one of the most valued Chinese medicinal products. This organism grows only in areas of high elevation on the Qinghai-Tibetan Plateau and features a complex life cycle. Studies have reported that *C. sinensis* comprises more than 90 intrinsic fungal species from more than 37 genera and at least 6 genotypes of *O. sinensis* fungi [7–23]. Of these, the most abundant culturable fungi are *Pseudogymnoascus roseus* in the sclerotia and

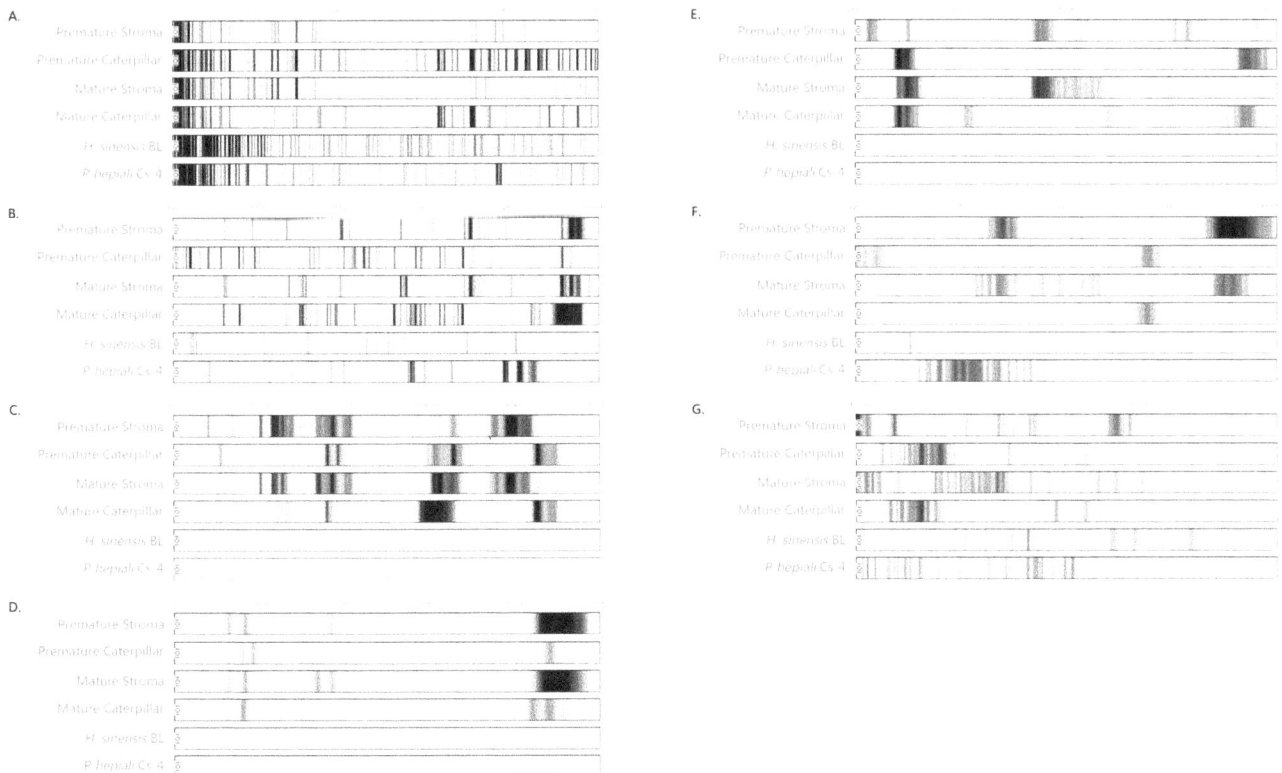

Figure 5. SELDI-TOF MS protein chromatograms to examine protein polymorphisms in the stroma and the caterpillar body specimens of premature and mature *C. sinensis* and the mycelial fermentation products *H. sinensis* Bailing and *P. hepiali* Cs-4. Total proteins were extracted from the caterpillar bodies or stromata of natural *C. sinensis* at 2 maturation stages and from the mycelial products *H. sinensis* Bailing (BL) or *P. hepiali* Cs-4. Panels A, B, C, D, E, F and G present proteins with molecular weights ranging from 0–5,000, 5,000–10,000, 10,000–15,000, 15,000–20,000, 20,000–30,000, 30,000–40,000 and 40,000 to>60,000 Daltons, respectively.

Table 2. Percentage similarities in proteomic polymorphisms between the *C. sinensis* compartment samples at 2 maturational stages and *H. sinensis* Bailing (BL) and *P. hepiali* Cs-4.

		Stroma		Caterpillar body		
		Premature	Mature	Premature	Mature	*H. sinensis* BL
Percentage similarity (%)	*H. sinensis* BL	0.69%	0.51%	1.04%	0.87%	—
	P. hepiali Cs-4	1.92%	2.33%	4.34%	4.18%	6.52%
Similarity Ratio (fold)	(*P. hepiali* Cs-4 *vs. H. sinensis* BL)	2.8-fold	4.5-fold	4.2-fold	4.8-fold	

cortices and *Penicillium chrysogenum* in the stromata, as reported by Zhang et al. [17]. Previously, we reported that *C. sinensis* maturation was associated with dynamic changes in the intrinsic fungal species and mutant *O. sinensis* genotypes along with significant changes in the RAPD molecular marker polymorphisms and component chemicals [13],[19–23],[29]. The fungal background of *C. sinensis* becomes even more complex when non-culturable fungal species are considered [18–23]. These findings reflect the altered fungal expression of multiple intrinsic fungi and support the hypothesis that *C. sinensis* is an integrated micro-ecosystem of multiple intrinsic fungi, as proposed by Liang *et al.* [27]. Density-weighted algorithms for similarity computations and cluster constructions were used in this study to analyze the mass spectrometry chromatograms of polymorphic proteomes, the downstream transcription/translation products of multiple fungal genomes. We observed different proteomic profiles with similar-

ities of 10.0% between the premature caterpillar bodies and stromata and 17.8% between the mature caterpillar bodies and stromata of *C. sinensis* (*cf.* Figure 4; Table 1), consistent with the mycological and molecular observations of diverse fungal populations in the two *C. sinensis* compartments [17–18]. However, considerably great proteomic polymorphism similarities of 42.8% and 57.4% were observed within the *C. sinensis* caterpillar body and stroma, respectively, at the 2 *C. sinensis* maturation stages (*cf.* Figure 4; Table 1). The differences in the across and within-compartment similarities between the proteomic profiles might possibly be derived from 2 major factors: (1) differential protein expression of the multiple fungal genomes (multiple mutant *O. sinensis* genotypes and the multiple intrinsic mesophilic and psychrophilic fungi), of which part or all undergo maturational alterations, and (2) protein species from the dead bodies of the *C. sinensis* ghost moth larvae, which are not merely a group of

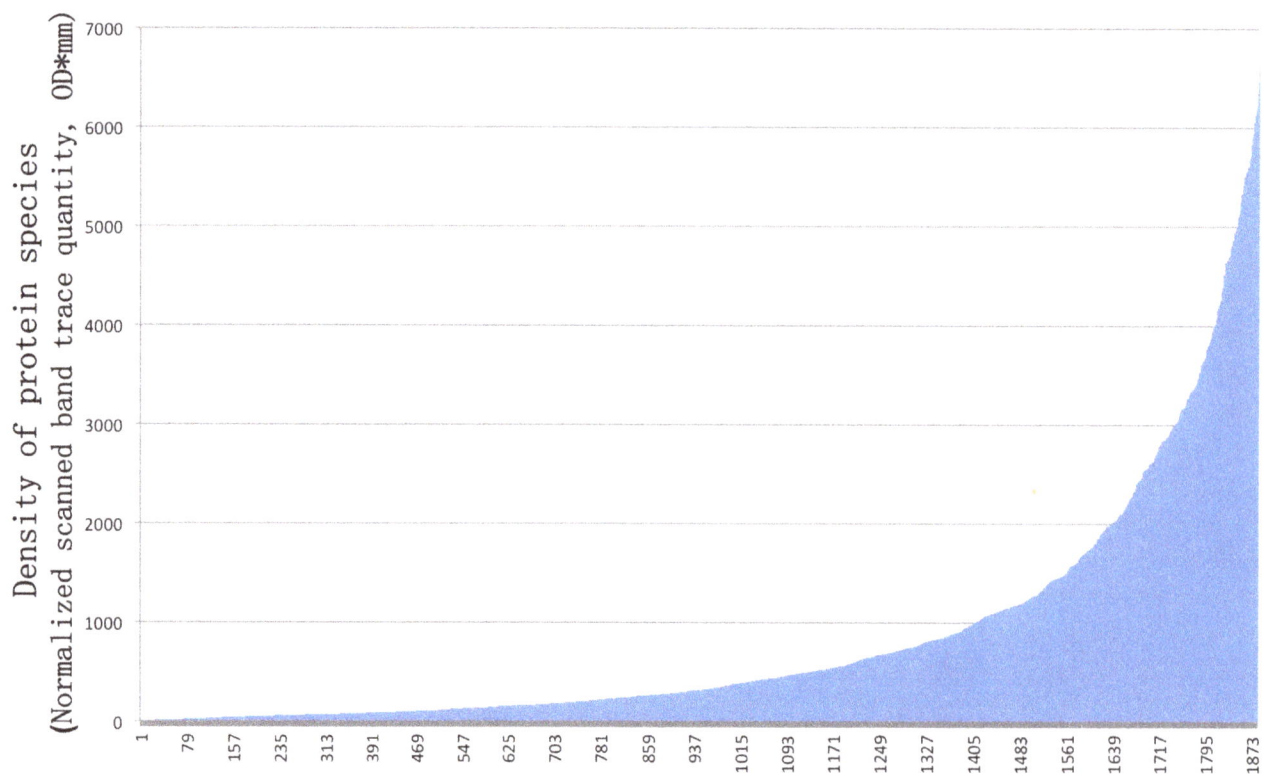

All protein species obtained from all mass spectrometry chromatograms

Figure 6. Distribution of the normalized scanned band trace quantities (OD*mm) of approximately 1,900 protein species from all segmented SELDI-TOF MS protein chromatograms in Figure 5.

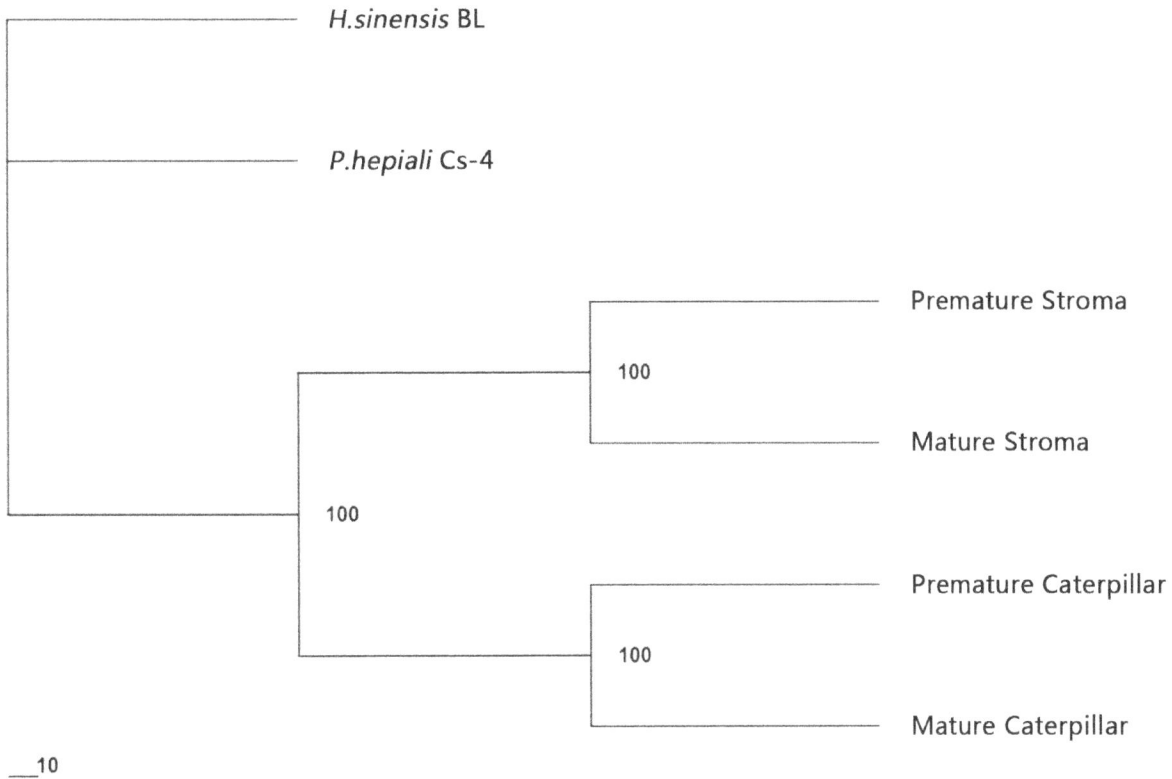

Figure 7. Integral cluster tree of all the proteomic chromatograms constructed with the semi-quantitative density-weighted algorithm. "*H.sinensis* BL" refers to the *H. sinensis* Bailing mycelial product, "*P.hepiali* Cs-4" refers to the *P. hepiali* Cs-4 mycelial product, and "Caterpillar" refers to the caterpillar body. Each protein species from all proteomic chromatograms in Figure 5 was assigned a score of 1–9 based on its density rank among the densities of all compared protein species; the missing protein band at the same molecular weight was assigned a score of 0. All protein species from the chromatograms in Figure 5 were entered into the cluster construction using the neighbor-joining distance method (bootstrap = 1000).

nutrients for fungal growth but also as a part of the species complex, along with all of the previously reported small chemical components [13],[38–47], contribute to the overall pharmacology of the natural medicinal product and partly explain the various therapeutic potencies of premature and mature *C. sinensis* that have been identified by traditional Chinese medicine quality grading system.

* * * * * * H I E R A R C H I C A L   C L U S T E R   A N A L Y S I S * * * * * *

Dendrogram using Complete Linkage

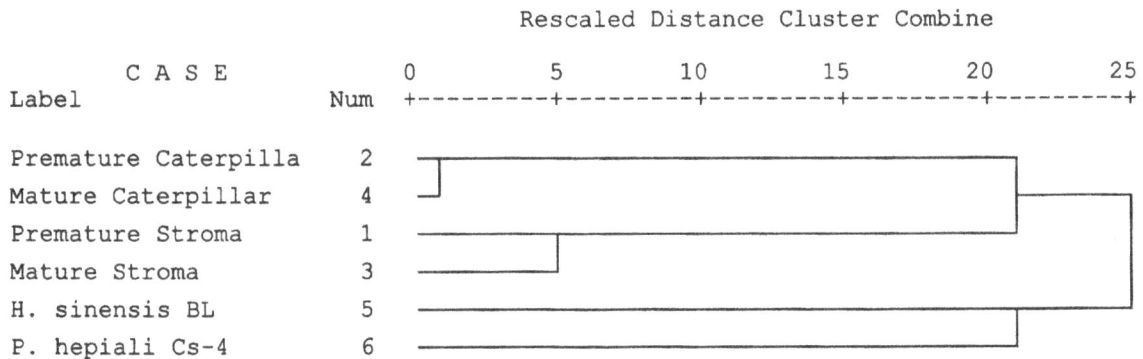

Figure 8. Integral cluster tree of all proteomic chromatograms constructed with the fully quantitative density-weighted algorithm. "H. sinensis BL" refers to the *H. sinensis* Bailing mycelial product; "Caterpillar" or "Caterpillar" refers to the caterpillar body. All protein species from all chromatograms in Figure 5 were entered into the cluster construction, using the furthest neighbor (Pearson correlation average linkage) method of hierarchical cluster analysis.

Density-weighted algorithms for similarity computations and cluster constructions were used to compare RAPD molecular marker polymorphisms in previous studies of *C. sinensis* [28–29]. Although density-unweighted arithmetic methods have been widely used in literature, these methods are only suitable for the analyses of "all-or-none" data. The density-weighted arithmetic methods used in this proteomic study are more mathematically general and sufficiently sensitive to capture all of the detailed information regarding dynamic changes in proteins expressed by the various intrinsic fungi during *C. sinensis* maturation [28–29]. These algorithms provide scientists with accurate analytical means with which to trace changes in the proteomic and molecular marker polymorphisms in natural *C. sinensis*.

In this first study testing the proteomic polymorphisms of different compartments of wild *C. sinensis* during maturation, the overall proteomic polymorphisms were compared in the pooled samples of 10 *C. sinensis* specimens of each maturation stage. The height of the *C. sinensis* stroma (*cf.* Figure 1) has been taken as the standard for the potency-quality grading of natural *C. sinensis* on the market. Such a common practice for potency grading can be explained by a *C. sinensis* mycology expert that the premature *C. sinensis* with a short stroma grows asexually, whereas the long-stroma *C. sinensis* with the formation of the ascocarp portion (the expanded portion close to the tip of the stroma) primarily grows sexually (personal communication with Prof. YL Guo). According to the comments of Prof. Guo regarding the asexual-sexual growth of *C. sinensis*, our previous molecular systematic studies demonstrated that the maturation of wild *C. sinensis* is a continuous biological course along with the weather during spring. However, there is no existing accurate method thus far to measure sample's maturation status. We also found previously large differences in the fungal activity of *H. sinensis*, biomasses of the fungus-specific DNA species, and small organic chemicals in wild *C. sinensis* during maturation, indicating large differences in fungal expression during *C. sinensis* maturation [13],[19–23],[28–29]. Therefore, we designed this study with 2 special sample arrangements of the test materials to minimize variations in individual specimens: (1) the selection of *C. sinensis* specimens with the clear morphological characters shown in Figure 1, i.e., very premature *C. sinensis* with a short possible stroma (1-2 cm) and mature *C. sinensis* with a long stroma (>5 cm) and with definite formation of the ascocarp portion; and (2) assessment of the pooled samples (10 specimens at each maturation stage; stromata and caterpillar bodies separated from the same specimens). In addition to the examination of maturational and compartmental group differences in proteomic polymorphism within a *C. sinensis* population, it is possible that there are individual differences in some degree in proteomic polymorphism within a *C. sinensis* population and within maturation groups. These individual differences are likely due to differences such as in the instar and nutrition status of the larvae of ghost moths within the family Hepialidae at the time of fungal infection, the growth location and environment (e.g., elevation and temperature, strength of plateau wind and sunshine in the growth area, amount of snow in winter and rain in spring, soil fertility, surrounding vegetation), the total weight and length of the *C. sinensis* specimens, and the weight ratio and height ratio of the caterpillar body and stroma of individual *C. sinensis* specimens. This study design of pooling samples, however, is limited regarding the exploration of such individual variations at each estimated maturation stage. However, there may also be population differences among the *C. sinensis* specimens collected from different production areas, likely due to the different species of larvae of ghost moths within the family Hepialidae, differences in latitude, possibly different local soil fungal flora or mycobiota

and other environment factors. All these considerations should encourage future studies to further explore variations in the molecular and proteomic polymorphisms and chemical profiles among the individual *C. sinensis* specimens collected within a production area or among the *C. sinensis* populations from various production areas. Perhaps prior to the future comparison of individual specimens, an accurate method for determining *C. sinensis* maturation stages may need to be established with the combined use of morphological characters and molecular markers to distinguish proteomic variations due to slightly different maturation stages of *C. sinensis* specimens or due to true differences in protein expression in individual specimens at the same maturation stage. To this end, fungal biomass ratios, for example, GC-biases *vs.* AT-biases of *O. sinensis*, may serve as a molecular marker to assist the morphological characterization when determining the *C. sinensis* maturation status [15–16],[19–23].

H. sinensis has been proposed as an anamorph of *O. sinensis*; natural *C. sinensis* is considered a single fungus product [4–5]. These hypotheses were proposed based on the aggregation of indirect evidence, such as morphological findings for the isolates from natural *C. sinensis*, ITS sequencing and results from microcycle conidiation of ascospores under particular culture conditions [4–5],[24]. No scientific studies to date have truly satisfied Koch's postulates, which have demonstrated the successful artificial induction of the *C. sinensis* sexual fruiting body and ascospores [5],[9],[24–27],[35]. Shen *et al.* [36] reported extremely slow growth (approximately 2 cm after 7 months) of artificial *Cephalosporium dongchongxiacae* (≡ *H. sinensis*; [4],[36]) fruiting bodies and observed regular, fine and deep twills on the surfaces of long, conically shaped fruiting bodies. The overall appearance of the artificial fruiting bodies, unfortunately, was distinct from that of natural *C. sinensis*, which has a long, round and cylindrical stroma with vertical fine wrinkles, as described in the Chinese Pharmacopoeia. Shen *et al.* [36] also reported the production of ascospores from one of the artificial *C. dongchongxiacae* fruiting bodies that featured no morphological formation of a *C. sinensis*-like ascocarp, the sexual organ of *C. sinensis*, thus indicating an overall teleomorphic morphology distinct from that of natural *C. sinensis*. Shen *et al.* [36] characterized in their paper the unique teleomorphic features of *C. dongchongxiacae*, and the results actually negated the anamorph-teleomorph connection between *C. dongchongxiacae* (≡ *H. sinensis*) and *O. sinensis* in accordance with Koch's postulates. In addition to the dramatic dissimilarities in the RAPD molecular marker polymorphisms between the *C. sinensis* ascocarp and *H. sinensis*, the drastically different proteomic polymorphisms in *C. sinensis* and *H. sinensis* might not support the "single-fungus" hypothesis of *C. sinensis* or the hypothesis of an anamorph-teleomorph connection between *H. sinensis* and *O. sinensis* [13],[25],[28–29],[37] (*cf.* Figure 5 and Table 2). Based on these microcosmic and macrocosmic studies, Liang *et al.* [27] hypothesized that *C. sinensis* is an integrated micro-ecosystem with varying compositions of multiple intrinsic fungi. In fact, the coexistence of these multiple fungi has been demonstrated in the culture-dependent and independent microbial communities or mycobiota present in natural *C. sinensis*, along with evidence of symbiotic interactions among the component fungi; these likely represent the key biological actions essential to the natural or artificial production of sexual fruiting bodies and ascospores [17–18],[31–32].

Studies of *C. sinensis* have detected several groups of chemical components, including carbohydrates; galactomannan; nucleosides; proteins, polypeptides, oligopeptides, polyamines, and diketopiperazines (cyclo-dipeptides); non-hormone sterols; fatty acids and other organic acids; and vitamins and inorganic elements [1],[38–40]. Other compounds such as verticiol, acid deoxyribonuclease, myriocin, 3-deoxyadenosine (coydycepin) and cordysinins A–E have also been detected [41–45]. Chemical constituent fingerprinting techniques have been used together with similarity comparisons and cluster constructions in *C. sinensis* studies to demonstrate the high similarity of *C. sinensis* samples collected from different production areas [13],[38],[40–47]. A cluster analysis of these small organic chemicals via capillary electrophoresis technology demonstrated that several mycelial fermentation products were situated in different clades outside of the cluster containing the natural *C. sinensis* samples collected from different production areas on Qinghai-Tibetan Plateau [40]. However, when analyzing the small chemicals of several natural *C. sinensis* samples collected from Tibet or Qinghai Provinces and of fermentation products (*P. hepiali* Cs-4 and *H. sinensis* Bailing) through HPLC fingerprinting, the clades of natural *C. sinensis* were much closer in rescaled distance to the clade containing the *P. hepiali* Cs-4 products than to the clade containing the *H. sinensis* Bailing products [38]. In the proteomic fingerprint analysis conducted in the current study, in which the bootstrap strategy (bootstrap = 1,000) was used in the density-weighting algorithm, the fermented products *P. hepiali* Cs-4 and *H. sinensis* Bailing were situated in an isolated clade outside of the *C. sinensis* cluster with the possibility that *P. hepiali* Cs-4 was closer to the *C. sinensis* cluster than was *H. sinensis* Bailing, as shown in Figure 7. This possibility was supported by the 2.8- to 4.8-fold higher similarities of the proteomic polymorphisms between the *P. hepiali* Cs-4 and *C. sinensis* compartments relative to those between the *H. sinensis* Bailing and *C. sinensis* compartments (*cf.* Table 2). The cluster relationship demonstrated by the semi-quantitative neighbor-joining algorithm was validated using the fully quantitative approach of the furthest neighbor algorithm without the bootstrap strategy provided by the SPSS software (*cf.* Figure 8).

Local herbal farmers in the *C. sinensis* production areas of the Qinghai-Tibetan Plateau have long recognized the temperature dependency of the *C. sinensis* maturational features and believe that "eating 'worms' in the winter and 'grass' in the summer" provides tonic herbal properties. Changes in the therapeutic properties of *C. sinensis* during its maturation have also been recognized by the field of traditional Chinese medicine, in which

natural *C. sinensis* is graded accordingly. We have reported maturational changes in the composition of multiple intrinsic fungal species of *C. sinensis* and at least 6 genotypes of *O. sinensis*, along with environmental changes (temperature, sunlight intensity, snow/rain, moisture and plateau wind) on the Qinghai-Tibetan Plateau [13],[19–23]. The altered fungal background of natural *C. sinensis* at various maturation stages causes large variations in (i) the RAPD molecular marker polymorphisms, (ii) the fingerprints of small organic compounds, and (iii) proteomic polymorphisms in the caterpillar bodies and stromata, as demonstrated in this and previous studies [13],[19–23],[28–29]. The integration of the component compounds that are differentially expressed in different compartments of *C. sinensis* and differentially altered during *C. sinensis* maturation constitutes the dynamic pharmacological base that is responsible for the varying potencies of the health benefits and therapeutic activities associated with *C. sinensis*.

In conclusion, SELDI-TOF MS proteomic profiling was used to macrocosmically detect the dynamic polymorphic alterations among differentially expressed proteins in the different *C. sinensis* compartments during maturation. The apparent proteomic polymorphism dissimilarity between *H. sinensis* and *C. sinensis* suggests different fungal backgrounds of these organisms and thus might not support the "single-fungus" hypothesis of *C. sinensis* or the hypothesis of an anamorph-teleomorph connection between *H. sinensis* and natural *C. sinensis*. However, the findings from this proteomic study, in corroboration with prior mycological and molecular observations, support the integrated micro-ecosystem hypothesis for natural *C. sinensis*.

Acknowledgments

The authors are grateful to Prof. P.Y. Xu, Mr. W. Chen, Ms. M. Yang and Mr. Y.C. Zhou for their assistance during the collection of the wild *C. sinensis* samples.

Author Contributions

Conceived and designed the experiments: IJZ JSZ. Performed the experiments: YZD IJZ. Analyzed the data: YZD LG YSY NZT LN. Contributed reagents/materials/analysis tools: ZMW. Wrote the paper: JSZ JYW. Formulated analytic algorithms: LN.

References

1. Zhu J-S, Halpern GM, Jones K (1998a) The scientific rediscovery of a precious ancient Chinese herbal regimen: *Cordyceps sinensis*: Part I. J Altern Complem Med 4(3): 289–303.
2. Zhu J-S, Halpern GM, Jones K (1998b) The scientific rediscovery of an ancient Chinese herbal medicine: *Cordyceps sinensis*: Part II. J Altern Complem Med 4(4): 429–457.
3. Tan NZ, Berger JL, Zhang Y, Prolla TA, Weindruch R, et al. (2011) The lifespan-prolonging effect of *Cordyceps sinensis* Cs-4 in normal mice and its molecular mechanisms. FASEB J 25(1): 599.1.
4. Wei XL, Yin XC, Guo YL, Shen NY, Wei JC (2006) Analyses of molecular systematics on *Cordyceps sinensis* and its related taxa. Mycosystema 25(2): 192–202.
5. Guo YL, Xiao PG, Wei JC (2010) On the biology and sustainable utilization of the Chinese Medicine treasure *Ophiocordyceps sinensis* (Berk.) G. H. Sung et al. Modern Chin Med 12(11): 3–8.
6. Sung GH, Hywel-Jones NL, Sung JM, Luangsa-Ard JJ, Shrestha B, et al. (2007) Phylogenetic classification of Cordyceps and the clavicipitaceous fungi. Stud Mycol 57(1): 5–59.
7. Zhao J, Wang N, Chen YQ, Li TH, Qu LH (1999) Molecular identification for the asexual stage of *Cordyceps sinensis*. Acta Sci Natural Univ Sunyatseni 38(1), 122–123.
8. Chen YQ, Wang N, Qu LH, Li TH, Zhang WM (2001) Determination of the anamorph of *Cordyceps sinensis* inferred from the analysis of the ribosomal DNA internal transcribed spacers and 5.8S rDNA. Biochem Syst Ecol 29: 597–607.
9. Jiang Y, Yao YJ (2003) A review for the debating studies on the anamorph of *Cordyceps sinensis*. Mycosistema 22(1): 161–176.
10. Chen YQ, Hu B, Xu F, Zhang WM, Zhou H, et al. (2004) Genetic variation of *Cordyceps sinensis*, a fruit-body-producing entomopathegenic species from different geographical regions in China. FEMS Microbiol Lett 230: 153–158.
11. Chen J, Zhang W, Lu T (2006) Morphological and genetic characterization of a cultivated *Cordyceps sinensis* fungus and its polysaccharide component possessing antioxidant property in H22 tumor-bearing mice. Life Sci 78(23): 2742–2748.
12. Leung PH, Zhang QX, Wu JY (2006) Mycelium cultivation, chemical composition and antitumour activity of a *Tolypocladium* sp. fungus isolated from wild *Cordyceps sinensis*. J Appl Microbiol 101(2): 275–283.
13. Zhu J-S, Guo YL, Yao YS, Zhou YJ, Lu JH, et al. (2007) Maturation of *Cordyceps sinensis* associates with co-existence of *Hirsutella sinensis* and *Paecilomyces hepiali* DNA and dynamic changes in fungal competitive proliferation predominance and chemical profiles. J Fungal Res 5(4): 214–224.
14. Stensrud Ø, Schumacher T, Shalchian-Tabrizi K, Svegardenib IB, Kauserud H (2007) Accelerated nrDNA evolution and profound AT bias in the medicinal fungus *Cordyceps sinensis*. Mycol Res 111: 409–415.
15. Yang JL, Xiao W, He HH, Zhu HX, Wang SF (2008) Molecular phylogenetic analysis of *Paecilomyces hepiali* and *Cordyceps sinensis*. Acta Pharmaceut Sinica 43(4): 421–426.
16. Xiao W, Yang JP, Zhu P, Cheng KD, He HX, et al. (2009) Non-support of species complex hypothesis of *Cordyceps sinensis* by targeted rDNA-ITS sequence analysis. Mycosystema 28(6): 724–730.

17. Zhang YJ, Sun BD, Zhang S, Wangmu, Liu XZ, et al. (2010a) Mycobiotal investigation of natural *Ophiocordyceps sinensis* based on culture-dependent investigation. Mycosistema 29(4): 518–527.
18. Zhang YJ, Zhang S, Wang M, Bai FY, Liu XZ (2010b) High Diversity of the Fungal Community Structure in Naturally-Occurring Ophiocordyceps sinensis. PLoS ONE 5(12): e15570. doi:10.1371/journal.pone.0015570.
19. Zhu J-S, Gao L, Li XH, Yao YS, Zhou YJ, et al. (2010) Maturational alterations of oppositely orientated rDNA and differential proliferations of CG:AT-biased genotypes of *Cordyceps sinensis* fungi and *Paecilomyces hepiali* in natural *Cordyceps sinensis*. Am J Biomed Sci 2(3): 217–238.
20. Gao L, Li XH, Zhao JQ, Lu JH, Zhu J-S (2011) Detection of multiple *Ophiocordyceps sinensis* mutants in premature stroma of *Cordyceps sinensis* by MassARRAY SNP MALDI-TOF mass spectrum genotyping. Beijing Da Xue Xue Bao 43(2): 259–266.
21. Yao YS, Zhou YJ, Gao L, Lu JH, Zhu J-S (2011) Dynamic alterations of the differential fungal expressions of *Ophiocordyceps sinensis* and its mutant genotypes in stroma and caterpillar during maturation of natural *Cordyceps sinensis*. J Fungal Res 9(1): 37–49,53.
22. Gao L, Li XH, Zhao JQ, Lu JH, Zhao JG, et al. (2012) Maturation of *Cordyceps sinensis* associates with alterations of fungal expressions of multiple *Ophiocordyceps sinensis* mutants with transition and transversion point mutations in stroma of *Cordyceps sinensis*. Beijing Da Xue Xue Bao 44(3): 454–463.
23. Zhu J-S, Zhao JG, Gao L, Li XH, Zhao JQ, et al. (2012) Dynamically altered expressions of at least 6 *Ophiocordyceps sinensis* mutants in the stroma of *Cordyceps sinensis*. J Fungal Res 10(2): 100–112.
24. Xiao YY, Chen C, Dong JF (2011) Morphological observation of ascospores of *Ophiocordyceps sinensis* and its anamorph in growth process. J Anhui Agricult Univ 38(4): 587–591.
25. Hu X, Zhang YJ, Xiao GH, Zheng P, Xia YL, et al. (2013) Genome survey uncovers the secrets of sex and lifestyle in caterpillar fungus. Chin Sci Bull 58: 2846–2854.
26. Jin ZX, Yang SH (2005) Progresses and trends of *Cordyceps sinensis* studies. J Tianjin Med Univ 11(1), 137–140.
27. Liang ZQ, Han YF, Liang JD, Dong X, Du W (2010) Issues of concern in the studies of *Ophiocordyceps sinensis*. Microbiol Chin 37(11): 1692–1697.
28. Ni LQ, Yao YS, Gao L, Wu ZM, Tan NZ, et al. (2014) Density-weighted algorithms for similarity computation and cluster tree construction in the RAPD analysis of natural *Cordyceps sinensis*. Am J Biomed Sci 6(2), 82–104.
29. Yao YS, Gao L, Li YL, Ma SL, Wu ZM, et al. (2014) Amplicon density-weighted algorithms analyze dissimilarity and dynamic alterations of RAPD polymorphisms in the integrated micro-ecosystem *Cordyceps sinensis*. Beijing Da Xue Xue Bao 46(4), 618–628.
30. Dai RQ, Lan JL, Chen WH, Li XM, Chen CT, et al. (1989) Discovery of a new fungus Paecilomyces hepiali Chen & Dai. Acta Agriculturea Universitatis Pekinensis 15(2), 221–224.
31. Ma SL, Li YL, Xu HF, Zhang ZH, Liu X (2010) Analyzing bacterial community in young *Hepialus* of intestinal tract of *Cordyceps sinensis* in Qinghai Province. Chin J Grassland 32(suppl.): 63–65.
32. Li YL, Yao YS, Ma SL, Xu HF, Li AP, et al. (2014) Inoculation potency enhancement using fungal complexes isolated from the intestine of *Hepialus armoricanus* larvae. (ABS0144) The 10th International Mycological Congress, Bangkok, Thailand 3–8 August 2014.
33. Issaq HJ, Veenstra TD, Conrads TP, Felschow D (2002) The SELDI-TOF MS Approach to Proteomics: Protein Profiling and Biomarker Identification. Biochem Biophys Res Comm 292, 587–592.
34. Zeidan BA, Cutress RI, Hastie C, Mirnezami AH, Packham G, et al. (2009) SELDI-TOF MS Proteomics in Breast Cancer. Clin Proteom (2009) 5:133–147.
35. Zhang P (2003) Advances in Cordyceps genus fungi research. J Biol 20(6): 43–45.
36. Shen NY, Zheng L, Zhang XC, Wei SL, Zhou ZR, et al. (1983) Anamorph of *Cordyceps sinensis* (Berk) Sacc. in: Monograph for Cordyceps studies 1980-1985. Xining: Qinghai Acad Animal Sci Veterin Med pp 1–13.
37. Li ZZ, Huang B, Li CR, Fan MZ (2000) Molecular evidence for anamorph determination of *Cordyceps sinensis* (Berk.) Sacc. Mycosystema 9(1): 60–64.
38. Wu YX, Zhou DL, Yan D, Ren YS, Fang YL, et al. (2008) HPLC fingerprint analysis of Cordyceps and mycelium of cultured Cordyceps. Chin J Chin Materia Med 33(19): 2212–2214.
39. Yue DC, Feng X, Liu H, Bao TT (1995) *Cordyceps sinensis*, Chapter 4. In: Institute of Materia Medica (Ed.) Advanced Studies in Traditional Chinese Herbal Medicine, Vol. 1, Beijing Med. Univ. and Peking Union University Press, Beijing, pp. 91–113.
40. Li SP, Song ZH, Dong TTX, Ji ZN, Lo CK, et al. (2004) Distinction of water-soluble constituents between natural and cultured Cordyceps by capillary electrophoresis. Phytomed 11, 684–690.
41. Hu Z, Xia FB, Wu XG, Wang Q, Xie JY, et al. (2004) New component analysis of essential oil from cultured *Cordyceps sinensis*. Edible Fungi Chin 23(5): 37–38.
42. Ye MQ, Hu Z, Fan Y, He L, Xia FB, et al. (2004) Purification and characterization of an acid deoxyribonuclease from the cultured mycelia of *Cordyceps sinensis*. J Biochem Mol Biol 37(4): 466–473.
43. Wang S, Yang FQ, Feng K, Li DQ, Zhao J, et al. (2009) Simultaneous determination of nucleosides, myriocin, and carbohydrates in Cordyceps by HPLC coupled with diode array detection and evaporative light scattering detection. J Sep Sci 32: 4069–4076.
44. Lu YY, Qiu XM, Jiang C, Liu SZ, Guo SW, et al. (2011) Analysis of nucleoside constituents in different parts of the artificial *Cordyceps sinensis*. Food Sci Technol 36(4): 250–256.
45. Yang ML, Kuo PC, Hwang TL, Wu TS (2011) Anti-inflammatory Principles from *Cordyceps sinensis*. J Nat Prod 74: 1996–2000.
46. Wu YS, Zhou DL, Yan D, Ren YS, Fang YL, et al. (2008) HPLC fingerprint analysis of Cordyceps and mycelium of cultured Cordy. Chin J Chin Materia Medica 33(19): 2212–2214.
47. Lai YH, Ruan GP, Xie YL, Chen HA (2008) Study on HPLC Fingerprint Characteristic Analysis of *Cordyceps sinensis* and Its Similar Products. J Chin Med Mater 31(8): 1142–114.

12

Angiogenesis Interactome and Time Course Microarray Data Reveal the Distinct Activation Patterns in Endothelial Cells

Liang-Hui Chu[1]*, Esak Lee[1], Joel S. Bader[1,2], Aleksander S. Popel[1,3]

1 Department of Biomedical Engineering, School of Medicine, Johns Hopkins University, Baltimore, Maryland, United States of America, **2** High-Throughput Biology Center, Johns Hopkins University, Baltimore, Maryland, United States of America, **3** Department of Oncology and Sidney Kimmel Comprehensive Cancer Center, School of Medicine, Johns Hopkins University, Baltimore, Maryland, United States of America

Abstract

Angiogenesis involves stimulation of endothelial cells (EC) by various cytokines and growth factors, but the signaling mechanisms are not completely understood. Combining dynamic gene expression time-course data for stimulated EC with protein-protein interactions associated with angiogenesis (the "angiome") could reveal how different stimuli result in different patterns of network activation and could implicate signaling intermediates as points for control or intervention. We constructed the protein-protein interaction networks of positive and negative regulation of angiogenesis comprising 367 and 245 proteins, respectively. We used five published gene expression datasets derived from in vitro assays using different types of blood endothelial cells stimulated by VEGFA (vascular endothelial growth factor A). We used the Short Time-series Expression Miner (STEM) to identify significant temporal gene expression profiles. The statistically significant patterns between 2D fibronectin and 3D type I collagen substrates for telomerase-immortalized EC (TIME) show that different substrates could influence the temporal gene activation patterns in the same cell line. We investigated the different activation patterns among 18 transmembrane tyrosine kinase receptors, and experimentally measured the protein level of the tyrosine-kinase receptors VEGFR1, VEGFR2 and VEGFR3 in human umbilical vein EC (HUVEC) and human microvascular EC (MEC). The results show that VEGFR1–VEGFR2 levels are more closely coupled than VEGFR1–VEGFR3 or VEGFR2–VEGFR3 in HUVEC and MEC. This computational methodology can be extended to investigate other molecules or biological processes such as cell cycle.

Editor: David D. Roberts, Center for Cancer Research, National Cancer Institute, United States of America

Funding: This work was supported by the National Institutes of Health (NIH) grants R01 CA138264, R01 HL101200, and R21 HL122721 (ASP), and U54 RR020839 and the Robert J. Kleberg, Jr. and Helen C. Kleberg Foundation (JSB). The funders had no role in study design, data collection and analysis, decision to publish, or preparation of the manuscript.

Competing Interests: The authors have declared that no competing interests exist.

* Email: lchu5@jhmi.edu

Introduction

Angiogenesis, the formation of new blood vessels from pre-existing vessels, is involved in both physiological (e.g. development, wound healing and exercise) and pathological conditions (e.g. cancer and ocular neovascularization, such as neovascular age-related macular degeneration). Numerous molecules are involved in angiogenesis: for example, vascular endothelial growth factors (VEGF) and their receptors, fibroblast growth factors (FGF) and their receptors, proteins in the matrix metalloproteinase (MMP) and Notch families. Other pro-angiogenic factors such as angiopoietin-1 and anti-angiogenic factors such as thrombospon-din-1 are also associated with regulation of angiogenesis. In order to integrate hundreds of angiogenesis-related molecules and infer angiogenesis-annotated genes, we have developed an algorithm to construct the angiome, a global protein-protein interaction network (PIN) relevant to angiogenesis [1].

Major regulators of angiogenesis for the endothelial cell, both ligands and their cell-surface receptors, were summarized in [2]. These regulators were classified as pro- or anti-angiogenic; such classification is important for application of our understanding of angiogenesis regulation to diseases. For example, suppression of major angiogenic regulators like VEGFA (conventionally referred to as VEGF), or release of endogenous anti-angiogenic factors like endostatin or thrombospondin can be used to inhibit tumor angiogenesis. An extended list of molecules involved in regulation of angiogenesis was constructed in [1], which included the families of VEGF, TGF (transforming growth factor), IGF (insulin-like growth factor), and PDGF (platelet-derived growth factor). Negative regulators of angiogenesis and associated proteins, including chemokines, angiopoietin, and serpin, were also considered.

Time course microarray data can help identify genes that are important in angiogenesis [1,3]. Cultured endothelial cells are widely used in angiogenesis research. The most commonly used

EC are human umbilical vein EC (HUVEC) and human microvascular EC (MEC); telomerase-immortalized human microvascular (TIME) EC are also used in functional genomics angiogenesis research [4]. Several time course microarray studies have been conducted to identify expressed genes in VEGF-treated HUVEC [5], MEC [6] and TIME cells [7]. The goal of this study is to combine the angiome with time-series gene expression data on VEGF-treated EC to investigate the dynamic responses of the key proteins and protein complexes in angiogenesis under different in vitro experimental conditions.

Materials and Methods

Constructing the networks of positive and negative regulation of angiogenesis

The flowchart of constructing the PIN of positive and negative regulation of angiogenesis is shown in Figure 1. We have constructed a gene search engine GeneHits described in [1] (accessible at http://sysbio.bme.jhu.edu). We constructed the angiome (the global protein-protein interaction network of angiogenesis) using the resources of SABiosciences, Gene Ontology (GO) and GeneCards [8]. The information on edges was downloaded from Michigan Molecular Interactions (MiMI) [9], which integrates eleven protein interaction data sources (BIND, CCSB, DIP, GRID, HPRD, IntAct, KEGG, MDC, MINT, PubMed and Reactome). The angiome network comprises 1,233 proteins and 5,726 interactions [1]. We will describe the new strategies, software and experimental datasets used in this study in the following sections.

Gene Ontology (GO) provides a rich resource of gene functions and locations in many different species [10]; positive regulation of angiogenesis (GO:0045766) and negative regulation of angiogenesis (GO:0016525) are included. Four genes are listed in both positive and negative regulators of angiogenesis: thrombospondin 1 (THBS1), angiopoietin 4 (ANGPT4), chemokine receptor 1 (CX3CR1), and serpin peptidase inhibitor member 1 (SERPINE1). However, THBS1 and SERPINE1 have been identified as anti-angiogenic [11–13]. Angiopoietin ANGPT4 is a protein that promotes angiogenesis [14]. Fractalkine (FKN)-induced activation of CX3CR1 in EC leads to in vivo angiogenesis through the induction of HIF-1alpha and VEGF-A gene expression by CX3CR1 activation and subsequent VEGF-A/KDR-induced angiogenesis [15]. Table 1 (A) and (B) presents 56 and 39 proteins annotated as positive and negative regulation of angiogenesis, respectively. We select the proteins in the extended angiome [1] which are linked to the 56 and 39 proteins in Table 1 (A) and (B) and their interactions to construct the two networks of positive and negative regulation of angiogenesis, respectively. Cytoscape is used to draw the PIN [16].

Microarray data analysis

We compiled five time-course microarray datasets at different experimental conditions on endothelial cells (Table 2). Schweighofer et al. [5] measured gene expression in HUVEC stimulated by VEGF and epidermal growth factor (EGF) (GSE10778). Glesne et al. [6] measured transcripts during proliferation and tubulogenesis in human MEC stimulated with VEGF (GSE2891). Mellberg et al. [7] cultured TIME cells (telomerase-immortalized human microvascular endothelial cells) in 3D collagen gels and on 2D

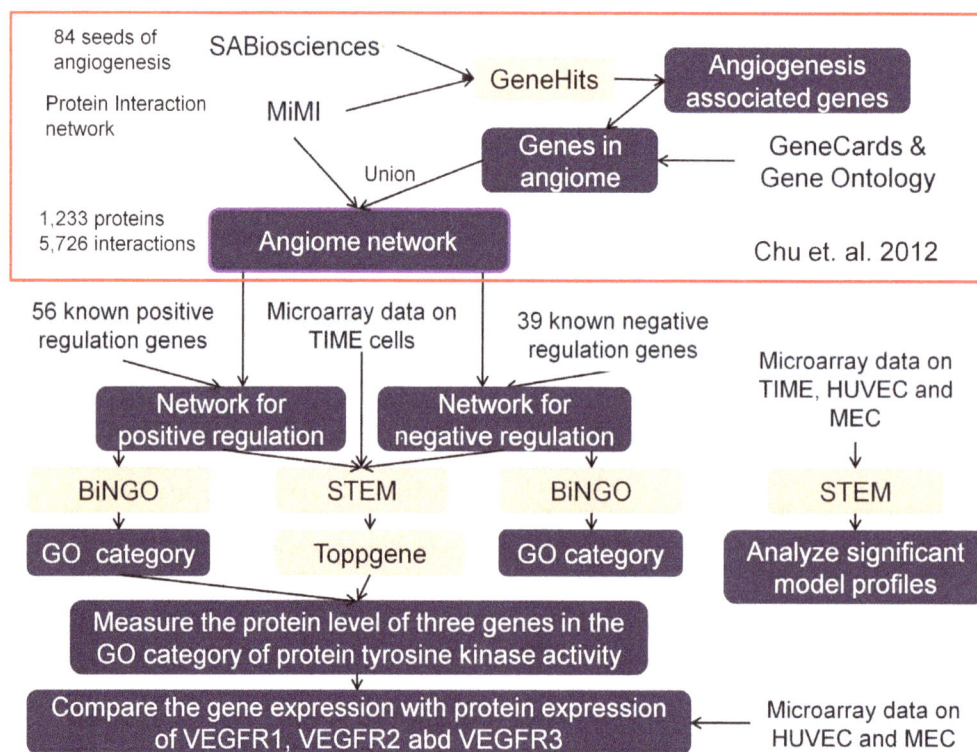

Figure 1. Flowchart of finding the protein complexes of angiome and merging time course gene expression data. We marked the methods used in the angiome study [1] with the red frame, and displayed the new methods in the lower part of the figure. These new strategies used in this study include software such as BiNGO (Biological Networks Gene Ontology) and STEM (Short Time-series Expression Miner), curated gene sets of positive and negative regulation of angiogenesis, use of microarray datasets and experimental design.

Table 1. List of genes in the angiome that are annotated as positive and negative regulators of angiogenesis shown in (A) and (B), respectively.

(A) 56 proteins annotated as positive regulators of angiogenesis: ADM, AGGF1, ANGPT4, ANGPTL3, ANXA3, AQP1, BTG1, C3, C3AR1, C5, CCL11, CCL24, CCL5, CCR3, CD34, CHRNA7, CTSH, CX3CR1, EPHA1, ERAP1, F3, FGF1, FGF2, FLT1, GATA2, GATA4, GATA6, HDAC9, HIF1A, HIPK1, HIPK2, HMOX1, IL1A, IL1B, KDR, MMP9, NOS3, PRKD1, PRKD2, PTGIS, PTGS2, RAMP2, RAPGEF3, RHOB, RRAS, RUNX1, SFRP2, SPHK1, TEK, TNFRSF1A, TNFSF12, TWIST1, UTS2R, VEGFA, VEGFB, WNT5A

(B) 39 proteins annotated as negative regulators of angiogenesis: AMOT, ANGPT2, APOH, BAI1, CCL2, CCR2, COL4A2, COL4A3, CXCL10, FASLG, FOXO4, GHRL, GTF2I, HDAC5, HHEX, HOXA5, HRG, KLF4, KLK3, KRIT1, LECT1, LIF, MAP2K5, NF1, NPPB, NPR1, PDE3B, PF4, PML, PTPRM, ROCK1, ROCK2, SERPINE1, SERPINF1, STAB1, THBS1, THBS2, THBS4, TIE1

fibronectin matrix, stimulated with VEGF and measured gene expression. Raw microarray data on TIME cells from Mellberg *et al.* [7] were kindly provided by the authors. We downloaded the time course microarray datasets from Gene Expression Omnibus (GEO) databases [5,6] and recovered the missing data from Mellberg *et al.* [7] using GenePattern 3.6.1 [17]. Gene Expression Omnibus (GEO) data were imported by GEOImporter version 5. Genes with missing values in Mellberg *et al.* [7] were recovered by the *k* nearest neighbors (KNN) algorithm in ImputeMissingValuesKNN version 13 module. We used the default settings in GenePattern software.

Temporal expression pattern

We use Short Time-series Expression Miner (STEM) [18] to identify significant temporal expression profiles and the genes associated with these profiles integrated with Gene Ontology (GO) database from microarray experiments. The clustering method of gene expression profiles is based on STEM clustering method; details of the algorithms are described in [19]. This clustering algorithm first selects several distinct and representative temporal expression profiles, called "model profiles". The model profile starts at the first time point, and then the profile between the two time points can be unchanged, increase or decrease with an integer number of time units. The model profiles are selected independently from the data to determine the significance of the different clusters. The STEM clustering algorithm assigns each gene to the model profile that matches the expression profile of genes most closely by the correlation coefficient. We set GO annotations as biological processes and molecular functions with minimum GO depth of 3, number of permutations per gene to 50, and significance level p-values to 0.05 by Bonferroni correction.

Functional enrichment of genes associated with positive and negative regulation of angiogenesis

We used BiNGO 2.44 (Biological Networks Gene Ontology tool) [20] on Cytoscape 2.8 [16] for the functional enrichment analysis of genes in the positive and negative regulation of angiogenesis PINs to identify pathways and biological processes. The p-values were computed by the hypergeometric test, and the Benjamini & Hochberg false discovery rate (FDR) correction was also computed at a significance level 0.05.

Cell culture

Human microvascular endothelial cells (MEC) and human umbilical vein endothelial cells (HUVEC) were purchased from Lonza (Walkersville, MD). MEC were propagated in microvascular endothelial cell growth medium-2 (EGM-2MV, Lonza). HUVEC were grown in endothelial cell growth medium-2 (EGM-2, Lonza). Cells were maintained under standard conditions of 37°C and 5% CO_2 and the passage numbers of the endothelial cells were kept between 3 and 6.

Western blot assay

MEC and HUVEC in passages 3 to 6 (Lonza) were plated in 75T tissue culture flasks at 1,000,000 cells/well in the normal growth media (EGM-2MV for MEC; EGM-2 for HUVEC, from Lonza). After 48 hr, normal growth media were replaced with serum-free media (EBM-2 without supplements) and incubation lasted 24 hr to starve the cells. Human $VEGF_{165}$ (50 ng/ml, R&D systems) in serum-free media was applied, and the flasks were incubated for 0, 1, 3, 6, 12, 24 hr at 37°C, and 5% CO_2. VEGF treatment was stopped by adding cold PBS and the cells were lysed in cold lysis buffer (150 mM NaCl, 1 mM EDTA, 1 l/ml protease inhibitors (Sigma Aldrich), 1 l/ml phosphatase inhibitors (Sigma) and 1% Triton X-100) for 2 hr at 4°C, then scraped to collect the lysates. Cell lysates were spun at 14,000 g for 30 min to remove

Table 2. Five VEGF-treated time-course microarray datasets with different experimental conditions on endothelial cells.

Treatment	Cells	Time	Resource	Ref
VEGFA	HUVEC	0,0.5,1,2.5,6 hr	GSE10778	(Schweighofer, et al., 2009)
VEGFA	MEC (proliferation)	0,0.5,1,2,4 hr	GSE3891	(Glesne, et al., 2006)
VEGFA	MEC (tubulogenesis)	0.5,1,2,4,8 hr	GSE3891	(Glesne, et al., 2006)
VEGFA	TIME (3D collagen I)	15 min,1,3 6,9,12, 18,24 hr	Provided by authors	(Mellberg, et al., 2009)
VEGFA	TIME (2D fibronectin)	15 min,1,3,6,9,12, 18,24 hr	Provided by authors	(Mellberg, et al., 2009)

dead cells and cell debris. Cell lysates were separated by sodium dodecyl sulfate polyacrylamide gel electrophoresis (SDS-PAGE) and transferred to nitrocellulose blots (Invitrogen, Carlsbad, CA), using the iBlot transfer module (Program 3, 14 min). We blocked the nitrocellulose membrane for 1 hr with 5% non-fat milk+1% BSA (bovine serum albumin, Sigma) in TBST (1X TBS with 0.1% Tween 20) at room temperature, and the membrane was probed with antibodies detecting human VEGFR1 or VEGFR2 or VEGFR3 at 1:1000 dilution (Cell Signaling Technology and Abcam). Glyceraldehyde 3-phosphate dehydrogenase (GAPDH, 1:2000, Cell Signaling) was used as a loading control. HRP-labelled secondary antibodies were added at 12000 dilution and protein bands were detected with the Amersham ECL Prime Western Blotting Detection Reagent (GE Healthcare). Western bands were analyzed by quantifying number of pixels in the band using ImageJ free software (NIH, Bethesda, MD). Full length VEGFR1 (180 kDa) in MEC and HUVEC were analyzed, as HUVEC does not show sVEGFR1 (110 kDa). Similarly full length VEGFR3 (195 kDa) was analyzed, as HUVEC do not show the unglycosylated precursor form (175 kDa). Each band was normalized by using the GAPDH level.

Results

Constructing the networks of positive and negative regulation of angiogenesis

Following the description in Methods and flowchart in Figure 1, we constructed the two networks of positive and negative regulation of angiogenesis. The PIN of positive regulation of angiogenesis comprises 367 proteins and 1,972 interactions (Table S1); the PIN of negative regulation of angiogenesis comprises 245 proteins and 1,154 interactions (Table S2). Some proteins in the positive regulation of angiogenesis are also connected to the proteins in the negative regulation of angiogenesis by physical interactions present in the MiMI [20] database and literature reports, such as anti-angiogenic thrombospondin (THBS1) directing binding to angiogenic proteins COL1A1 (collagen type I) [21] and MMP9 (matrix metallopeptidase 9) [22]. Details of protein interaction types and resources of interactions are provided in Table S1 and S2. The list of repeated proteins included in both positive and negative regulation of angiogenesis is provided in Table S2. We used BiNGO 2.44 (Biological Networks Gene Ontology tool) [23] for the functional enrichment of genes in the two angiogenesis PINs (Table S3 and S4).

Temporal gene expression pattern

Among microarray datasets shown in Table 2, Mellberg's dataset on TIME cells [7] contains the most time points at 15 min and 1, 3, 6, 9, 12, 18, and 24 h. We used the STEM [18] to identify significant temporal gene expression profiles and the genes associated with these profiles integrated GO database. We found the temporal gene expression pattern of all the genes in the raw microarray data. We normalized the microarray data to the first time point in each of the set [5,6] except Mellberg's data [7] which have been normalized to the untreated conditions. The maximum number of model profiles was set as 20 and also compared with the maximum number of model profiles as 10 and 40 in temporal gene expression profiles of TIME cells on 2D fibronectin and 3D collagen I (Table S5). The genes with absolute log2 fold change between the maximum and minimum values of any two over all time points less than 1 are removed in the analysis.

We show the four statistically significant (adjusted p-value<0.05 by Bonferroni correction) temporal gene expression profiles of TIME cells on 2D fibronectin, and sort the four profiles by their p-values in Figure 2 (A). The p-value was calculated by the number of genes assigned to the model profile, compared to the expected number of assigned genes. The number on top left represents the assigned profile number by STEM, and the number on bottom left represents the significance level before the Bonferroni correction. The box is colored if the statistically significant number of genes, based on the adjusted p-value<0.05 by Bonferroni correction, are assigned to the model profile. The black and red lines in the individual profile boxes indicate the assigned pattern, e.g. the sequence (0,1,2,3,4,5,6,7,8) over the eight time points and initial points in profile #16, and the gene expression of genes assigned in that profile. We compare the four statistical significant profiles on 2D fibronectin in Figure 2 (A) with 3D type I collagen, and plot the four profiles (#16, #4, #5, #9) in Figure 2 (B). We found statistically significant patterns of continuous up- and down-regulation depicted by profiles #16 and #4 (shown on the top-left corner of each profile box) exist for both matrices on TIME cells, but fluctuation patterns depicted by profiles #5 and #9 are only exhibited on 2D fibronectin (Figure 2A).

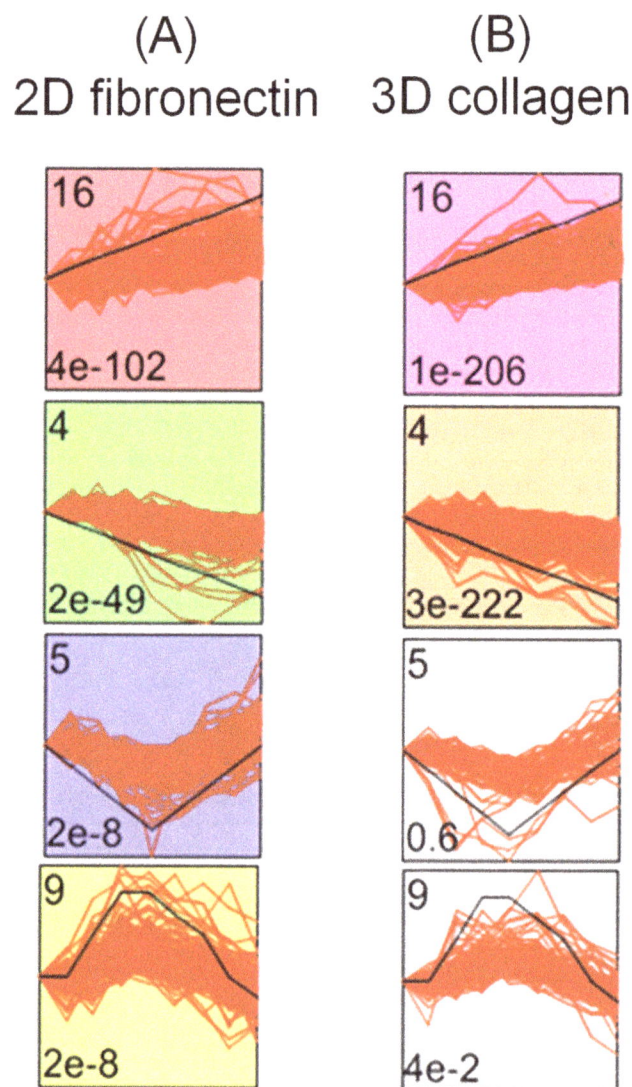

Figure 2. Temporal gene expression profiles on 2D fibronectin and 3D collagen I for TIME cells in (A) and (B), respectively.

We found that the top five most significant GOs on 2D fibronectin of profile #16 (Bonferroni corrected p-value<0.008 by using a randomization test) in Figure 2A are angiogenesis (GO:0001525), vasculature development (GO:0001944), extracellular matrix organization (GO:0030198), extracellular structure organization (GO:0043062) and system development (GO:0048731). The top five most significant GOs (corrected p-value<0.001) on 3D type I collagen matrix of the monotonically increasing profile #16 in Figure 2B are angiogenesis (GO:0001525), blood vessel morphogenesis (GO:0048514), vasculature development (GO:0001944), blood vessel development (GO:0001568) and extracellular matrix organization (GO:0030198). We found several significant GOs with similar GO functions, such as angiogenesis (GO:0001525), vasculature development (GO:0001944), blood vessel development (GO:0001568), and blood vessel morphogenesis (GO:0048514). The clusters of significant profiles, GOs and genes in the profile are listed in Table S6.

We compared the statistically significant model profiles between 2D fibronectin (#16, 4, 5 and 9 in Figure 2A) and 3D type I collagen (#4 and 16 in Figure 2B) by the maximum uncorrected intersection p-value as 0.005 in Table S7. Table S7 also lists the genes in the pairwise comparisons with at least one significant GO category by the corrected p-value<0.05. The p-value was calculated based on the hypergeometric distribution of the intersection of genes assigned to the two profiles, one profile from the original data set and the other from the comparison data set [18]. 115 and 112 genes are assigned to the intersection of the increasing profile #16 and the decreasing profile #4 on two matrices, respectively. Among these 115 genes in pattern #16 of the intersection, fourteen genes annotated as GO:0001525 angiogenesis (corrected p-value = 0.004) include ADM, CDH13, COL4A1, EPHA2, HSPG2, ISL1, ITGAV, MMP2, NOTCH4, RAMP2, RGC32, RHOB, VASH1 and VEGFC. We found three genes, the LIM-homeobox transcription factor islet-1 (ISL1), Response gene to complement 32 (RGC32), and vasohibin-1 (VASH1) annotated as angiogenesis, which were not included in the angiome in our previous study [1]. We also compare the different activation pattern pairs between the two substrates. Profile #5 on 2D fibronectin and #4 in 3D collagen I share some genes in the GO:0048584 positive regulation of response to stimulus (corrected p-value = 0.006), including BCAR1, DAB2, DAPK3, DBNL, DUSP7, F2RL1, FZD4, GPR177, IGKC and TNFSF10.

Significant temporal gene expression profiles in VEGFA-treated MEC and HUVEC

We used the STEM software [18] to find significant temporal gene expression profiles in previously reported datasets of VEGFA-treated MEC (GSE3891) [6] and HUVEC (GSE10778) [5]; the results are presented in Tables S8–S9 listing the significant profiles (p-value<5E-2). The GOs for the significantly decreasing profiles #4 (p-value = 6.40E-40) in VEGF-treated MEC during proliferation [7] include translational initiation, termination and elongation. We compare the genes in the intersection of significant profiles during proliferation and tubulogenesis in VEGFA-treated MEC (Table S8). The GO categories for the genes in the intersection of increasing profile #17 during tubulogenesis and decreasing profile #4 during proliferation include translational termination and elongation (p-value<0.001). This analysis shows that some genes involved in protein translation behave differently during endothelial cell proliferation and tubulogenesis. The temporal gene expression profiles in HUVEC [5] in Table S9 show more diverse patterns than TIME cells shown in Figure 2.

Activation patterns of the receptor protein tyrosine kinase

We used the defined gene set of the 367 and 245 genes in the positive and negative regulation of angiogenesis in STEM clustering, respectively (Table S10). We found 21 and 19 genes in the positive regulation of angiogenesis which were assigned to the increasing profile #16 on 3D type I collagen and 2D fibronectin, respectively. We further used Toppgene [24] to analyze the functional enrichment of proteins included in the increasing profile #16 of positive regulation of angiogenesis (Table S10). One of the top significant GO molecular functions for 21 and 19 genes in the increasing profile #16 of positive regulation of angiogenesis on 3D type I collagen is protein tyrosine kinase activity (p-value = 9.568E-5 and 2.648E-3).

Since tyrosine kinase activity is of great interest in translational applications, we scrutinize the genes annotated as protein tyrosine kinase activity in the functional enrichment of positive regulation of angiogenesis in Table S3. The GO category "transmembrane receptor protein tyrosine kinase activity" (adjusted p-value = 1.64E-13) contains eighteen proteins ALK, EGFR, EPHA1, EPHB2, FGFR1, FGFR2, FGFR3, FGFR4, FGFRL1, FLT1 (VEGFR1), FLT4 (VEGFR3), IGF1R, KDR (VEGFR2), NRP1, NRP2, NTRK2, TEK and TIE1. We merge proteomic and genomic data based on the 2007 protocol [25] by Cytoscape [16]. The proteins with gene transcripts on 3D type I collagen for TIME cells in Figure 3 show that FLT1 is activated consistently after 6 h, KDR only activated at 24 h, and FLT4 decreased from 15 min to 9 hr then increased from 12 hr to 24 hr. The VEGF ligands family and their receptors play important roles in the development, maintenance, and remodeling of the vasculature [26,27]. Thus, we select three VEGF receptor tyrosine kinases VEGFR1 (FLT1), VEGFR2 (KDR), and VEGFR3 (FLT4) to perform protein-level time series in vitro experiments.

Comparison between gene transcripts and protein expression

We explored time-dependent VEGF receptor expression in blood endothelial cells after VEGF treatment, as the VEGF-VEGFR axis is pivotal in endothelial cell growth and maintenance. The experimental results for FLT1 (VEGFR1), KDR (VEGFR2) and FLT4 (VEGFR3) are shown in Figure 4. Briefly, one million of HUVEC or MEC were starved overnight, after which we treated the cells with 50 ng/ml of human $VEGF_{165}$ and incubated them for 0, 1, 3, 6, 12 and 24 hr at 37°C. Total protein levels of VEGFR1/2/3 and glyceraldehyde 3-phosphate dehydrogenase (GAPDH) were obtained for normalization in data analyses. The number of pixels of each western band was analyzed using ImageJ (NIH, Bethesda). In VEGFR1 and VEGFR3 analyses, the full length VEGFR1 (180 kDa) and the full length VEGFR3 (195 kDa) in MEC and HUVEC were analyzed, as HUVEC do not express some isoforms. Each band was finally normalized by the GAPDH level.

We plotted the ratio of measured protein levels of VEGFR1, VEGFR2 and VEGFR3 to the GAPDH in HUVEC and MEC in Figure 5 (A) and (B), respectively, normalized to the first time point. We observed that VEGFR1 and VEGFR2 levels are more closely coupled than VEGFR1-VEGFR3 or VEGFR2-VEGFR3 in HUVEC and MEC. Interestingly, VEGFR1 and VEGFR2 were restored to the initial protein level after the mid-time point of VEGF treatment, suggesting that some downstream signaling of VEGF pathway may induce VEGFR1 and VEGFR2 expression. Figure 5 (B) shows that VEGFR1 drops after VEGF treatment in MEC, showing its minimum level at 3 hr, then, VEGFR1 recovers

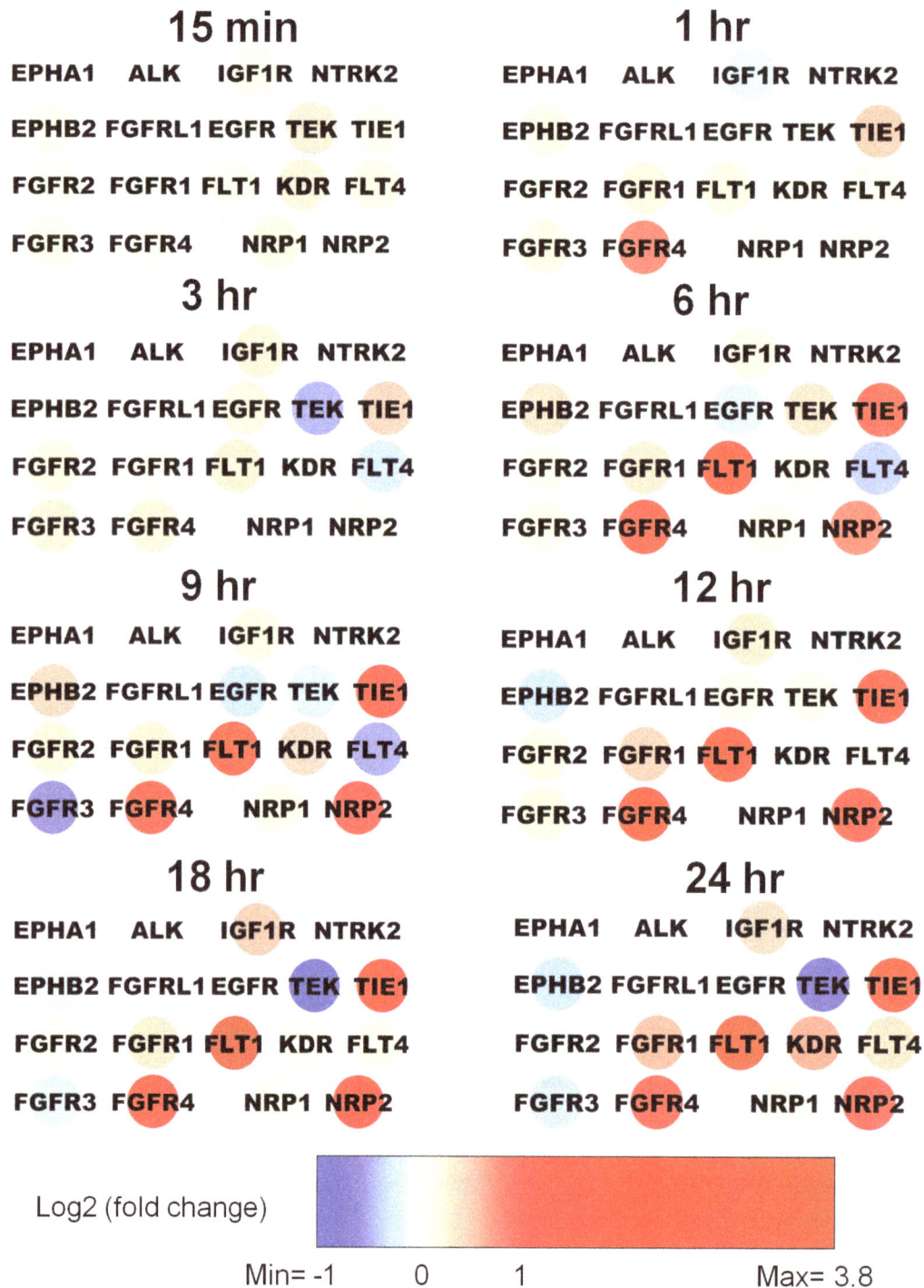

Figure 3. Activation pattern of receptor tyrosine kinases on 3D collagen I for TIME cells.

continuously. Similarly, VEGFR2 shows minimum level at 6 hr, and recovered after that time point. This "drop and recovery" pattern is not shown in VEGFR3. There might indicate different regulation mechanisms for VEGFR1/2 and VEGFR3 in endothelial cells.

VEGFR1 and VEGFR2 levels increase after 12 hr in VEGF treated MEC and HUVEC. This confirms that VEGF is a potent mitogen for blood endothelial cells, and that VEGF-treated HUVEC and MEC may not be significantly involved in VEGFR3

signaling, which is known as a lymphangiogenic receptor, though these endothelial cells express VEGFR3. Also crosstalk between VEGFR1 and VEGFR2 in the presence of VEGF, especially their reciprocal mitogenic signaling is an important topic in the endothelial cell biology. VEGFR1 has three isoforms: full length/membrane bound form (180 kDa), and two soluble forms (110, 75 kDa) [28]. These soluble VEGFR1 lack the intracellular tyrosine kinase and the transmembrane domains, and play as scavenger molecules of VEGF. VEGFR3 has three isoforms:

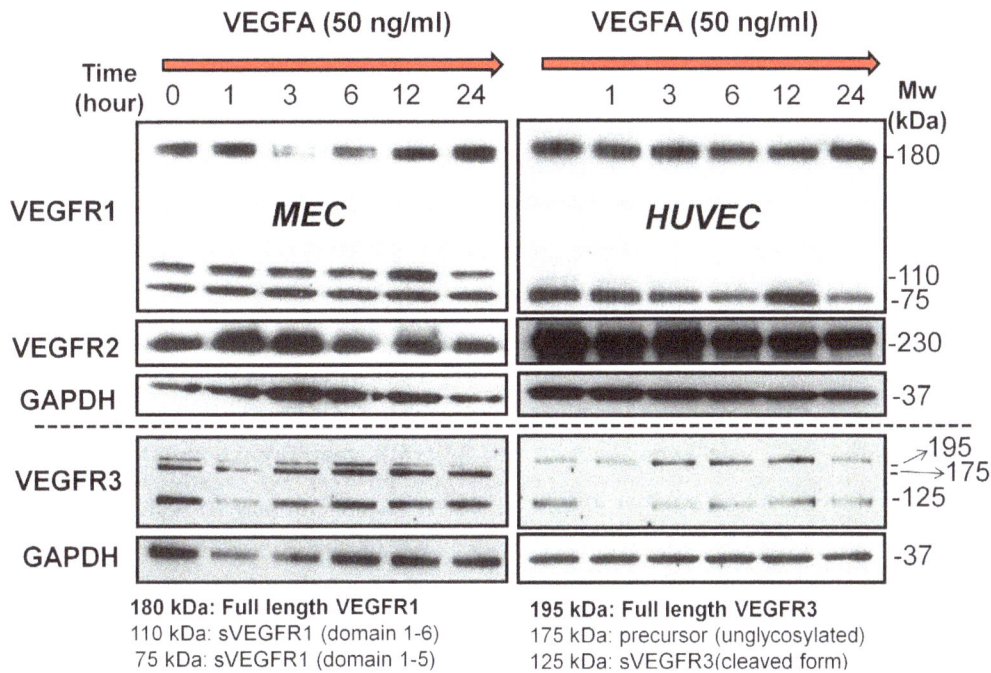

Figure 4. Experiments of VEGFR1, VEGFR2 and VEGFR3 for MEC and HUVEC.

glycosylated (195 kDa), unglycosylated precursor (175 kDa), and their cleaved form (125 kDa). Interestingly, HUVEC did not show larger soluble form of VEGFR1 (110 kDa) and unglycosylated precursor form of VEGFR3 (175 kDa) compared to MEC that express all three isoforms of VEGFR1 and VEGFR3 (Figure 4) [29]. Different expression of the VEGFR isoforms, their processes of cleavage, and biological functions of these isoforms under VEGF treatment in different endothelial cells need to be further investigated.

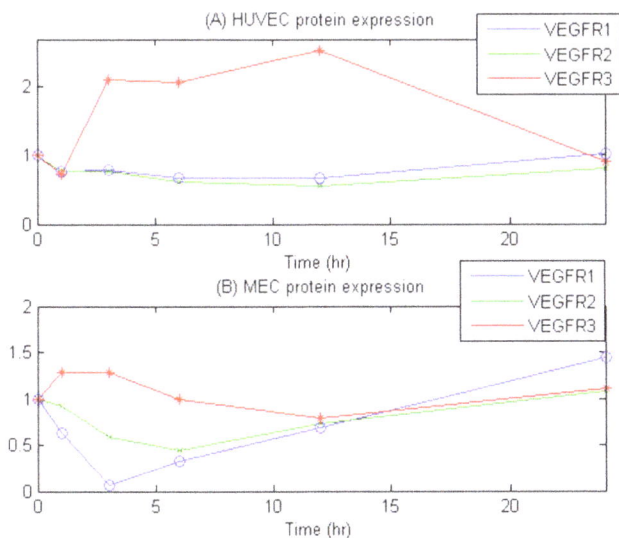

Figure 5. Normalized protein level measurement of VEGFR1, VEGFR2 and VEGFR3 to GAPDH on HUVEC and MEC in (A) and (B), respectively.

The study of time-specific differences in gene expression in EC could provide important insights into their role in normal physiology and diseases. In normal physiology, temporal gene expression can result in EC heterogeneity [30]. EC heterogeneity is observed in different organs and different stages of development [31]. Our study may enable understanding of angiogenesis processes in different location or stages of organs by identifying crucial genes in each context. We particularly identified time-specific activation patterns of genes in VEGF-treated TIME cells, HUVEC and MEC. VEGF is pivotal for life: if it is abolished, it results in the embryonic death and impaired tissue maintenance and regeneration [32]. As our study was based on microarray data, it could be applied to the experimental design with time-dependent quantitative RT-PCR to identify other genes that regulate EC proliferation and migration in the presence of VEGF. Our analyses, however, need to be further explored at protein expression levels. The time-dependent approach for the gene expression in angiogenesis is also important for diseases. Plasma concentrations of VEGF and its receptors vary in a time-dependent manner before and after lung cancer surgery [33]. Proangiogenic plasma alterations such as Angiopoietin-1 (Ang1), VEGF and soluble VEGFR1 may result in cancer patients developing recurrent disease after surgery. Time-dependent changes of plasma VEGF levels and VEGFR1 in acute lung injury in the rat sepsis model revealed the pulmonary VEGF and the signaling pathways [33]. The VEGF and VEGFR1 levels are increased in liver tissues in lipopolysaccharide (LPS)-induced endotoxemia in a time-dependent manner [34]. Therefore, studies in VEGF-dependent diseases and associated abnormalities in blood endothelium can benefit from the current study.

In summary, we investigated the different activation patterns of genes in VEGF treated human endothelial cells (TIME cells, HUVEC and MEC). This computational methodology can be extended to investigate various VEGF dependent biological processes. All the files including the Cytoscape and microarray datasets for STEM simulations are provided on our laboratory website http://pages.jh.edu/~apopel/software.html.

Conclusions

Combining gene expression data and protein interactions could reveal the dynamics of positive and negative regulation of angiogenesis in different endothelial cells and under different experimental conditions. We constructed two protein interaction networks representing positive and negative regulation of angiogenesis and found several clusters from gene ontology annotations and network properties. These findings capture the dynamics of protein interactions in regulation of angiogenesis, and can serve as a guide for experimental design related to activation patterns of important proteins in angiogenesis.

Supporting Information

Table S1 Proteins and interactions in positive regulation of angiogenesis.

Table S2 Proteins and interactions in negative regulation of angiogenesis.

Table S3 Functional enrichment analysis of genes in the positive regulation of angiogenesis.

Table S4 Functional enrichment analysis of genes in the negative regulation of angiogenesis.

Table S5 Comparison of number of model profiles as 10, 20 and 40 in STEM analysis.

Table S6 Significant temporal profiles on 2D fibronectin and 3D type I collagen for TIME cells.

Table S7 Comparisons of significant profiles in 2D fibronectin and 3D type I collagen.

Table S8 Significant temporal profiles for MEC.

Table S9 Significant temporal profiles for HUVEC.

Table S10 Significant temporal profiles by defined gene sets of positive and negative regulation of angiogenesis in 2D fibronectin and 3D type I collagen.

Acknowledgments

The authors thank Dr. Sofie Mellberg and Dr. Lena Claesson-Welsh for providing the raw data from their microarray experiments [7].

Author Contributions

Conceived and designed the experiments: LHC EL. Performed the experiments: EL. Analyzed the data: LHC. Contributed reagents/materials/analysis tools: LHC EL. Contributed to the writing of the manuscript: LHC EL JSB ASP.

References

1. Chu LH, Rivera CG, Popel AS, Bader JS (2012) Constructing the angiome: a global angiogenesis protein interaction network. Physiol Genomics 44: 915–924.
2. Hagedorn M, Bikfalvi A (2000) Target molecules for anti-angiogenic therapy: from basic research to clinical trials. Crit Rev Oncol Hematol 34: 89–110.
3. Rivera CG, Mellberg S, Claesson-Welsh L, Bader JS, Popel AS (2011) Analysis of VEGF-A regulated gene expression in endothelial cells to identify genes linked to angiogenesis. PLoS One 6: e24887.
4. van Beijnum JR, van der Linden E, Griffioen AW (2008) Angiogenic profiling and comparison of immortalized endothelial cells for functional genomics. Exp Cell Res 314: 264–272.
5. Schweighofer B, Testori J, Sturtzel C, Sattler S, Mayer H, et al. (2009) The VEGF-induced transcriptional response comprises gene clusters at the crossroad of angiogenesis and inflammation. Thromb Haemost 102: 544–554.
6. Glesne DA, Zhang W, Mandava S, Ursos L, Buell ME, et al. (2006) Subtractive transcriptomics: establishing polarity drives in vitro human endothelial morphogenesis. Cancer Res 66: 4030–4040.
7. Mellberg S, Dimberg A, Bahram F, Hayashi M, Rennel E, et al. (2009) Transcriptional profiling reveals a critical role for tyrosine phosphatase VE-PTP in regulation of VEGFR2 activity and endothelial cell morphogenesis. FASEB J 23: 1490–1502.
8. Stelzer G, Dalah I, Stein TI, Satanower Y, Rosen N, et al. (2011) In-silico human genomics with GeneCards. Hum Genomics 5: 709–717.
9. Gao J, Ade AS, Tarcea VG, Weymouth TE, Mirel BR, et al. (2009) Integrating and annotating the interactome using the MiMI plugin for cytoscape. Bioinformatics 25: 137–138.
10. Ashburner M, Ball CA, Blake JA, Botstein D, Butler H, et al. (2000) Gene ontology: tool for the unification of biology. The Gene Ontology Consortium. Nat Genet 25: 25–29.
11. Karagiannis ED, Popel AS (2008) A systematic methodology for proteome-wide identification of peptides inhibiting the proliferation and migration of endothelial cells. Proc Natl Acad Sci U S A 105: 13775–13780.
12. Tolsma SS, Volpert OV, Good DJ, Frazier WA, Polverini PJ, et al. (1993) Peptides derived from two separate domains of the matrix protein thrombospondin-1 have anti-angiogenic activity. J Cell Biol 122: 497–511.
13. Filleur S, Volz K, Nelius T, Mirochnik Y, Huang H, et al. (2005) Two functional epitopes of pigment epithelial-derived factor block angiogenesis and induce differentiation in prostate cancer. Cancer Res 65: 5144–5152.
14. Sillen A, Brohede J, Lilius L, Forsell C, Andrade J, et al. (2010) Linkage to 20p13 including the ANGPT4 gene in families with mixed Alzheimer's disease and vascular dementia. J Hum Genet 55: 649–655.
15. Ryu J, Lee CW, Hong KH, Shin JA, Lim SH, et al. (2008) Activation of fractalkine/CX3CR1 by vascular endothelial cells induces angiogenesis through VEGF-A/KDR and reverses hindlimb ischaemia. Cardiovasc Res 78: 333–340.
16. Smoot ME, Ono K, Ruscheinski J, Wang PL, Ideker T (2011) Cytoscape 2.8: new features for data integration and network visualization. Bioinformatics 27: 431–432.
17. Reich M, Liefeld T, Gould J, Lerner J, Tamayo P, et al. (2006) GenePattern 2.0. Nat Genet 38: 500–501.
18. Ernst J, Bar-Joseph Z (2006) STEM: a tool for the analysis of short time series gene expression data. BMC Bioinformatics 7: 191.
19. Ernst J, Nau GJ, Bar-Joseph Z (2005) Clustering short time series gene expression data. Bioinformatics 21 Suppl 1: i159–168.
20. Tarcea VG, Weymouth T, Ade A, Bookvich A, Gao J, et al. (2009) Michigan molecular interactions r2: from interacting proteins to pathways. Nucleic Acids Res 37: D642–646.
21. Galvin NJ, Vance PM, Dixit VM, Fink B, Frazier WA (1987) Interaction of human thrombospondin with types I-V collagen: direct binding and electron microscopy. J Cell Biol 104: 1413–1422.
22. Bein K, Simons M (2000) Thrombospondin type 1 repeats interact with matrix metalloproteinase 2. Regulation of metalloproteinase activity. J Biol Chem 275: 32167–32173.
23. Maere S, Heymans K, Kuiper M (2005) BiNGO: a Cytoscape plugin to assess overrepresentation of gene ontology categories in biological networks. Bioinformatics 21: 3448–3449.
24. Chen J, Bardes EE, Aronow BJ, Jegga AG (2009) ToppGene Suite for gene list enrichment analysis and candidate gene prioritization. Nucleic Acids Res 37: W305–311.
25. Cline MS, Smoot M, Cerami E, Kuchinsky A, Landys N, et al. (2007) Integration of biological networks and gene expression data using Cytoscape. Nat Protoc 2: 2366–2382.
26. Vempati P, Popel AS, Mac Gabhann F (2013) Extracellular regulation of VEGF: Isoforms, proteolysis, and vascular patterning. Cytokine Growth Factor Rev.
27. Mac Gabhann F, Popel AS (2008) Systems biology of vascular endothelial growth factors. Microcirculation 15: 715–738.
28. Cebe-Suarez S, Zehnder-Fjallman A, Ballmer-Hofer K (2006) The role of VEGF receptors in angiogenesis; complex partnerships. Cell Mol Life Sci 63: 601–615.
29. Bando H, Brokelmann M, Toi M, Alitalo K, Sleeman JP, et al. (2004) Immunodetection and quantification of vascular endothelial growth factor receptor-3 in human malignant tumor tissues. Int J Cancer 111: 184–191.

30. Minami T, Aird WC (2005) Endothelial cell gene regulation. Trends Cardiovasc Med 15: 174–184.

31. Nolan DJ, Ginsberg M, Israely E, Palikuqi B, Poulos MG, et al. (2013) Molecular signatures of tissue-specific microvascular endothelial cell heterogeneity in organ maintenance and regeneration. Dev Cell 26: 204–219.

32. Kajdaniuk D, Marek B, Borgiel-Marek H, Kos-Kudla B (2011) Vascular endothelial growth factor (VEGF) - part 1: in physiology and pathophysiology. Endokrynol Pol 62: 444–455.

33. Jesmin S, Zaedi S, Islam AM, Sultana SN, Iwashima Y, et al. (2012) Time-dependent alterations of VEGF and its signaling molecules in acute lung injury in a rat model of sepsis. Inflammation 35: 484–500.

34. Zaedi S, Jesmin S, Yamaguchi N, Shimojo N, Maeda S, et al. (2006) Altered expression of endothelin, vascular endothelial growth factor, and its receptor in hepatic tissue in endotoxemic rat. Exp Biol Med (Maywood) 231: 1182–1186.

13

Vesicular LL-37 Contributes to Inflammation of the Lesional Skin of Palmoplantar Pustulosis

Masamoto Murakami[1]*, Takaaki Kaneko[2], Teruaki Nakatsuji[3], Kenji Kameda[4], Hidenori Okazaki[1], Xiuju Dai[1], Yasushi Hanakawa[1], Mikiko Tohyama[1], Akemi Ishida-Yamamoto[2], Koji Sayama[1]

1 Department of Dermatology, Ehime University Graduate School of Medicine, Ehime, Japan, 2 Department of Dermatology, Asahikawa Medical College, Asahikawa, Japan, 3 Division of Dermatology, University of California San Diego, and VA San Diego Healthcare Center, San Diego, California, United States of America, 4 Integrated Center for Science, Ehime University Graduate School of Medicine, Ehime, Japan

Abstract

"Pustulosis palmaris et plantaris", or palmoplantar pustulosis (PPP), is a chronic pustular dermatitis characterized by intraepidermal palmoplantar pustules. Although early stage vesicles (preceding the pustular phase) formed in the acrosyringium contain the antimicrobial peptides cathelicidin (hCAP-18/LL-37) and dermcidin, the details of hCAP-18/LL-37 expression in such vesicles remain unclear. The principal aim of the present study was to clarify the manner of hCAP-18/LL-37 expression in PPP vesicles and to determine whether this material contributed to subsequent inflammation of lesional skin. PPP vesicle fluid (PPP-VF) induced the expression of mRNAs encoding IL-17C, IL-8, IL-1α, and IL-1β in living skin equivalents, but the level of only IL-8 mRNA decreased significantly upon stimulation of PPP vesicle with depletion of endogenous hCAP-18/LL-37 by affinity chromatography (dep-PPP-VF). Semi-quantitative dot-blot analysis revealed higher concentrations of hCAP-18/LL-37 in PPP-VF compared to healthy sweat (2.87 ± 0.93 μM vs. 0.09 ± 0.09 μM). This concentration of hCAP-18/LL-37 in PPP-VF could upregulate expression of IL-17C, IL-8, IL-1α, and IL-1β at both the mRNA and protein levels. Recombinant hCAP-18 was incubated with dep-PPP-VF. Proteinase 3, which converts hCAP-18 to the active form (LL-37), was present in PPP-VF. Histopathological and immunohistochemical examination revealed that early stage vesicles contained many mononuclear cells but no polymorphonuclear cells, and the mononuclear cells were CD68-positive. The epidermis surrounding the vesicle expresses monocyte chemotactic chemokine, CCL2. In conclusion, PPP-VF contains the proteinase required for LL-37 processing and also may directly upregulate IL-8 in lesional keratinocytes, in turn contributing to the subsequent inflammation of PPP lesional skin.

Editor: Paul Proost, University of Leuven, Rega Institute, Belgium

Funding: This study was supported by a Grant-in-Aid for Scientific Research (c) 25461670, Japan Society for the Promotion of Science (JSPS), and a grant for creative bioscience research at Ehime University, 2012–2013. The funders had no role in study design, data collection and analysis, decision to publish, or preparation of the manuscript.

Competing Interests: The authors have declared that no competing interests exist.

* Email: mamuraka@m.ehime-u.ac.jp

Introduction

"Pustulosis palmaris et plantaris", or palmoplantar pustulosis (PPP), is a chronic pustular dermatitis characterized by intraepidermal palmoplantar pustules [1]. On careful observation in the clinic, a PPP lesion exhibits several unique characteristics including vesicles, pustules, erythema, lichenification, and abnormal desquamation. Although PPP is a common skin disease that is often recalcitrant to available treatments, the pathogenesis of the condition remains unknown. Prior to pustule formation, vesicles form early in the acrosyringium, and antimicrobial peptides found in human sweat, hCAP-18/LL-37 and dermcidin are present in vesicles of the palms and soles [2]. In eccrine sweat, these components protect the body surface via the innate immune system. Dermcidin is continuously secreted in eccrine sweat but is not induced during inflammation [3,4]. In contrast, hCAP-18/LL-37 is induced in inflammatory conditions such as psoriasis and wound healing [5,6].

Later, secondary leukocyte accumulation in vesicles is associated with expression of complement and/or IL-8 in the stratum corneum or the surrounding epidermal keratinocytes [7]. Furthermore, interleukin (IL)-17-positive cells infiltrate around the acrosyringium [8]. Although the mechanism of abnormal desquamation remains unclear, aberrant expression of kallikrein-related peptidases (KLK-5, -7, and -14) in lesional skin may be important in this context [9].

Human skin contains two major classes of antimicrobial peptides: the cathelicidins [10–12] and the β-defensins [13–15]. Like many other antimicrobial peptides, cathelicidins are synthesized as preproproteins [12]. The only human cathelicidin is hCAP-18 [5,16], expressed in leukocytes and on a variety of epithelial surfaces. hCAP-18 is processed by a proteinase, principally proteinase-3, to the mature form, LL-37, which exhibits antimicrobial activity [17]. hCAP-18/LL-37 has been detected in human keratinocytes, but only at sites of inflammation, suggesting that the peptide functions primarily in response to injury. Though main role of LL-37 is "antibacterial" but several studies reported that LL-37 is chemotactic in vitro, inducing selective migration of human peripheral blood monocytes, neutrophils, and CD4-positive T cells [18,19]. Recent evidence indicates that skin antimicrobial peptides, including cathelicidin, are chemotactic for PMNs [20]. LL-37, the mature form of cathelicidin, plays an important role in skin barrier function and contributes to inflammation of skin lesions [21–23]. In addition, LL-37 can be processed to physiological fragments such as RK-31, KR-30, and KS-20, after secretion in sweat. They exhibit antimicrobial activity as LL-37 shows [24] However, several additional LL-37 fragments are found in the pathogenesis of rosacea, one the inflammatory skin disorders, and they contribute the inflammatory cytokines up-regulations [25]. Hence, LL-37 regarded as a double-edged sword for skin defense barrier and regeneration.

We have observed that lesions do not develop pustules or scales if vesicle/pustule ruptures occur, suggesting that the vesicle/pustule contains some heretofore-undefined factor causing subsequent inflammation. As mentioned above, hCAP-18/LL-37 occurs in PPP vesicles, and may be the factor triggering inflammatory changes.

In the present study, we sought to detail the manner of hCAP-18/LL-37 expression in PPP vesicular fluid (PPP-VF) and to determine whether this material contributed to subsequent inflammation of lesional skin.

Materials and Methods

Ethics Statement

All procedures that involved human subjects except the skin biopsy received prior approval from the Ethics Committees of Asahikawa Medical College and Ehime University Graduate School of Medicine. The skin biopsy procedure from the patient was approved by the Ethics Committee of Asahikawa Medical University. We have already got an approval by the Ethics Committee of Ehime University and confirmation of written informed consent from the donor's patient for the collection and generation of the cell lines described [20]. This study was conducted according to the principles of the Declaration of Helsinki. All subjects provided written informed consent.

PPP vesicle and sweat collection

Fifteen volunteers (13 females and 2 males; mean age: 62.7 ± 18.5 years, range: 33–82 years) with 2–10-year histories of PPP were recruited. Additional clinical information at the first visit to our hospital is as follows; smoking history (10/15, over 20 yrs), sternoclavicular joint pain (3/15), anti-streptlysin O test (5/15), periodontitis (3/15). All subjects had used only topical steroid ointments and had not been treated previously with any systemic therapy. Two board-certified dermatologists performed clinical diagnoses. To obtain early stage material, lesional vesicles were carefully observed using a dermatoscope and samples were collected as follows: The skin lesions were cleaned with 70% (v/v) alcohol; lids were removed using 18-G needles; and vesicle fluid was collected immediately (using a micropipette) from several small vesicles. About 3 µl of vesicle fluid were diluted in 30-µl double-distilled water (DDW) and stored in a microtube at –80°C prior to evaluation. Healthy sweat served as a control. After 30 min of exercise, eccrine sweat was collected from the forearms of 14 healthy volunteers (students of Ehime University; three females and eight males; age, 19–23 years) using tissue paper, as described previously [24]. After collection, crude sweat was centrifuged at 17,000×g for 10 min, and the supernatants collected and stored at –80°C prior to use.

Tissue sampling

Punch biopsies (~5 mm in diameter) were taken from the palmar vesicular lesions of five PPP cases at Asahikawa Medical University with their written informed consent, and subjected to pathological diagnosis at our clinic. Specimens were fixed in 10% (v/v) buffered formalin overnight and next embedded in paraffin blocks for routine pathological diagnosis. Sections 4 µm in thickness were prepared for hematoxylin-and-eosin (H&E) staining and immunostaining.

Cell cultures and stimulation

Primary normal human keratinocytes (NHKs) were isolated from surgically discarded neonatal skin samples (dactylosymphysis, 0 M) and cultured in MCDB153 medium supplemented with insulin (5 µg/ml), hydrocortisone (5×10^{-7} M), ethanolamine (0.1 mM), phosphoethanolamine (0.1 mM), bovine pituitary extract (50 mg/ml), and Ca^{2+} (0.03 mM), as described previously [26]. Subconfluent keratinocyte cultures that had been passaged four times were used in stimulation experiments. Cells were incubated with 3 µM synthetic LL-37 peptide (this level was shown to be appropriate in a pilot study) for various times (0, 2, 4, 8, 20, and 24 h) at 37°C.

Preparation of human living skin equivalents (LSEs) and stimulation thereof

The LSE preparation method has been described previously [20,27]. Briefly, a collagen gel was prepared by mixing 6 volumes of ice-cold porcine collagen type I solution (Nitta Gelatin, Osaka, Japan) with 1 volume of 8× DMEM (Gibco, Auckland, NZ), 10 volumes of 1× Dulbecco's Minimal Essential Eagle's Medium (DMEM) supplemented with 20% (v/v) FCS, and 1 volume of 0.1 M NaOH. The final collagen concentration was 0.8 mg/ml. One milliliter amounts of the mixture were added to culture inserts (Transwel-COL, membrane pore size 3 µm; Costar, Corning, NY, NY) in a six-well Costar culture plate (Corning). After gel polymerization at 37°C, two volumes of a fibroblast suspension (5×10^5 cells/ml in 1× DMEM supplemented with 10% [v/v] FCS) were added to eight volumes of collagen solution, and 3.5 ml of the mixture were applied to each insert. When the fibroblast-containing gel had polymerized, DMEM supplemented with 10% (v/v) FCS and ascorbic acid (50 ng/ml final concentration) was added. The culture medium was changed twice weekly. Five days after dermal components were prepared, 6.0×10^5 keratinocytes in 60-µl MCDB 153 type II medium were seeded onto the concave surfaces of contracted gels. The keratinocytes were kept submerged in culture medium for 2 days. When the keratinocytes attained confluence, the LSE was lifted to form an air-liquid interface and cornification medium [27] was added. This medium was changed every other day. Ten days after airlift, the LSEs were used in PPP vesicle stimulation experiments. At the end of each experiment, LSEs were fixed in 20% (v/v) formalin and embedded in paraffin for histological evaluation of morphology.

To stimulate LSEs with PPP-VF, 10 µl PPP-VF was diluted in 100 µl 1% (w/v) agarose gel (SeaPlaque agarose; FMC Bioproducts, Rockland, ME). The heat-melted gel was placed into the cap of a 1.5-ml tube and cooled to RT. Next, a gel cylinder was excised using a disposable biopsy punch (6 mm in diameter, Kai Medical, Kyoto, Japan) and placed on the surface of an LSE, followed by incubation for 12 h at 37°C in a humidified incubator under 5% (v/v) CO_2. As a control, 100 µl of a 1% (w/v) agarose gel cylinder containing 10 µl of eccrine sweat was prepared. After depletion of endogenous hCAP-18/LL-37 in PPP-VF (as described below), 10-μl amounts of such depleted PPP-VF diluted in 100 μl of 1% (w/v) agarose gel were incubated with LSEs.

Synthetic LL-37 and native human proteinase 3 peptides

Both authentic and scrambled LL-37 were commercial preparations (Peptide Institute Inc., Osaka, Japan; and Eurogentec, Seraing, Belgium, respectively). Peptides were purified to over 95% by HPLC and their identities confirmed by mass spectrometry as described previously [2]. Native human proteinase 3 was purchased from Cell Sciences (Canton, MA).

Preparation of the hCAP-18 plasmid

A PCR product encoding full-length hCAP-18 was generated using cDNA obtained from NHKs stimulated with vitamin D; the forward PCR primer was (5′-TAAGGCCTCTGTCGAC-CAGGTCCTCAGCTACAAGGAAGC-3′) and the reverse primer (5′-CAGAATTCGCAAGCTTCTAG-GACTCTGTCCTGGGTACAAG-3′); both primers contained restriction enzyme sites. Amplified DNA was digested with *Sal*I and *Hind*III, inserted into the In-Fusion Ready pEcoli-Nterm 6×HN vector (Clontech Laboratories, Mountain View, CA), and transformed into competent cells [*Escherichia coli* BL21 (DE3); Invitrogen, Carlsbad, CA].

Preparation of recombinant hCAP-18 (rhCAP-18) with a GST tag

The rhCAP-18 peptide was prepared as full-length human cathelicidin using a cell-free protein synthesis system employing wheat germ ribosomal RNA [28,29]. Briefly, the hCAP-18 cDNA was inserted into a pEUE01-GST-N2 expression vector containing a GST tag region. GST-hCAP-18 proteins were automatically synthesized using a Robotic Protein Synthesizer Protemist DT II (CellFree Sciences, Matsuyama, Japan). The transcription mixture (250 μl) containing 25 μg plasmid DNA, 80 mM HEPES-KOH (pH 7.8), 16 mM magnesium acetate, 2 mM spermidine, 10 mM dithiothreitol, 2.5 mM each of the four nucleoside triphosphates, 250 U of SP6 RNA polymerase (Promega, Madison, WI), and 250 U RNasin (Promega), was incubated for 6 h at 37°C. Next, the solution was mixed with 250-μl wheat germ extract WE-PRO7240G (CellFree Sciences) supplemented with 1 μl of 20 mg/ml creatine kinase in a single well of a six-well plate. The substrate mixture (5.5 ml; 30 mM HEPES-KOH [pH 7.8], 100 mM potassium acetate, 2.7 mM magnesium acetate, 0.4 mM spermidine, 2.5 mM dithiothreitol, 0.3 mM amino acid mix, 1.2 mM ATP, 0.25 mM GTP and 16 mM creatine phosphate [CellFree Sciences]) was placed on top of the translation mix and incubated at 17°C for 20 h. Synthetic GST-rhCAP-18 was purified on a glutathione Sepharose 4B column (GE Healthcare, Uppsala, Sweden). The protein solution was loaded onto the column and the column next washed with PBS. We sought to remove the GST tag from the recombinant peptide using Turbo-TEV protease (GST fusion) employing a standard procedure, but the peptide remained attached to the resin and could not be eluted. Therefore, we used the GST-rhCAP-18 peptide in processing experiments.

Antibodies

A mouse anti-LL-37 monoclonal antibody was purchased from Santa Cruz Biotechnology, Inc. (Santa Cruz, CA; catalog no. Sc-166770). Chicken polyclonal antibodies against the cathelin-domain (CATH) peptides were produced by Aves Labs (Tigard, OR), as described [30]. Mouse anti-PR3 monoclonal antibody, mouse anti-CD68 (KP1) antibody, and mouse-anti MCP1 (CCL2)

antibody were purchased from Abcam (Tokyo, Japan) (catalog nos. ab91181, ab955, and ab9669, respectively). Mouse anti-CD56 (1B6) antibody was purchased from Nichirei Biosciences Inc. (Tokyo, Japan).

Antibody column depletion of endogenous hCAP-18/LL-37 in PPP-VF (dep-PPP-VF)

To deplete endogenous vesicular hCAP-18/LL-37, an anti-hCAP-18/LL-37 antibody column was prepared using the anti-LL-37 monoclonal antibody and a Protein G HP SpinTrap/Ad Spin Trap (GE Healthcare Life Sciences, Tokyo, Japan) according to the manufacturer's instructions. Ten microliter amounts of PPP-VF were diluted in 50 μl of DDW, and endogenous hCAP-18/LL-37 removed by affinity chromatography, as confirmed by Western blotting (described below).

Processing experiment using depleted PPP vesicles

After preparing dep-PPP-VF as described above, 20-μl amounts of depleted solution were immediately applied to GST-rhCAP-18-Sepharose columns. The column resin was incubated with dep-PPP-VF, eccrine sweat, or without any addition, at 37°C for 6 h. All samples were subsequently centrifuged at 1,000 rpm for 15 s; eluent (flowthrough) and resin collected; and 5-μl amounts of both eluent and resin boiled in Laemmli sample buffer and subjected to SDS-PAGE gel electrophoresis as described below. The experiments were repeated three times.

Dot-blot analysis and densitometry

To estimate the concentrations of hCAP-18/LL-37 in sweat and PPP-VF, we performed semi-quantitative dot-blot analysis. Briefly, 10-μl PPP-VF was diluted in 100-μl DDW and spotted onto nitrocellulose membranes containing serial dilutions of the LL-37 synthetic peptide using a Bio-Dot apparatus (Bio-Rad Laboratories, Hercules, CA). As a negative control, 10 μM of scrambled LL-37 peptide was applied to a membrane. After air-drying, membranes was treated with blocking solution (0.1% TTBS: 5% [w/v] nonfat milk in 0.1% [v/v] Tween 20/Tris-buffered saline [TBS] with 150 mM NaCl, and 10 mM Tris Base [pH 7.6]) for 60 min at room temperature (RT), and mouse anti-LL37 monoclonal antibody (1:100 in blocking solution) was next added, followed by overnight incubation at 4°C. After three washes with 0.1% (v/v) TTBS, the membranes were incubated with horseradish peroxidase (HRP)-labeled rabbit anti-mouse IgG polyclonal antibody (1:2,000 in blocking solution, Bio-Rad) for 60 min at RT. After washing with 0.1% TTBS, membranes were immersed in ECL solution (Western Lightning Chemiluminescence Reagents Plus; GE Healthcare, Buckinghamshire, UK) for 60 s and visualized using the LAS-4000 imaging system (GE Healthcare) according to the manufacturer's instructions. To estimate hCAP-18/LL-37 concentrations, the intensity of dot staining was compared to that of a standard curve constructed by spotting concentrations of the synthetic peptide onto the nitrocellulose membrane. Dot staining intensities were measured in triplicate experiments using NIH-image software, as described previously [9].

Western blotting

Proteins in PPP-VF (10 μl), and in GST-hCAP18 recombinant protein preparations after incubation with various factors, were resolved on 15% (w/v) Tris-HCl gels (BIO CRAFT, Tokyo, Japan) and transferred to PVDF membranes (Immobilon-P; Millipore, Darmstadt, Germany); 3.2 pmol of the LL-37 synthetic peptide served as a positive control. Membrane bands were visualized as described above. To confirm the identities of bands detected by Western blotting, the filter was stripped of antibody using WB Stripping Solution (Nacalai Tesque, Kyoto, Japan) following the manufacturer's instructions, and next reacted with a chicken anti-cathelin polyclonal antibody (1:2,000) and an HRP-conjugated-anti-chicken IgY goat antibody (1:2,000, Promega), as described previously [30].

To detect proteinase 3 in PPP-VF, 10-μl PPP-VF were subjected to 15% (w/v) Tris-HCl gel electrophoresis using native human proteinase 3 as a positive control. A mouse anti-proteinase 3 monoclonal antibody (1:30) was used as primary antibody to detect the protein, as described above.

Immunohistochemistry

Tissue sections were immersed in PBS after deparaffinization and endogenous peroxidase activity was blocked by incubation with 0.3% (v/v) H_2O_2 in methanol for 30 min. After washing with phosphate buffered saline (PBS), immunostaining with mouse anti-CD68 (1:200) and mouse anti-CD56 antibodies (ready-to-use) was performed using a Histofine SAB-PO kit (Nichirei Biosciences Inc.) according to the manufacturer's instructions. As a negative control, the polyclonal and monoclonal antibodies were replaced by normal mouse preimmune IgG diluted with PBS containing 3% (w/v) BSA to the protein concentrations of the primary antibodies. Nuclear counterstaining was performed using hematoxylin. All procedures were conducted at RT, except for the initial antibody incubation (4°C, overnight).

Enzyme-linked immunosorbent assays (ELISAs)

NHK monolayers were treated with 3 μM of an LL37 synthetic peptide (which was shown to be appropriate in a pilot experiment) at 37°C for 12 h, and the culture supernatants collected to measure protein expression levels. ELISA kits detecting IL-1α, IL-1β, IL-8, and IL-17C were purchased from R&D Systems (Minneapolis, MN) and used according to the manufacturer's protocols. Optical densities at 450 nm were measured using an Immuno Mini NJ-2300 microplate reader. All assays were performed in triplicate.

RNA preparation and quantitative real-time PCR (qRT-PCR)

Total RNA was isolated using Isogen (Nippon Gene, Tokyo, Japan) and cDNA was prepared from 1-μg amounts of total RNA using the iScript cDNA synthesis kit (Bio-Rad Laboratories) according to the manufacturer's instructions. TaqMan probes for glyceraldehyde-3-phosphate dehydrogenase (GAPDH), IL-1α, IL-1β, IL-8, IL-22, and IL-17 (A, F, and C) were obtained from Applied Biosystems (Foster city, CA). qRT-PCR was performed on an ABI PRISM 7700 sequencing platform (Applied Biosystems) according to the manufacturer's instructions. The mRNA levels of target genes were normalized to that of GAPDH. Gene expression levels in treated cells were compared to those in untreated cells. Data were analyzed using the Comparative Ct Method, where Ct indicates the number of cycles required to reach an arbitrary threshold [31].

Statistical analysis

Quantitative data are expressed as means ± standard deviations. The data from dot-blot analysis and real-time PCR were evaluated using the STATFLEX software (version 6.0, ARTEC Inc., Osaka, Japan). The paired Student's t-test was used to compare between-group differences, and p-values<0.05 were considered to indicate statistical significance.

Results

PPP-VF induces inflammatory cytokine mRNA expression in LSEs

After incubation with PPP-VF, the expression levels of cytokine mRNAs in LSEs were evaluated via qRT-PCR (Fig. 1A). All of IL-17C (2.00±1.79-fold), IL-8 (1.8±1.7-fold), IL-1α (3.47±1.28-fold), and IL-1β (18.67±11.72-fold) were upregulated compared to non-treated LSEs (control). None of IL-22, IL-17A, or IL-17F was detected. For IL-8, IL-1α, and IL-1β, the increases were statistically significant (Fig. 1A).

Confirmation of LL-37 expression in PPP-VF and preparation of depleted PPP-VF using an anti-LL-37 antibody column

Expression of hCAP-18 in PPP vesicles was confirmed using Western blotting. hCAP-18 (18 kDa; full-length), an intermediate-sized fragment (~14 kDa), and mature LL-37 (4.5 kDa), were detected, as were additional fragments of ~6 and ~8 kDa (Fig. 1B, lane b). Endogenous vesicle hCAP-18/LL-37 was successfully depleted using an anti-LL-37 antibody column (Fig. 1B, lane c). No processing was noted upon incubation with a crude rhCAP-18 peptide (Fig. 1B, lane a), whereas the positive control (a synthetic LL-37 peptide) yielded a 4.5-kDa band (Fig. 1B, lane d).

Endogenous hCAP-18/LL-37 directly contributes to upregulation of IL-8 mRNA

After depleting endogenous hCAP-18/LL-37 via specific antibody column purification, LSEs were incubated with the depleted PPP-VF using the method described above. The IL-8 mRNA level upon stimulation with depleted PPP-VF was significantly lower than that after incubation with non-depleted PPP-VF (Fig. 1C). However, no significant differences in the levels of IL-17C, IL-1α, or IL-1β mRNAs were evident (Fig. 1C).

PPP-VF contains a higher concentration of hCAP-18/LL-37 than it in eccrine sweat

To determine hCAP-18/LL-37 concentrations in PPP-VF, we performed dot-blot analysis and densitometry using sweat as a control. All 15 PPP-VF samples and all 14 samples of healthy sweat contained hCAP-18/LL-37 (Fig. 2). The average concentration of hCAP-18/LL-37 in PPP-VF was 2.87±0.93 μM, but the average concentration in sweat was only 0.09±0.09 μM, consistent with the results of a previous study reporting an average hCAP-18/LL-37 concentration ~1 μM in crude sweat collected from normal volunteers [30]. The difference was significant ($p<$ 0.05).

LL-37 (the mature form of hCAP-18) upregulates PPP-associated cytokines in NHKs

As hCAP-18/LL-37 was present in PPP-VF, we next explored whether LL-37 stimulated proinflammatory cytokine expression in NHKs. Pilot experiments indicated that stimulation with 3 μM LL-37 was appropriate. NHKs were incubated with 3 μM synthetic LL-37 for 0, 2, 4, 8, 20 and 24 h at 37°C, and mRNA levels next evaluated via qRT-PCR. LL-37 stimulated upregulation of mRNAs encoding IL-17C, IL-8, IL-1α and IL-1β; but not IL-22, IL-17A, or IL-17F (Fig. 3A). Upregulation of mRNAs encoding IL-8, IL-1α and IL-1β was defined by increases from the values at 0 h. However, IL-17C mRNA was not expressed at 0 h and upregulation was thus defined by reference to the level at 2 h. All mRNA expression levels increased in a time-dependent

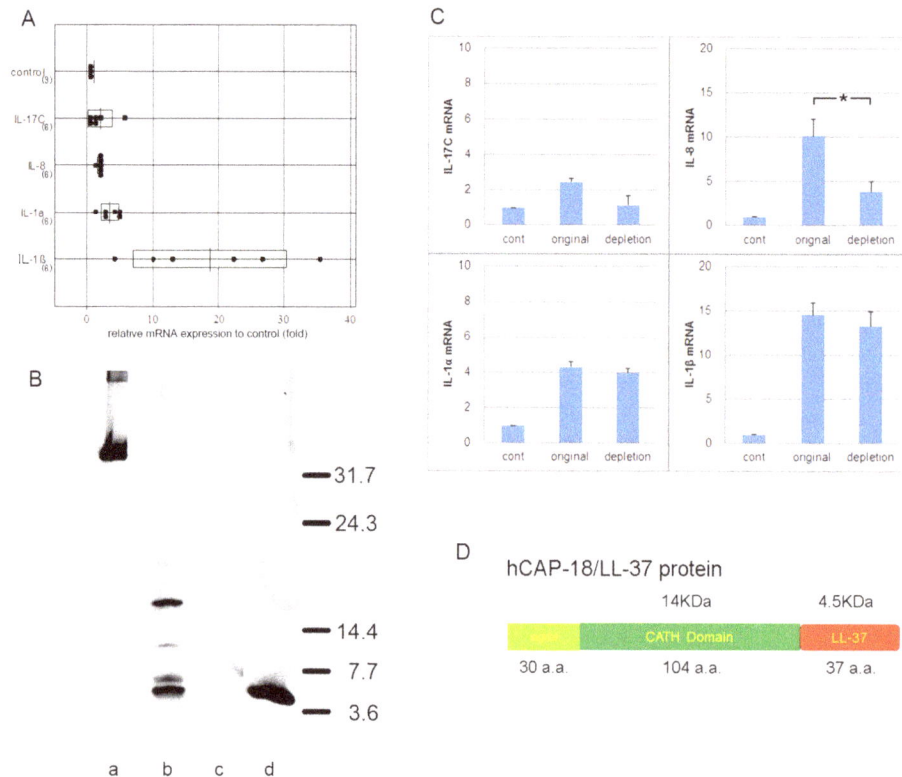

Figure 1. PPP vesicles stimulate expression of cytokine-encoding mRNAs in LSEs. PPP vesicles were suspended in 1% (w/v) agar and mRNAs measured using qRT-PCR. **A)** All of IL-17C (2.00 ± 1.79-fold), IL-8 (1.8 ± 1.7-fold), IL-1α (3.47 ± 1.28-fold), and IL-1β (18.67 ± 11.72-fold) were upregulated compared to the levels in non-treated LSEs (controls). The levels of cathelicidin, IL-8, IL-1α, and IL-1β differed significantly from those in control sweat. **B)** Western blotting showed that the hCAP-18/LL-37-depleted PPP-VF sample contained bands equivalent to hCAP-18 (18 kDa; full-length); an intermediate-sized fragment ($\sim$14 kDa); mature LL-37 (4.5 kDa); and two additional bands (lane b). Lane a: the GST-hCAP18 peptide prior to incubation; Lane b: PPP-VF before depletion; Lane c: PPP-VF after depletion of endogenous hCAP-18/LL-37; Lane d: synthetic LL-37 peptide (3.2 pmol). **C)** mRNA expression levels in LSEs stimulated by original and depleted PPP-VF, as calculated via qRT-PCR. $*p<0.05$. **D)** The illustration of structure of hCAP-18/LL-37.

manner, but to significantly different extents. To confirm protein expression, culture media were assessed by ELISAs (Fig. 3B). IL-17C, IL-8, IL-1α, and IL-1β protein levels in media increased in parallel with the corresponding mRNA levels, suggesting that PPP-VF contained a level of LL-37 sufficient to induce IL-17C, IL-8, IL-1α, and IL-1β mRNA expression in NHKs, consistent with data on mRNA expression levels in PPP lesional skin biopsy samples [32].

hCAP-18 is processed to LL-37 in PPP-VF

To determine whether hCAP-18 in PPP-VF could be processed to LL-37 by proteinases in PPP-VF, GST-rhCAP-18 was incubated with dep-PPP-VF from individual two cases described above. Western blotting revealed several processed bands, and the hCAP-18 (18 kDa; full-length) protein. Derived bands included an intermediate-sized fragment ($\sim$14 kDa), mature LL-37 (4.5 kDa), and two additional bands of $\sim$6 and $\sim$8 kDa (Fig. 4A, lanes F [flowthrough] and R [resin-bound] material from dep-PPP-VF (1); and in lane F of material from dep-PPP-VF (2); detected using an anti-LL37 Ab). In a separate experiment, we found that GST-rhCAP-18 incubated in eccrine sweat was cleaved into hCAP-18 (18 kDa; full-length) protein, an intermediate-sized fragment ($\sim$14 kDa) and two additional bands of $\sim$6 and $\sim$8 kDa, but mature LL-37 (4.5 kDa)was not detected (Fig. 4A, lane R of sweat tr; detected using an anti-LL-37 Ab). In addition, 18-kDa bands were detected using anti-CATH Ab in both dep-PPP-VF (1) (F and

R) and dep-PPP-VF (2). These results suggest that PPP-VF contains a proteinase that can process LL-37. However, although sweat also contains proteinases, these cannot directly process hCAP-18 to LL37.

PPP-VF contains proteinase 3

We used Western blotting to determine whether proteinase 3 [21], which processes LL-37, was present in PPP-VF. A band 29 kDa in size, thus that of native proteinase 3, was present in PPP-VF but not in the control sweat sample (Fig. 4B).

PPP-VF contains CD68-positive cells

H&E staining revealed that vesicle contained many mononuclear cells but no polymorphonuclear cells (Fig. 5a). The mononuclear cells were positive for CD68 (Fig. 5c, d) but not for CD56 (Fig. 5e, f), suggesting that the cells were monocytes/macrophages and not NK or plasma cells. No signal was detected from pre-immune mouse IgG (Fig. 5b).

Lesion epidermis expresses monocyte chemotactic chemokine (MCP-1) protein

The epidermis surrounding the PPP vesicle expressed the MCP-1 protein (Fig. 6a–c, e, f). No signal was detected from pre-immune mouse IgG (Fig. 6d).

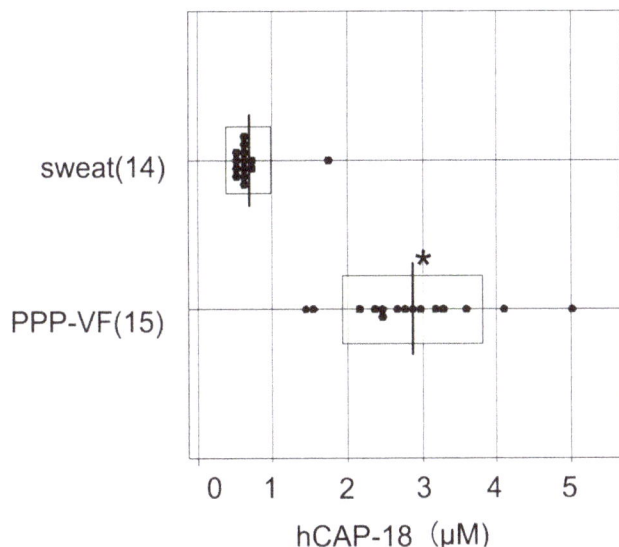

Figure 2. Quantification of hCAP-18/LL-37 in PPP-VF and eccrine sweat samples. Dot-blot analyses and densitometry were performed on PPP vesicles (15 samples), eccrine sweat samples (14 samples), a serially diluted LL-37 synthetic peptide solution, and a 10 μM solution of scrambled LL-37 synthetic peptide (negative control). hCAP-18/LL-37 was confirmed to be present in all PPP-VF and eccrine sweat samples, but not in the scrambled peptide control. The average concentrations of hCAP-18/LL-37 in PPP-VF and control sweat were 2.87±0.93 and 0.09±0.09 μM, respectively. *$p<0.05$ compared to sweat.

Discussion

Previously, we reported that IL-17 (A, C, and F), IL-22, and IL-8 mRNAs were upregulated in the lesional skin of PPP patients [32]; confirmed that PPP-VF could upregulate the expression of IL-17C, IL-8, IL-1α, and IL-1β in LSE keratinocytes; and that endogenous hCAP-18/LL-37 induced high-level IL-8 synthesis (Fig. 1). PPP-VF contains a higher concentration of hCAP-18/LL-

37 than does eccrine sweat (Fig. 2), sufficient to induce IL-17C, IL-8, IL-1α, and IL-1β mRNA and protein expression in monolayer NHKs (Fig. 3). In addition, GST-rhCAP-18 was incubated with depleted PPP-VF, and the presence of processed LL-37 confirmed thereafter (Fig. 4A). This suggests that PPP-VF contains the proteinase responsible for cleavage of hCAP-18 to LL-37, and, indeed, Western blotting confirmed that proteinase 3 was present in PPP-VF (Fig. 4B).

This is the first report to show that PPP-VF induces IL-17C, IL-8, IL-1α, and IL-1β synthesis in keratinocytes of LSEs, and to quantify the hCAP-18/LL-37 concentration in PPP-VF. The average concentration of LL-37 in PPP-VF was 2.87±0.93 μM, and this was sufficient to induce IL-17C, IL-8, IL-1α, and IL-1β mRNA and protein expression in monolayer NHKs.

We have shown that 3 μM LL-37 was sufficient to stimulate IL-8 release from cultured keratinocytes, without toxic effects [22,24], and that LL-37 at 13.5 or 45 μg/ml (3 or 10 μM, respectively) increased production of IL-6 and secretion of IL-1α [22]. LL-37 enhanced IL-8 release by human airway smooth muscle (HASM) cells, via a process dependent on ERK1/2 activation [23]. IL-17 also induced release of IL-8 and IL-6 from human keratinocytes, but the amounts released were lower than noted upon LL-37 induction. However, the Th17 cytokines (IL-17 and IL-22) synergistically upregulated IL-8 and IL-6 expression in keratinocytes [25]. Upon PPP-VF stimulation in LSE, we could not confirm IL-22 upregulation in keratinocytes, but IL-17C upregulation was indeed evident. However, the level of IL-17C mRNA in LSE was not greatly reduced in dep-PPP-VF. The big difference between monolayer culture cells and LSE is that LSE has piled-up keratinocytes covered with stratum corneum and also has a lot of differentiated keratinocyte such as granular cells and prickle cells. Though LSE was stimulated with LL-37 through the stratum corneum with a gel cylinder, NHK were stimulated with LL-37 in the culture medium which they were covered. Under inflammatory skin condition, stratum corneum could be easily disrupted so that keratinocyte could not have such a protection no more. In addition, from the view of cell differentiation, many of keratinocytes in monolayer culture might be in proliferative stage, but it in LSE might be constructed with many differentiated keratinocytes

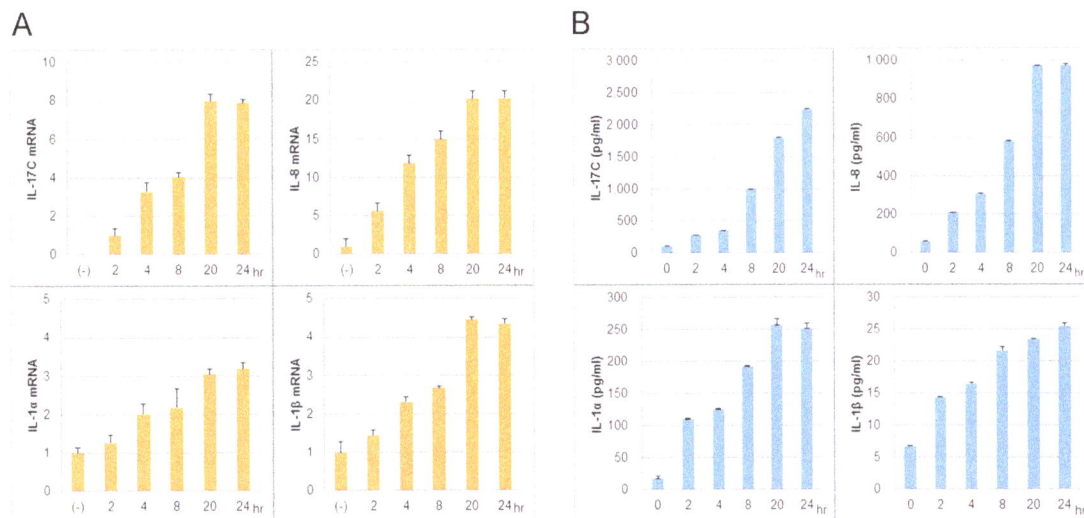

Figure 3. Cytokine induction in NHKs by the synthetic LL-37 peptide. To assess the ability of LL-37 to induce cytokines, NHKs were incubated with 3 μM LL-37 for 0, 2, 4, 8, 20, and 24 h at 37°C. (A) Expression levels of mRNAs encoding IL-17C, IL-8, IL-1α, and IL-1β mRNA, measured by qRT-PCR. The relative mRNA levels are expressed as means ±SDs (in -fold changes). Enzyme-linked immunoassays (ELISAs) were performed on culture media. All later values yielded by both qRT-PCR and ELISA were significantly different from those at 0 h (A, B, $p<0.05$).

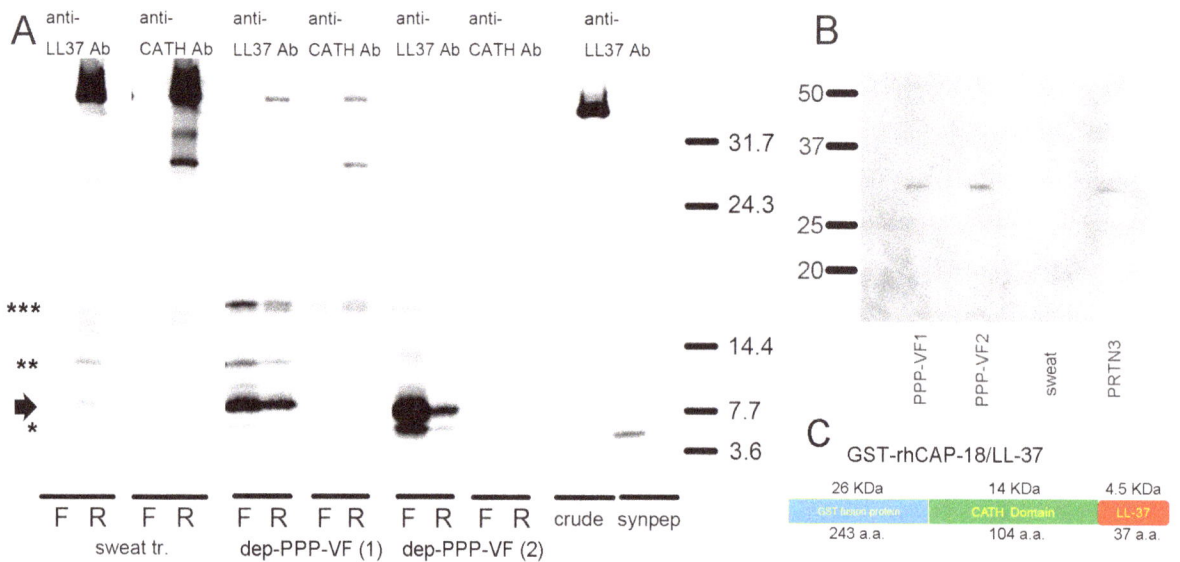

Figure 4. LL-37 is synthesized from GST-rhCAP18 in depleted PPP-VF containing proteinase 3. A) Several bands derived from GST-rhCAP18 were evident with dep-PPP-VF incubation. These were hCAP-18 (18 kDa; full length, indicated with ***), an intermediate-sized fragment (~14 kDa, indicated with **), mature LL-37 (4.5 kDa, indicated with *), and two additional bands of ~6 and 8 kDa (indicated with right arrow). In addition, 18-kDa bands reacting with anti-CATH Ab were present in PPP-VF tr-1 (F and R) and PPP-VF tr-2 (F). No mature LL-37 was detected in the sweat treated sample (lane R: sweat tr; anti-LL37 Ab staining). Abbreviations: α-LL37, anti-LL-37 antibody; α-CATH, anti-CATH antibody; F, peptide in flowthrough; R, peptide binding to resin; Sweat tr, eccrine sweat-treated peptide; PPP-VF tr, depleted PPP-VF component-treated peptide; crude, non-treated GST-hCAP-18 peptide (binding to resin); syn pep, LL-37 synthetic peptide (3.2 pmol). **B)** Proteinase 3 expression in concentrated PPP-VF was confirmed by Western blotting. Both depleted samples, PPP-VFs 1 and 2, exhibited single bands 29 kDa in size, thus that of PRTN3. Authentic PRTN3 (10 ng) served as a positive control. PPP-VF tr, depleted PPP-VF component-treated peptide; sweat, sweat sample 1; PRTN3, native proteinase 3. **C)** The illustration of structure of GST-rhCAP-18/LL-37.

after air-lifting. We are speculating that these differences are responsible, so far. However the detail of the reason is still unclear so that we are now continuing to elucidate this reason from the view of the skin barrier problem and keratinocyte differentiation.

We earlier reported that PPP-VF was formed in the acrosyringium, and thus originally contains components of eccrine sweat [2], and sweat contains IL-1α and IL-1β [33]. Thus, baseline IL-1α and IL-1β mRNA levels reflect not only the action of the endogenous hCAP-18/LL-37 system but also the fact that sweat

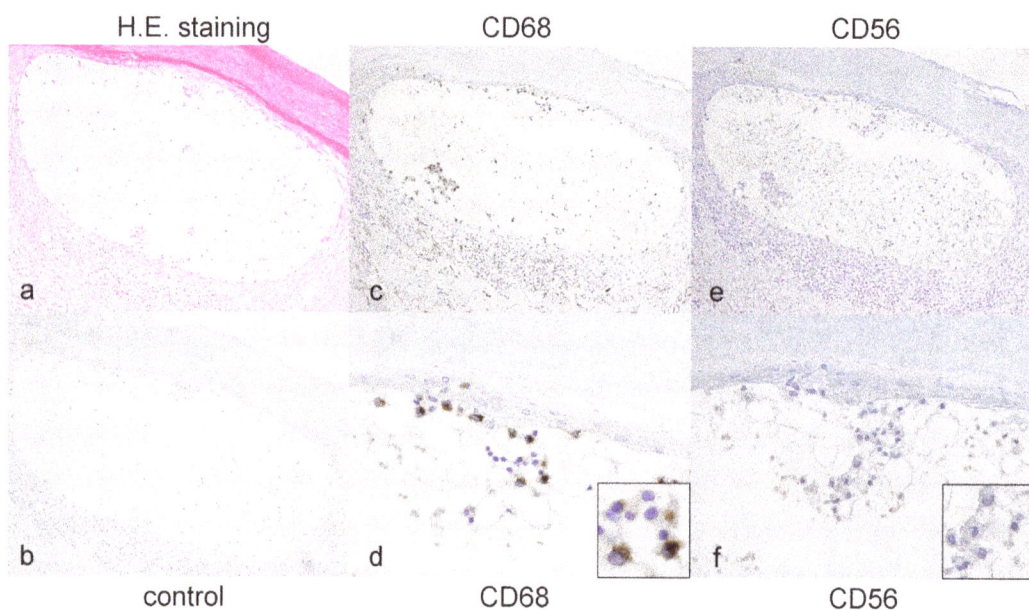

Figure 5. Monocytes in PPP-VF. Hematoxylin-eosin staining revealed many mononuclear cells, but no polymorphonuclear cells, in vesicles (Fig. 5a). The mononuclear cells were positive for CD68 (Figs. 5c, d) but not CD56 (Figs. 5e, f) in all five instances. Pre-immune anti-mouse IgG did not stain the sections. (Fig. 5d). (Original magnifications: a, b, c, e: 100×, d, f: 400×).

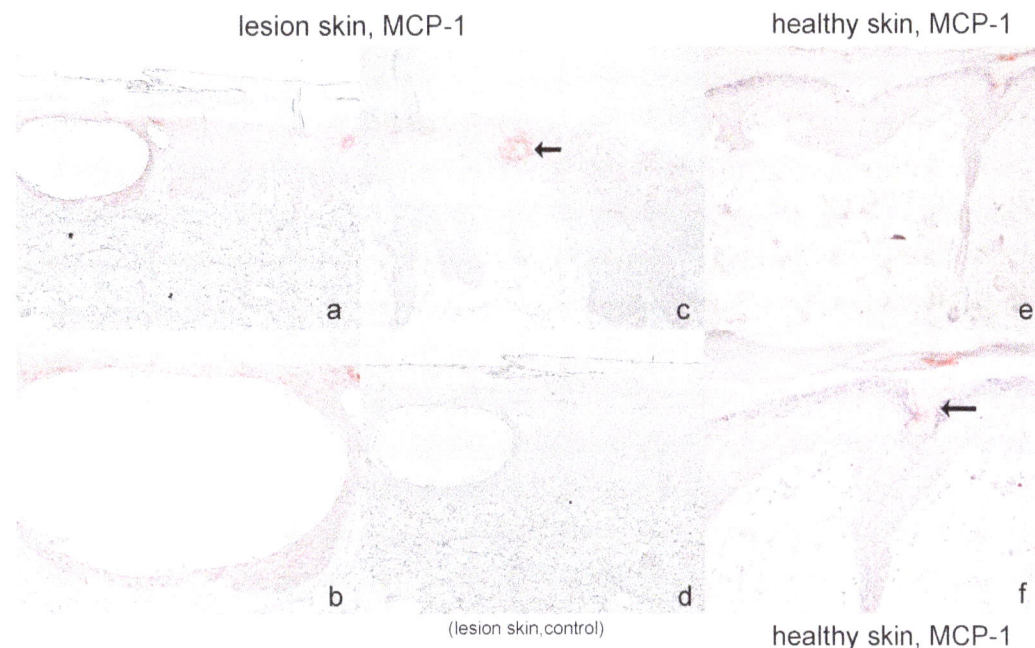

Figure 6. MCP-1 expression in PPP lesion skin and healthy skin. In lesion skin, strong expression of MCP-1 was detected around the PPP vesicle in the epidermis (Figs. 6a, b). In addition, acrosyringium in the lesions skin also showed the protein expression (Fig. 6c), but not in healthy skin (Fig. 6e, f). The eccrine pore at the surface of skin showed weak positive staining locating (Fig. 6f, arrowhead). (Original magnifications: a, c, d, e: 40×, b, f: 100×).

components are included in vesicles. We have also reported that sweat IL-1α induced IL-8 expression in keratinocytes [33]. This may explain why baseline IL-8 mRNA expression remained high even after endogenous hCAP18/LL-37 was removed from PPP-VF.

From the result of LSE stimulation with dep-PPP-VF experiment, IL-8 mRNA expression was significantly decreased in dep-PPP-VF. This suggests that IL-8 expression in keratinocytes is directly upregulated by LL-37, consistent with our previous data showing that LL-37 potently stimulated IL-8 release from keratinocytes [24]. IL-8 is regarded as a major factor triggering pustule formation [7]. IL-8 action has been implicated in a number of inflammatory diseases involving neutrophil activation. A human mAb directed against IL-8 (HuMab 10F8), which neutralizes IL-8-dependent human neutrophil activation and migration, has been evaluated (in a Scandinavian Phase 2 trial) as a candidate for PPP treatment [34]. This trial was well-tolerated and significantly reduced clinical disease activity at all five endpoints evaluated; fresh pustule formation was reduced by ≥ 50%. This result strongly suggests that IL-8 is involved in the pathogenesis of PPP pustulation. IL-8 is barely detectable in healthy skin, but it is rapidly induced (by 10–100-fold) in response to several proinflammatory cytokines (including tumor necrosis factor and IL-1), bacterial and viral products, cellular stress [35], and IL-17 [36]. As mentioned above, we confirmed that the average concentration of hCAP-18/LL-37 was 2.87±0.93 μM, which can directly induce expression of both IL-8 mRNA and protein. This suggests that IL-8 in keratinocytes was induced principally via LL-37 stimulation. However, it has been reported that several factors, including IL-17A/C/F, IL-1α, and IL-1β, can upregulate IL-8 expression in keratinocytes. The mechanism of IL-8 upregulation in PPP requires further elucidation, but it is clear that LL-37 may play a role and may also contribute to subsequent inflammation of PPP lesional skin.

LL-37, mature form of human cathelicidin, can be processed after secretion in sweat physiologically, and fragments derived from LL-37, including RK-31, KR-30, and KS-20, exhibit antimicrobial activity as LL-37 shows [24]. Additional LL-37 fragments are found in the pathogenesis of several inflammatory skin disorder, such as rosacea, and they contribute the inflammatory cytokines up-regulations [25]. Hence, we speculated that fragments derived from LL-37 might be processed in early stage PPP vesicles which could trigger subsequent inflammation. However, western blotting revealed no fragment smaller than LL-37 in PPP-VF.

PPP vesicle originally contains additional 6 and 8 kDa in addition to 18 and 4.5 kDa (figure 1B, lane b), and GST-rhCAP-18 treated with dep-PPP-VF showed all of them and another band about 10 kDa (Fig. 4 A). With sweat treatment, those bands were detected without 4.5 kDa. This result suggested that not only the proteinase for LL-37 processing was included but also other proteinase which could process or degrade the GST-rhCAP-18 protein. And it was also suggested that control sweat we collected did not contain enough concentration of the responsible proteinase, such as proteinase 3 [21], to process LL-37.

The three known serine proteases in azurophil granules (elastase, cathepsin G, and proteinase 3) cleave many of the same substrates, and hCAP-18 is susceptible to cleavage by all three enzymes in vitro. However, proteinase 3 is solely responsible for cleavage of hCAP-18 after exocytosis [21]. Proteinase 3 is a 29-kDa serine protease normally transcribed during myelopoiesis, and is thought to be turned off in both mature leucocytes [21,37] and monocytes [38–40]. We thus doubted that PPP-VF contained proteinase 3, and Western blotting showed that the proteinase was indeed present.

We very carefully collected PPP-VF using a dermoscope not to collect polymorphonuclear cells such as neutrophils. When infiltrating cells were evaluated histopathologically, early stage vesicles contained principally mononuclear cells suggestive of monocytes/macrophages with only a few polymorphic nuclear cells. Upon immunostaining for CD68, many vesicular cells were positive, especially near the lids of vesicles, suggesting that proteinase 3 in early stage vesicles is secreted from monocytes that had infiltrated the vesicles. For further study to know if the epidermis expressed CCL2 expression, MCP-1 immunohisto-chemistry was additionally performed because CCL2 is a strong monocyte/macrophage chemoattractant. The epidermal kerati-nocyte surrounding PPP vesicle showed MCP-1 protein expression with immunohistochemical examination. The recruitment of CD68 positive mononuclear cells in the early vesicles could be induced by this phenomenon (Fig. 5). However, if so, the exact mechanism by which monocytes/macrophages infiltrate the PPP-VF in the early stages of disease remains still unclear, and need to elucidate further.

In the present study, we thus report a new role for the LL-37 of PPP-VF. The protein contributes to the pathophysiology of

subsequent inflammation. However, further studies are required to clarify how PPP is initiated.

Acknowledgments

The authors thank Ms. K. Nishikura at Asahikawa Medical University for her excellent pathological assistance. We also thank Ms. T. Tsuda, Ms. E. Tan, and Ms. K. Ozaki at Ehime University for their excellent technical assistance and Ms. Y. Tanaka at INCS, Ehime University for preparing the GST-hCAP-18 protein. The English in this document has been checked by at least two professional editors, both native speakers of English. For a certificate, please see: http://www.textcheck.com/certificate/MeULuw.

Author Contributions

Conceived and designed the experiments: MM. Performed the experiments: MM TK TN KK HO XD. Analyzed the data: MM YH MT AIY KS. Contributed reagents/materials/analysis tools: MM YH MT AIY KS. Contributed to the writing of the manuscript: MM.

References

1. Uehara M, Ofuji S (1974) The morphogenesis of pustulosis palmaris et plantaris. Arch Dermatol 109: 518–520.
2. Murakami M, Ohtake T, Horibe Y, Ishida-Yamamoto A, Morhenn VB, et al. (2010) Acrosyringium is the main site of the vesicle/pustule formation in palmoplantar pustulosis. J Invest Dermatol 130: 2010–2016.
3. Rieg S, Garbe C, Sauer B, Kalbacher H, Schittek B (2004) Dermcidin is constitutively produced by eccrine sweat glands and is not induced in epidermal cells under inflammatory skin conditions. Br J Dermatol 151: 534–539.
4. Schittek B, Hipfel R, Sauer B, Bauer J, Kalbacher H, et al. (2001) Dermcidin: a novel human antibiotic peptide secreted by sweat glands. Nat Immunol 2: 1133–1137.
5. Frohm M, Agerberth B, Ahangari G, Stahle-Backdahl M, Liden S, et al. (1997) The expression of the gene coding for the antibacterial peptide LL-37 is induced in human keratinocytes during inflammatory disorders. J Biol Chem 272: 15258–15263.
6. Gallo RL, Murakami M, Ohtake T, Zaiou M (2002) Biology and clinical relevance of naturally occurring antimicrobial peptides. J Allergy Clin Immunol 110: 823–831.
7. Ozawa M, Terui T, Tagami H (2005) Localization of IL-8 and complement components in lesional skin of psoriasis vulgaris and pustulosis palmaris et plantaris. Dermatology 211: 249–255.
8. Hagforsen E, Hedstrand H, Nyberg F, Michaelsson G (2010) Novel findings of Langerhans cells and interleukin-17 expression in relation to the acrosyringium and pustule in palmoplantar pustulosis. Br J Dermatol 163: 572–579.
9. Kaneko T, Murakami M, Kishibe M, Brattsand M, Morhenn VB, et al. (2012) Over-expression of kallikrein related peptidases in palmoplantar pustulosis. J Dermatol Sci 67: 73–76.
10. Gallo RL, Kim KJ, Bernfield M, Kozak CA, Zanetti M, et al. (1997) Identification of CRAMP, a cathelin-related antimicrobial peptide expressed in the embryonic and adult mouse. J Biol Chem 272: 13088–13093.
11. Nizet V, Ohtake T, Lauth X, Trowbridge J, Rudisill J, et al. (2001) Innate antimicrobial peptide protects the skin from invasive bacterial infection. Nature 414: 454–457.
12. Zanetti M, Gennaro R, Romeo D (1995) Cathelicidins: a novel protein family with a common proregion and a variable C-terminal antimicrobial domain. FEBS Lett 374: 1–5.
13. Ali RS, Falconer A, Ikram M, Bissett CE, Cerio R, et al. (2001) Expression of the peptide antibiotics human beta defensin-1 and human beta defensin-2 in normal human skin. J Invest Dermatol 117: 106–111.
14. Harder J, Bartels J, Christophers E, Schroder JM (1997) A peptide antibiotic from human skin. Nature 387: 861.
15. Stolzenberg ED, Anderson GM, Ackermann MR, Whitlock RH, Zasloff M (1997) Epithelial antibiotic induced in states of disease. Proc Natl Acad Sci U S A 94: 8686–8690.
16. Agerberth B, Gunne H, Odeberg J, Kogner P, Boman HG, et al. (1995) FALL-39, a putative human peptide antibiotic, is cysteine-free and expressed in bone marrow and testis. Proc Natl Acad Sci U S A 92: 195–199.
17. Sorensen O, Bratt T, Johnsen AH, Madsen MT, Borregaard N (1999) The human antibacterial cathelicidin, hCAP-18, is bound to lipoproteins in plasma. J Biol Chem 274: 22445–22451.
18. Agerberth B, Charo J, Werr J, Olsson B, Idali F, et al. (2000) The human antimicrobial and chemotactic peptides LL-37 and alpha- defensins are expressed by specific lymphocyte and monocyte populations. Blood 96: 3086–3093.
19. De Y, Chen Q, Schmidt AP, Anderson GM, Wang JM, et al. (2000) LL-37, the neutrophil granule- and epithelial cell-derived cathelicidin, utilizes formyl

peptide receptor-like 1 (FPRL1) as a receptor to chemoattract human peripheral blood neutrophils, monocytes, and T cells. J Exp Med 192: 1069–1074.
20. Yang L, Shirakata Y, Tokumaru S, Xiuju D, Tohyama M, et al. (2009) Living skin equivalents constructed using human amnions as a matrix. J Dermatol Sci 56: 188–195.
21. Sorensen OE, Follin P, Johnsen AH, Calafat J, Tjabringa GS, et al. (2001) Human cathelicidin, hCAP-18, is processed to the antimicrobial peptide LL-37 by extracellular cleavage with proteinase 3. Blood 97: 3951–3959.
22. Braff MH, Hawkins MA, Di Nardo A, Lopez-Garcia B, Howell MD, et al. (2005) Structure-function relationships among human cathelicidin peptides: dissociation of antimicrobial properties from host immunostimulatory activities. J Immunol 174: 4271–4278.
23. Zuyderduyn S, Ninaber DK, Hiemstra PS, Rabe KF (2006) The antimicrobial peptide LL-37 enhances IL-8 release by human airway smooth muscle cells. J Allergy Clin Immunol 117: 1328–1335.
24. Murakami M, Lopez-Garcia B, Braff M, Dorschner RA, Gallo RL (2004) Postsecretory processing generates multiple cathelicidins for enhanced topical antimicrobial defense. J Immunol 172: 3070–3077.
25. Chen X, Takai T, Xie Y, Niyonsaba F, Okumura K, et al. (2013) Human antimicrobial peptide LL-37 modulates proinflammatory responses induced by cytokine milieus and double-stranded RNA in human keratinocytes. Biochem Biophys Res Commun 433: 532–537.
26. Dai X, Sayama K, Tohyama M, Shirakata Y, Hanakawa Y, et al. (2011) Mite allergen is a danger signal for the skin via activation of inflammasome in keratinocytes. The Journal of allergy and clinical immunology 127: 806–814 e801–804.
27. Yang L, Shirakata Y, Tamai K, Dai X, Hanakawa Y, et al. (2005) Microbubble-enhanced ultrasound for gene transfer into living skin equivalents. J Dermatol Sci 40: 105–114.
28. Madin K, Sawasaki T, Ogasawara T, Endo Y (2000) A highly efficient and robust cell-free protein synthesis system prepared from wheat embryos: plants apparently contain a suicide system directed at ribosomes. Proc Natl Acad Sci U S A 97: 559–564.
29. Sawasaki T, Ogasawara T, Morishita R, Endo Y (2002) A cell-free protein synthesis system for high-throughput proteomics. Proc Natl Acad Sci U S A 99: 14652–14657.
30. Murakami M, Ohtake T, Dorschner RA, Schittek B, Garbe C, et al. (2002) Cathelicidin anti-microbial peptide expression in sweat, an innate defense system for the skin. J Invest Dermatol 119: 1090–1095.
31. Dorschner RA, Lin KH, Murakami M, Gallo RL (2003) Neonatal skin in mice and humans expresses increased levels of antimicrobial peptides: innate immunity during development of the adaptive response. Pediatr Res 53: 566–572.
32. Murakami M, Hagforsen E, Morhenn V, Ishida-Yamamoto A, Iizuka H (2011) Patients with palmoplantar pustulosis have increased IL-17 and IL-22 levels both in the lesion and serum. Exp Dermatol 20: 845–847.
33. Dai X, Okazaki H, Hanakawa Y, Murakami M, Tohyama M, et al. (2013) Eccrine sweat contains IL-1alpha, IL-1beta and IL-31 and activates epidermal keratinocytes as a danger signal. PLoS One 8: e67666.
34. Skov L, Beurskens FJ, Zachariae CO, Reitamo S, Teeling J, et al. (2008) IL-8 as antibody therapeutic target in inflammatory diseases: reduction of clinical activity in palmoplantar pustulosis. J Immunol 181: 669–679.
35. Hoffmann E, Dittrich-Breiholz O, Holtmann H, Kracht M (2002) Multiple control of interleukin-8 gene expression. J Leukoc Biol 72: 847–855.
36. Mirandola P, Gobbi G, Micheloni C, Vaccarezza M, Di Marcantonio D, et al. (2011) Hydrogen sulfide inhibits IL-8 expression in human keratinocytes via MAP kinase signaling. Lab Invest 91: 1188–1194.

37. Cowland JB, Borregaard N (1999) Isolation of neutrophil precursors from bone marrow for biochemical and transcriptional analysis. J Immunol Methods 232: 191–200.

38. Just J, Moog-Lutz C, Houzel-Charavel A, Canteloup S, Grimfeld A, et al. (1999) Proteinase 3 mRNA expression is induced in monocytes but not in neutrophils of patients with cystic fibrosis. FEBS Lett 457: 437–440.

39. Braun MG, Csernok E, Gross WL, Muller-Hermelink HK (1991) Proteinase 3, the target antigen of anticytoplasmic antibodies circulating in Wegener's granulomatosis. Immunolocalization in normal and pathologic tissues. Am J - Pathol 139: 831–838.

40. Charles LA, Falk RJ, Jennette JC (1992) Reactivity of antineutrophil cytoplasmic autoantibodies with mononuclear phagocytes. J Leukoc Biol 51: 65–68.

14

Proteome Profile of Swine Testicular Cells Infected with Porcine Transmissible Gastroenteritis Coronavirus

Ruili Ma[1,2], Yanming Zhang[1]*, Haiquan Liu[3], Pengbo Ning[1]

1 College of Veterinary Medicine, Northwest Agriculture & Forestry University, Yangling, Shaanxi, China, 2 College of Life Sciences, Northwest Agriculture & Forestry University, Yangling, Shaanxi, China, 3 School of Computer Science and Engineering, Xi'an Technological University, Xi'an, Shaanxi, China

Abstract

The interactions occurring between a virus and a host cell during a viral infection are complex. The purpose of this paper was to analyze altered cellular protein levels in porcine transmissible gastroenteritis coronavirus (TGEV)-infected swine testicular (ST) cells in order to determine potential virus-host interactions. A proteomic approach using isobaric tags for relative and absolute quantitation (iTRAQ)-coupled two-dimensional liquid chromatography-tandem mass spectrometry identification was conducted on the TGEV-infected ST cells. The results showed that the 4-plex iTRAQ-based quantitative approach identified 4,112 proteins, 146 of which showed significant changes in expression 48 h after infection. At 64 h post infection, 219 of these proteins showed significant change, further indicating that a larger number of proteomic changes appear to occur during the later stages of infection. Gene ontology analysis of the altered proteins showed enrichment in multiple biological processes, including cell adhesion, response to stress, generation of precursor metabolites and energy, cell motility, protein complex assembly, growth, developmental maturation, immune system process, extracellular matrix organization, locomotion, cell-cell signaling, neurological system process, and cell junction organization. Changes in the expression levels of transforming growth factor beta 1 (TGF-β1), caspase-8, and heat shock protein 90 alpha (HSP90α) were also verified by western blot analysis. To our knowledge, this study is the first time the response profile of ST host cells following TGEV infection has been analyzed using iTRAQ technology, and our description of the late proteomic changes that are occurring after the time of vigorous viral production are novel. Therefore, this study provides a solid foundation for further investigation, and will likely help us to better understand the mechanisms of TGEV infection and pathogenesis.

Editor: Volker Thiel, University of Berne, Switzerland

Funding: This work was supported by the National Natural Science Foundation of China (No. 31172339). The funder had no role in study design, data collection and analysis, decision to publish, or preparation of the manuscript.

Competing Interests: The authors have declared that no competing interests exist.

* Email: zhangym@nwsuaf.edu.cn

Introduction

Porcine transmissible gastroenteritis coronavirus (TGEV) is an animal coronavirus that causes severe gastroenteritis in young TGEV-seronegative pigs. Various breeds of pigs, regardless of age, are susceptible to TGEV; however, the mortality rate for piglets under 2 weeks of age is the highest, reaching almost 100%. Diseased pigs often present with vomiting, dehydration, and severe diarrhea. Further, the disease is known to affect pigs in many countries throughout the world and an outbreak can cause enormous losses in the pig industry [1,2]. The pathogen, TGEV, which belongs to the *Alphacoronavirus* genus of the *Coronavirinae* subfamily within the family *Coronaviridae*, is an enveloped, non-segmented, single-stranded positive-sense RNA virus [3,4]. The envelop, core, and nucleocapsid of the TGEV virion contain four major structural proteins: the nucleocapsid (N) protein, the membrane (M) glycoprotein, the small envelope (E) protein, and the spike (S) protein [5]. The tropism and pathogenicity of the virus are influenced by the S protein, which has four major antigenic sites, A, B, C, and D, with site A being the major inducer of antibody neutralization [3,5]. The M protein, which plays a central role in virus assembly by interacting with viral ribonucleoprotein

(RNP) and S glycoproteins [6], is embedded within the virus membrane and interacts with the nucleocapsid, forming the core of TGEV virion. In addition, the N-terminal domain of the M protein is essential for interferon alpha (IFN-α) induction [7], which is involved in the host's innate immune response. The E protein, a transmembrane protein that acts as a minor structural component in TGEV and affects virus morphogenesis, is essential for virion assembly and release [8].

TGEV RNA, along with the N protein, is infectious and invades the organism through the digestive and respiratory tracts, resulting in infection of the small intestinal enterocytes, villous atrophy, and severe watery diarrhea. These changes in intestinal health are known to be important during the pathogenesis of TGEV infection [9]. Furthermore, corresponding to these pathologic changes observed in vivo, TGEV can also propagate and cause cytopathic effects (CPEs) in multiple types of cultured cells, such as swine testicular (ST) cells, PK-15 cells, and villous enterocytes. Notably, ST cells are more susceptible to TGEV, and higher levels of virus replication have been observed in this cell line [10,11].

The full RNA genome of TGEV is approximately 28.5 kb in length and has a 5′-cap structure and a poly(A) tail at the 3′ end. The 9 open reading frame (ORF) genes included in the TGEV genome are arranged in the following order 5'-la- lb-S-3a-3b-E-M-N-7-3'. The first gene at the 5' end consists of two large ORFs, ORF la and ORF lb, which constitute the replicase gene, known for its RNA-dependent RNA-polymerase and helicase activities, as well as other enzymes, such as endoribonuclease, 3'–5'exoribonuclease, 2'-O-ribose methyltransferase, ribose ADP 1'' phosphatase, etc. [12]. ORF2, ORF4, ORF5, and ORF6 encode the S, E, M, and N proteins, respectively, while ORF3a, ORF3b, and ORF7 encode non-structural proteins [13]. Some investigators have suggested that ORF3 may be related to viral virulence and pathogenesis [12], while ORF7 may interact with host cell proteins and play a role in TGEV replication [14]. In fact, a recent study indicates that plasmid-transcribed small hairpin (sh) RNAs targeting the ORF7 gene of TGEV is capable of inhibiting virus replication and expression of the viral target gene in ST cells in vitro [15]. Although we have some knowledge concerning the translation and function of these viral proteins, the interactions that occur between these proteins and host cell proteins are not fully understood.

Importantly, recent advances in proteomic technology have allowed for more in depth investigation of virus-host interactions, and different techniques have been successfully applied to identify altered proteins in infected host cells and tissues. For example, Sun et al. [16] have identified 35 differentially expressed proteins in PK-15 cells infected with classical swine fever virus (CSFV) using two-dimensional polyacrylamide gel electrophoresis (2D PAGE) followed by matrix-assisted laser desorption-ionization time-of-flight tandem mass spectrometry (MALDI-TOF-MS/MS). In addition, two-dimensional fluorescence difference gel electrophoresis (2D-DIGE) and MS/MS proteomic approaches have been applied to characterize protein changes occurring in host cells in response to porcine circovirus type 2 (PCV2) infection [17]. The same methods have also been studied for many other pathogenic animal viruses, including porcine reproductive and respiratory syndrome virus (PRRSV) [18], coronavirus infectious bronchitis virus (IBV) [19], severe acute respiratory syndrome-associated coronavirus (SARS-CoV) [20], and TGEV [21]. However, these conventional approaches based on 2D gel electrophoresis are not suitable for detecting low abundance, hydrophobic, or very acidic/basic proteins. On the other hand, the isobaric tags for relative and absolute quantitation (iTRAQ) technique, in association with liquid chromatograph (LC), is a more advanced method for proteomic research, and is capable of detecting a much larger number of proteins, even those with low abundance, in addition to identifying and quantifying the proteins simultaneously [22]. To this end, Lu et al. [23] previously used the iTRAQ method to identify 160 significantly altered proteins in pulmonary alveolar macrophages (PAMs) infected with PRRSV. Similarly, this method has been used to investigate influenza virus infection in primary human macrophages [24], human immunodeficiency virus 1 (HIV-1) infection in CD4+ T cells [25], and Epstein–Barr virus (EBV) infection in nasopharyngeal carcinoma cell line [26].

Here, we report the first differential proteomic analysis of TGEV-infected and uninfected ST cells using iTRAQ labeling followed by 2D-LC-MS and bioinformatic analyses. The proteomic data obtained in this study will help to enhance our understanding of the host response to TGEV infection, but also provide new insights on the mechanisms of disease onset.

Materials and Methods

Cell culture and viral replication

ST cells were obtained from the American Type Culture Collection (ATCC). The cells were cultured in high-glucose Dulbecco's modified Eagle's medium (DMEM; GIBCO, UK) containing 1% L-glutamine and 10% fetal bovine serum (FBS) (Hyclone, Logan, UT) at 37°C in 5% CO_2. Culture medium was replaced two to three times per week. The TGEV TH-98 strain was isolated from a suburb of Harbin, Heilongjiang province, China. The virus was propagated in ST cells and preserved at −70°C in our laboratory.

TGEV infection

The monolayer of confluent ST cells was dispersed with 0.25% trypsin and 0.02% ethylenediaminetetraacetic acid (EDTA) and seeded in 6-cm cell culture flasks. After a 24 h incubation period, the culture medium was removed and the ST cells were washed with phosphate buffered saline (PBS, pH 7.4). The cells were then infected with the TGEV TH-98 strain at a 50% tissue culture infectious dose ($TCID_{50}$) of $1 \times 10^{3.53}$ viruses per well, with absorption for 2 h at 37°C. Maintenance medium (DMEM medium supplemented with 2% FBS) was then added to the cells. A mock group of ST cells that were not infected with TGEV was used as a negative control for each of the following experiments. Three replicates of virus-infected and mock-infected cultures with different passage numbers were prepared at each time point. The morphological changes were observed under the light microscope at 24, 40, 48, and 64 hours post infection (hpi).

Reverse transcription polymerase chain reaction (RT-PCR) and real time quantitative PCR (qRT-PCR)

To determine the extent of TGEV infection, conventional RT-PCR and qRT-PCR assays were performed to detect the viral N gene. Monolayers of ST cells were infected with TGEV as described above. Cells were collected from 24 to 80 hpi at 8 h intervals, and the total RNA of the infected cells was extracted using Trizol (Invitrogen). RNA samples were reverse-transcribed using PrimeScript RT reagent Kit (Takara Bio, Dalian, China), according to the manufacturer's instructions. The RT reaction was incubated at 37°C for 15 min followed by 85°C for 5 s. A mixture of oligo dT primers and random 6 mers was used in the RT step. The cDNA was stored at −20°C until further use.

PCR was performed for the TGEV N gene in a 25 μl reaction mixture containing 1 μl of the cDNA, 0.5 μl of each forward (F) and reverse (R) primer, 12.5 μl of Premix Taq (Takara Bio, Dalian, China), and 10.5 μl DEPC water, starting with a 5 min denaturation at 95 C followed by 32 cycles of 30 s denaturation at 95 C, 30 s annealing at 56 C, and 40 s extension at 72 C. A final extension step was carried out at 72 C for 10 min. RT-PCR products were resolved on a 15 g/L agarose gel. The following PCR primers were used in this study: TGEV N (F, 5'-GAGC-AGTGCCAAGCATTACCC-3' and R, 5'-GACTTCTAT CT-GGTCGCCATCTTC-3') and β-actin (F, 5'-GCAAGGACCTC-TACGCCAA-3' and R, 5'-CTGGAAGGTGGACAGCGAG-3').

The mRNA expression level of the TGEV N gene was quantified using a SYBR Green assay on a Bio-Rad iQ5 real time PCR detection system as described previously [27]. We used the same primers listed above for qRT-PCR. Reactions were carried out in 50 μl volumes containing 0.5 μl of 20 × SYBR Green I, 2 μl of cDNA template, 1 μl of each F and R primer, 25 μl of 2 × PCR buffer, and 20.5 μl DEPC water. The cycling conditions were 94°C for 4 min, followed by 35 cycles of 94°C for 20 s, 60°C for 30 s, 72°C for 30 s, and then a final extension of 10 min at 72°C. The relative gene expression was determined with the $2^{(-\Delta\Delta Ct)}$ method [28], and the tests were performed in triplicate.

Protein isolation, digestion, and labeling with iTRAQ reagents

Following ST cell infection, cells were collected at 48 and 64 hpi by centrifugation at 3,000 rpm for 5 min at 4°C, washed twice with PBS, and 1 mL of iTRAQ lysis solution (8 M urea, 1% (w/v) dithiothreitol (DTT)) containing protease inhibitor was added. Then, the cells were put in an ice bath and broken up by sonication. The solution was then mixed for 30 min at 4°C. The soluble protein fraction was harvested by centrifugation at 40,000 × g for 30 min at 4°C and the debris was discarded. The protein concentration was determined with the Bradford protein assay (2-D Quant Kit, Bestbio, China). A 100 µg aliquot of protein from each sample was reduced, alkylated, and trypsin-digested as described in the iTRAQ protocol (AB Sciex, American), followed by labeling with the 4-plex iTRAQ Reagents Multiplex Kit according to the manufacturer's instructions (AB Sciex, American). Two virus-free samples at 48 h and 64 h were labeled with iTRAQ tags 114 and 115, while two TGEV-infected samples at 48 h and 64 h were labeled with tags 116 and 117. The labeled digests were then pooled, dried using a vacuum freeze drier (Christ RVC 2−25, Germany), and preserved at −20°C for later use.

2D LC-MS/MS analysis

The combined peptide mixtures were separated by reversed phase high-performance liquid chromatography (HPLC) (Ekspert ultraLC 100, AB Sciex, USA) on a Durashell-C18 reverse phase column (4.6 mm × 250 mm, 5 µm 100 Å, Agela). The mobile phases used were composed of 20 mM ammonium formate (pH 10) in water (labeled mobile phase A) and 20 mM ammonium formate (pH 10) in acetonitrile(ACN) (mobile phase B). The flow rate was 0.8 mL/min, and the elutant was collected into 48 centrifuge tubes at each minute after the first 5 min. Each aliquot was then dried by vacuum freezing.

The peptides were then analyzed with a nanoflow reversed-phase liquid chromatography-tandem mass spectrometry (nano-RPLC-MS/MS) system (TripleTOF 5600, AB Sciex, USA). The above 48 tubes were merged into 10 components dissolved in 2% ACN and 0.1% formic acid (FA), then centrifuged at 12,000 × g for 10 min. The supernatant (8 µl) was used for loading at a rate of 2 µl/min, with a separation rate of 0.3 µl/min. The mobile phase A used in this analysis was composed of 2% ACN and 0.2% FA, while mobile phase B was composed of 98% ACN and 0.1% FA. The following MS parameters were utilized: source gas parameters (ion spray voltage: 2.3 kV, GS1:4, curtain gas: 30 or 35, DP: 100 or 80); TOF MS (m/z: 350–1250, accumulation time: 0.25 s); and product ion scan (IDA number: 30, m/z: 100–1500, accumulation time: 0.1 s, dynamic exclusion time: 25 s, rolling CE: enabled, adjust CE when using iTRAQ reagent: enabled, CES: 5).

Data analysis and bioinformatics

Protein identification and quantification were performed with the ProteinPilot software (version 4.0, AB Sciex) using the Paragon algorithm. Each MS/MS spectrum was searched against a database of *Sus scrofa* protein sequences (NCBI nr, released in March 2011, downloaded from ftp://ftp.ncbi.nih.gov/genomes/Sus_scrofa/protein/). The following search parameters were used: iTRAQ 4-plex (peptide labeled), cysteine alkylation with methyl methanethiosulfonate(MMTS), trypsin digestion, biological modifications allowed, a thorough search, a detected protein threshold of 95% confidence (unused Protscore ≥1.3), and a critical false discovery rate (FDR) of 1%. The peptide and protein selection criteria for relative quantitation were performed as described previously, whereby only peptides unique for a given protein were

considered [29]. In addition, proteins with an iTRAQ ratio higher than 20 or lower than 0.05 as well as proteins in reverse database were removed [30].

To assign enriched Gene Ontology (GO) terms to the identified proteins, the differentially expressed proteins identified from iTRAQ experiments and all of the 4,112 measured proteins were classified based on their GO annotations using QuickGO (http://www.ebi.ac.uk/QuickGO/), with UniProt ID (http://www.uniprot.org/?tab=mapping) as the data source. GO enrichment analysis of the differentially regulated proteins was evaluated using all of the 4,112 quantified proteins as background with hypergeometric distribution [31]. Categories belonging to biological processes, molecular functions, and cellular components that were identified at a confidence level of 95% were included in the analysis. The protein-protein interaction network for a select group of proteins was analyzed using the STRING 9.1 database (http://string-db.org/). Network analysis was set at medium confidence (STRING score >0.4).

Western blot analysis

Following ST cell infection with TGEV, the culture medium was removed after incubating for 48 h and 64 h; then, the cells were washed with cold PBS and collected after centrifugation at 3,000 rpm for 10 min. Cells were then lysed in RIPA lysis buffer with protease inhibitors (Applygen Technologies Inc., China). Cellular debris was removed by centrifugation at 12,000 × g for 5 min at 4°C, and the protein concentration was measured by Coomassie blue G250 staining. An equal amount (20 µg) of cell lysate from each sample was separated using 10% SDS-PAGE and then transferred to polyvinyl difluoride (PVDF) membranes (Millipore, Bedford, USA). The PVDF membranes were then blocked with 5% (w/v) de-fatted milk powder dissolved in tris buffered saline and tween 20 (TBST) buffer (150 mM NaCl, 50 mM Tris, 0.05% Tween 20) for 1 h at 37°C. After blocking, membranes were incubated with anti-glyceraldehyde 3-phosphate dehydrogenase (GAPDH) mouse monoclonal antibody (1:3000; Western Biotechnology, China), anti-heat shock protein 90 alpha (Hsp90α/HSP90AA1) antibody (1:300; Abcam, Cambridge, UK), anti-caspase 8 antibody (1:300; Abcam, Cambridge, UK), or anti-transforming growth factor β 1 (TGF-β1/TGFB1) antibody (1:300; Abcam, Cambridge, UK) overnight at 4°C, followed by HRP-conjugated secondary antibody (1:5000; Western Biotechnology, China) for 1.5 h at 37°C. The membranes were then washed four times in TBST buffer for 5 min each time. Protein band detection was performed using ECL reagents (Applygen Technologies Inc., China), and the band intensities were analyzed using Labworks 4.6 software.

Results

Confirmation of TGEV infection in ST Cells

After introducing TGEV into the ST cells, we observed the induction of typical CPEs, including cell rounding, swelling, granular degeneration of the cytoplasm, cell detachment, and severely diseased cell morphology, from 40 to 64 h after inoculation (Figure 1 A–D) compared to the non-infected control cells (Figure 1 E–H). Virus infection at 48 and 64 h was also confirmed by RT-PCR detection of the viral N gene in the sample (Figure 2A).

Dynamic changes in viral gene expression in infected cells

To further identify the extent of TGEV infection, the mRNA expression levels of viral genes in infected cells were determined

TGEV infection

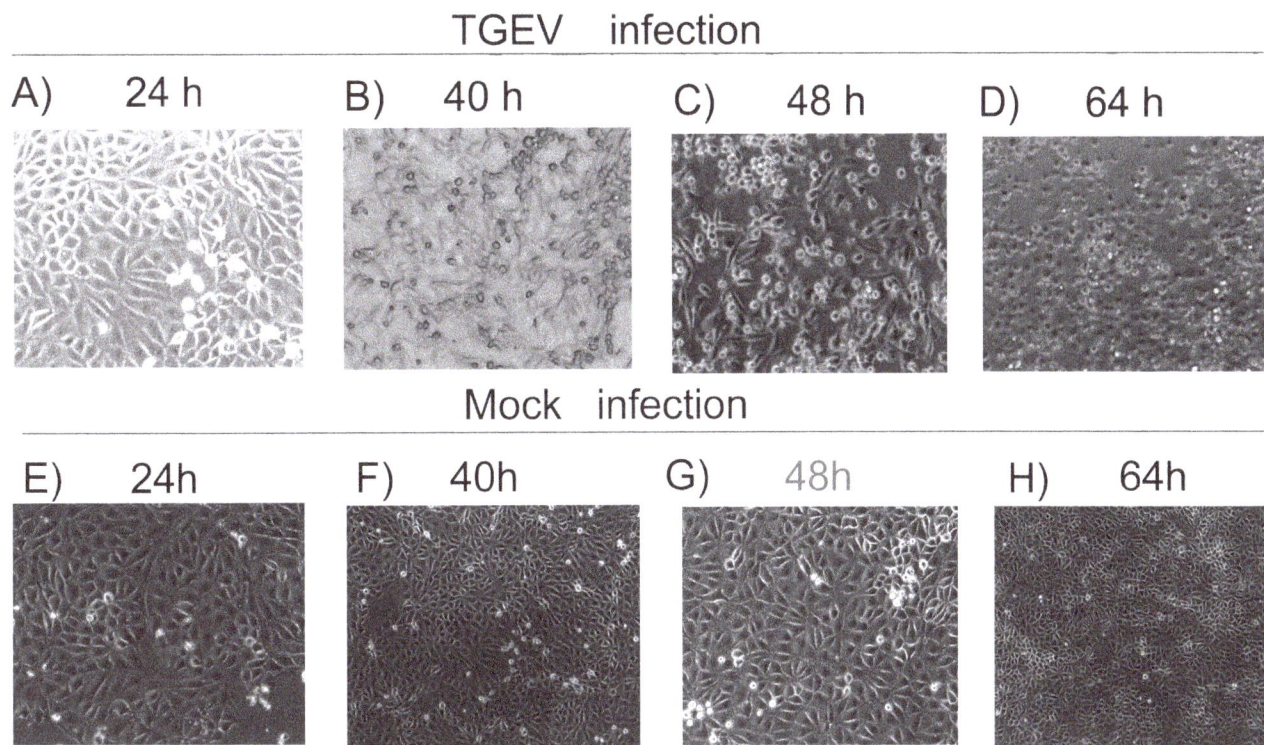

A) 24 h   B) 40 h   C) 48 h   D) 64 h

Mock infection

E) 24h   F) 40h   G) 48h   H) 64h

Figure 1. Morphological changes in TGEV-infected cells. ST cells were seeded into 6-cm culture plates, infected with TGEV, and the cytopathic effects (CPEs) were imaged at 24 (A), 40 (B), 48 (C), and 64 (D) hours following infection. Images of non-infected cells (mock infection) are shown for comparison at each time point (E, F, G, H).

using qRT-PCR. Comparative threshold (Ct) cycle values in three independent experiments were calculated and the results indicated that the average Ct value for the TGEV N gene ranged from 25.2 to 27.5. Correspondingly, the average Ct value observed for the β-actin control gene ranged from 19.6 to 21.0. The relative expression of TGEV N mRNA was calculated using the $2^{(-\Delta\Delta CT)}$ method [28], and the change in expression at each time point is indicated in Figure 2B. These data show that, following infection, the viral mRNA levels increased gradually over time, and reached a peak at 48 hpi. Following this time point, the viral mRNA levels appear to decrease.

Protein identification by MS

In the infected ST cells, a total of 29,214 peptides and 4,364 proteins were detected (Table S1); however, only 4,112 proteins were quantified reliably (Table S2). Notably, the abnormal proteins, such as the proteins with iTRAQ ratio higher than 20 or lower than 0.05, which are not quantifiable [30], were removed and only proteins with reasonable ratios across all channels were investigated further. Figure 3A depicts the scatter plots for the $\log_{10}$ 116/114 and $\log_{10}$ 117/115 ratios in the iTRAQ experiment. Linear regression analysis showed that correlation (R^2) was 0.58, with a p-value less than 0.05. These results suggest that the alterations in protein abundance due to virus infection were near-linear dependency between the two time points. In order to

(A)   (B)

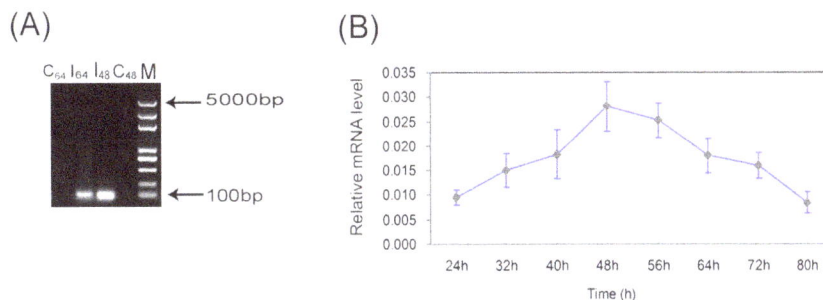

Figure 2. Validation of TGEV virus infection of ST cells. (A) RT-PCR validation of TGEV infection in ST cells at 48 hpi (I_{48}) and 64 hpi (I_{64}) compared to the control at 48 h (C_{48}) and 64 h (C_{64}). A marker (M) was used to identify fragment size. (B) qRT-PCR analysis of changes in TGEV mRNA expression levels in the ST cells over time. The changes in mRNA expression level at the various time points is indicated, and show that the expression level of TGEV increased gradually, reaching a peak at 48 h, then decreased dramatically. Values are the means of three repeated experiments. The error bars in the graphs represent the standard deviation.

(A)

(B)

(C)

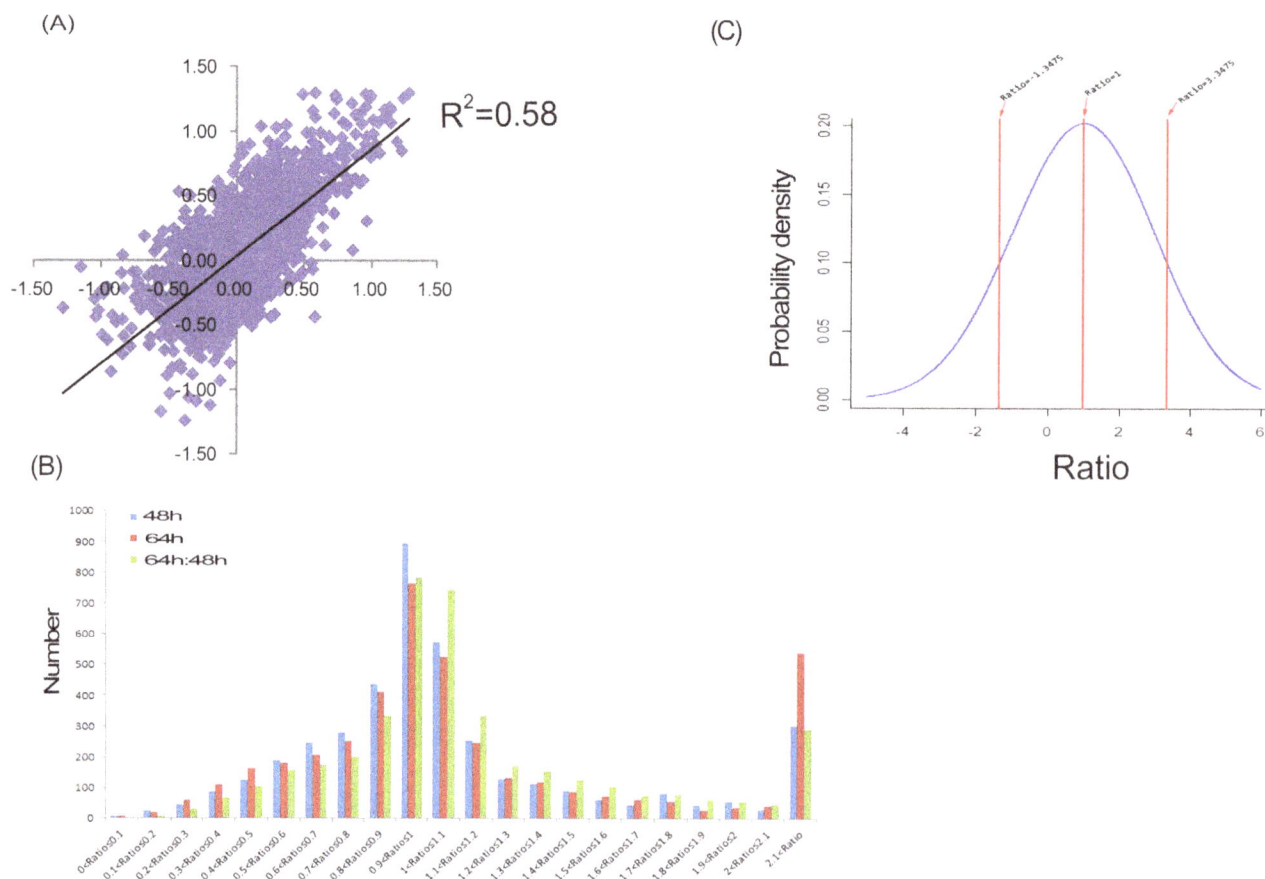

Figure 3. Results of the iTRAQ ratios analysis. (A) A scatter plot showing the correlation between the $\log_{10}$ infection/mock ratios at 48 hpi and 64 hpi for the 4,112 reliably quantified proteins in the iTRAQ experiment. Linear regression analysis shows that correlation (R^2) was 0.58, with a p-value less than 0.05. (B) Histograms showing the distribution of protein ratios identified at 48 and 64 hpi. (C) The distribution range of differentially expressed proteins identified at 48 hpi. iTRAQ ratios higher than 3.3475 (p = 0.975) or lower than -1.3475 (p = 0.025) were defined as statistically significant.

identify the proteins that were significantly different at each time point (infected/uninfected) or between the different time points, we analyzed the distribution of ratios for the identified proteins as shown in the Figure 3B. For the distribution range of the differentially expressed proteins identified at 48 hpi, shown in Figure 3C, a ratio higher than 3.35 or lower than -1.35 was defined as a statistically significant difference in protein expression. At 64 hpi, a ratio higher than 4.55 or lower than -2.15 was defined as a statistically significant difference in protein expression. According to analyses, the differentially expressed proteins identified were considered to show a significant upward or downward trend if their expression ratios were greater than 4.0 or less than 0.25 compared to the control group.

Using the criterion listed above, the expression of 146 proteins was significantly changed at 48 hpi (95 upregulated and 51 downregulated), while 219 proteins were significantly changed at 64 hpi (172 upregulated and 47 downregulated). Further, 72 proteins were identified to be significantly different between the two time points (54 upregulated and 18 downregulated), resulting in a total of 316 unique proteins being significantly altered during TGEV infection, including 162 predicted proteins (Table S3 and Table 1 (excluding the predicted proteins)). Because the current pig genome database is poorly annotated compared to the human genome database, there were numerous proteins that were unassigned or uncharacterized, resulting in a large number of

predicted proteins in our analysis. However, our ability to detect the unannotated proteins by MS demonstrates that they do existence in this species, and additional research concerning their function is warranted.

GO enrichment analysis

Biological process-based enrichment analysis of the differentially expressed proteins revealed that six common GO terms were significantly enriched in this set of proteins (p<0.05). Thus, it appears that in TGEV-infected ST cells at 48 and 64 hpi there are expression changes in proteins that are related to cell adhesion, neurological system processes, extracellular matrix organization, locomotion, cell junction organization, and cell-cell signaling. Moreover, at the later time point, 64 hpi, our GO term analysis also indicated that a significant number of the differentially expressed proteins were related to cellular stress (p = 8.18E-4), generation of precursor metabolites and energy (p = 2.74E-3), cell motility (p = 6.71E-3), protein complex assembly (p = 4.69E-2), growth (p = 3.87E-2), developmental maturation (p = 1.53E-2), and immune system processes (p = 4.67E-2) (Table 2).

To further investigate the localization pattern of these differentially expressed genes, a cellular component-based enrichment analysis was performed. At 48 hpi, we observed the significant enrichments in extracellular region (p = 1.29E-4), proteinaceous extracellular matrix (p = 1.62E-4), and extracellular

Table 1. Differentially expressed proteins identified by iTRAQ analysis of ST cells infected with TGEV.

Accession number	Protein name	Gene symbol	Unused ProtScore	Infected/uninfected (48 h)		Infected/uninfected (64 h)	
				Ratio	P-value	Ratio	P-value
Upregulated proteins							
gi|359811347	60 kDa heat shock protein, mitochondrial	-	139.52	3.16	0.00	5.86 ↑	0.00
gi|227430407	Keratin, type II cytoskeletal 8	KRT8	110.35	4.02 ↑	0.00	6.49 ↑	0.00
gi|347300243	Glutamate dehydrogenase 1, mitochondrial	GLUD1	102.28	1.79	0.18	4.17 ↑	0.00
gi|297591975	ATP synthase subunit alpha, mitochondrial	ATP5A1	96.61	1.17	0.67	4.66 ↑	0.00
gi|417515796	Hypoxia up-regulated protein 1 precursor	-	92.84	3.91	0.01	6.92 ↑	0.00
gi|349732227	Heterogeneous nuclear ribonucleoprotein M	-	89.74	7.66 ↑	0.00	9.64 ↑	0.00
gi|56748897	Heat shock 70 kDa protein 1B	HSPA1B	62.37	4.33 ↑	0.21	4.02 ↑	0.12
gi|47522630	Aspartate aminotransferase, mitochondrial precursor	GOT2	60.36	1.43	0.01	4.66 ↑	0.00
gi|387912908	Calreticulin	CALR	55.58	2.40	0.11	4.61 ↑	0.00
gi|346421378	Serpin H1 precursor	-	52.10	3.22	0.00	4.06 ↑	0.00
gi|2506849	Malate dehydrogenase, mitochondrial	MDH2	49.19	3.28	0.00	6.98 ↑	0.00
gi|148230268	Galectin-3	LGALS3	48.39	3.08	0.15	5.06 ↑	0.01
gi|417515899	2-oxoglutarate dehydrogenase, mitochondrial	-	45.01	2.99	0.02	5.25 ↑	0.00
gi|745552	Voltage-dependent anion channel 1	VDAC1	43.46	6.19 ↑	0.01	11.59 ↑	0.00
gi|330417958	Phosphoenolpyruvate carboxykinase [GTP], mitochondrial	PCK2	42.89	1.96	0.09	5.75 ↑	0.00
gi|353468887	Signal transducer and activator of transcription 1	STAT1	42.79	1.80	0.33	6.98 ↑	0.00
gi|21264506	Succinyl-CoA ligase [GDP-forming] subunit beta, mitochondrial	SUCLG2	41.68	1.53	0.00	4.06 ↑	0.00
gi|47716872	Galectin-1	-	41.49	5.20 ↑	0.05	4.66 ↑	0.06
gi|342349346	Lon peptidase 1, mitochondrial	-	41.02	2.49	0.03	6.03 ↑	0.00
gi|210050415	Mx2 protein	Mx2	40.44	3.08	0.79	18.88 ↑ *	0.00
gi|342349319	Calnexin precursor	-	37.71	4.79 ↑	0.00	6.14 ↑	0.00
gi|72535198	Histone H1.3-like protein	-	36.51	1.69	0.38	8.32 ↑ *	0.12
gi|347300207	Nucleobindin-1 precursor	NUCB1	35.30	3.40	0.00	5.01 ↑	0.00
gi|347800693	Ferredoxin reductase	FDXR	33.41	1.56	0.05	4.57 ↑	0.00
gi|417515788	Prolow-density lipoprotein receptor-related protein 1 precursor	-	32.44	1.60	0.13	5.65 ↑	0.00
gi|297747350	FAT tumor suppressor homolog 1	-	32.04	6.08 ↑	0.00	7.24 ↑	0.00
gi|298104076	Enoyl-CoA hydratase, mitochondrial	-	30.50	2.21	0.23	6.55 ↑	0.00
gi|7939586	Dihydrolipoamide succinyltransferase	DLST	30.42	1.71	0.17	4.06 ↑	0.00
gi|7404364	Hydroxyacyl-coenzyme A dehydrogenase, mitochondrialPrecursor	HADH	29.29	1.92	0.00	5.50 ↑	0.00
gi|346644866	Coiled-coil-helix-coiled-coil-helix domain-containing protein 3, mitochondrial	CHCHD3	28.98	2.51	0.01	4.61 ↑	0.00
gi|47522814	Dihydrolipoyllysine-residue acetyltransferase component of pyruvate dehydrogenase complex, mitochondrial precursor	-	28.30	1.98	0.32	4.92 ↑	0.00

Table 1. Cont.

Accession number	Protein name	Gene symbol	Unused ProtScore	Infected/uninfected (48 h)		Infected/uninfected (64 h)	
				Ratio	P-value	Ratio	P-value
gi|6165556	Long-chain 3-ketoacyl-CoA thiolase	LCTHIO	26.88	3.98	0.00	7.59 ↑	0.00
gi|156720190	Mx1 protein	Mx1	26.26	3.80 ↑	0.98	19.41 ↑ *	0.00
gi|47522770	Clusterin precursor	CLU	25.75	13.80 ↑	0.00	14.59 ↑	0.00
gi|347300323	Thioredoxin-dependent peroxide reductase, mitochondrial	PRDX3	24.24	2.72	0.00	6.92 ↑	0.00
gi|47522610	Succinyl-CoA:3-ketoacid coenzyme A transferase 1, mitochondrial precursor	OXCT1	23.94	1.72	0.22	6.37 ↑	0.00
gi|346986361	Electron-transfer-flavoprotein, alpha polypeptide	ETFA	22.41	2.42	0.34	5.55 ↑	0.00
gi|172072653	Lactadherin precursor	MFGE8	22.33	10.19 ↑	0.00	9.64 ↑	0.00
gi|56417363	Cathepsin D protein	—	21.95	0.54	0.0	2.38* ↑	0.02
gi|87047636	ATP synthase H+-transporting mitochondrial F1 complex O subunit	ATP5O	21.83	1.11	0.66	7.66 ↑ *	0.00
gi|89573851	Succinate dehydrogenase complex subunit B	SDHB	21.18	2.00	0.04	5.97 ↑	0.00
gi|5921142	Amyloid precursor protein	APP	20.29	13.43 ↑	0.00	15.14 ↑	0.00
gi|347658971	ATP synthase, H+ transporting, mitochondrial Fo complex, subunit d	—	20.26	3.34	0.01	9.82 ↑	0.00
gi|75052621	Transcription factor A, mitochondrial	TFAM	19.14	2.01	0.00	5.50 ↑	0.00
gi|312283580	Superoxide dismutase [Mn], mitochondrial	—	18.40	1.79	0.18	5.35 ↑	0.00
gi|6093657	Propionyl-CoA carboxylase beta chain, mitochondria	PCCB	17.92	2.33	0.13	6.85 ↑	0.00
gi|346716275	DnaJ homolog subfamily B member 11 precursor	DNAJB11	17.48	2.63	0.01	4.74 ↑	0.00
gi|118403762	Extracellular superoxide dismutase precursor	—	16.78	11.27 ↑	0.01	11.48 ↑	0.01
gi|150251019	Adenylate kinase 3-like 1	AK3L1	15.85	2.05	0.12	4.06 ↑	0.00
gi|158517860	Thymosin beta-10	TMSB10	13.45	6.25 ↑	0.30	7.52 ↑	0.30
gi|47522698	Cathepsin L1 precursor	CTSL	12.72	4.45 ↑	0.01	5.11 ↑	0.01
gi|329744622	Low-density lipoprotein receptor precursor	LDLR	12.69	3.50	0.04	4.06 ↑	0.01
gi|346644882	Reticulocalbin 2, EF-hand calcium binding domain precursor	RCN2	12.05	2.00	0.04	4.09 ↑	0.00
gi|284519712	**Caspase-8**	—	**11.38**	**7.11 ↑**	**0.13**	**16.14 ↑**	**0.00**
gi|211578396	Nitrogen fixation 1-like protein	LOC100156145	11.29	3.60	0.02	5.45 ↑	0.00
gi|346644830	Sulfide:quinone oxidoreductase, mitochondrial	SQRDL	10.92	1.87	0.38	4.53 ↑	0.02
gi|417515419	Semaphorin-3C precursor	—	10.76	4.92 ↑	0.08	3.94	0.24
gi|75064988	Syndecan-4	SDC4	10.29	18.88 ↑	0.00	19.59 ↑	0.00
gi|346716228	Histidine triad nucleotide-binding protein 2, mitochondrial isoform 2 precursor	HINT2	10.06	3.77	0.20	12.71 ↑	0.03
gi|85720739	Beta-enolase 3	ENO3	9.83	15.42 ↑	0.20	8.32 ↑	0.25
gi|223634702	Succinyl-CoA ligase [ADP/GDP-forming] subunit alpha, mitochondrial	SUCLG1	9.76	4.74 ↑	0.00	9.04 ↑	0.00

Table 1. Cont.

Accession number	Protein name	Gene symbol	Unused ProtScore	Infected/uninfected (48 h)		Infected/uninfected (64 h)	
				Ratio	P-value	Ratio	P-value
gi|4579751	130 kDa regulatory subunit of myosin phosphatase, partial	–	9.64	8.39 ↑	0.00	3.94	0.27
gi|76781337	ADAMTS1	ADAMTS1	9.61	6.79 ↑	0.00	7.38 ↑	0.00
gi|417515625	Interferon-induced protein with tetratricopeptide repeats 2	–	9.52	1.53	0.68	10.76 ↑*	0.00
gi|47522640	CD97 antigen	–	8.47	5.40 ↑	0.01	5.20 ↑	0.02
gi|55247591	Granulin precursor	GRN	8.43	12.71 ↑	0.00	14.59 ↑	0.01
gi|8347147	Inflammatory response protein 6	RSAD2	8.22	0.72	0.88	4.06 ↑*	0.00
gi|148234138	Cytochrome c oxidase subunit 6B1	COX6B	8.11	2.65	0.11	4.97 ↑	0.01
gi|343790890	Acyl-CoA dehydrogenase family, member 8	–	8.06	1.56	0.30	5.55 ↑	0.02
gi|9957597	Probable ATP-dependent RNA helicase DDX58	DDX58	7.83	1.22	0.52	8.02 ↑*	0.00
gi|347300255	DAZ-associated protein 1	DAZAP1	7.32	7.52 ↑	0.01	4.29 ↑	0.05
gi|148887343	ATP synthase subunit e, mitochondrial	ATP5I	7.00	1.11	0.97	4.21 ↑	0.01
gi|297632426	Signal sequence receptor, alpha	–	6.36	4.49 ↑	0.03	5.20 ↑	0.03
gi|6919844	Transforming growth factor-beta-induced protein ig-h3	TGFBI	6.12	4.92 ↑	0.01	3.56	0.20
gi|47523704	Double stranded RNA-dependent protein kinase	PKR	6.07	5.65 ↑	0.19	6.67 ↑	0.15
gi|339895859	Lipase, endothelial precursor	LIPG	5.14	4.06 ↑	0.04	3.25	0.05
gi|6226834	2'-5'-oligoadenylate synthase 1	OAS1	5.03	1.96	0.09	10.47 ↑*	0.01
gi|21636588	ATP synthase gamma subunit 1	–	4.61	2.78	0.16	4.49 ↑	0.05
gi|56392985	Asparagine-linked glycosylation 2	ALG2	4.31	2.65	0.30	4.57 ↑	0.23
gi|52346216	Fibroleukin precursor	FGL2	4.22	3.13	0.11	4.33 ↑	0.07
gi|154147577	Interferon-induced helicase C domain-containing protein 1	MDA5	4.20	2.09	0.78	6.67 ↑	0.06
gi|343098453	Chromatin target of PRMT1 protein	CHTOP	4.10	8.47 ↑	0.05	6.79 ↑	0.24
gi|343478189	Tubulin beta-2B chain	TUBB2B	4.04	5.25 ↑	0.30	5.40 ↑	0.24
gi|47523638	Nexin-1 precursor	PN-1	4.01	5.97 ↑	0.16	8.95 ↑	0.14
gi|346716354	Protein lunapark	–	4.00	10.76 ↑	0.17	7.94 ↑	0.23
gi|87047624	C-C motif chemokine 5	CCL5	3.80	5.35 ↑	0.31	18.71 ↑	0.12
gi|75056555	Integral membrane protein 2B	ITM2B	3.70	12.82 ↑	0.20	12.94 ↑	0.18
gi|264681460	Acyl carrier protein, mitochondrial	NDUFAB1	3.13	2.21	0.25	4.33 ↑	0.09
gi|456752927	Lectin, galactoside-binding, soluble, 3 binding protein	–	2.94	1.04	0.13	6.43 ↑*	0.06
gi|116175255	Regulator of differentiation 1	ROD1	2.79	2.68	0.23	4.29 ↑	0.14
gi|164664468	ATP synthase subunit epsilon, mitochondrial	ATP5E	2.74	3.66	0.14	14.06 ↑	0.02
gi|47522704	Vascular cell adhesion protein 1 precursor	–	2.72	3.56	0.11	6.79 ↑	0.02
gi|417515517	Solute carrier family 2,facilitated glucose transporter member 1	–	2.52	4.06 ↑	0.17	2.83	0.23
gi|346644790	Eukaryotic translation initiation factor 4E-binding protein 1	–	2.15	11.48 ↑	0.05	6.73 ↑	0.17

Table 1. Cont.

Accession number	Protein name	Gene symbol	Unused ProtScore	Infected/uninfected (48 h)		Infected/uninfected (64 h)	
				Ratio	P-value	Ratio	P-value
gi|346644828	Nuclear ubiquitous casein and cyclin-dependent kinases substrate	NUCKS1	2.01	5.70 ↑	0.24	3.60	0.37
gi|35208827	Macrophage colony-stimulating factor 1 precursor	MCSF alpha	2.01	6.37 ↑	0.24	7.73 ↑	0.21
gi|158726687	IGFBP-6	–	2.00	9.29 ↑	0.11	9.20 ↑	0.11
gi|146345485	Plasminogen	PLG	2.00	7.94 ↑	0.12	13.30 ↑	0.10
gi|63809	**Transforming growth factor beta-1**	**TGFB1**	**2.00**	**8.32 ↑**	**0.31**	**13.43 ↑**	**0.21**
gi|239504564	Claudin-4	CLDN4	1.97	4.92 ↑	0.27	8.63 ↑	0.16
gi|75049861	C-X-C motif chemokine 16	CXCL16	1.96	3.02	0.22	4.92 ↑	0.15
gi|158514029	ATP synthase lipid-binding protein, mitochondrial	ATP5G1	1.45	1.38	0.49	5.81 ↑*	0.34
gi|872313	Monocyte chemoattractant protein 1	CCL2	1.32	3.44	0.25	4.79 ↑	0.19
gi|81295909	Mitochondrial aldehyde dehydrogenase 2	ALDH2	34.88	0.72	0.14	3.16*	0.00
gi|224593280	Tyrosine-protein phosphatase non-receptor type 1	PTPN1	12.65	0.33	0.01	1.37*	0.11
gi|83415439	MHC class I antigen	PD1	7.05	0.45	0.43	3.13*	0.04
gi|148747492	Keratin, type II cytoskeletal 2 epidermal	KRT2A	6.68	0.67	0.98	3.40*	0.10
gi|75054309	N-acetylgalactosamine-6-sulfatase	GALNS	6.61	0.34	0.05	1.72*	0.11
gi|343791025	Lysosomal protective protein precursor	–	5.84	0.81	0.80	3.25*	0.05
gi|262204920	Peroxisomal trans-2-enoyl-CoA reductase	PECR	5.77	0.26	0.13	1.25*	0.32
gi|75063982	Alpha-crystallin B chain	CRYAB	4.92	0.37	0.19	3.40*	0.07
gi|456753359	Mevalonate (diphospho) decarboxylase, partial	–	4.01	0.26	0.44	1.41*	0.77
gi|343478257	Peptidase M20 domain containing 1	–	3.19	0.31	0.36	1.34*	0.69
gi|90024980	Peroxisomal enoyl coenzyme A hydratase 1	ECH1	17.09	0.79	0.88	3.37*	0.00
Downregulated proteins							
gi|346986428	Heat shock 90kD protein 1, beta	HSPCB	130.10	0.70	0.52	0.21 ↓	0.00
gi|48675927	Tropomyosin alpha-3 chain	TPM3	91.83	0.53	0.01	0.20 ↓	0.00
gi|28948618	Chain A, structure of full-length annexin A1 in the presence of calcium	ANXA1	72.35	0.42	0.00	0.06 ↓*	0.00
gi|6016267	**Heat shock protein HSP 90-alpha**	**HSP90AA1**	**53.06**	**0.74**	**0.10**	**0.19 ↓***	**0.00**
gi|47523720	Glucose-6-phosphate isomerase	GPI	50.00	0.54	0.00	0.18 ↓	0.00
gi|575527982	Radixin	RDX	44.08	0.53	0.00	0.22 ↓	0.00
gi|51702768	Peptidyl-prolyl cis-trans isomerase A	PPIA	41.51	0.75	0.35	0.24 ↓	0.00
gi|7650140	Gag-pol precursor	–	40.78	0.23 ↓	0.00	0.82	0.04
gi|262263205	Triosephosphate isomerase 1	TPI1	37.70	0.47	0.02	0.13 ↓	0.00
gi|1927	Cardiac alpha tropomyosin	TPM1	36.76	0.50	0.01	0.08 ↓*	0.00
gi|75074817	Peroxiredoxin-6	PRDX6	35.65	0.90	0.03	0.16 ↓*	0.00
gi|94962086	Aldo-keto reductase family 1 member C4	AKR1C4	34.49	0.12 ↓	0.00	0.37	0.00

Table 1. Cont.

Accession number	Protein name	Gene symbol	Unused ProtScore	Infected/uninfected (48 h)		Infected/uninfected (64 h)	
				Ratio	P-value	Ratio	P-value
gi|473575	Lactate dehydrogenase-B	LDHB	24.97	0.63	0.01	0.08↓*	0.00
gi|164414678	Alternative pig liver esterase	APLE	23.65	0.19↓	0.06	0.64	0.43
gi|343780946	D-dopachrome decarboxylase	DDT	19.05	0.15↓	0.26	0.60*	0.61
gi|347300176	Peroxiredoxin-2	PRDX2	24.03	0.74	0.30	0.24↓	0.01
gi|302372516	Heart fatty acid-binding protein	FABP3	23.79	0.60	0.00	0.23↓	0.00
gi|343887360	Proteasome (prosome, macropain) subunit, alpha type	-	21.65	0.42	0.00	0.25↓	0.00
gi|47522644	Acylamino-acid-releasing enzyme	APEH	20.10	0.38	0.01	0.15↓	0.00
gi|346716148	Importin-5	-	18.84	0.83	0.27	0.22↓	0.00
gi|47523046	Acyl-CoA-binding protein	DBI	18.36	0.44	0.05	0.23↓	0.00
gi|47523158	Glutathione S-transferase A2	-	15.78	0.31	0.00	0.09↓	0.00
gi|297591959	Farnesyl pyrophosphate synthase precursor	FDPS	15.29	0.67	0.32	0.16↓*	0.00
gi|56384247	Ribosomal protein L7	-	15.34	0.09↓	0.01	0.37	0.05
gi|347300398	Core histone macro-H2A.1 isoform 1	H2AFY	14.24	0.25↓	0.24	1.16*	0.58
gi|417515487	Collectin sub-family member 12	-	14.04	0.19↓	0.00	0.48	0.14
gi|94471896	signal transducer and activator of transcription 3	STAT3	13.42	0.21↓	0.00	0.58	0.14
gi|417515866	KIAA0196	-	12.91	0.39	0.00	0.14↓	0.00
gi|584724	Aminoacylase-1	ACY1	12.26	0.30	0.00	0.13↓	0.00
gi|158514030	60S ribosomal protein L14	RPL14	10.79	0.14↓	0.00	0.79*	0.85
gi|187606917	40S ribosomal protein S26	RPS26	6.00	0.19↓	0.07	0.45	0.15
gi|89257972	Protein phosphatase 1 catalytic subunit beta isoform	PPP1CB	5.27	0.11↓	0.25	0.61*	0.52
gi|417515889	FK506-binding protein 15	-	4.06	0.05↓	0.12	0.43*	0.25
gi|48474224	Scavenger receptor class B member 1	SCARB1	2.98	0.24↓	0.08	0.41	0.13
gi|83778524	Beta-tropomyosin	TPM2	2.55	0.28	0.21	0.07↓*	0.04
gi|298104074	Protein FAM54A	-	2.08	0.46	0.30	0.21↓	0.16
gi|342349308	Calmegin precursor	-	2.00	0.16↓	0.14	0.21↓	0.16
gi|1016311	Cytochrome P450 2C33v3, partial	-	1.96	0.10↓	0.11	0.41*	0.26
gi|346716298	Heterogeneous nuclear ribonucleoprotein G	RBMX	32.46	3.63	0.00	0.86*	0.41
gi|262263201	Squalene epoxidase	SQLE	2.02	1.37	0.67	0.34*	0.32

Table 2. Biological process-based GO term enrichment analysis.

GO term	Gene symbol or protein name (48 hpi)	P-value(48 hpi)	Gene symbol or protein name (64 hpi)	P-value(64 hpi)
Cell adhesion	MFGE8, CYR61, ITGA5, FN1, TGFB1, PN-1, CCL5, APP, PPP1CB, SCARB1	2.57E-3	MFGE8, CYR61, ITGA5, FN1, TGFB1, CALR, APP, TACSTD2, PN-1, Vascular cell adhesion molecule, CCL5	2.57E-3
Response to stress	NUDT9, CYR61, ITGA5, FN1, PLG, TGFB1, Extracellular superoxide dismutase precursor, CLU, CCL5, HSPA1B, PN-1, VDAC1	7.98E-1	CCL2, NUDT9, CYR61, ITGA5, FN1, LOC100516779, Mitochondrial heat shock 60 kDa protein 1, CCDC47, PLG, TGFB1, CALR, Extracellular superoxide dismutase precursor, OAS1, HSPA1B, PN-1, DDX58, VDAC1, RSAD2, HSP90AA1, HSPCB, DBI, PRDX2, PRDX3, CLU, CCL5, CXCL16, PRDX6	8.18E-4
Generation of precursor metabolites and energy	ENO3, SUCLG1, PPP1CB	6.51E-1	ENO3, TPI1, IDH3A, SUCLG1, LDHB, MDH2, GPI, SUCLG2, SDHB, DLST	2.74E-3
Extracellular matrix organization	CYR61, TGFB1, APP	1.22E-2	LGALS3, CYR61, TGFB1, APP	1.51E-3
Locomotion	TGFB1, APP, CCL5	1.22E-2	TGFB1, APP, CCL5, CXCL16	1.51E-3
Cell motility	STAT3, CYR61, ITGA5, TUBB2B, TGFB1, CCL5	2.07E-1	CCL2, TACSTD2, CYR61, ITGA5, TUBB2B, TGFB1, CALR, CCL5, CXCL16, DDX58	6.71E-3
Cell-cell signaling	ITPR3, APP, PN-1, VDAC1, CCL5	2.58E-2	GLUD1, APP, PN-1, VDAC1, CCL5	2.58E-2
Neurological system process	ITPR3, ITGA5, APP, VDAC1, PN-1	7.91E-3	ITGA5, APP, PN-1, VDAC1	3.34E-2
Protein complex assembly	H2AFY, SLAIN2, HIST1H2BF, TUBB2B, HIST1H2BJ, TMSB10, TGFB1, CLU, CCL5	2.66E-1	SLAIN2,HIST1H2BF,TUBB2B,HIST1H2BJ, TGFB1, TMSB10, RDX, CALR,CLU, CCL5, Histone H1.3-like protein, TFAM	4.69E-2
Cell junction organization	ITGA5, FN1, TGFB1	1.22E-2	ITGA5, FN1, TGFB1	1.22E-2
Growth	STAT3, CYR61, TGFB1, APP, PN-1	1.01E-1	COL9A1, CYR61, TGFB1, PN-1, APP, CXCL16	3.87E-2
Developmental maturation	APP	1.13E-1	ARCN1, APP	1.53E-2
Immune system process	TGFB1, CCL5	9.81E-1	CCL2, HSPCB, PRDX3, LOC100516779, TGFB1, CALR, OAS1, CCL5, CXCL16, DDX58, RSAD2	4.67E-2

Note: P-values were calculated in the hypergeometric test. Gene symbols were retrieved from UniProt. The significantly common processes affected are highlighted in bold.

space (p = 1.52E-2) (Table S4). In addition, 37 differentially expressed proteins were also significantly enriched (p = 8.65E-3) in mitochondrion at 64 hpi (Table S5).

The final step of our GO enrichment analysis consisted of investigating the mechanistic role these genes play in the cell. To do so, we performed a molecular function-based enrichment analysis. This analysis showed that two GO terms, unfolded protein binding (p = 2.67E-2) and transmembrane transporter activity (p = 3.55E-2), were significantly enriched at 64 hpi (Table S5). Further GO analysis of the differentially expressed proteins between the two time points indicated that there were no significant enriched terms.

Protein–protein interaction analysis

In order to understand the interactions between TGEV and host cell proteins, we further analyzed the differentially expressed proteins by searching the STRING 9.1 database (http://string-db.org/) for protein-protein interactions (Figure 4). In this STRING analysis, the interactions (edges) of the submitted proteins (nodes) were scored according to known and predicted protein-protein interactions. We created three protein network maps: one for proteins changed significantly at 48 hpi (30 nodes and 15 edges; Figure 4A), one for proteins changed significantly at 64 hpi (66 nodes and 70 edges; Figure 4B), and one for the proteins that were significantly changed when the viral infection was prolonged from 48 to 64 h (24 nodes and 9 edges; Figure 4C). Notably, the protein network constructed for the 64 hpi time point is clearly much more extensive than the two other networks, and these protein-protein interactions suggest the existence of reported functional linkages. GO enrichment analysis for the STRING protein network at 64 hpi showed that several biological processes were significantly affected (p<0.05 based on the FDR correction) in this network, including the regulation of viral genome replication, the innate immune response, negative regulation of viral genome replication, positive and negative regulation of viral processes, and ATP biosynthetic processes (Table 3). However, at 48 hpi, the most enriched biological process was related to cell recognition during phagocytosis(p = 8.02E-1). In Figure 4C, we have shown that the majority proteins in these protein networks, such as radical S-adenosyl methionine domain containing protein 2 (RSAD2), Mx dynamin-like GTPase 1 (Mx1), 2′-5′-oligoadenylate synthetase 1 (OAS1), Mx dynamin-like GTPase 2 (Mx2), are involved in the innate immune response. These data suggest that some entirely different host proteins, interactions, or processes, including the immune response, were perturbed at these times during TGEV infection.

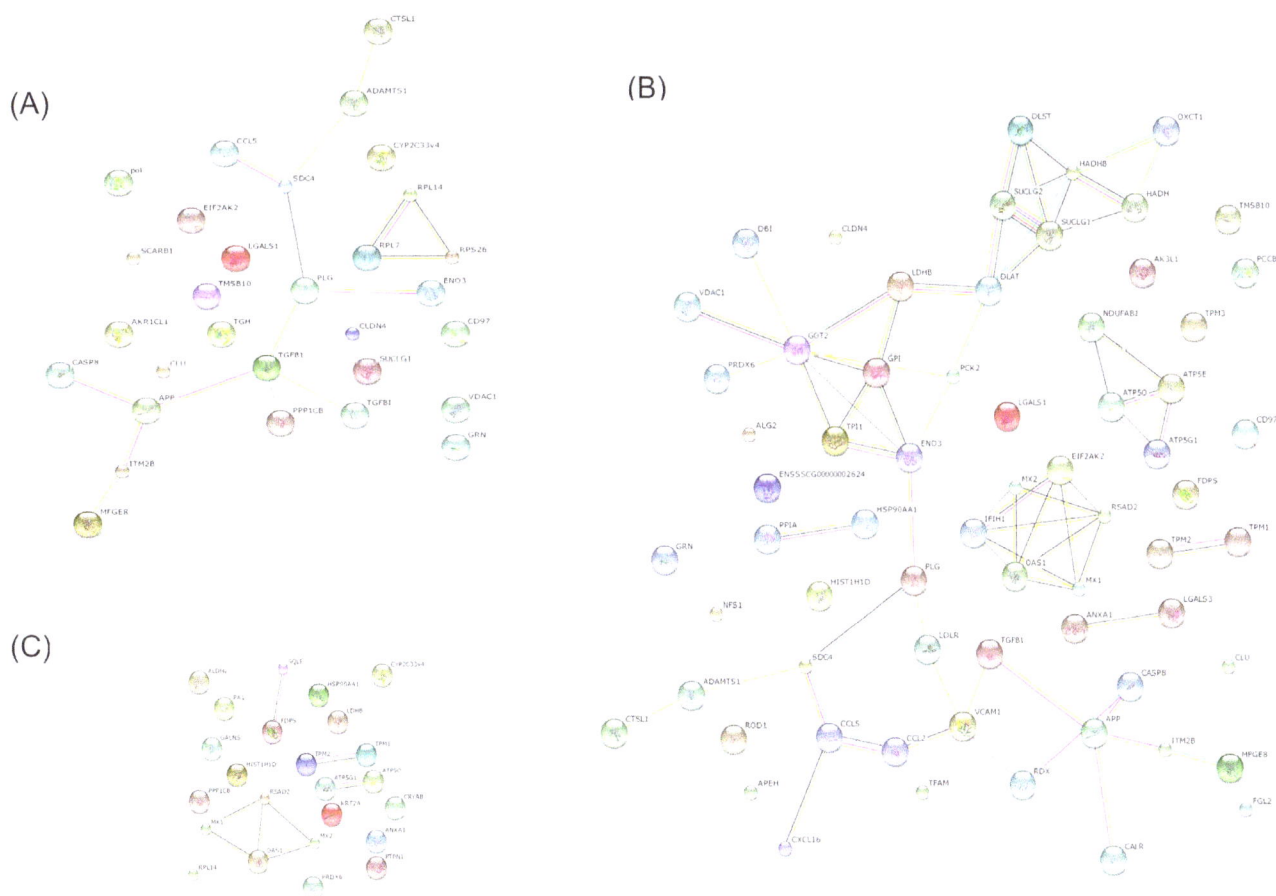

Figure 4. Protein-protein interaction network created using the STRING database. (A) Network of the differentially expressed proteins at 48 hpi. The network includes 30 nodes (proteins) and 15 edges (interactions). (B) Network of differentially expressed proteins at 64 hpi. The network includes 66 nodes and 70 edges. (C) Network of differentially expressed proteins between the two time points. The network includes 24 nodes and 9 edges. Network analysis was set at medium confidence (STRING score = 0.4). Seven different colored lines were used to represent the types of evidence for the association: green, neighborhood evidence; red, gene fusion; blue, co-occurrence; black, co-expression; purple, experimental; light blue, database; yellow, text mining.

Table 3. List of the GO biological processes enriched for the proteins present in the STRING protein network.

GO biological process	P-value
Regulation of viral genome replication	1.33E-2
Innate immune response	1.35E-2
Negative regulation of viral genome replication	2.36E-2
Regulation of viral process	2.70E-2
Negative regulation of viral process	2.83E-2
ATP biosynthetic process	2.89E-2

Note: The significance of the GO biological process is derived from the network in Figure 4B and was determined using the FDR correction (p<0.05).

Western blot confirmation of altered expression for three of the differentially expressed proteins

To further confirm the proteomic data for three of the proteins, western blot analysis was performed to investigate the changes in the expression of HSP90α, caspase 8, and TGF-β1. The proteins were selected based on three criteria: 1) the expression of the protein was increased or decreased during TGEV infection according to our proteomics data; 2) the protein is known to be relevant during viral infection; and 3) each protein analyzed needs to be involved in a special biological process as determined by our GO enrichment analysis [32]. HSP90α, caspase 8, and TGF-β1 all filled these criteria and their protein expression was analyzed via western blot analysis of the cell lysate. As shown in Figure 5, the expression of HSP90α was significantly downregulated in TGEV-infected cells at 64 hpi, while the expression of caspase-8 was upregulated from 48 to 64 hpi in these cells. The expression of TGF-β1 was also significantly induced in TGEV-infected cells following infection. Thus, these results confirm the altered expression observed in the proteomic data for these three representative proteins during TGEV infection.

Discussion

The interactions between a virus and a host cell during a viral infection are complex, involving numerous genes and signaling pathways. ST cells are known to be sensitive to TGEV, resulting in increased viral multiplication and CPEs [15]. In order to better understand the interactions between the host proteome and TGEV, we adopted an iTRAQ quantitative proteomic approach to investigate the altered cellular proteins of the ST cells during TGEV infection in vitro. Compared with the 2-DE and 2D-DIGE methods often used, the 2D-LC-MS/MS method utilized here provides more quantitative and qualitative information about the proteins, and can also detect membrane proteins, hydrophobic proteins, higher molecular weight proteins, and low-abundance proteins, which are often missed by other methods. iTRAQ also has more advantages compared to isotope-coded affinity tags (ICAT) and stable isotope labeling by amino acids in cell culture (SILAC) methods, which both allow multiple labeling and quantitation of four to eight samples simultaneously with high sensitivity [22,33,34]. Further, the iTRAQ technique has been widely used for quantitative proteomics, including protein expression analysis and biomarker identification [23–26,35].

Prior to proteomic analysis, we determined which time points to investigate following infection by observing the morphological changes and analyzing viral gene expression dynamics in the TGEV infected cells. The results indicated that TGEV induced significant CPEs from 40 to 64 hpi in infected cells compared to the mock infected cells. At 40 hpi, less than 50% of the infected cells were morphologically altered, while at 48 hpi more than 80% infected cells showed rounding and granular degeneration. Further, the mRNA level of the viral N gene in ST cells continuously increased in the infected cells until 48 h, at which time we observed the highest viral replication level. At 64 hpi, the morphological effects observed were much more pronounced, characterized by even more cellular rounding and detachment. However, the mRNA levels of the viral N gene decreased rapidly from 48 to 64 h, a phenomenon we believe may be attributed to the host's immune response or a decrease in infected cell viability as the TGEV infection progressed. Based on our qRT-PCR and CPE analyses, we choose to more deeply investigate the proteomic changes occurring in the TGEV-infected ST cells at 48 hpi and 64 hpi using a 4-plex iTRAQ analysis.

In our analysis, we observed a statistically significant change in the expression of 316 proteins during TGEV infection in vitro. This number includes protein changes that were unique for a specific time point as well as those shared at these different time conditions. For example, the expression level of HSP90α expression was unchanged at 48 hpi, but decreased at 64 hpi,

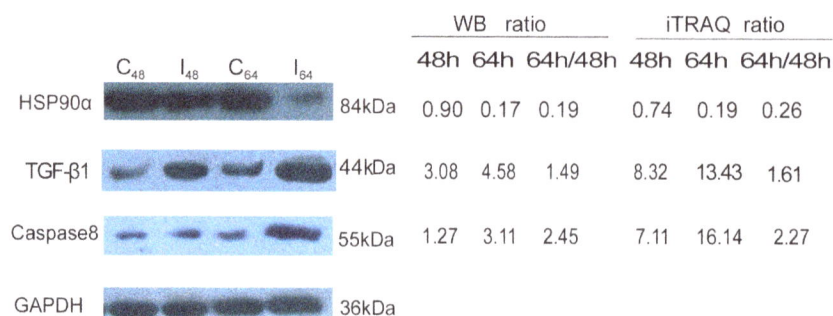

					WB ratio			iTRAQ ratio		
	C₄₈	I₄₈	C₆₄	I₆₄	48h	64h	64h/48h	48h	64h	64h/48h
HSP90α					84kDa					
					0.90	0.17	0.19	0.74	0.19	0.26
TGF-β1					44kDa					
					3.08	4.58	1.49	8.32	13.43	1.61
Caspase8					55kDa					
					1.27	3.11	2.45	7.11	16.14	2.27
GAPDH					36kDa					

Figure 5. Western blot confirmation for three differentially expressed proteins (caspase-8, HSP90α, and TGF-β1). Following TGEV and mock infection of the ST cells, equal amounts of protein were separated by SDS-PAGE and transferred to PVDF membranes. The membranes were then probed with the specified antibody, and the identified bands were visualized. GAPDH was used as an internal control to normalize the quantitative data. The representative images shown are typical of two independent experiments. At 48 hpi (I₄₈), integrated optical density (IOD) analysis showed an upregulation of caspase-8 (1.27 fold) and TGF-β1 (3.08 fold), but HSP90α was almost unchanged (0.90 fold). At 64 hpi (I₆₄), we observed an upregulation in both caspase-8 (3.11 fold) and TGF-β1 (4.58 fold), but a 5.82 fold downregulation of HSP90α. The IOD was normalized against GAPDH.

making this change unique for the latter time point. On the other hand, TGF-β1 was observed to increase at both of the time points, and was thus labeled a shared protein change. Moreover, the 316 altered proteins also includes proteins that changed from 48 hpi to 64 hpi, rather than one of these time points compared to non-infected cells. For example, mitochondrial aldehyde dehydrogenase 2 (ALDH2) and MHC class I antigen (PD1) were not changed at 48 or 64 hpi compared to the control group, but increased at 64 hpi compared with 48 hpi. We also observed a larger proteomic shift at 64 hpi compared to the 48 hpi time point in the infected ST cells.

Further, some proteins previously reported to play a role in virus-induced host cell death, such as caspase-8, caspase-3, caspase-9, and porcine aminopeptidase-N (pAPN) [36–38], were also identified using this iTRAQ technique. These caspase proteins are known to be involved in TGEV-induced cell apoptosis processes, while pAPN is the cell receptor for TGEV. Our results indicate that TGEV infection caused significant upregulation of caspase-8 expression at two time points (approximately 7-fold at 48 hpi and 16-fold at 64 hpi) in the virus-infected ST cells, and this change was verified by western blotting analysis. However, the expression of caspase-3, caspase-9, and pAPN was not significantly altered, indicating that the pathways involving these genes are not altered or that other proteins are compensating for their lack of change. In this regard, we identified an additional 15 proteins involved in cell death pathways that had significantly altered expression levels (p = 4.46E-2) (Table S6), including melanoma differentiation associated protein-5 (MDA5), monocyte chemoattractant protein 1 (CCL2), thioredoxin- dependent peroxide reductase, mitochondrial (PRDX3), peroxiredoxin-2 (PRDX2), predicted protein CYR61 (CYR61), keratin, type II cytoskeletal 8 (KRT8), predicted bcl-2-like protein 13 (BCL2L13), predicted integrin alpha-5 isoform 1 (ITGA5), TGF-β1, amyloid beta A4 protein (APP), clusterin (CLU), C–C motif chemokine 5 (CCL5), heat shock 70 kDa protein 1B (HSPA1B), alpha-crystallin B chain (CRYAB), voltage-dependent anion-selective channel protein 1 (VDAC1), all of which, with the exception of PRDX2 and BCL2L13 were upregulated at one or two time points. Regulation of cell death is known to be important for replication and pathogenesis in various coronaviruses [39], and we believe that further research on these proteins will lead to a better understanding of cell death regulation during TGEV infection.

In order to determine what other processes, in addition to cell death, were affected by TGEV infection, we performed a GO enrichment analysis for the different temporal conditions. This analysis indicated that six biological processes were significantly affected at 48 and 64 hpi, and the differentially expressed proteins involved in these processes were almost the same. The large overlap between the two time points suggests that some of the same sets of host proteins or processes were disturbed at these times. However, it is also likely that some processes were affected solely at one time point or the other. At 48 hpi, serine/threonine-protein phosphatase PP1-beta-catalytic subunit (PPP1CB), scavenger receptor class B member 1 (SCARB1), transforming growth factor-beta-induced protein ig-h3 (TGFBI), and predicted inositol 1,4,5-trisphosphate receptor type 3 (ITPR3) were uniquely altered, likely indicating changes in cell adhesion and/or cell-cell signaling processes. At 64 hpi, on the other hand, calreticulin (CALR), predicted tumor- associated calcium signal transducer 2-like (TACSTD2), vascular cell adhesion molecule, galectin-3 (LGALS3), glutamate dehydrogenase 1 (GLUD1), and C–X–C motif chemokine 16 (CXCL16) were uniquely changed, also indicating changes in cell adhesion and/or cell-cell signaling as well as extracellular matrix organization and locomotion. We

believe that these uniquely altered proteins reflect changes in specific/specialized processes at each time point that are tightly linked to the temporal changes observed in the host cell morphology and gene/protein expression after TGEV infection.

The most significantly enriched GO category related to the differentially expressed proteins was stress, which included 12 differentially expressed proteins at 48 hpi and 27 different proteins at 64 hpi. The increased number of proteins association with this GO term at 48 hpi likely highlights the initial upregulation of the cellular stress response, while the higher number at 64 hpi indicates that the stress response to TGEV infection is likely more fully induced at this later stage. HSPs, also known as stress proteins, are often involved in the cellular response to stress, influencing changes in the state or activity of the cell or organism. HSP90, which has two isoforms (HSP90α and HSP90β), is one of the most abundant molecular chaperones that is induced in response to cellular stress, and it functions to stabilize proteins involved in cell growth and anti-apoptotic signaling [40]. The expression of HSP90α has been reported to play an important role in the replication of some viruses, such as Ebola virus (EBOV) [41], hepatitis C virus (HCV) [42], influenza virus [43], and Japanese encephalitis virus [44]. On the other hand, the reduction of HSP90β has been reported to decrease the correct assembly of human enterovirus 71 viral particles [40]. In this study, HSP90α and heat shock 90kD protein 1, beta (HSPCB/HSP90β) were significantly downregulated at 64 hpi in the TGEV-infected ST cells, but were unchanged at 48 hpi, indicating that they may play a similar role in TGEV infection. Interestingly, a member of the HSP70 protein family, heat shock 70 kDa protein 1B (HSPA1B), as well as mitochondrial 60 kDa heat shock protein (HSP60) were both upregulated in infected ST cells at 48 and/or 64 hpi. HSP60 is a mitochondrial chaperonin protein involved in protein folding and a number of extracellular immunomodulatory activities. Elevated expression of HSP60 is associated with a number of inflammatory disorders [45]. HSP70 plays an important role in multiple processes within cells, including protein translation, folding, intracellular trafficking, and degradation. A previous study has revealed that HSP70 is involved in all steps of the viral life cycle, including replication, and is highly specific in regards to viral response, differing from one cell to another for any given virus type [46]. For example, silencing HSP70 expression has been associated with an increase in viral protein levels, while an increase in HSP70 has been suspected to be the initial cellular response to protect against viral infection in rotavirus-infected cells [47]. Further, a recent study showed that HSP70 is an essential host factor for the replication of PRRSV as the silence of HSP70 significantly reduced PRRSV replication [48]. Our results provide new experimental evidence relating the expression of HSP90, HSP70, and HSP60 to TGEV infection, and we speculate that these proteins play a potential role in TGEV replication. Additional work is required to investigate the detailed role of these proteins during TGEV infection.

Furthermore, another significantly enriched GO process we observed that 11 significantly altered proteins was immune system processes. Most of these proteins were significantly upregulated at 64 hpi in response to the viral infection, while some were first upregulated at 48 hpi, including CCL5 and TGF-β1. Chemokines, such as CCL2, CCL5, and CXCL16, whose main function is macrophage recruitment and activation, are potentially involved in host-mediated immunopathology. A recent study showed that coronavirus infection of transgenic mice expressing CCL2 led to a dysregulated immune response without effective virus clearance and enhanced death [49]. In additional, TGEV-infection can induce the expression of proinflammatory genes, including CCL2,

CCL5, and probable ATP-dependent RNA helicase DDX58 (DDX58/RIG-1), in cell culture and in vivo in the absence of viral protein 7 [50]. In this study, we observed an upregulation of CCL2, CCL5, CXCL16, TGF-β1, and DDX58 expression. TGF-β1 is a multifunctional cytokine, secreted from various cells, and, in immunology, it regulates cellular proliferation, differentiation, and other cellular functions for a variety of cell types, especially regulatory T cells [51]. Some research has indicated that SARS-CoV papain-like protease (PLpro) increases TGF-β1 mRNA expression and protein production in human promonocytes [52]. Further, Gomez-Laguna et al. [53] inferred that the upregulation of the TGF-β may impair the host immune response during PRRSV infection by limiting the overproduction of proinflammatory cytokines necessary to decrease PRRSV replication. In response to viral infection, DDX58 plays important roles in the recognition of RNA viruses in various cells, and has been identified as a candidate for a cytoplasmic viral dsRNA receptor [54]. Further, upregulation of this gene activates cells to produce type I interferons, which may increase the antiviral status of cells to protect against viral infection. In this regard, we found that interferon-inducible antiviral proteins, RSAD2, OAS1, were also upregulated in the period of late infection, suggesting that many of the proteins identified in this study are associated with inflammation, IFN activation, and the innate immune response. Increased expression of these proteins may help the virus enter the cell as well as potentially enhance TGEV replication or the host response against the virus, during the late stages of infection.

In conclusion, we used the iTRAQ method to identify 316 significantly altered proteins in TGEV-infected ST cells. A larger number of protein expression changes occurred at 64 hpi compared to 48 hpi, indicating a larger shift in the proteome in the later stages of infection. GO analysis of these differentially expressed proteins indicated that a number of diverse biological processes are affected. In addition, many of the significant immune response related changes in protein expression we discovered are novel and, to our knowledge, have not been detected in previous proteome study. Results from this study complement the previous proteomics data obtained concerning the host response to a viral infection, and further facilitates a better understanding of the pathogenic mechanisms of TGEV infection and molecular responses of host cells to this virus.

Supporting Information

Table S1   Total proteins (4,364) identified and quantified by iTRAQ.

Table S2   List of the 4,112 reliably quantified proteins selected from Table S1.

Table S3   Differentially expressed proteins identified under different conditions.

Table S4   GO enrichment analysis of differentially expressed proteins identified at 48 hpi.

Table S5   GO enrichment analysis of differentially expressed proteins identified at 64 hpi.

Table S6   GO enrichment of all the differentially expressed proteins.

Acknowledgments

We thank Qiangqiang Zhao, Chen Lou, Wulong Liang, and Helin Li for their technical support, and Shuo Chen for his valuable advice.

Author Contributions

Conceived and designed the experiments: RM YZ. Performed the experiments: RM. Analyzed the data: RM HL. Contributed reagents/materials/analysis tools: PN YZ. Contributed to the writing of the manuscript: RM. Drafted the work or revised it critically: RM YZ HL PN.

References

1. Jones T, Pritchard G, Paton D (1997) Transmissible gastroenteritis of pigs. Vet Rec 141: 427–428.
2. Wesley RD, Lager KM (2003) Increased litter survival rates, reduced clinical illness and better lactogenic immunity against TGEV in gilts that were primed as neonates with porcine respiratory coronavirus (PRCV). Vet Microbiol 95: 175–186.
3. Kim L, Hayes J, Lewis P, Parwani AV, Chang KO, et al. (2000) Molecular characterization and pathogenesis of transmissible gastroenteritis coronavirus (TGEV) and porcine respiratory coronavirus (PRCV) field isolates co-circulating in a swine herd. Arch Virol 145: 1133–1147.
4. Vlasova AN, Halpin R, Wang S, Ghedin E, Spiro DJ, et al. (2011) Molecular characterization of a new species in the genus Alphacoronavirus associated with mink epizootic catarrhal gastroenteritis. J Gen Virol 92: 1369–1379.
5. Spaan W, Cavanagh D, Horzinek MC (1988) Coronaviruses: structure and genome expression. J Gen Virol 69: 2939–2952.
6. Neuman BW, Kiss G, Kunding AH, Bhella D, Baksh MF, et al. (2011) A structural analysis of M protein in coronavirus assembly and morphology. J Struct Biol 174: 11–22.
7. Baudoux P, Carrat C, Besnardeau L, Charley B, Laude H (1998) Coronavirus pseudoparticles formed with recombinant M and E proteins induce alpha interferon synthesis by leukocytes. J Virol 72: 8636–8643.
8. Curtis KM, Yount B, Baric RS (2002) Heterologous gene expression from transmissible gastroenteritis virus replicon particles. J Virol 76: 1422–1434.
9. Weingartl HM, Derbyshire JB (1993) Binding of porcine transmissible gastroenteritis virus by enterocytes from newborn and weaned piglets. Vet Microbiol 35: 23–32.
10. Weingartl HM, Derbyshire JB (1994) Evidence for a putative second receptor for porcine transmissible gastroenteritis virus on the villous enterocytes of newborn pigs. J Virol 68: 7253–7259.
11. Sirinarumitr T, Paul PS, Kluge JP, Halbur PG (1996) In situ hybridization technique for the detection of swine enteric and respiratory coronaviruses, transmissible gastroenteritis virus (TGEV) and porcine respiratory coronavirus (PRCV), in formalin-fixed paraffin-embedded tissues. J Virol Methods 56: 149–160.
12. Galán C, Sola I, Nogales A, Thomas B, Akoulitchev A, et al. (2009) Host cell proteins interacting with the 3′end of TGEV coronavirus genome influence virus replication. Virology 391: 304–314.
13. Penzes Z, Gonzalez JM, Calvo E, Izeta A, Smerdou C, et al. (2001) Complete genome sequence of transmissible gastroenteritis coronavirus PUR46-MAD clone and evolution of the purdue virus cluster. Virus Genes 23: 105–118.
14. Ortego J, Sola I, Almazán F, Ceriani JE, Riquelme C, et al. (2003) Transmissible gastroenteritis coronavirus gene 7 is not essential but influences in vivo virus replication and virulence. Virology 308: 13–22.
15. He L, Zhang YM, Dong LJ, Cheng M, Wang J, et al. (2012) In vitro inhibition of transmissible gastroenteritis coronavirus replication in swine testicular cells by short hairpin RNAs targeting the ORF 7 gene. Virol J 9: 176–184.
16. Sun J, Jiang Y, Shi Z, Yan Y, Guo H, et al. (2008) Proteomic alteration of PK-15 cells after infection by classical swine fever virus. J Proteome Res 7: 5263–5269.
17. Zhang X, Zhou J, Wu Y, Zheng X, Ma G, et al. (2009) Differential proteome analysis of host cells infected with porcine circovirus type 2. J Proteome Res 8: 5111–5119.
18. Yang Y, An T, Gong D, Li D, Peng J, et al. (2012) Identification of porcine serum proteins modified in response to HP-PRRSV HuN4 infection by two-dimensional differential gel electrophoresis. Vet Microbiol 158: 237–246.
19. Cao Z, Han Z, Shao Y, Liu X, Sun J, et al. (2012) Proteomics analysis of differentially expressed proteins in chicken trachea and kidney after infection with the highly virulent and attenuated coronavirus infectious bronchitis virus in vivo. Proteome Sci 10: 24.
20. Jiang XS, Tang LY, Dai J, Zhou H, Li SJ, et al. (2005) Quantitative analysis of severe acute respiratory syndrome (SARS)-associated coronavirus-infected cells using proteomic approaches implications for cellular responses to virus infection. Mol Cell Proteomics 4: 902–913.

21. Zhang X, Shi HY, Chen JF, Shi D, Lang HW, et al. (2013) Identification of cellular proteome using two-dimensional difference gel electrophoresis in ST cells infected with transmissible gastroenteritis coronavirus. Proteome Sci 11: 31.

22. Wu WW, Wang G, Baek SJ, Shen RF (2006) Comparative study of three proteomic quantitative methods, DIGE, cICAT, and iTRAQ, using 2D gel- or LC-MALDI TOF/TOF. J Proteome Res 5: 651–658.

23. Lu Q, Bai J, Zhang L, Liu J, Jiang Z, et al. (2012) Two-dimensional liquid chromatography-tandem mass spectrometry coupled with isobaric tags for relative and absolute quantification (iTRAQ) labeling approach revealed first proteome profiles of pulmonary alveolar macrophages infected with porcine reproductive and respiratory syndrome virus. J Proteome Res 11: 2890–2903.

24. Lietzen N, Ohman T, Rintahaka J, Julkunen I, Aittokallio T, et al. (2011) Quantitative subcellular proteome and secretome profiling of influenza A virus-infected human primary macrophages. PLoS Pathog 7: e1001340.

25. Navare AT, Sova P, Purdy DE, Weiss JM, Wolf-Yadlin A, et al. (2012) Quantitative proteomic analysis of HIV-1 infected CD4+ T cells reveals an early host response in important biological pathways: protein synthesis, cell proliferation, and T-cell activation. Virology 429: 37–46.

26. Feng X, Zhang J, Chen WN, Ching CB (2011) Proteome profiling of Epstein-Barr virus infected nasopharyngeal carcinoma cell line: identification of potential biomarkers by comparative iTRAQ-coupled 2D LC/MS-MS analysis. J proteomics 74: 567–576.

27. Liu W, Saint DA (2002) Validation of a quantitative method for real time PCR kinetics. Biochem Biophys Res Commun 294: 347–353.

28. Livak KJ, Schmittgen TD (2001) Analysis of relative gene expression data using real-time quantitative PCR and the 2(-Delta Delta C(T)) Method. Methods 25: 402–408.

29. Ruppen I, Grau L, Orenes-Pinero E, Ashman K, Gil M, et al. (2010) Differential protein expression profiling by iTRAQ-two-dimensional LC-MS/MS of human bladder cancer EJ138 cells transfected with the metastasis suppressor KiSS-1 gene. Mol Cell Proteomics 9: 2276–2291.

30. Sun L, Bertke MM, Champion MM, Zhu G, Huber PW, et al. (2014) Quantitative proteomics of Xenopus laevis embryos: expression kinetics of nearly 4000 proteins during early development. Sci Rep 4: 4365.

31. Rivals I, Personnaz L, Taing L, Potier MC (2007) Enrichment or depletion of a GO category within a class of genes: which test? Bioinformatics 23: 401–407.

32. Chiu HC, Hannemann H, Heesom KJ, Matthews DA, Davidson AD (2014) High-Throughput Quantitative Proteomic Analysis of Dengue Virus Type 2 Infected A549 Cells. PloS One 9: e93305.

33. Ross PL, Huang YN, Marchese JN, Williamson B, Parker K, et al. (2004) Multiplexed protein quantitation in Saccharomyces cerevisiae using amine-reactive isobaric tagging reagents. Mol Cell Proteomics 3: 1154–1169.

34. Munday DC, Surtees R, Emmott E, Dove BK, Digard P, et al. (2012) Using SILAC and quantitative proteomics to investigate the interactions between viral and host proteomes. Proteomics 12: 666–672.

35. Li H, DeSouza LV, Ghanny S, Li W, Romaschin AD, et al. (2007) Identification of candidate biomarker proteins released by human endometrial and cervical cancer cells using two-dimensional liquid chromatography/tandem mass spectrometry. J Proteome Res 6: 2615–2622.

36. Ding L, Xu X, Huang Y, Li Z, Zhang K, et al. (2012) Transmissible gastroenteritis virus infection induces apoptosis through FasL- and mitochondria-mediated pathways. Vet Microbiol 158: 12–22.

37. Eleouet JF, Chilmonczyk S, Besnardeau L, Laude H (1998) Transmissible gastroenteritis coronavirus induces programmed cell death in infected cells through a caspase-dependent pathway. J Virol 72: 4918–4924.

38. Delmas B, Gelfi J, L'Haridon R, Vogel LK, Sjostrom H, et al. (1992) Aminopeptidase N is a major receptor for the enteropathogenic coronavirus TGEV. Nature 357: 417–420.

39. Tan YJ, Lim SG, Hong W (2007) Regulation of cell death during infection by the severe acute respiratory syndrome coronavirus and other coronaviruses. Cell Microbiol 9: 2552–2561.

40. Wang RYL, Kuo RL, Ma WC, Huang HI, Yu JS, et al. (2013) Heat shock protein-90-beta facilitates enterovirus 71 viral particles assembly. Virology 443: 236–247.

41. Smith DR, McCarthy S, Chrovian A, Olinger G, Stosselet A, et al. (2010) Inhibition of heat-shock protein 90 reduces Ebola virus replication. Antivir Res 87: 187–194.

42. Okamoto T, Nishimura Y, Ichimura T, Suzuki K, Miyamura T, et al. (2006) Hepatitis C virus RNA replication is regulated by FKBP8 and Hsp90. EMBO J 25: 5015–5025.

43. Momose F, Naito T, Yano K, Sugimoto S, Morikawa Y, et al. (2002) Identification of Hsp90 as a stimulatory host factor involved in influenza virus RNA synthesis. J Biol Chem 277: 45306–45314.

44. Hung CY, Tsai MC, Wu YP, Wang RY (2011) Identification of heat-shock protein 90 beta in Japanese encephalitis virus-induced secretion proteins. J Gen Virol 92: 2803–2809.

45. Johnson BJ, Le TTT, Dobbin CA, Banovic T, Howard CB, et al. (2005) Heat shock protein 10 inhibits lipopolysaccharide-induced inflammatory mediator production. J Biol Chem 280: 4037–4047.

46. Lahaye X, Vidy A, Fouquet B, Blondel D (2012) Hsp70 protein positively regulates rabies virus infection. J virol 86: 4743–4751.

47. Broquet AH, Lenoir C, Gardet A, Sapin C, Chwetzoff S, et al. (2007) Hsp70 negatively controls rotavirus protein bioavailability in caco-2 cells infected by the rotavirus RF strain. J virol 81: 1297–1304.

48. Gao J, Xiao S, Liu X, Wang L, Ji Q, et al. (2014) Inhibition of HSP70 reduces porcine reproductive and respiratory syndrome virus replication in vitro. BMC microbiol 14: 64.

49. Trujillo JA, Fleming EL, Perlman S (2013) Transgenic CCL2 expression in the central nervous system results in a dysregulated immune response and enhanced lethality after coronavirus infection. J virol 87: 2376–2389.

50. Cruz JLG, Becares M, Sola I, Oliveros JC, Enjuanes L, et al. (2013) Alphacoronavirus protein 7 modulates host innate immune response. J virol 87: 9754–9767.

51. Yang Y, Zhang N, Lan F, Crombruggen K, Fang L, et al. (2014) Transforming growth factor-beta 1 pathways in inflammatory airway diseases. Allergy 69: 699–707.

52. Li SW, Yang TC, Wan L, Lin YJ, Tsai FJ, et al. (2012) Correlation between TGF-β1 expression and proteomic profiling induced by severe acute respiratory syndrome coronavirus papain-like protease. Proteomics 12: 3193–3205.

53. Gomez-Laguna J, Rodriguez-Gomez IM, Barranco I, Pallares FJ, Salguero FJ, et al. (2012) Enhanced expression of TGFβ protein in lymphoid organs and lung, but not in serum, of pigs infected with a European field isolate of porcine reproductive and respiratory syndrome virus. Vet Microbiol 158: 187–193.

54. Takeuchi O, Akira S (2008) MDA5/RIG-I and virus recognition. Curr Opin Immunol 20: 17–22.

15

A Disordered Region in the EvpP Protein from the Type VI Secretion System of *Edwardsiella tarda* is Essential for EvpC Binding

Wentao Hu[1], Ganesh Anand[1], J. Sivaraman[1], Ka Yin Leung[2,3], Yu-Keung Mok[1]*

1 Department of Biological Sciences, 14 Science Drive 4, National University of Singapore, Singapore, Singapore 117543, 2 Department of Biology, Faculty of Natural and Applied Sciences, Trinity Western University, Langley, British Columbia, Canada V2Y 1Y1, 3 State Key Laboratory of Bioreactor Engineering, East China University of Science and Technology, Shanghai, China 200237

Abstract

The type VI secretion system (T6SS) of pathogenic bacteria plays important roles in both virulence and inter-bacterial competitions. The effectors of T6SS are presumed to be transported either by attaching to the tip protein or by interacting with HcpI (haemolysin corregulated protein 1). In *Edwardsiella tarda* PPD130/91, the T6SS secreted protein EvpP (*E. tarda* virulent protein P) is found to be essential for virulence and directly interacts with EvpC (Hcp-like), suggesting that it could be a potential effector. Using limited protease digestion, nuclear magnetic resonance heteronuclear Nuclear Overhauser Effects, and hydrogen-deuterium exchange mass spectrometry, we confirmed that the dimeric EvpP (40 kDa) contains a substantial proportion (40%) of disordered regions but still maintains an ordered and folded core domain. We show that an N-terminal, 10-kDa, protease-resistant fragment in EvpP connects to a shorter, 4-kDa protease-resistant fragment through a highly flexible region, which is followed by another disordered region at the C-terminus. Within this C-terminal disordered region, residues Pro143 to Ile168 are essential for its interaction with EvpC. Unlike the highly unfolded T3SS effector, which has a lower molecular weight and is maintained in an unfolded conformation with a dedicated chaperone, the T6SS effector seems to be relatively larger, folded but partially disordered and uses HcpI as a chaperone.

Editor: Shannon Wing-Ngor Au, The Chinese University of Hong Kong, China

Funding: This project was supported by Academic Research Fund, National University of Singapore grant R-154-000-498-112 to Y.K.M. K.Y.L. was supported by a grant from the National Science and Engineering Research Council (NSERC) Discovery Grant (372373-2010), Canada, and the Open Funding Project of the State Key Laboratory of Bioreactor Engineering of China. The funders had no role in study design, data collection and analysis, decision to publish, or preparation of the manuscript.

Competing Interests: The authors have declared that no competing interests exist.

* Email: dbsmokh@nus.edu.sg

Introduction

The type VI secretion system (T6SS) was first discovered as an essential bacterial secretion system for bacterial pathogenesis [1,2], and its activity is tightly controlled by both environmental and bacterial regulatory factors [3]. Accumulated evidence now shows that the T6SS of pathogenic bacteria have dual roles in both pathogenicity and inter-bacterial competition, and that the T6SS is an effective weaponry against other Gram-negative bacteria in the living habitat [4]. As the prey will usually fight back by secreting toxins from its own T6SS, this process of inter-bacterial competition is termed as "T6SS dueling" [5].

The T6SS core apparatus assembles from 13 "core components" proteins, from TssA-M (type six sub-units), in addition to Hcp (haemolysin corregulated protein), VgrG (valine-glycine repeat protein G), the ClpV AAA+ ATPase and the T4SS IcmF- and IcmH-like proteins [6–8]. These proteins are arranged in the form of two sub-assemblies: a dynamic bacteriophage-like structure and a cell-envelope-spanning membrane-associated assembly [9]. The phage-like complex is functionally related to the phage tail and comprises the Hcp (haemolysin corregulated protein) tube, the trimeric VgrG (valine-glycine repeat protein G) tip protein with a PAAR (proline-alanine-alanine-arginine) sharpened tip [10], and the sheath-containing TssB and TssC proteins [11,12]. Extension and contraction of the sheath will drive the Hcp tube and the VgrG tip toward the host cell/bacteria to export T6SS effectors [13]. The trans-envelope complex contains a large base-plate with inner membrane components, TssL and TssM, and the outer membrane lipoprotein, TssJ, for anchoring the phage-like structure to the membrane [14–16]. TssK is a trimer that connects the two systems by directly interacting with TssL, HcpI and TssC [17].

T6SS has two currently known routes for effector proteins delivery. The first route is through the "evolved VgrG proteins", where the VgrG tip protein harbors an additional effector domain fused to the C-terminus of the gp-5-like β-helix needle [18]. Such additional effector domains include an actin cross-linking domain, which fuses to VgrG [19]. Alternatively, the RhsA effector nuclease, a PAAR protein that recognizes the VgrG trimer and decorates the tip of the T6SS injectisome for export [20]. The second route is through the delivery of "classic toxins" into the host cell by passing through the HcpI channel [4,21]. Recent findings, however, showed an alternate mechanism in which effector form a complex with the HcpI ring and injected as such. The hexameric HcpI ring is proposed to act as a chaperone and receptor for these and other effectors molecules [22]. In general, this second class of effector binds to residues on the inside of the HcpI hexamers [21].

Edwardsiella tarda is a recognized fish pathogen that has been shown to also cause gastrointestinal infections in human [23]. The EvpP (*E. tarda* virulent protein P) protein is essential for *E. tarda* virulence and has been identified as one of the three secreted proteins—together with EvpC (hexameric and HcpI-like) [24] and EvpI (VgrG-like)—from the T6SS of *E. tarda*. The secretion of EvpC and EvpI is mutually dependent, and both are required for the secretion of EvpP, with evidence to show binding between EvpP and EvpC [25]. In the T6SS gene cluster, EvpP is located at the first ORF and its expression is tightly regulated by Fur and the PhoB-PhoR two-component system, which sense iron and phosphate, respectively [26]. The promoter of *evpP* binds directly to the transcription activator EsrC, which in turn is inhibited by the direct interaction between EsrC and Fur protein. EvpP has orthologs in only two other genomes (*E. italuri* and *Vibrio* MED222 [27]) and has no known function or sequence homology with other existing proteins. EvpP is not an anti-bacterial peptidoglycan peptidase like many other known T6SS effectors and could represent a potential novel class of T6SS effector.

Using limited protease digestion, nuclear magnetic resonance (NMR) experiments and amide proton exchange mass spectrometry, we found that EvpP is highly disordered with the disordered region mapping to the middle of the protein and shown to link two relatively folded regions. The highly disordered nature of the protein and the known interaction between EvpP and EvpC suggests that EvpP is a T6SS secreted substrate which translocates through interaction with EvpC. Further experiments are needed to confirm whether EvpP is a *bone fide* T6SS effector.

Materials and Methods

Protein expression and purification

Wild type and mutant *evpP* sequences were cloned between BamH I and EcoR I sites of the pET-M vector (derived from pET32-a, Novagen; Madison, WI) and then transformed into *E. coli* BL21 (DE3) competent cells. The cells were allowed to grow in LB with ampicillin at 37°C until the O.D. reached 0.6. IPTG (0.4 mM) was added into the media to induce protein expression and the cells were grown overnight at 20°C. The cells were then harvested in 20 mM Tris buffer, pH 7.0, sonicated and centrifuged. The supernatant was collected and passed through a Ni-NTA affinity column. Expression of ^{15}N- or ^{13}C-labeled protein was carried out in similar conditions as mentioned above, except that the cells were grown in M9 minimal medium supplemented with ^{15}N-labeled ammonium chloride and/or ^{13}C-labeled D-glucose. To obtain a single labeled sample (either ^{15}N-labeled or ^{13}C-labeled sample), cells were grown in 1×M9 minimal medium in the presence of ^{15}N-ammonium chloride (Cambridge Isotope Laboratories; Andover, MA) or with ^{13}C-glucose (Cambridge Isotope Laboratories) as the sole nitrogen or carbon source. For double labeled ^{15}N- and ^{13}C-sample, 1×M9 minimal medium with ^{15}N-ammonium chloride and ^{13}C-glucose (Cambridge Isotope Laboratories) was used as the sole nitrogen and carbon sources, respectively.

Dynamic light scattering

Dynamic light scattering was carried out for both EvpP and its mutant EvpP-P143T using DynaPro Dynamic Light Scattering

(Protein Solutions LLC; Joplin, MO) machine linked to a Temperature Controlled MicroSampler (Protein Solutions). Twenty microliters of protein sample (50 µM) was loaded into a quartz cuvette with a path length of 15 mm (Protein Solutions). Data were recorded and analyzed using the Instrument Control Software for Molecular Research (Protein Solutions).

Limited protease digestion

Limited protease digestion (enzymes used were trypsin, chymotrypsin, elastase, thermolysin) was carried out for both EvpP and its mutant EvpP-P143T using protein samples in their native buffer (20 mM Tris-Cl, pH 7) with different enzymes at different ratios (usually 1:50, 1:100 or 1:200). Both the protein sample and the enzyme stock were prepared as 1 mg/ml. Ten microliters of enzyme solution was added into 1 ml of protein sample to form the reaction mixture. Aliquots were removed from the reaction mixture every 10 or 15 min and mixed with an equal volume of 2×SDS sample buffer and boiled at 100°C for 5 min. All of the aliquots were analyzed on SDS-PAGE.

Mass spectrometry and N-terminal sequencing

Trypsin-digested EvpP samples were blotted onto a MALDI target plate with sinapinic acid as a matrix. The molecular weight of each component in the complex was determined using Voyager-DE STR Mass Spectrometer (GE Healthcare; Buckinghamshire, UK). A voltage of 2100 V and laser intensity of 2500 was used for the experiment. For N-terminal sequencing, protein samples were separated on a 15% SDS-PAGE or 15% Tricine-SDS-PAGE gels, and then blotted onto PVDF membranes (EMD Millipore, Billerica, MA) using a Mini Trans-blot Electrophoretic Transfer Cell (Bio-Rad; Hercules, CA) at 4°C at a voltage of 90 V for 90 min or until complete transfer of protein bands onto the membrane. Protein bands on the membrane were visualized by coomassie blue staining and excess stain on the membrane was removed by washing with a 50% methanol solution. Membrane-bound protein bands were excised and N-terminal residues were fragmented using Procise Protein sequencing system (Applied Biosystems, Life Technologies; Carlsbad, CA). Data were collected and analyzed using SequencePro Data Analysis Application software v2.1 (Applied Biosystems, Life Technologies).

Nuclear magnetic resonance and backbone assignment

3D and 4D heteronuclear NMR experiments were carried out to assign the backbone chemical shifts for EvpP-P143T. All NMR experiments were carried out on a Bruker AVANCE 800 MHz spectrometer (Madison, WI) equipped with a cryoprobe. Samples were loaded into a 5-mm NMR tube and all experiments were carried out at 297 K. All data acquired were processed using NMRPipe [28] and analyzed using NMRDraw [28], SPARKY [29] and NMRspy (Zheng, Yu; Yang, Daiwen. NMRspy, National University of Singapore).

HDX-MS (hydrogen-deuterium exchange mass spectrometry)

Sample protein was exposed to 100% D_2O buffer for various incubation times and the reaction quenched by trifluoroacetic acid prior to mass spectrometry. In order to increase the resolution in terms of amino acid sequence, the quenched sample was subjected to pepsin digestion. Prior to ESI mass spectrometry, the peptides were seperated on an HPLC column to further increase resolution. The whole process was carried out in Waters nanoACQUITY UPLC Hydrogen Deuterium Exchange (HDS) System (Waters Corporation; Milford, MA). Data were recorded and analyzed by

ProteinLynx software (ProteinLynx Global Server, Waters Corporation). The peptide spectra assignment for each identified peptide in the undeuterated sample as well as in the subsequent deuterated samples was carried out using the DynamX software (Waters Corporation). The output file was converted to a Microsoft Excel format (Microsoft, Redmond, WA) for further data analysis. The back-exchange constant used for calculating the exchange rate was 1.49.

Glutathione-S-transferase (GST) pull-down assay

The buffer for all EvpP and its truncation mutants was exchanged to PBS before GST pull down assay. Glutathione-sepharose 4B beads (50 μl) (Amersham Biosciences, GE Healthcare, Piscataway, NJ) was added into an Eppendorf tube (Eppendorf, Hamburg, Germany) and was washed at least three times with 1 ml of 1×PBS. GST-EvpC (100 μg) or GST protein alone as a control was added into the tube and the mixture was topped up to 500 μl with 1×PBS. The protein:bead mixture was incubated at 4°C for at least 1 h. The beads were washed with 1 ml of 1×PBS three times. EvpP or its truncation mutants (200 μg) was added into the tube and the reaction volume was topped up to 500 μl by 1×PBS. The mixture was incubated at 4°C for at least 2 h with mixing. Finally, the beads were washed with 1 ml of 1×PBS to remove unbound or loosely bound proteins, and the beads suspended in 100 μl of 2×SDS sample buffer. The suspension was boiled for 5 min, centrifuged at 13,000 rpm for 10 min, and 20 μl of the supernatant was analyzed on SDS-PAGE.

Results

EvpP purified as a dimeric protein

EvpP from *E. tarda* strain PPD130/91 is a 185-amino acid protein with a molecular weight of 20.3 kDa and a theoretical isoelectric point of 9.47. 6× His-tagged EvpP was over-expressed in *E. coli* BL21 (DE3) as a soluble protein and purified through Ni-NTA affinity column. The purified EvpP protein eluted from the Superdex 75 gel filtration column (HiLoad 16/60, GE Healthcare) at a volume of ~72 ml. Molecular weight standards ovalbumin (44 kDa) and myoglobulin (17 kDa) on the same column eluted at 63 ml and 83 ml, respectively. This elution comparison suggests that EvpP could be a dimer and the slight increased elution volume may be due to the flexible conformation of the protein (Figure 1A). To confirm the dimeric nature of EvpP, we performed dynamic light scattering (DLS) experiments to study the oligomeric state of proteins in solution. At 20°C, EvpP showed a profile corresponding to a dimeric molecule. The hydro-dynamic radius of EvpP was determined to be from 2.92 nm to 3.09 nm (mean, 3.01 nm; n = 11 readings) and the estimated corresponding molecular weight ranged from 41.2 kDa to 47.3 kDa (mean, 44.4 kDa; n = 11 readings); this measured molecular weight is twice that of the theoretical molecular weight for 6× His-tagged EvpP (22.0 kDa) (Figure 1B). For most of the readings, the polydispersity index (PDI) lies within the range from 0.0 to 0.1 (% polydispersity from 0% to 31%), suggesting that EvpP is a mono-dispersed dimeric species under the conditions used (Figure S1 in File S1). In addition, cross-linking experiments using glutaraldehyde were performed showing that the dimeric form of EvpP is the main cross-linked species (Figure S2 in File S1).

EvpP limited protease digestion generated a folded core comprising two fragments

To ascertain if EvpP contained any disordered regions and to further map these regions, we performed limited protease digestion on EvpP using trypsin. At a protein:trypsin ratio of 100:1, trypsin reproducibly digested EvpP to two fragments of 10 kDa and 4 kDa as observed on SDS PAGE (Figure 2A). This digestion was completed within 20 to 30 min and further digestion was not observed even after 60 min. When analysed on gel filtration columns (HiLoad 16/60 Superdex 75), the two fragments were found to associate with each other and both eluted at a volume similar to that of native EvpP (Figure 2B). This suggested that the trypsin digestion retained EvpP as a dimer (~28 kDa) and that each monomer (~14 kDa) contained folded regions from two different parts of the protein. Freshly digested EvpP also showed a ^{1}H-^{15}N HSQC NMR spectrum typical of a folded protein (Figure 3B) and the line widths of the cross-peaks are comparable to that of native EvpP, although the protein tended to denature after approximately 3 days (data not shown).

We next determined the boundaries of the 10-kDa and 4-kDa protease-resistant regions of EvpP by performing N-terminal sequencing on both fragments. The 10-kDa fragment N-terminal sequence was "GSELMSILN", where "GSEL" is from the vector and "MSILN" corresponds to the first five residues of EvpP; this indicates that the N-terminal region of EvpP is resistant to tryptic digestion. Using electrospray mass spectrometry, we determined the molecular weight of this 10-kDa fragment as 9,614 Da. Combining this molecular weight and N-terminal sequencing data, the 10-kDa trypsin-resistant fragment was able to be mapped to residues Met1 to Arg85 of EvpP (Figure 3C). For the shorter 4-kDa fragment, our attempts at N-terminal sequencing were not successful but the molecular weight of the fragment was determined as 4,121 Da. Instead of N-terminal sequencing, we performed peptide identification on the 4-kDa fragment using MALDI-TOF-TOF tandem mass spectrometry with trypsin digestion. Two peptides were chosen and their molecular weights (1,796 Da and 1,270.7 Da) were found to match the EvpP sequences "ENWSSLDHLLEIVLK" (residues Glu108 to Lys122) and "VPVYAWFGGFK" (residues Val123 to Lys133), respectively. Based on the determined molecular weight of the fragment, we deduced the boundaries of this 4-kDa fragment to be from residues Ile106 to Gly141 of EvpP (Figure 3C).

NMR backbone assignment of P143T EvpP

The wild type EvpP contained too many disordered regions to be crystallized for structure determination. In addition, the NMR sample of EvpP at room temperature tended to precipitate steadily. Thus, we sought to obtain a more stable mutant of EvpP to carry out NMR experiments. Sequence alignment of EvpP from *E. tarda* PPD130/91 [2] with homologues from *E. tarda* EIB202, *E. tarda* 080813, *Aeromonas hydrophila* [30] and *E. tarda* FL6–60 [31] showed that residues Pro20, Leu34, Pro58, Leu91, Pro143 and Gly156 are unique to *E. tarda* PPD130/91. The corresponding residues in other homologues are Ala22, Gln36, Ser60, Gln93, Thr145 and Ser158, respectively (Figure S3 in File S1). The contrasting branching and flexibility properties of these residues would likely contribute to the differences in stability between EvpP from *E. tarda* PPD130/91 and that of other strains. Therefore, we performed site-directed mutagenesis at these locations (P20A, L34Q, P58S, L91Q, P143T and G158S) and found that one of the mutants, P143T, had a higher stability than the wild type EvpP, as determined with urea denaturation experiments (Figure S4 in File S1). The EvpP P143T mutant showed a similar digestion pattern as wild type EvpP in limited protease digestion using trypsin, suggesting that the same disordered regions still exist in the mutated protein (Figure 2A). The ^{1}H-^{15}N HSQC NMR spectrum of EvpP P143T was also similar to that of wild type EvpP (Figure 3A); however, the mutant was stable at room temperature,

A

B

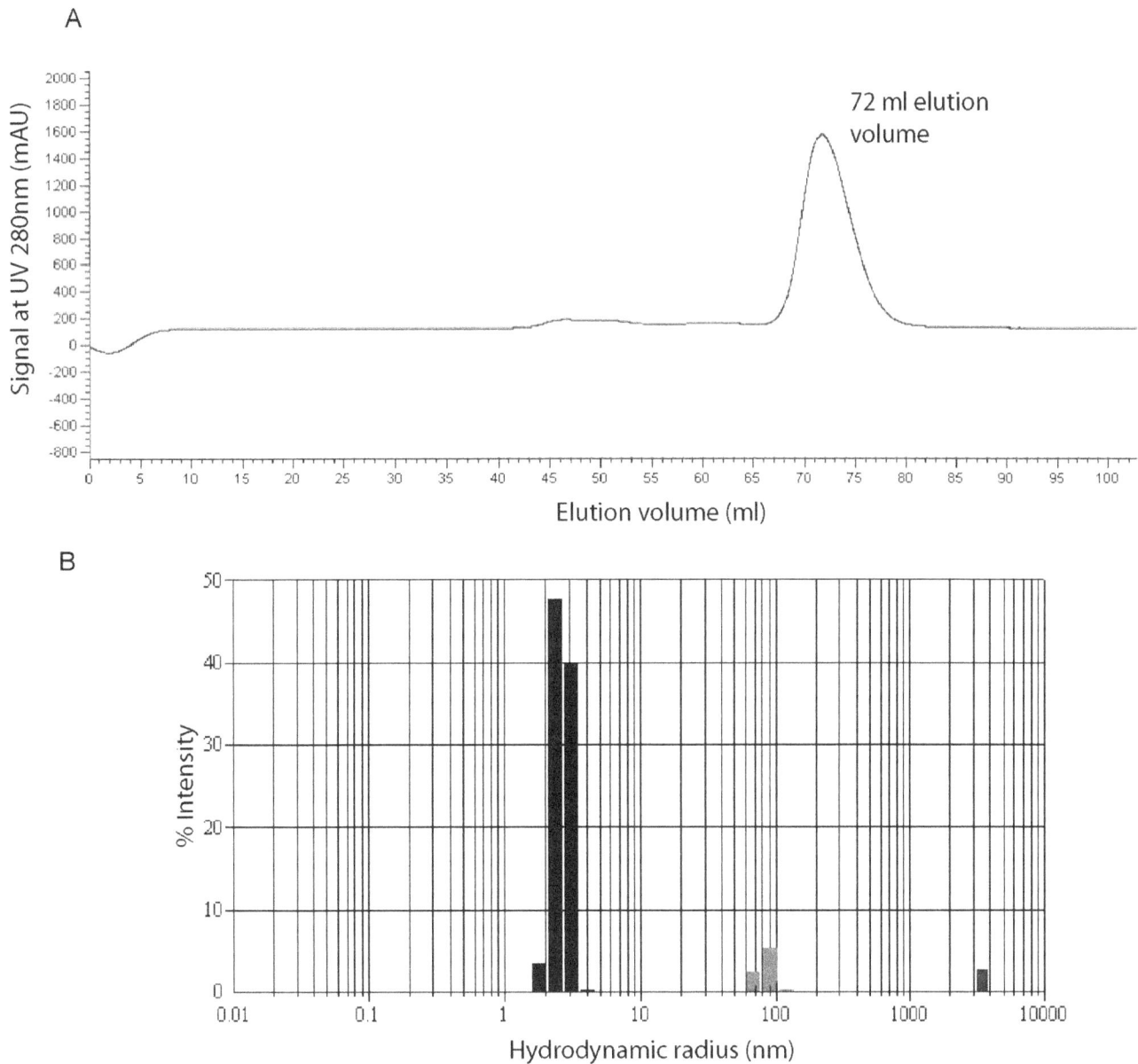

Figure 1. EvpP is a dimeric protein in solution. (A) The recombinant wild type EvpP purified as a dimeric protein with an elution volume of 72 ml from the Superdex 75 gel filtration column (HiLoad 16/60, GE Healthcare). (B) Dynamic light scattering data showing that the hydrodynamic radius of EvpP ranged from 2.92 nm to 3.09 nm, with a mean value of 3.01 nm. The estimated molecular weight ranged from 41.2 kDa to 47.3 kDa, with mean value of 44.4 kDa (data not shown).

which thus allowed us to carry out sequential backbone assignment through various 3D and 4D NMR experiments.

The theoretical number of backbone amide cross-peaks in the ^{1}H-^{15}N HSQC spectrum of His-tagged EvpP should be 190 peaks; however, only ~50% of these peaks were able to be observed in the ^{1}H-^{15}N HSQC spectra of wild type EvpP (90 peaks) and P143T EvpP (96 peaks) (Figure 3A). This large number of missing peaks suggested that EvpP contains an extensive number of disordered regions, which undergo medium time scale conformational exchange. We attempted to assign P143T EvpP using standard through-bond triple resonance experiments; however, this was not successful because of the exceptionally short relaxation time of the protein. Subsequently, we adopted an NMR strategy that utilized both through-bond and through-space NMR

experiments for resonance assignment. A set of five experiments were performed: ^{15}N-edited HSQC, ^{15}N-^{13}C-edited HNCA, HN(CO)CA, 4D-NOESY and ^{13}C-edited CCH-TOCSY. This strategy was previously designed for backbone and side chain assignment of large proteins without deuteration [32]. Using this method, we are able to assign 89 amide proton cross-peaks out of 96 detected cross-peaks from the ^{1}H-^{15}N HSQC of P143T EvpP (Figure 4A). The assigned residues were located mainly at the more-ordered and protease-resistant 10-kDa and 4-kDa regions (Figure 3C). This agrees with the notion that regions lacking HSQC peaks are likely to be flexible and more susceptible to protease digestion. One short region from residues Gly51 to Pro59 could not be assigned, suggesting that this region may also be undergoing medium time scale conformational exchange even

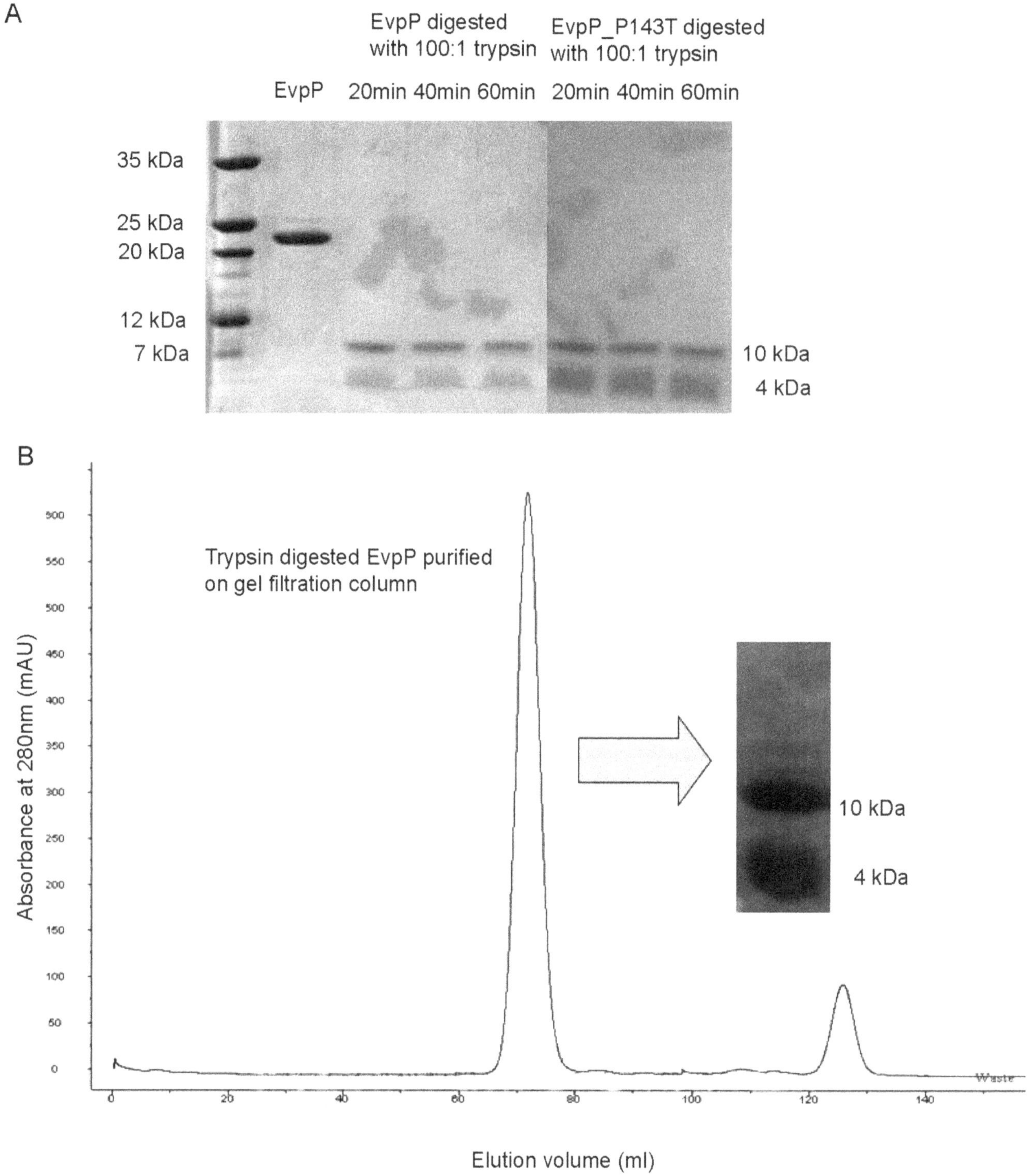

A

B

Figure 2. Limited trypsin digestion of EvpP and EvpP P143T. (A) Limited trypsin digestion of both wild type and P143T EvpP resulted in two protease-resistant fragments of 10 kDa and 4 kDa consistently within a digestion period of 20–60 min. (B) Gel filtration elution profile showed that the 10-kDa and 4-kDa fragments remain associated with each other after protease digestion.

though it is located inside the 10-kDa, protease-resistant fragment. Prediction of disordered regions in EvpP using the software RONN [33] and PrDOS [34] also picked up this unassigned region and another unassigned region from residues Met142 to Gly161 between the more ordered 10-kDa and 4-kDa regions (Figure 3C).

Heteronuclear nuclear Overhauser effects (NOE) experiments verify ordered regions on P143T EvpP

To further verify whether the cross-peaks shown on the ^{1}H-^{15}N HSQC spectrum account for residues in the ordered regions, we performed heteronuclear NOE experiments using ^{15}N-labeled P143T EvpP. The two HSQC-like spectra, with or without proton

A

B

C

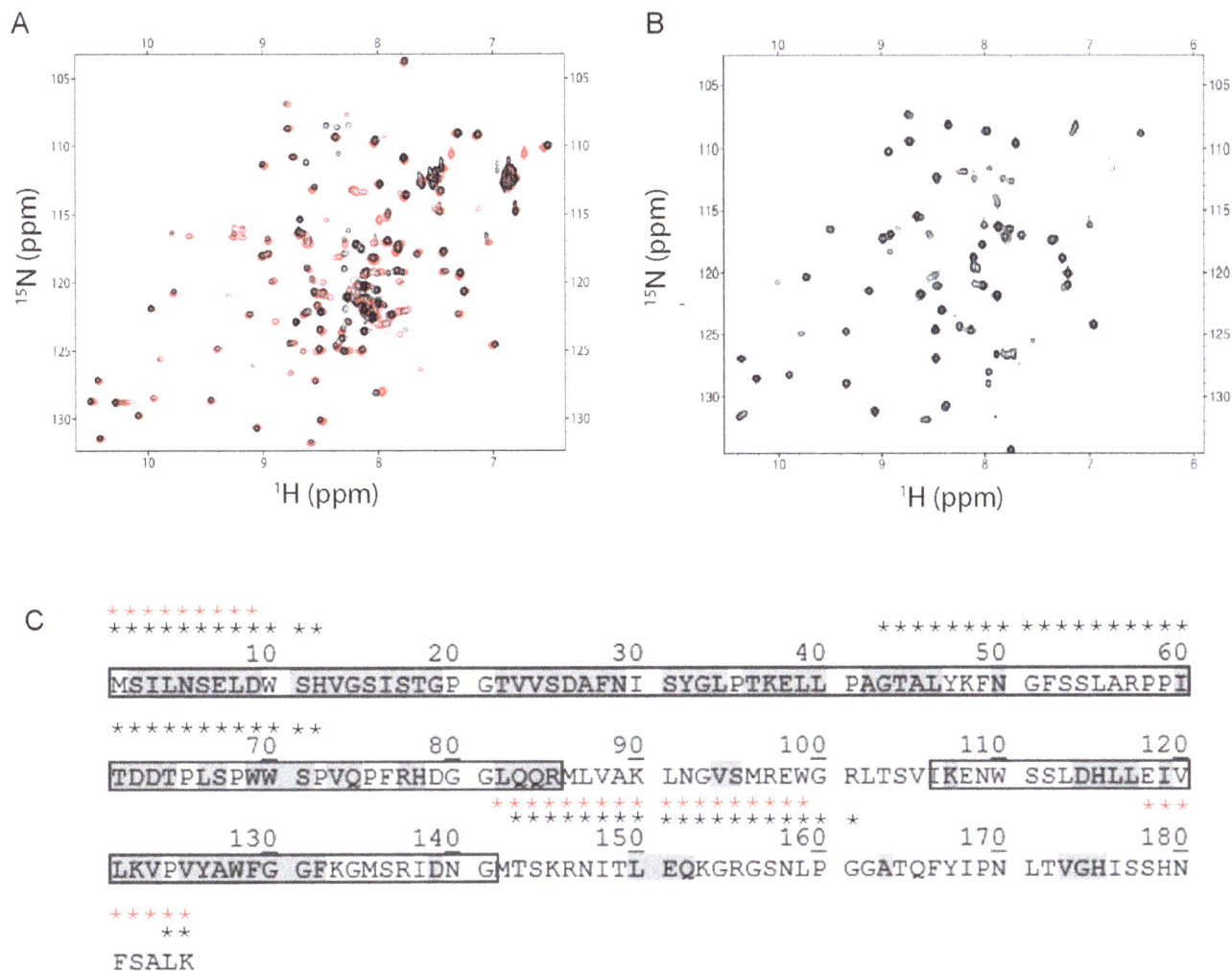

Figure 3. **¹H-¹⁵N HSQC spectra of wild type, P143T and trypsin digested EvpP.** (A) Overlay of the ¹H-¹⁵N HSQC spectra of wild type EvpP (black) and P143T EvpP (red). (B) ¹H-¹⁵N HSQC spectrum of trypsin digested EvpP. (C) Sequence of P143T EvpP from *E. tarda* PPD130/91, with the boundaries of the 10-kDa (residues Met1 to Arg85) and 4-kDa (residues Ile106 to Gly141) protease-resistant fragments boxed. Residues that can be assigned by NMR experiments are shaded in grey. Residues that are predicted to be disordered by the software RONN [33] and PrDOS [34] are marked by black and red asterisks, respectively, above the sequence.

saturation, were superimposed for peak intensity comparisons (Figure 4A). A ratio of the peak intensity from the spectrum after proton saturation over that before proton saturation was obtained for each assigned residue. A ratio close to 1.0 represents a residue with an ordered backbone conformation located in a more rigid region. Residues with intensity ratios of ~0.7–0.9 (as indicated in boxed regions 1, 2 and 3; Figure 4B) represent residues within more ordered regions. Residues within regions 1 and 2 are located within the 10-kDa and 4-kDa fragments, respectively, and are thus resistant to limited protease digestion. Residues in region 3 may represent a very short ordered region that cannot be resolved on SDS-PAGE, which is also resistant to limited protease digestion. Indeed, we observed two long-range NOEs from residues Val173 and Gly174 to residues Asp62 and Thr64 at the N-terminal ordered region (data not shown). Most of the assigned residues located outside of these three boxed regions have a peak intensity ratio lower than 0.6, which is in agreement with our supposition that these regions have a disordered conformation and are thus more susceptible to protease digestion (Figure 4B).

HDX-MS experiment to map disordered regions on EvpP

Hydrogen-deuterium exchange mass spectrometry (HDX-MS) experiments [35–37] provide an alternative approach with which to probe the disordered regions in EvpP. The wild type EvpP generated a library of 117 peptide fragments with sequence coverage of 99.5%. In order to better depict the H/D exchange patterns of EvpP, a subset of 12 peptides with the highest mass spectrometry signal intensities was chosen from the peptide library to plot a "heat map" spanning the entire protein sequence with the least overlapping residues. The extent of deuteration for each peptide was plotted in a time-dependent manner, and the peptide was mapped against the sequence of the protein to help visualize the global pattern (Figure 5). The extensively deuterated peptides all pointed to three regions of EvpP that could be disordered. The most disordered of these stretched from residues Leu87 to Leu102. This region connects the 10-kDa and 4-kDa fragments and is susceptible to limited protease digestion. Extensive hydrogen/deuterium exchange was found among the two peptides in this region, with one of them showing as much as 80% deuteration. A second flexible region was observed from residues Arg136 to

A

B

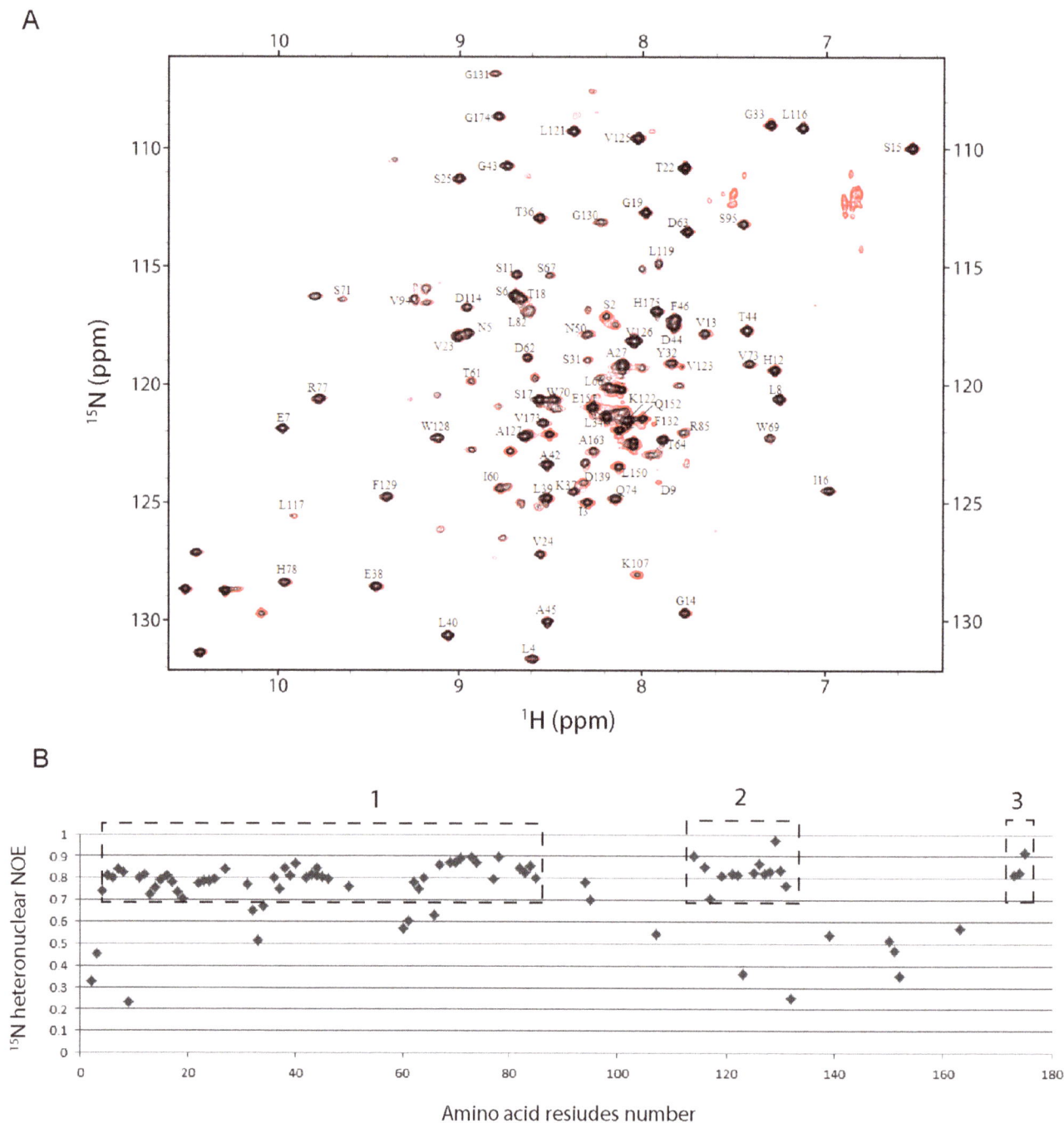

Figure 4. Heteronuclear NOE experiment on P143T EvpP. (A) Overlay spectra from heteronuclear NOE experiments showing cross-peaks from an experiment without proton saturation (red) and an experiment with proton saturation (black). The label next to each peak corresponds to sequential assignment of the residue. (B) A plot showing the relative ratio of peak intensities with proton saturation against those without proton saturation for all residues that can be assigned. Residue with a low value for the ratio is located at a disordered region. Boundaries of the three identified ordered regions are boxed by dashed lines and labeled Box 1, 2 and 3.

Phe166, which represents the region C-terminal to the 4-kDa fragment obtained from protease digestion. This region is also susceptible to limited protease digestion and has missing cross-peaks in the HSQC spectrum. The third flexible region ran from residues Lys48 to Ser54. This short stretch is located within the 10-kDa fragment obtained from protease digestion of EvpP. Residues from this region have missing cross-peaks from the HSQC

spectrum and were unable to be assigned, suggesting conformational flexibility.

The heat maps also pointed to three regions that are more protected from deuteration. Firstly, a protected and presumably more-ordered region was found from residues Thr103 to Lys133, a region which overlaps with the 4-kDa fragment. The region spanning residues Val120 to Tyr126 represents the most ordered/buried region throughout the entire EvpP sequence, located at the

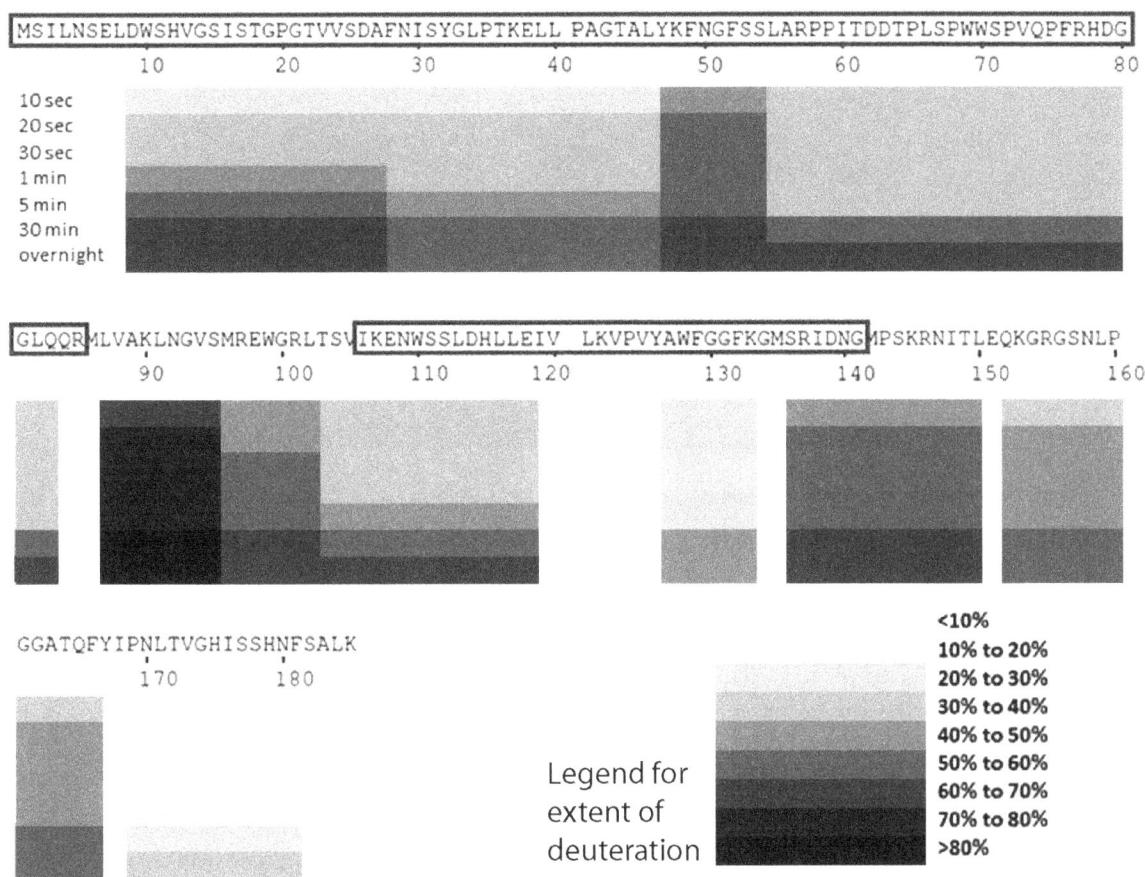

Figure 5. Hydrogen-deuterium (H/D) exchange heat map of wild type EvpP. The peptides from 12 subsets with the highest signal intensities were chosen from respective mass spectra. The extent of deuteration for the peptides exchanged for various time periods (from 10 sec to overnight) is plotted along the amino acid sequence of EvpP. The boundaries of the 10-kDa and 4-kDa protease-resistant fragments are boxed in the sequence of EvpP above the heat map.

middle of the 4-kDa fragment. A second region, corresponding to the 10-kDa fragment, was found in the N-terminal half of EvpP, running from residues Asp9 to Gln83. With the exception of residues Lys48 to Ser54, this region has a lower overall extent of hydrogen deuterium exchange as compared with the disordered regions, particularly during the initial time period of 10–30 sec. The last protected region was found from residues Pro169 to Phe181 at the C-terminus of EvpP. The extent of deuteration in this region was even less than most of the N-terminal regions. The presence of this protected region agrees with our findings from the heteronuclear NOE experiment that residues Val173 to His175 are ordered (Figure 4B, boxed region 3). Residues within this region could be interacting with residues from the N-terminal ordered region of EvpP through tertiary folding and provide stability.

Interaction of truncation mutants of EvpP with GST-EvpC

To determine if the disordered region is important for the binding of EvpP to EvpC, we generated C-terminal truncation mutants of EvpP and determined its binding to GST-EvpC (Figure 6). The GST-EvpC is neither in dimeric nor hexameric form and likely assumed an aggregated form. Removal of the C-terminal ordered region of EvpP (EvpP$_{1-168}$) did not affect its binding to EvpC significantly, suggesting that this C-terminal ordered region is not essential for EvpC interaction. Further deletion up to residue Pro143 (EvpP$_{1-142}$) caused a significant

reduction in the interaction between EvpP and EvpC (Figure 6). This result suggests that the C-terminal disordered region between residues Pro143 and Ile168 is essential for the interaction between EvpP and EvpC. The ordered region containing both the 10-kDa and 4-kDa protease resistant fragments, as well as the disordered region between them, is not sufficient for the interaction between EvpP and EvpC.

Discussion

EvpP is a potential T6SS effector with substantial disorder

Many proteins are intrinsically disordered or equipped with disordered regions so that they can decouple specificity from affinity and achieve faster rates of association and dissociation [38,39]. The disordered nature of the protein also allows multiple protein targets through the induced-fit model [40] and the disordered region usually undergoes folding or becomes ordered upon interaction [41]. In the case of T3SS effectors, a partially disordered conformation is necessary to enable passage through the narrow channel of the secretion needle [42]. A specific chaperone is required for each T3SS effector protein to keep it in a non-native or unfolded conformation ready for secretion. For instance, the disordered chaperone binding region of YopE becomes ordered upon binding to the SycE chaperone [43]. YopE is kept in an extended and non-native conformation which

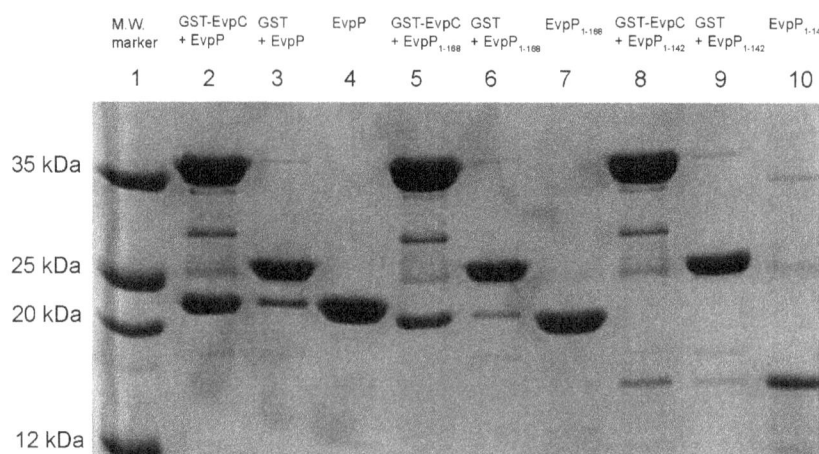

Figure 6. GST-EvpC pull-down assay for truncation mutants of EvpP. SDS-PAGE showing pull-down results using GST-EvpC and truncation mutants of EvpP. Lane 1: M.W. marker; lane2: GST-EvpC+EvpP; lane 3: GST+EvpP; lane 4: EvpP; lane 5: GST-EvpC+EvpP$_{1-168}$; lane 6: GST+EvpP$_{1-168}$; lane 7: EvpP$_{1-168}$; lane 8: GST-EvpC+EvpP$_{1-142}$; lane 9: GST+EvpP$_{1-142}$; and lane 10: EvpP$_{1-142}$. The region from residues Pro143 to Ile168 is essential for the interaction between EvpC and EvpP.

wraps around the chaperone and readies the protein for secretion through the narrow channel of the YscF needle [44].

In *E. tarda*, the T6SS gene cluster harbors a secreted protein EvpP, which could be a potential effector protein essential for bacterial virulence. As most T3SS effectors contain a substantial proportion of disordered regions in the absence of the chaperone, we sought to determine whether this is also the case for EvpP and, if in existence, whether these disordered regions are involved in its interaction with EvpC, the only known interaction partner of EvpP. Indeed, a significant proportion of EvpP is disordered (~40%). A specific disordered region at the C-terminal region of EvpP, spanning from residues Pro143 to Ile168, was found to be involved in its interaction with EvpC, which also suggests that EvpP could be a potential classic type of T6SS effector that requires interaction with EvpC for secretion [24]. EvpC may also act as a chaperone to stabilize EvpP through these interactions.

Up to date, most identified T6SS effectors are anti-bacterial effectors targeting cell wall of neighboring bacterial cells. In *Pseudomonas aeruginosa*, peptidoglycan degrading effectors, Tse1-3 (type VI secretion exported 1–3) toxins are known to kill a broad range of bacteria through secretion by the T6SS [45]. Four families of cell wall-targeting peptidoglycan amidase enzymes, Tae1-4 (type VI amidase effector), have also been identified as substrates of the T6SS-1 in *Burkholderia thailandensis* using mass spectrometry [46]. These anti-bacterial effectors are relatively ordered as many of them can be crystallized for structure determination [47–50]. They carry out their function by enzymatically digesting bacterial peptidoglycan without the need to interact with proteins of the host bacteria. In contrast, no anti-bacterial T6SS effectors were identified in *E. tarda* (unpublished results) and no homologues of Tse1 (*Pseudomonas*), Tse3 (*Psuedomonas*), Tae3 (*Ralstonia*) and Tae4 (*Salmonella*) were found in *E. tarda* as confirmed by BLAST search. EvpP is likely a potential effector against eukaryotic host with no known enzymatic activity. The substantial amount of disordered regions may allow EvpP to interact easily with host targets to manifest virulence. To confirm that EvpP is a *bone fide* T6SS effector, further experiments, such as TEM-1 β-lactamase translocation assay or western blots using anti-EvpP antibody, are needed to verify that EvpP is transferred into the target host cells.

Differences between T3SS and T6SS effector in chaperone requirement and degree of folding

Unlike many T3SS effectors, the currently identified classic type T6SS effectors, for example, Tse1-3, do not appear to require a dedicated chaperone to keep them soluble and in an unfolded conformation. Their interaction with the HcpI, however, is still required for secretion [22]. During the course of our study, we also noticed that there are significant differences in the degree of unfolding between EvpP and known T3SS-secreted substrates. The ^{1}H-^{15}N HSQC spectrum of EvpP showed that the ordered region of EvpP contains residues with NH chemical shifts, widely ranging from 7.0 to 10.0 ppm, which suggests a folded conformation with varying molecular environments. In contrast, the ^{1}H-^{15}N HSQC of a classic T3SS substrate, ExsE, showed a typical unfolded conformation with peaks ranging narrowly from 7.8 to 8.6 ppm in the absence or presence of the chaperone, ExsC [51]. The crystal structure of the ExsC-ExsE complex showed that ExsE wrapped around the dimeric ExsC chaperone in a way similar to that of the SycE-YopE complex [52].

One probable explanation for this difference could be the variation in the internal diameter of the HcpI ring and that of the T3SS secretion needle. Six units of HcpI form a ring with an internal diameter of ~40 Å [53]. This diameter is twice that of the T3SS needle tunnel, which has an internal diameter of ~20 Å. The T3SS needle also contains a narrower channel section, with an internal diameter of only 10 Å [42]. The wider diameter of the HcpI ring allows passage of substrates of a larger molecular weight and more folded conformation than does the T3SS. The T3SS substrate, ExsE, on the other hand is smaller (8.7 kDa) and is in a highly unfolded conformation in the absence or presence of the chaperone ExsC [51]. Thus, given the differences in chaperone requirement and the degree of effector folding, it is likely T3SS and T6SS use different mechanisms for substrate transport.

An effector or a regulator?

EvpP is essential for virulence of *E. tarda* PPD130/91 and the *evpP* ORF is regulated by a dedicated promoter. Interestingly, EvpP lacks homology with many of the currently identified, peptidoglycan peptidase T6SS effectors. Despite this, the disordered nature of EvpP as well as its interaction with EvpC strongly suggests that EvpP is a secreted substrate of T6SS. These findings, however, cannot prove whether EvpP is an effector or a regulator. The intrinsically disordered substrate, ExsE, secreted from the T3SS of *P. aeruginosa*, forms a tight complex with the ExsC chaperone under non-inducing conditions [51], whereas the transcription factor ExsA forms a complex with the anti-activator ExsD. Under inducing conditions, ExsE is secreted by the T3SS; this causes a reduction in intracellular ExsE levels, which in turn favors the formation of the ExsD-ExsC complex and dissociation of the ExsD-ExsA complex, thus releasing ExsA for binding to promoter to up-regulate T3SS expression [51]. In this case, T3SS-secreted ExsE acts more like a regulator than an effector. Future experiments, such as translocation assays, will be required to confirm that EvpP is indeed transferred into the host cells. The structure of the complex formed between EvpC and EvpP would provide insight into the mechanism of EvpP transport and aid in the design of small molecule inhibitors that could disrupt the secretion of EvpP from T6SS. This therapeutic approach could be applied not only against *E. tarda* but also other pathogenic bacteria to prevent bacterial infections and diseases.

Supporting Information

File S1   Figure S1. Raw data for dynamic light scattering experiment on EvpP. The polydispersity index is ranged from 0.0 to 0.1. **Figure S2.** SDS PAGE showing cross-linking of EvpP with glutaraldehyde at room temperature for differnet time periods. **Figure S3.** Sequence alignment of EvpP protein from different bacterial strains. **Figure S4.** Urea denaturation curves of wild type and mutant EvpP monitored by fluorescence.

Acknowledgments

This project was supported by Academic Research Fund, National University of Singapore grant R-154-000-498-112 to Y.K.M. K.Y.L. was supported by a grant from the National Science and Engineering Research Council (NSERC) Discovery Grant (372373-2010), Canada, and the Open Funding Project of the State Key Laboratory of Bioreactor Engineering of China.

Author Contributions

Conceived and designed the experiments: YKM. Performed the experiments: WH. Analyzed the data: WH YKM. Contributed reagents/materials/analysis tools: KYL GA JS. Wrote the paper: YKM.

References

1. Pukatzki S, Ma AT, Sturtevant D, Krastins B, Sarracino D, et al. (2006) Identification of a conserved bacterial protein secretion system in *Vibrio cholerae* using the *Dictyostelium* host model system. Proc Natl Acad Sci USA 103: 1528–1533.
2. Srinivasa Rao PS, Yamada Y, Tan YP, Leung KY (2004) Use of proteomics to identify novel virulence determinants that are required for *Edwardsiella tarda* pathogenesis. Mol Microbiol 53: 573–586.
3. Leung KY, Siame BA, Snowball H, Mok Y-K (2011) Type VI secretion regulation: crosstalk and intracellular communication. Curr Opin Microbiol 14: 9–15.
4. Kapitein N, Mogk A (2013) Deadly syringes: type VI secretion system activities in pathogenicity and interbacterial competition. Curr Opin Microbiol 16: 52–58.
5. Basler M, Ho BT, Mekalanos JJ (2013) Tit-for-Tat: Type VI secretion system counterattack during bacterial cell-cell interactions. Cell 152: 884–894.
6. Bingle LEH, Bailey CM, Pallen MJ (2008) Type VI secretion: a beginner's guide. Curr Opin Microbiol 11: 3–8.
7. Cascales E (2008) The type VI secretion toolkit. EMBO Rep 9: 735–741.
8. Filloux A, Hachani A, Bleves S (2008) The bacterial type VI secretion machine: yet another player for protein transport across membranes. Microbiology 154: 1570–1583.
9. Silverman JM, Brunet YR, Cascales E, Mougous JD (2012) Structure and regulation of the type VI secretion system. Annu Rev Microbiol 66: 453–472.
10. Shneider MM, Buth SA, Ho BT, Basler M, Mekalanos JJ, et al. (2013) PAAR-repeat proteins sharpen and diversify the type VI secretion system spike. Nature 500: 350–353.
11. Zhang XY, Brunet YR, Logger L, Douzi B, Cambillau C, et al. (2013) Dissection of the TssB-TssC interface during type VI secretion sheath complex formation. PLoS ONE 8: e81074.
12. Lossi NS, Manoli E, Förster A, Dajani R, Pape T, et al. (2013) The HsiB1C1 (TssB-TssC) complex of the *Pseudomonas aeruginosa* type VI secretion system forms a bacteriophage tail sheathlike structure. J Biol Chem 288: 7536–7548.
13. Brunet YR, Espinosa L, Harchouni S, Mignot T, Cascales E (2013) Imaging type VI secretion-mediated bacterial killing. Cell Rep 3: 36–41.
14. Felisberto-Rodrigues C, Durand E, Aschtgen M-S, Blangy S, Ortiz-Lombardia M, et al. (2011) Towards a structural comprehesion of bacterial type VI secretion systems: Characterization of the TssJ-TssM complex of an *Escherichia coli* pathovar. PLoS Pathogens 7: e1002386.
15. Durand E, Zoued A, Spinelli S, Watson PJ, Aschtgen MS, et al. (2012) Structural characterization and oligomerization of the TssL protein, a component shared by bacterial type VI and type IVb secretion systems. J Biol Chem 287: 14157–14168.
16. Robb CS, Assmus M, Nano FE, Boraston AB (2013) Structure of the T6SS lipoprotein TssJ1 from *Pseudomonas aeruginosa*. Acta Crystallogr F Struct Biol Cryst Commun 69: 607–610.
17. Zoued A, Durand E, Bebeacua C, Brunet YR, Douzi B, et al. (2013) TssK is a trimeric cytoplasmic protein interacting with components of both phage-like and membrane anchoring complexes of the type VI secretion system. J Biol Chem 288: 27031–27041.
18. Leiman PG, Basler M, Ramagopal UA, Bonanno JB, Sauder JM, et al. (2009) Type VI secretion apparatus and phage tail-associated protein complexes share a common evolutionary origin. Proc Natl Acad Sci U S A 106: 4154–4159.
19. Pukatzki S, Ma AT, Revel AT, Sturtevant D, Mekalanos JJ (2007) Type VI secretion system translocates a phage tail spike-like protein into target cells where it cross-links actin. Proc Natl Acad Sci USA 104: 15508–15513.
20. Koskiniemi S, Lamoureux JG, Nikolakakis KC, de Roodenbeke CtK, Kaplan MD, et al. (2013) Rhs proteins from diverse bacteria mediate intercellular competition. Proc Natl Acad Sci U S A 110: 7032–7037.
21. Ho BT, Dong TG, Mekalanos JJ (2014) A view to a kill: the bacterial type VI secretion system. Cell Host & Microbe doi: 10.1016/j.chom.2013.11.008.
22. Silverman JM, Agnello DM, Zheng H, Andrews BT, Li M, et al. (2013) Haemolysin coregulated protein is an exported receptor and chaperone of type VI secretion substrates. Mol Cell 51: 584–593.
23. Leung KY, Siame BA, Tenkink BJ, Noort RJ, Mok Y-K (2012) *Edwardsiella tarda* - Virulence mechanisms of an emerging gastroenteritis pathogen. Microbes Infect 14: 26–34.
24. Jobichem C, Chakraborty S, Li M, Zheng J, Joseph L, et al. (2010) Structural basis for the secretion of EvpC: a key type VI secretion system protein from *Edwardsiella tarda*. PLoS ONE 5: e12910.
25. Zheng J, Leung KY (2007) Dissection of a type VI secretion system in *Edwardsiella tarda*. Mol Microbiol 66: 1192–1206.
26. Chakraborty S, Sivaraman J, Leung KY, Mok Y-K (2011) Two-component PhoB-PhoR regulatory system and ferric uptake regulator sense phosphate and iron to control virulence genes in type III and VI secretion systems of *Edwardsiella tarda*. J Biol Chem 286: 39417–39430.
27. Records AR (2011) The type VI secretion system: a multipurpose delivery system with a phage-like machinery. Mol Plant Microbe Interact 24: 751–757.
28. Delaglio F, Grzesiek S, Vuister GW, Zhu G, Pfeifer J, et al. (1995) NMRPipe: a multidimensional spectral processing system based on UNIX pipes. J Biomol NMR 6: 277–293.
29. Goddard TD, Kneller DG SPARKY 3. University of California, San Francisco.
30. Wang X, Wang Q, Xiao J, Liu Q, Wu H, et al. (2009) *Edwardsiella tarda* T6SS component *evpP* is regulated by *esrB* and iron, and plays essential roles in the invasion of fish. Fish Shellfish Immunol 27: 469–477.
31. van Soest JJ, Stockhammer OW, Ordas A, Bloemberg GV, Spaink HP, et al. (2011) Comparison of static immersion and intravenous injection systems for exposure of zebrafish embryos to the natural pathogen *Edwardsiella tarda*. BMC Immunol 12: 58.
32. Xu Y, Zheng Y, Fan J-S, Yang D (2006) A new strategy for structure determination of large proteins in solution without deuteration. Nat Methods 3: 931–937.
33. Yang ZR, Thomson R, McNeil P, Esnouf RM (2005) RONN: the bio-basis function neural network technique applied to the detection of natively disordered regions in proteins. Bioinformatics 21: 3369–3376.
34. Ishida T, Kinoshita K (2007) PrDOS: prediction of disordered protein regions from amino acid sequence. Nucleic Acids Res 35: W460–W464.
35. Abzalimov RR, Kaplan DA, Easterling ML, Kaltashov IA (2009) Protein conformations can be probed in top-down HDX MS experiments utilizing

electron transfer dissociation of protein ions without hydrogen scrambling. J Am Soc Mass Spectrom 20: 1514–1517.

36. Kaltashov IA, Bobst CE, Abzalimov RR (2009) H/D exchange and mass spectrometry in the studies of protein conformation and dynamics: Is there a need for a top-down approach? Anal Chem 81: 7892–7899.

37. Konermann L, Pan J, Liu Y-H (2011) Hydrogen exchange mass spectrometry for studying protein structure and dynamics. Chem Soc Rev 40: 1224–1234.

38. Radivojac P, Iakoucheva LM, Oldfield CJ, Obradovic Z, Uversky VN, et al. (2007) Intrinsic disorder and functional proteomics. Biophy J 92: 1439–1456.

39. Teilum K, Olsen JG, Kragelund BB (2009) Functional aspects of protein flexibility. Cell Mol Life Sci 66: 2231–2247.

40. Dunker AK, Silman I, Uversky VN, Sussman JL (2008) Function and structure of inherently disordered proteins. Curr Opin Struct Biol 18: 756–764.

41. Dyson HJ, Wright PE (2005) Intrinsically unstructured proteins and their functions. Nat Rev Mol Cell Biol 6: 197–208.

42. Radics J, Königsmaier L, Marlovits TC (2014) Structure of a pathogenic type 3 secretion system in action. Nat Struct Mol Biol 21: 82–89.

43. Rodgers L, Gamez A, Riek R, Ghosh P (2008) The type III secretion chaperone SycE promotes a localized disorder-to-order transition in the natively unfolded effector YopE. J Biol Chem 283: 20857–20863.

44. Birtalan SC, Phillips RM, Ghosh P (2002) Three-dimensional secretion signals in chaperone-effector complexes of bacterial pathogens. Mol Cell 9: 971–980.

45. Wang T, Ding J, Zhang Y, Wang DC, Liu W (2013) Complex structure of type VI peptidoglycan muramidase effector and a cognate immunity protein. Acta Crystallogr D Biol Crystallogr 69: 1889–1900.

46. Russell AB, Singh P, Brittnacher M, Bui NK, Hood RD, et al. (2012) A widespread bacterial type VI secretion effector superfamily identified using a heuristic approach. Cell Host & Microbe 11: 538–549.

47. Chou S, Bui NK, Russell AB, Lexa KW, Gardiner TE, et al. (2012) Structure of a peptidoglycan amidase effector targeted to Gram-negative bacteria by the type VI secretion system. Cell Rep 1: 656–664.

48. Lu D, Shang G, Zhang H, Yu Q, Cong X, et al. (2014) Structural insights into the T6SS effector protein Tse3 and the Tse3-Tsi3 complex from *Pseudomonas aeruginosa* reveal a calcium-dependent membrane-binding mechanism. Mol Microbiol 92: 1092–1112.

49. Dong C, Zhang H, Gao Z-Q, Wang W-J, She Z, et al. (2013) Structural insights into the inhibition of type VI effector Tae3 by its immunity protein Tai3. Biochem J 454: 59–68.

50. Zhang H, Zhang H, Gao Z-Q, Wang W-J, Liu G-F, et al. (2013) Structure of the type VI effector-immunity complex (Tae4-Tai4) provides novel insights into the inhibition mechanism of the effector by its immunity protein. J Biol Chem 288: 5928–5939.

51. Zheng Z, Ma D, Yahr TL, Chen L (2012) The transiently ordered regions in intrinsically disordered ExsE are correlated with structural elements involved in chaperone binding. Biochem Biophy Res Commun 417: 129–134.

52. Vogelaar NJ, Jing X, Robinson HH, Schubot FD (2010) Analysis of the crystal structure of the ExsC.ExsE complex reveals distinctive binding interactions of the *Pseudomonas aeruginosa* type III secretion chaperone ExsC with ExsE and ExsD. Biochemistry 49: 5870–5879.

53. Mougous JD, Cuff ME, Raunser S, Shen A, Zhou M, et al. (2006) A virulence locus of *Pseudomonas aeruginosa* encodes a protein secretion apparatus. Science 312: 1526–1530.

16

Anomalies in Network Bridges Involved in Bile Acid Metabolism Predict Outcomes of Colorectal Cancer Patients

Sunjae Lee[1]⁹, **KiYoung Lee**[2]⁹, **Seyeol Yoon**[1], **Jae W. Lee**[3]*, **Doheon Lee**[1]*

1 Department of Bio and Brain Engineering, KAIST, Yuseong-gu, Daejeon, Republic of Korea, 2 Department of Medical Informatics, School of Medicine, Ajou University, Yeongtong-gu, Suwon-si, Republic of Korea, 3 Neuroscience Section, Papé Family Pediatric Research Institute, Department of Pediatrics, Oregon Health and Science University, Portland, Oregon, United States of America

Abstract

Biomarkers prognostic for colorectal cancer (CRC) would be highly desirable in clinical practice. Proteins that regulate bile acid (BA) homeostasis, by linking metabolic sensors and metabolic enzymes, also called bridge proteins, may be reliable prognostic biomarkers for CRC. Based on a devised metric, "bridgeness," we identified bridge proteins involved in the regulation of BA homeostasis and identified their prognostic potentials. The expression patterns of these bridge proteins could distinguish between normal and diseased tissues, suggesting that these proteins are associated with CRC pathogenesis. Using a supervised classification system, we found that these bridge proteins were reproducibly prognostic, with high prognostic ability compared to other known markers.

Editor: Antonio Moschetta, IRCCS Istituto Oncologico Giovanni Paolo II, Italy

Funding: This work was supported by grants from the NIH (DK064678 to J.W.L.). K.L. was supported by Basic Science Research Program through the National Research Foundation of Korea (NRF) funded by the Ministry of Education, Science and Technology (NRF-2013R1A2A2A04013317). D.L. was supported by grants of the Korea Health Technology R&D Project, Ministry of Health and Welfare, Republic of Korea (A112022); and also by the Bio-Synergy Research Project (NRF-2012M3A9C4048758) of the Ministry of Science, ICT and Future Planning through the National Research Foundation. The funders had no role in study design, data collection and analysis, decision to publish, or preparation of the manuscript.

Competing Interests: The authors have declared that no competing interests exist.

* Email: dhlee@kaist.ac.kr (DL); leejae@ohsu.edu (JWL)

⁹ These authors contributed equally to this work.

Introduction

Colorectal cancer (CRC) is the third leading cause of cancer deaths worldwide, with 746,000 persons dying from this disease in 2012 [1]. Prognostic biomarkers would improve treatment strategies through risk stratifications [2]. To date, however, few indicators of patient prognosis have been identified, impeding the selection and timing of adjuvant therapy for at-risk patients.

Prognostic biomarkers should be mechanistically relevant to disease pathogenesis. Although current data-driven expression-signatures, where gene expression patterns are highly correlated with patient prognosis, have shown substantial prognostic ability, they have not revealed underlying mechanism and thus obscured proper therapeutic interventions [3]. Biological hypotheses have provided a priori evidence of mechanistic relevance [4], but existing targeted hypothesis-driven approaches are likely to miss out numerous genes related to the biological hypotheses, requiring new alternative approaches to find many hypothesis-relevant genes.

Bile acids (BAs) are carcinogenic [5,6], with high-fat diets modulating BA homeostasis and altered levels of BAs leading to

CRC pathogenesis. For example, a BA-supplemented diet in mice has been shown to induce CRCs directly, suggesting that BAs are carcinogenic [7]. However, although BAs lead to CRC pathogenesis, BAs were not utilized as practical markers. At in vivo levels, they were weak and indistinctive between patients with CRC and matched controls across studies [8] since changed BA levels by food intake are temporary and weak, thus difficult to detect. Anomalies in genes regulating cellular BA homeostasis are more of determinate factors to develop CRCs.

Proteins involved in the regulation of the homeostasis of not only BAs but all metabolites include metabolic sensors and metabolic enzymes. Metabolic sensors recognize the metabolic information during the regulation of homeostasis by detecting the levels of intracellular metabolites [9–11]. For example, the farnesoid X receptor (FXR, also known as NR1H4) detect the level of intracellular BAs, with this information utilized during the regulation of cellular BA homeostasis. Metabolic enzymes catalyze the reactions of metabolites, altering their intracellular levels. Anomalies in these sensors and enzymes would therefore alter BA homeostasis [12,13] and ultimately affect CRC pathogenesis. For example, genetic defects in BA regulating enzymes or sensor proteins were found to lead to CRC pathogenesis [14,15]. However, these genes also were not prognostic markers due to the low incidence of mutations in CRCs.

Interestingly, additional factors that are neither metabolic sensors nor enzymes were shown to modulate BA homeostasis [16]. As an alternative method of identifying reliable prognostic markers, we hypothesized that these factors may relay information on metabolic status between metabolic sensors and enzymes, functionally linking these two classes of molecules. These factors, called bridge proteins, may serve as reliable prognostic markers in patients with CRC, because anomalies in these proteins would disturb the delivery of metabolic information and the proper regulation of BA homeostasis. Current targeted approaches would be ineffective in probing relay proteins specifically between metabolic sensors and enzymes, due in large part to the lack of a method to quantify the relay degree of proteins. Systematic approaches, using information about known molecular interactions and the proteins connecting sensors and enzymes may identify and distinguish bridge proteins implicated in cellular signaling networks.

Here, we propose a network-based approach that identifies prognostic markers among proteins that play a critical role possibly linking sensors and enzymes of BA metabolism, relating to known biological hypothesis. These proteins, referred to as bridge proteins, can be assessed systematically based on information about molecular interactions recorded in several databases. To this end, we have defined a "bridgeness" metric, representing the degrees of connection between sensors and enzymes, and propose key bridge proteins as network markers for prognosis in patients with CRC. Using this "hypothesis-initiated" approach, we identified a set of markers that could better predict outcomes in patients with CRC than previously identified prognostic markers. A network-based investigation of biomarkers based on their bridgeness property may identify prognostic biomarkers implicated in cellular networks.

Results

Bridge networks and bridge proteins for bile acid metabolism

Our network-based approach identified 50 bridge proteins as reliable prognostic markers (**Table S1**). Top-ranked bridge proteins included peroxisome proliferator-activated receptor gamma, coactivator 1 alpha (PPARGC1A), hepatocyte nuclear factor 4 alpha (HNF4A), glycogen synthase kinase 3 beta (GSK3B), retinoid X receptor gamma (RXRG), caspase 8, apoptosis-related cysteine peptidase (CASP8), CREB binding protein (CBP), peroxisome proliferator-activated receptor alpha (PPARA), p53 (also known as TP53), E1A binding protein p300 (EP300) and retinoid X receptor alpha (RXRA). Notably, RXRA, forming a heterodimer with a BA sensor, FXR, participates in the regulation of BA homeostasis [17]. Also, p53 regulates BA homeostasis by linking between a BA sensor and BA enzymes, leading to abnormal BA accumulation by its defect [16,18]. Likewise, some bridge proteins that function in regulating BA homeostasis are summarized in **Table S2**, showing evidence that bridge proteins, though they are computationally selected, may participate in the regulation of BA homeostasis.

To investigate these bridge proteins, we constructed a reference network for BA metabolism (**Figure 1**), a network composed of metabolic sensors, metabolic enzymes and proteins linking sensors and enzymes. Pivotal bridge proteins that regulate given metabolic pathways were investigated by first integrating previous knowledge and interactome data. To date, 53 enzymes, including transport-

ers, have been reported to be involved in BA metabolism and recorded in the EHMN database (**Table S3**) [19]. As detecting BAs and regulating their levels by altering downstream pathways for BAs, FXR has been found in vivo and in vitro to be a sensor for BAs [11]. Based on previous knowledge and the database, the sensor and enzymes were included in a BA bridge network. Large-scale interactome data from the databases, including HPRD [20] and TRANSFAC [21], were integrated to identify proteins that link sensors and enzymes (**Figure 1B**). We found that 10,805 genes or gene products were responsible for 110,741 interactions; of these gene products, we extracted only the sensors, enzymes and related intermediate proteins. All proteins responsible for direct and indirect interactions between sensors and enzymes were considered, with any intermediate protein being a possible bridge protein.

Constraints were subsequently imposed on both proteins and their interactions by considering the tissue-specific context of metabolism (**Figure 1C; Materials and Methods**). Despite abundant information on large-scale interactome data, there may be selection biases and tissue-specific variations. As a result of imposing constraints, we obtained a final reference network of 63,070 edges and 7,011 nodes, with sensors and enzymes constituting 23 nodes (**Figure 1D**, see **Figure S1** for the final reference network).

From the reference network, we selected bridge proteins, among intermediate proteins, that better link BA sensors and BA enzymes, using a "bridgeness" metric, assuming that the highly linking proteins critically regulate BA homeostasis through delivering metabolic information (**Figure 1E; Methods**). Compared with other existing centralities, including degree, closeness and betweenness centralities (see **Text S1**), our method was better able to focus on a particular protein's connections in specific paths between sensors and enzymes, regardless of the connections in other unrelated paths on the network. As expected, locally dense proteins among paths between BA sensors and BA enzymes contribute significantly to the regulation of BA metabolism; thus, these proteins may be associated with CRC carcinogenesis. We therefore focused on the prognostic potential of bridge proteins with high bridgeness scores.

Biological characteristics of bridge proteins

Before investigating their prognostic potentials, we examined the biological characteristics of bridge proteins that were computationally selected by bridgeness scores in CRCs. First, we identified expression patterns of bridge proteins embedded in CRCs; we examined discriminative patterns of bridge proteins at the transcriptomic level, using gene-expression profiles of CRC patients, as described previously [22]. Using univariate Student t-tests, we checked the ability of individual bridge proteins to distinguish between normal colon ($N = 54$) and primary CRC tissue samples ($N = 186$) at the transcriptomic level. Of the top-50 proteins, 42 (84%) were significantly discriminative (two sided $P < 0.01$). Gene ontology enrichment analysis of these 42 proteins revealed that most were enriched in terms such as "regulation of transcription from RNA polymerase II promoter" and "transcription regulator activity", which are related to regulatory roles in cellular processes (**Table S4**). They were also enriched in CRC pathogenic pathway-related terms, such as "canonical Wnt receptor signaling pathway" and "axin-APC-beta-catenin-GSK3B complex", suggesting the relevance of these bridge proteins to CRC pathogenesis.

Next, we compared the p-value distributions of i) bridge proteins, ii) a sensor and an enzyme, and iii) a combined group of i) and ii) (**Figure 2**). Compared with the background distribution

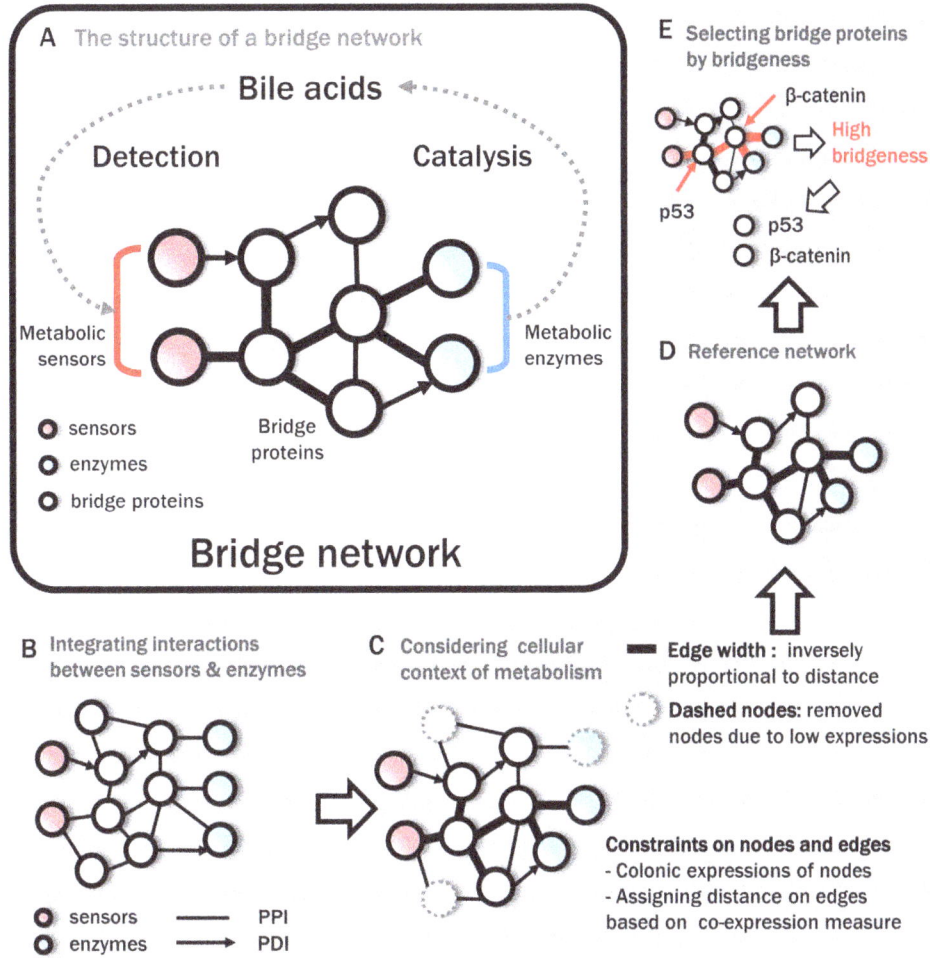

Figure 1. A bridge network for bile acid metabolism for determining bridge proteins. The overall process of the network construction is described in (**B–E**). (**A**) Structure of a bridge network, composed of a metabolic sensor (red), a metabolic enzyme (blue) and a bridge protein (gray). Metabolic enzymes catalyze the reactions of metabolites. Metabolic sensors detect the levels of intracellular metabolites. Bridge proteins link metabolic sensors and metabolic enzymes. (**B**) Integration of possible interactions between sensors and enzymes using protein-protein interactions (PPI) and protein-DNA interactions (PDI). Information on sensors and enzymes was collected from published studies and databases. (**C**) Imposing constraints on nodes and edges of an integrated network. (**D**) A final reference network to identify bridge proteins. (**E**) Selection of bridge proteins from the reference network by their bridgeness scores.

of p-values from overall gene products detected in a microarray ($N = 12,752$), the p-value distribution of the combined group was somewhat right-shifted (Kolmogorov-Smirnov (KS) test, one-sided $P = 7.89 \times 10^{-2}$). However, when we focused only on the bridge proteins, they showed high statistical significance in the KS test ($P = 2.93 \times 10^{-3}$), indicating that the discriminative power of bridge proteins, at the transcriptome level, was significantly greater than that of overall gene products in the microarray. Interestingly, sensor and enzyme proteins showed similar distributions relative to background ($P = 0.812$), indicating that sensor and enzyme proteins are less informative than bridge proteins in distinguishing between normal and diseased colon tissues.

We also investigated whether the top-50 bridge proteins are a feasible number of selections showing high statistical significance. We therefore compared the p-value distributions of selections with various numbers of bridge proteins, using the KS-test. The top-50 bridge proteins showed the lowest p-value on this comparison (**Figure S2**), with the statistical significance of selected bridge proteins being lower. Hence, we focused on the top-50 bridge

proteins in further analysis. We also included other constraints used in network construction in a similar fashion (**Figure S3**).

We next compared the discriminative power of selected bridge proteins from different networks, through multivariate classification (**Figure 3A**) (**See detailed process in Materials and Methods**). The generated networks for comparisons were: (i) a bridge network developed from BA metabolism, (ii) a bridge network developed from glucose metabolism (i.e., glycolysis pathway) and (iii) a whole protein network without confining by sensors and enzymes in certain metabolic pathways. We also compared randomly selected proteins regardless of their interactions. Glycolysis was chosen for comparison to BA metabolism due to its relevance to common cancer progression [23,24]. As expected, the discriminative power of a BA bridge network at the transcriptome level exceeded that of a glycolysis bridge network because glycolysis is not specifically involved in CRCs. The ability of components of the BA bridge network to classify a sample as normal colon or primary CRC tissue (**Figure 3B**) largely exceeded that of randomly selected gene products. In contrast, components of other networks, including that involved in

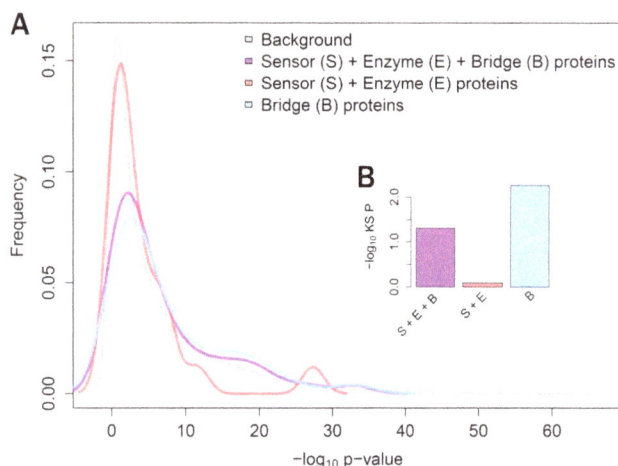

Figure 2. p-value distributions of components of a bridge network for bile acid metabolism. (**A**) p-value distributions of (i) sensor, enzyme and bridge proteins (S + E + B), (ii) sensor and enzyme proteins (B) and (iii) bridge proteins (B). (**B**) Comparisons of those p-value distributions with background p-value distribution. The statistical significance levels of shifted p-value distributions were determined by one-sided Kolmogorov Smirnov tests.

glycolysis, were equal to or barely exceeded randomly selected gene products in discriminative ability. That is, only gene expression levels of bridge proteins selected from a BA bridge network according to bridgeness were informative in distinguishing between normal colon and CRC.

We then examined CRC stage-specific expression patterns of selected bridge proteins. Most sporadic CRCs develop from normal colon via adenomatous polyps, with the sequence involving accumulated genetic anomalies in a stepwise manner [25]. To identify stage-specific variations in bridge proteins, we performed multivariate classifications between normal colons and adenomatous polyps and between polyps and primary CRCs. We found substantial variations in gene expressions of bridge proteins between normal colons and polyps (**Figure 3C and D**). Namely, bridge proteins associated with BA metabolism varied substantially during early stages of CRC pathogenesis, suggesting that these bridge proteins may be initiators of CRC tumorigenesis. We also found that bridge proteins from BA metabolism and glycolysis exhibited inverse patterns between polyps and primary CRCs, showing weaker, but substantial, variations during later stage of CRC pathogenesis, as if these changes were followers of CRC development (**Figure 3C and D**). Together, these findings showed that bridge proteins from BA metabolism and glycolysis behaved commutatively during CRC progression.

Furthermore, using pathway enrichment tests, we observed other meaningful biological characteristics of bridge proteins. Bridge proteins involved in BA metabolism were enriched in CRC-related pathways, including the Wnt (KEGG ID: hsa04310; false discovery rate-adjusted, hypergeometric $P = 4.47 \times 10^{-5}$), CRC (KEGG ID: hsa05210; $P = 2.80 \times 10^{-5}$) and common cancer (KEGG ID: hsa05200; $P = 6.94 \times 10^{-10}$) pathways (**Table S5**). This finding indicates that most bridge proteins are involved in CRC pathogenesis-related pathways and have the potential to promote CRCs through these pathways. Thus, characteristics determined from discriminative patterns and enrichment tests indicate that bridge proteins selected by bridgeness are associated with CRC pathogenesis.

Potential of bridge proteins as prognostic markers

To assess the prognostic ability of computationally-selected bridge proteins, we assessed their expression patterns in patients classified as having a good or poor prognosis. First, we clustered patients in an unsupervised way, based on similarities of expression patterns, and compared survival outcomes among patients in clusters. Total 178 patients from previous dataset [26] were clustered into three subgroups using a hierarchical clustering algorithm: BA-m1 ($N = 106$), BA-m2 ($N = 28$) and BA-m3 ($N = 44$) (**Figure 4A**). The Kaplan-Meier method with the log-rank test showed that among three subgroups of patients, the relapse-free survival was significantly different, indicating their substantial prognostic potential ($P = 2.37 \times 10^{-3}$) (**Figure 4B**). Then, we assessed the prognostic potential of other known expression-signature markers in the same way. Using expression patterns of genes selected in Wang et al [27] and ColoPrint [28], we classified patients into three subgroups and compared survival outcomes among their subgroups (ColoPrint's subgroups: col-m1 ($N = 20$), col-m2 ($N = 1$) and col-m3 ($N = 157$); Wang's subgroups: wang-m1 ($N = 19$), wang-m2 ($N = 3$) and wang-m3 ($N = 156$)). As a result, subgroups of patients clustered by ColoPrint's genes can distinguish between good and poor prognoses ($P = 2.75 \times 10^{-8}$), though just a single patient found in the poorest prognosis group (col-m2), but Wang's genes were not prognostic ($P = 0.258$) (**Figure 4C and D**). In addition, known molecular markers, including p53 mutations ($P = 0.233$), mismatch repair gene status ($P = 9.8 \times 10^{-2}$), KRAS mutations ($P = 5.75 \times 10^{-2}$), and BRAF mutations ($P = 0.338$), were not also substantially prognostic in this dataset (**Figure 4E–H**).

To assess the prognostic reproducibility of these bridge proteins and other expression-signature markers, we then classified patients in an independent dataset [22] as having good or bad prognoses, through a supervised classification system, using previous dataset [26] as the training dataset (**Figure 5**). Patients in the test data were classified, using their expression levels, based on correlation coefficients to mean expression levels of poor-prognosis-group patients in the training data, like previously performed [29]; we assigned patients into a poor-prognosis group if their correlation coefficients were high. We obtained thresholds of correlation coefficients to decide poor-prognosis patients with the highest statistical significance, through cross-validation procedures on the training data (See Materials and Methods). Noteworthy, patients in the test data can be significantly distinguished between good and poor prognoses when we used expression levels of bridge proteins as features for correlation coefficients; survival outcomes, i.e., CRC-specific survivals, of classified groups by the bridge proteins were significantly different when the Kaplan-Meier method with the log-rank test was used ($P = 2.70 \times 10^{-2}$) (**Figure 5A**). Other expression signatures, including ColoPrint ($P = 0.210$) and Wang's ($P = 0.558$) (**Figure 5B and C**), were not prognostic in the independent test dataset, suggesting that only bridge proteins were reproducibly prognostic. These results underline the potential and reliability of bridge proteins as prognostic markers.

Discussion

By investigating genes involved in the regulation of BA homeostasis, this study has identified numerous genes for prognostic biomarkers of CRC, with showing mechanistic relevance to CRC pathogenesis. Although various prognostic biomarkers have been proposed based on biological hypotheses [4], these biomarkers have shown limited clinical usefulness. The hypothesis, that BAs play pivotal roles in CRC, provides clues to

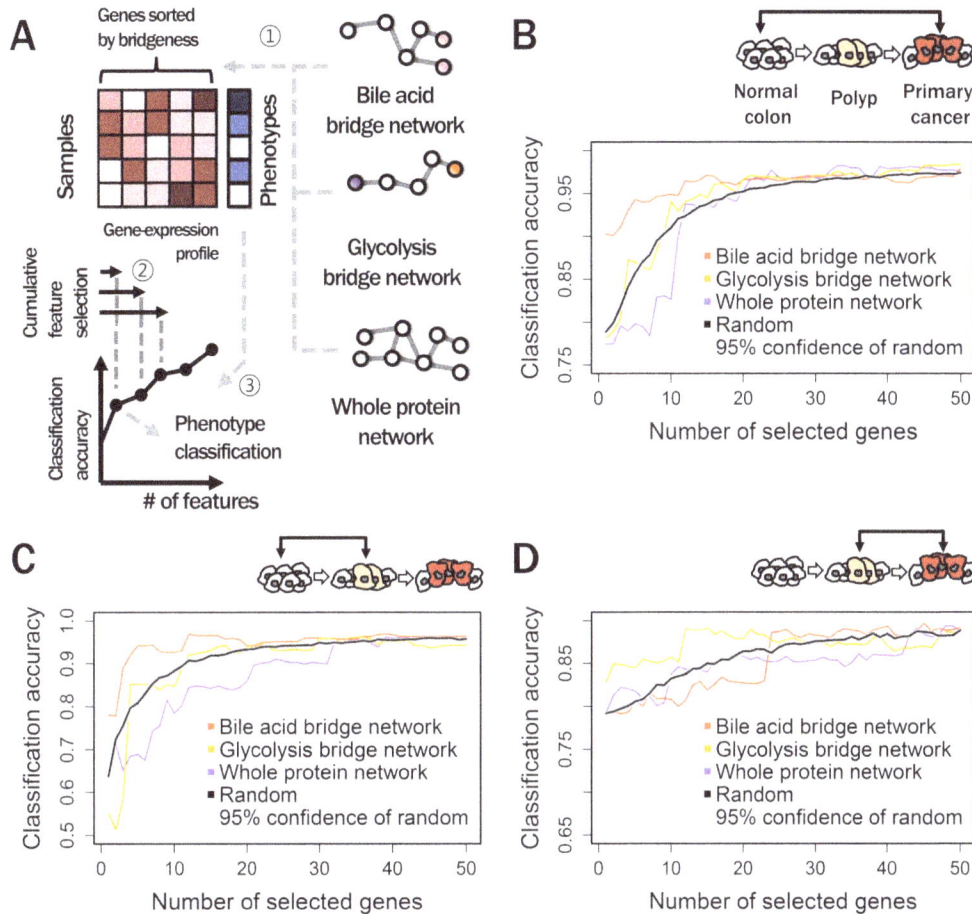

Figure 3. Multivariate analysis of bridge proteins from different networks. (A) Overall process of multivariate classifications using features from bridge proteins of different networks. After sorting bridge proteins by their bridgeness (①), features were extracted cumulatively from top-ranked bridge proteins (②). Samples were subsequently classified by cumulatively selected features and calculated classification accuracies (③). **(B)** Accuracies of classifications between normal colon and primary CRC tissues. For classifications, bridge proteins were obtained from (i) a bile acid bridge network (red), (ii) a glycolysis bridge network (yellow) and (iii) an whole protein network (purple). Classification accuracies were also calculated using randomly selected proteins (black) with 95% confidence intervals (gray) on the mean classification accuracies of repeated random selections **(C)** Accuracies of classifications between normal colon and polyp tissues. **(D)** Accuracies of classifications between polyp and primary CRC tissues.

understanding the pathogenesis of this disease. However, rather than focusing on BAs themselves, we focused on the genes involved in regulating BA metabolism by linking metabolic sensors and metabolic enzymes. Based on a devised metric, "bridgeness", numerous bridge proteins were selected from a reference, or bridge, network, and their prognostic abilities were analyzed. Bridge proteins could distinguish between normal and diseased tissues and are therefore relevant to the pathogenesis of CRC. These bridge proteins had greater and reproducible prognostic ability, as shown by statistical significance, than previously identified prognostic markers, suggesting that they are reliable prognostic markers in patients with CRC.

Interestingly, however, neither sensor nor enzyme proteins could significantly distinguish between normal colon tissue and CRC, a finding that may result from the housekeeping roles of these sensor and enzyme proteins for cell survival. Cells lack proteins with molecular functions similar to those of most of these sensor and enzyme proteins; thus, defects in their expression would have detrimental effects on cellular functions. Thus, evolutionarily, genetic anomalies in bridge proteins may have survival advantages over anomalies in sensor and enzyme proteins. Indeed, some bridge proteins, including caspase 8, apoptosis-related cysteine

peptidase (CASP8), p53 and catenin (cadherin-associated protein) beta 1, 88 kDa (CTNNB1, also known as β-catenin), showed high mutational frequencies in CRC samples, whereas sensor and enzymes proteins for BA metabolism did not [30]. This evolutionary pressure, including during CRC tumorigenesis, would accelerate the acquisition of anomalies by bridge proteins.

In previous studies, notably, one bridge protein, STK11, was shown to have particular mechanistic potential to promote colorectal tumorigenesis [31–33]. STK11 has been associated with Peutz-Jeghers syndrome (PJS), a condition that enhances the formation of gastric adenomatous polyps and hepatocellular carcinoma [31]. In most PJS patients, one allele of STK11 is mutated, causing multiple gastric adenomatous polyps or hepato-cellular carcinoma [32,33]. Similarly, STK11 may have the mechanistic potential to promote colorectal tumorigenesis. Other bridge proteins may also have prognostic value in CRC pathogenesis.

STK11 is also associated with energy metabolism, either alone or by interacting with AMPK, making it a potential bridge protein involved in the regulation of energy metabolism [34,35]. Among the other bridge proteins involved in energy metabolism are PPARGC1A, GSK3B, PPARA, peroxisome proliferator-activated

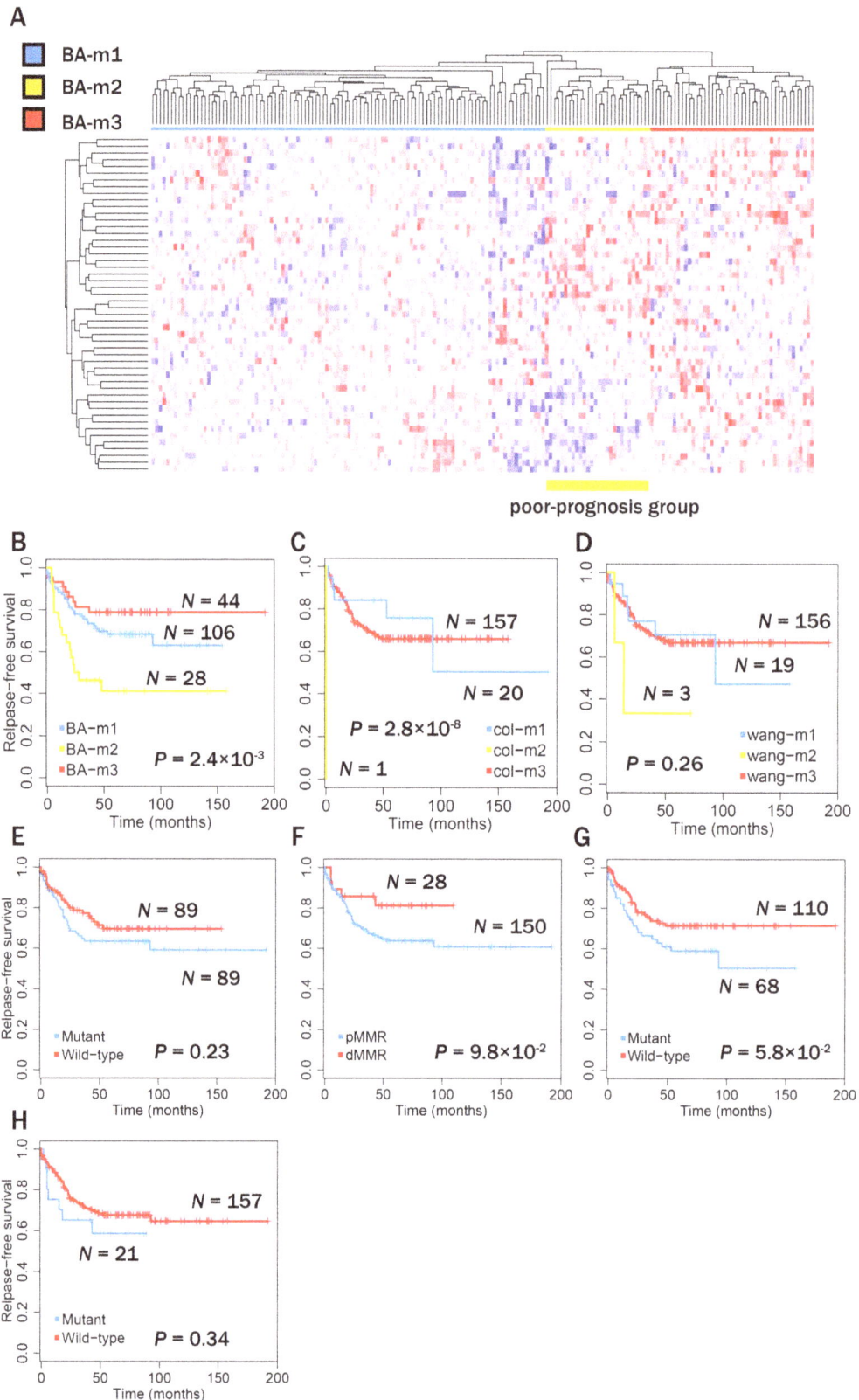

Figure 4. Identification of the prognostic ability of markers. Their prognostic ability was examined using a dataset of tissue samples from patients with CRC [26]. (**A**) Heatmap of CRC tumor samples with subgroups classified by the expression patterns of bridge proteins: BA-m1 (blue), BA-m2 (yellow) and BA-m3 (red). Prognostic ability was assessed by Kaplan-Meier survival analyses. The BA-m2 group showed the poorest prognosis. (**B**) Prognostic ability of our bridge proteins. (**C**) Prognostic ability of the ColoPrint gene set [28], with subgroups classified as col-m1 (blue), col-m2 (yellow) and col-m3 (red). (**D**) Prognostic ability of the Wang et al. signature gene set [27], with subgroups classified as wang-m1 (blue), wang-m2 (yellow) and wang-m3 (red). (**E**) Prognostic ability of p53 mutation status, mutant and wild-type. (**F**) Prognostic ability of mismatch repair gene (MMR)

status, deficient (dMMR) and proficient (pMMR). (**G**) Prognostic ability of KRAS mutation status, mutant and wild-type. (**H**) Prognostic ability of BRAF mutation status, mutant and wild-type.

receptor gamma (PPARG), solute carrier family 2 (facilitated glucose transporter) member 4 (SLC2A4, also known as GLUT4), glyceraldehyde-3-phosphate dehydrogenase (GAPDH), and lactate dehydrogenase A (LDHA), all important regulators of or enzymes involved in energy metabolism. Thus, their molecular functions may explain the activities of BAs that increase energy expenditure [36]. Assessments of the molecular functions of bridge proteins may provide novel insights on their as yet unidentified roles in BA homeostasis.

Despite bridge proteins showing prognostic potential, BA bridge networks show limited ability to identify other known CRC-susceptibility genes. For example, we found that a BA bridge network was unable to identify several well-known CRC-susceptibility genes, such as APC, KRAS, and BRAF. Inaccuracies originating from large-scale interactome data could impede in-depth analysis of bridge networks. Also, the interrelations of metabolic pathways, such as lipid, cholesterol, and glucose

metabolism, would extend the ability to investigate all risk factors for CRC pathogenesis. This approach could also be applied to other diseases vulnerable to metabolic anomalies, including obesity, type-2 diabetes and Alzheimer's disease once metabolic sensors, enzymes and proper interactome data are generated for these diseases. The determination of proper and accurate bridge networks for metabolic pathways can allow the identification of disease-susceptibility genes and their clinical use as prognostic markers.

In summary, we found that bridge proteins, which are involved in the regulation of BA metabolism, have prognostic potential in patients with CRC. Despite their potential to promote CRC pathogenesis, bridge proteins had not been systematically investigated in previous studies. Based on a devised metric for "bridgeness", we computationally selected bridge proteins from a reference network and examined their prognostic potential in CRC. We also tested whether differences in their discriminative

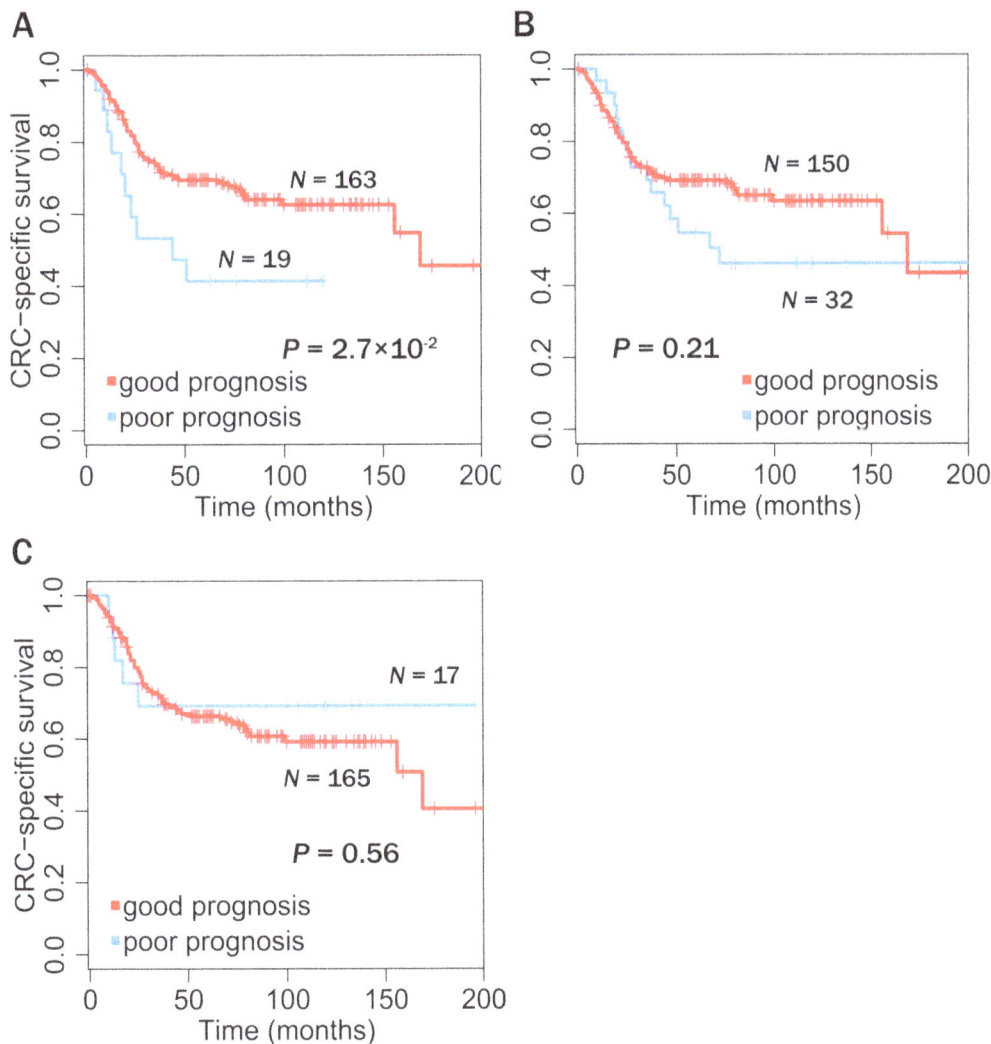

Figure 5. Identification of the prognostic reproducibility of markers. Their prognostic ability was examined in an independent test data [22] by supervised classifications and thus confirmed their prognostic reproducibility. (**A**) Prognostic ability of our bridge proteins (**B**) Prognostic ability determined by the ColoPrint gene set in reference [28] (**C**) Prognostic ability determined by the Wang et al. gene set in reference [27].

expression patterns in normal colon and CRC made them relevant to CRC pathogenesis. The findings indicate that bridge proteins involved in the regulation of BA metabolism may be reliable prognostic markers for CRC patients.

Materials and Methods

Bridge network construction

The reference network for BA metabolism consisted of metabolic sensors, metabolic enzymes and proteins interacting with both. The selected BA sensor was FXR and the BA enzymes were those designated in the EHMN human metabolic network database as enzymes involved in the "bile acid biosynthesis" pathway [19]. Possible interactions between the sensor and the enzymes were integrated using protein-protein interactions (PPI) described in the HPRD human protein information database [20] and protein-DNA interactions (PDI) from the commercial TF binding site database, TRANSFAC (Ver. 11.1) [21]. PPIs were regarded as bidirectional interactions and PDIs as unidirectional interactions from TFs to target genes. Next, we imposed constraints on the integrated network. On edges, we assigned distance values using a co-expression measure (i.e., the distance d_{ij} between genes i and j was defined as $d_{ij} = 1 - r^2_{ij}$ where r_{ij} is Pearson's correlation coefficient for the correlation in expression between genes i and j). Co-expression, defined as the functional relationship of a pair of proteins [37], was calculated using recently published FACS-sorted cell expression profiles from 52 patients with CRC [38], obtained from the public gene expression profile database, GEO (ID: GSE39397). On nodes, we imposed constraints regarding colonic gene expression. Using human whole-tissue gene expression data obtained from the public database, BioGPS [39], we determined the colonic expression of individual genes and compared the colonic and tissue-wide expression of each (total 176 samples with 84 tissue types; two samples for a colon tissue). If the average ratio of colonic to tissue-wide expression was lower than our criterion, that gene was removed. The criterion for gene removal was determined by comparing the p-value distribution of 50 bridge proteins with a background p-value distribution, as described in Results (**Figure S3**). In those comparisons, a 40th percentile cutoff produced the highest significance of the shifted p-value distribution.

Similarly, we constructed bridge networks relative to glycolysis and all proteins without specification for network comparisons. All the processes were identical to those used to construct the BA bridge network, except for the selection of metabolic sensors and enzymes. For glycolysis, we selected the metabolic sensors egl nine homolog 2 (C. elegans) (EGLN2, also known as PHD1), egl nine homolog 1 (C. elegans) (EGLN1, also known as PHD2), egl nine homolog 3 (C. elegans) (EGLN3, also known as PHD3) and hypoxia inducible factor 1 alpha subunit inhibitor (HIF1AN, also known as FIH). Their sensing of glycolysis metabolites was determined in vitro and in vivo [40]. Metabolic enzymes for glycolysis pathway were obtained from the "glycolysis and gluconeogenesis" pathway in the EHMN database [19]. For the whole protein network, we regarded metabolic sensors and enzymes as all the genes in the network in order to avoid specification by certain types of metabolism.

Bridgeness score

The bridgeness metric of a gene i with a set of sensors S and a set of enzymes T was calculated as:

$$B_{i,S,T} = \frac{1}{|S| \times |T|} \sum_{s \in S, t \in T, s \neq t} \frac{d(s,t)}{d_i(s,t)}$$

$$= \frac{1}{|S| \times |T|} \sum_{s \in S, t \in T, s \neq t} \frac{d(s,t)}{d(s,i) + d(i,t)}$$

where $d(s,t)$ represents the distance of the shortest path between a sensor s and an enzyme t, and $d_i(s,t)$ represents the distance of the shortest path between node s and node t via node i. If gene i in the network is far from the shortest path between sensors and enzymes (i.e., $d_i(s,t) \gg d(s,t)$), then the addend tends to zero. Therefore, the bridgeness of gene i would be high if it is located near the shortest paths between sensors and enzymes, thus avoiding unrelated paths in cellular signaling networks. All calculations of network features and bridgeness were determined using R language and the *igraph* package [41].

Univariate and multivariate analysis

For univariate and multivariate analyses, we used a gene expression profile from CRC patients [22], which we obtained from the GEO database (ID: GSE41258). This dataset includes gene expression in 54 normal colons, 49 adenomatous polyps and 186 primary CRC tissue samples. Before using gene expression profiles to distinguish among tissue types, we performed gene-wise normalization on the profile using Z score transformation. In univariate analysis, the ability of each gene's expression to distinguish normal colon and primary CRC tissues was assessed by Student's t-test. We also calculated the statistical significance of the shifted p-value distribution of genes of interest against a background p-value distribution using the two-sample Kolmogorov-Smirnov one-sided test with the support of R package, *stats*. In multivariate analysis, we identified a bridge protein's discriminative ability, at the transcriptome level, using a logistic regression model with the support of java machine learning API, *Weka* [42]. Multivariate features were cumulatively selected from top-ranked bridge proteins of networks. The ability of each selection to classify samples as normal colon or primary CRC was evaluated using the five-fold cross-validation method with five repeats. The ability of each to distinguish between normal colon and polyp tissues, and between polyps and primary CRCs tissues, was assessed using the same features. We also simultaneously evaluated randomly selected proteins with an equal number of features. At each evaluation step, classification accuracy (i.e. accuracy = $\frac{TP + TN}{TP + FP + TN + FN}$) was measured and averaged after five repeats. In assessing features of randomly selected proteins, we calculated the mean classification accuracy after 100 repeats of selections and afterward calculated a 95% confidence interval of mean classification accuracies.

Survival analysis

First, the prognostic ability of bridge proteins was determined using related information from the Marisa et al. dataset [26] in the GEO database (ID: GSE39582). Information was available about gene expression; CRC recurrence-free survival event and time; treatment status; molecular marker status, including p53, KRAS, and BRAF mutations; and mismatch repair gene status. In this dataset, we used 178 tumor samples of patients to assess the prognostic ability; samples from treated patients or with missing information about survival outcomes or molecular status were excluded, avoiding unexpected effects of treatment on survival outcomes or unknown information. Identifying prognostic ability, we clustered patients, based on Euclidean distances between gene expressions of patients, by an unsupervised hierarchical clustering algorithm and measured the difference of survival outcomes among the patient clusters by the Kaplan-Meier method with the log-rank test. To compare prognostic abilities with other gene-expression signature markers, we used two gene sets, Wang's ($N = 21$) [27] and ColoPrint ($N = 15$) [28], and assessed their prognostic ability using their expression profiles from patients with CRC. In the two comparative gene sets [27,28], we only utilized genes detected in microarray data that we applied.

We also assessed the prognostic reproducibility of bridge proteins through a supervised classification system (**Figure S4**). In this classification system, the previous dataset [26] were used as a training data and the Sheffer et al. dataset [22] were used as an independent test data during supervised classifications. Total 182 tumor samples of patients from the Sheffer et al. data were used, after excluding samples that were not used in the original study [22]. This dataset contains information about gene expression and CRC-specific survival event and time. Performing supervised classification, we first determined a patient group with the poorest prognosis from the training data, after clustering patients by a hierarchical clustering and comparing survival probabilities among patient clusters. Referencing mean expression levels of patients in the poorest prognosis group (i.e. BA-m2 in **Figure 4A**) as a criterion, we classified patients of the test data into poor prognoses if their correlations of gene expressions with the reference expression levels are higher than a threshold, like existing study [29]. We calculated the correlations based on Pearson's correlation coefficients. A threshold of a correlation coefficient deciding prognosis was obtained through cross-validated procedures using the training data [26]. In this data set, we performed supervised classifications through five-fold cross-validations with various thresholds and selected the best threshold that can distinguish patients into a good or poor prognosis group with the most statistical significance. The statistical significance was measured by the Kaplan-Meier method with the log-rank test. We repeated cross-validations 100 times and averaged best thresholds in all repeats as a final threshold to use. Based on the final threshold, at last, we classified patients in an independent test data with learning a training data. We performed supervised classifications by other expression signatures in a similar way. All the statistical analyses, including Kaplan-Meier survival analysis, were performed by R packages.

Supporting Information

Figure S1 A final reference network for bile acid metabolism. This network is composed of a metabolic sensor (red), metabolic enzymes (blue) and interplay proteins (the outer layer of the largest circle). The Top-50 bridge proteins (black) are also shown. The edges representing the shortest paths between a sensor or an enzyme and a top-50 bridge protein are underlined (red edges).

Figure S2 p-value distributions of bridge proteins with varying numbers of selections. They stand for p-value distributions of the (**A**) top-10, (**B**) top-20, (**C**) top-30, (**D**) top-40, (**E**) top-50, (**F**) top-60, (**G**) top-70, (**H**) top-80, (**I**) top-90, and (**J**) top-100 bridge proteins. Statistical significance was highest for the top-50 bridge proteins when the shifted degrees of background (gray) and selected (blue) p-value distributions were measured using the one-sided Kolmogorov-Smirnov test.

Figure S3 p-value distribution of bridge proteins according to imposed constraints. (**A**) without node or edge constraints, (**B**) without node constraints, (**C–F**) with node constraints of (**C**) 10%, (**D**) 20%, (**E**) 30%, and (**F**) 40% removal criteria. Node removals within 40% were the most feasible for network construction.

Figure S4 An overview of a supervised classification system. The pipeline of supervised classification system was demonstrated. We used Marisa et al. dataset as a training data and Sheffer et al. dataset as a test data, after filtering out samples of patients in undesired conditions (1). Supervised classifications were based on correlations of gene expressions between the reference from the training data and samples from the test data. To select the threshold of correlation coefficients for deciding prognosis, we performed cross-validation procedure; we repeated five-fold cross-validation 100 times and averaged best threshold in all repeats (2). Based on the threshold obtained, we classified patients in the test data (3) and compared survival outcomes among classified patient groups, having a good or poor prognosis, through the Kaplan-Meier method with the log-rank test.

Table S1 Top-50 bridge protein information. We showed statistics of each bridge protein about discriminative power (T-score and T-test P) using datasets of Sheffer et al.

Table S2 Evidence of bridge proteins involved in the regulation of bile acid homeostasis. Shown was previous literature that identified bridge proteins as being involved in the regulation of bile acid homeostasis. In the second column, we provided literature with definitive evidence that defects of some bridge proteins cause abnormal changes of bile acid levels. In the third and fourth columns, we provided literature with indirect evidence: studies in the third column showing that bridge proteins were regulated by or co-activated with a bile acid sensor; studies in the fourth column showing that bridge proteins regulated enzymes in bile acid metabolism.

Table S3 Sensor and enzyme proteins in bile acid or glucose metabolism.

Table S4 Enriched GO terms under corrected p-value<0.01.

Table S5 Enriched KEGG non-metabolic pathways under FDR-adjusted hypergeometric p-value<0.01.

Dataset S1 A source code and a dataset for extracting bridge proteins involved in bile acid metabolism. Performing a source code with a dataset will provide an output file showing top-50 bridge proteins we used.

Dataset S2 A source code and a dataset for survival analyses in Figure 4 and 5. Performing a source code with a dataset will provide figures shown in our manuscripts.

Text S1 Characteristics of bridgeness scores.

Author Contributions

Conceived and designed the experiments: SL KL JWL DL. Performed the experiments: SL SY. Analyzed the data: SL KL SY JWL DL. Wrote the paper: SL KL.

References

1. Ferley J, SoerjomataramI I, Ervik M, Dikshit R, Eser S, et al. (2013) GLOBOCAN 2012 v1.0, Cancer Incidence and Mortality Worldwide: IARC CancerBase No. 11 [Internet]. Lyon, Fr Int Agency Res Cancer. Available from: http://globocan.iarc.fr, accessed on day/month/year.
2. Joensuu H (2008) Risk stratification of patients diagnosed with gastrointestinal stromal tumor. Hum Pathol 39: 1411–1419.
3. Chang HY, Sneddon JB, Alizadeh AA, Sood R, West RB, et al. (2004) Gene expression signature of fibroblast serum response predicts human cancer progression: similarities between tumors and wounds. PLoS Biol 2: E7.
4. Walther A, Johnstone E, Swanton C, Midgley R, Tomlinson I, et al. (2009) Genetic prognostic and predictive markers in colorectal cancer. Nat Rev Cancer 9: 489–499.
5. Willett WC, Stampfer MJ, Colditz GA, Rosner BA, Speizer FE (1990) Relation of meat, fat, and fiber intake to the risk of colon cancer in a prospective study among women. N Engl J Med 323: 1664–1672.
6. Bernstein H (2009) Bile acids as endogenous etiologic agents in gastrointestinal cancer. World J Gastroenterol 15: 3329.
7. Bernstein C, Holubec H, Bhattacharyya AK, Nguyen H, Payne CM, et al. (2011) Carcinogenicity of deoxycholate, a secondary bile acid. Arch Toxicol 85: 863–871.
8. Chey WD, Camilleri M, Chang L, Rikner L, Graffner H (2012) Response to Drs Trivedi and Ward. Am J Gastroenterol 107: 140–141.
9. Lage R, Diéguez C, Vidal-Puig A, López M (2008) AMPK: a metabolic gauge regulating whole-body energy homeostasis. Trends Mol Med 14: 539–549.
10. Guarani V, Potente M (2010) SIRT1 - a metabolic sensor that controls blood vessel growth. Curr Opin Pharmacol 10: 139–145.
11. Makishima M (1999) Identification of a Nuclear Receptor for Bile Acids. Science 284: 1362–1365.
12. Sinal CJ, Tohkin M, Miyata M, Ward JM, Lambert G, et al. (2000) Targeted disruption of the nuclear receptor FXR/BAR impairs bile acid and lipid homeostasis. Cell 102: 731–744.
13. Schwarz M, Russell DW, Dietschy JM, Turley SD (2001) Alternate pathways of bile acid synthesis in the cholesterol 7alpha-hydroxylase knockout mouse are not upregulated by either cholesterol or cholestyramine feeding. J Lipid Res 42: 1594–1603.
14. Wertheim BC, Smith JW, Fang C, Alberts DS, Lance P, et al. (2012) Risk modification of colorectal adenoma by CYP7A1 polymorphisms and the role of bile acid metabolism in carcinogenesis. Cancer Prev Res 5: 197–204.
15. Maran RRM, Thomas A, Roth M, Sheng Z, Esterly N, et al. (2009) Farnesoid X receptor deficiency in mice leads to increased intestinal epithelial cell proliferation and tumor development. J Pharmacol Exp Ther 328: 469–477.
16. Kim D-H, Lee JW (2011) Tumor suppressor p53 regulates bile acid homeostasis via small heterodimer partner. Proc Natl Acad Sci U S A 108: 12266–12270.
17. Goodwin B, Jones SA, Price RR, Watson MA, McKee DD, et al. (2000) A regulatory cascade of the nuclear receptors FXR, SHP-1, and LRH-1 represses bile acid biosynthesis. Mol Cell 6: 517–526.
18. Kim D-H, Kim J, Lee JW (2011) Requirement for MLL3 in p53 regulation of hepatic expression of small heterodimer partner and bile acid homeostasis. Mol Endocrinol 25: 2076–2083.
19. Hao T, Ma H-W, Zhao X-M, Goryanin I (2010) Compartmentalization of the Edinburgh Human Metabolic Network. BMC Bioinformatics 11: 393.
20. Keshava Prasad TS, Goel R, Kandasamy K, Keerthikumar S, Kumar S, et al. (2009) Human Protein Reference Database–2009 update. Nucleic Acids Res 37: D767–72.
21. Matys V, Kel-Margoulis O V, Fricke E, Liebich I, Land S, et al. (2006) TRANSFAC and its module TRANSCompel: transcriptional gene regulation in eukaryotes. Nucleic Acids Res 34: D108–10.
22. Sheffer M, Bacolod MD, Zuk O, Giardina SF, Pincas H, et al. (2009) Association of survival and disease progression with chromosomal instability: a genomic exploration of colorectal cancer. Proc Natl Acad Sci U S A 106: 7131–7136.
23. Vander Heiden MG, Cantley LC, Thompson CB (2009) Understanding the Warburg effect: the metabolic requirements of cell proliferation. Science 324: 1029–1033.
24. Koppenol WH, Bounds PL, Dang C V (2011) Otto Warburg's contributions to current concepts of cancer metabolism. Nat Rev Cancer 11: 325–337.
25. Davies RJ, Miller R, Coleman N (2005) Colorectal cancer screening: prospects for molecular stool analysis. Nat Rev Cancer 5: 199–209.
26. Marisa L, de Reyniès A, Duval A, Selves J, Gaub MP, et al. (2013) Gene expression classification of colon cancer into molecular subtypes: characterization, validation, and prognostic value. PLoS Med 10: e1001453.
27. Wang Y, Jatkoe T, Zhang Y, Mutch MG, Talantov D, et al. (2004) Gene expression profiles and molecular markers to predict recurrence of Dukes' B colon cancer. J Clin Oncol 22: 1564–1571.
28. Salazar R, Roepman P, Capella G, Moreno V, Simon I, et al. (2011) Gene expression signature to improve prognosis prediction of stage II and III colorectal cancer. J Clin Oncol 29: 17–24.
29. Van de Vijver MJ, He YD, van't Veer LJ, Dai H, Hart AAM, et al. (2002) A gene-expression signature as a predictor of survival in breast cancer. N Engl J Med 347: 1999–2009.
30. The Cancer Genome Atlas Network (2012) Comprehensive molecular characterization of human colon and rectal cancer. Nature 487: 330–337.
31. Giardiello FM, Brensinger JD, Tersmette AC, Goodman SN, Petersen GM, et al. (2000) Very High Risk of Cancer in Familial Peutz-Jeghers Syndrome. Gastroenterology 119: 1447–1453.
32. Nakau M, Miyoshi H, Seldin MF, Imamura M, Oshima M, et al. (2002) Hepatocellular carcinoma caused by loss of heterozygosity in Lkb1 gene knockout mice. Cancer Res 62: 4549–4553.
33. Thorgeirsson SS (2003) Hunting for tumor suppressor genes in liver cancer. Hepatology 37: 739–741.
34. Shackelford DB, Shaw RJ (2009) The LKB1-AMPK pathway: metabolism and growth control in tumour suppression. Nat Rev Cancer 9: 563–575.
35. Gurumurthy S, Xie SZ, Alagesan B, Kim J, Yusuf RZ, et al. (2010) The Lkb1 metabolic sensor maintains haematopoietic stem cell survival. Nature 468: 659–663.
36. Watanabe M, Houten SM, Mataki C, Christoffolete MA, Kim BW, et al. (2006) Bile acids induce energy expenditure by promoting intracellular thyroid hormone activation. Nature 439: 484–489.
37. Xulvi-Brunet R, Li H (2010) Co-expression networks: graph properties and topological comparisons. Bioinformatics 26: 205–214.
38. Calon A, Espinet E, Palomo-Ponce S, Tauriello DVF, Iglesias M, et al. (2012) Dependency of colorectal cancer on a TGF-β-driven program in stromal cells for metastasis initiation. Cancer Cell 22: 571–584.
39. Su AI, Wiltshire T, Batalov S, Lapp H, Ching K a, et al. (2004) A gene atlas of the mouse and human protein-encoding transcriptomes. Proc Natl Acad Sci U S A 101: 6062–6067.
40. Chen N, Rinner O, Czernik D, Nytko KJ, Zheng D, et al. (2011) The oxygen sensor PHD3 limits glycolysis under hypoxia via direct binding to pyruvate kinase. Cell Res 21: 983–986.
41. Csardi G, Nepusz T (2006) The igraph software package for complex network research. InterJournal, Complex Syst.
42. Hall M, Frank E, Holmes G, Phahringer B, Reuteman P, et al. (2009) The WEKA data mining software: an update. ACM SIGKDD Explor 11.

17

Analysis of Alpha-Synuclein in Malignant Melanoma – Development of a SRM Quantification Assay

Charlotte Welinder[1,2]*, **Göran B. Jönsson**[1], **Christian Ingvar**[3,4], **Lotta Lundgren**[1,3], **Bo Baldetorp**[1], **Håkan Olsson**[1,3,5], **Thomas Breslin**[1], **Melinda Rezeli**[6], **Bo Jansson**[1], **Thomas E. Fehniger**[1,2], **Thomas Laurell**[2,6], **Elisabet Wieslander**[1], **Krzysztof Pawlowski**[1,7], **György Marko-Varga**[2,6,8]

1 Division of Oncology and Pathology, Clinical Sciences, Lund University, Lund, Sweden, **2** Centre of Excellence in Biological and Medical Mass Spectrometry, Lund University, Lund, Sweden, **3** Skåne University Hospital, Lund, Sweden, **4** Dept. of Surgery, Clinical Sciences, Lund University, Skåne University Hospital, Lund, Sweden, **5** Dept. of Cancer Epidemiology, Clinical Sciences, Lund University, Lund, Sweden, **6** Clinical Protein Science & Imaging, Biomedical Center, Biomedical Engineering, Lund University, Lund, Sweden, **7** Dept. of Experimental Design and Bioinformatics, Faculty of Agriculture and Biology, Warsaw University of Life Sciences, Warszawa, Poland, **8** First Department of Surgery, Tokyo Medical University, Tokyo, Japan

Abstract

Globally, malignant melanoma shows a steady increase in the incidence among cancer diseases. Malignant melanoma represents a cancer type where currently no biomarker or diagnostics is available to identify disease stage, progression of disease or personalized medicine treatment. The aim of this study was to assess the tissue expression of alpha-synuclein, a protein implicated in several disease processes, in metastatic tissues from malignant melanoma patients. A targeted Selected Reaction Monitoring (SRM) assay was developed and utilized together with stable isotope labeling for the relative quantification of two target peptides of alpha-synuclein. Analysis of alpha-synuclein protein was then performed in ten metastatic tissue samples from the Lund Melanoma Biobank. The calibration curve using peak area ratio (heavy/light) versus concentration ratios showed linear regression over three orders of magnitude, for both of the selected target peptide sequences. In support of the measurements of specific protein expression levels, we also observed significant correlation between the protein and mRNA levels of alpha-synuclein in these tissues. Investigating levels of tissue alpha-synuclein may add novel aspect to biomarker development in melanoma, help to understand disease mechanisms and ultimately contribute to discriminate melanoma patients with different prognosis.

Editor: Benjamin Edward Rich, Cellcuity, United States of America

Funding: This work was supported by The Kamprad Foundation (http://familjenkampradsstiftelse.se), (to CW GBJ CI LL BB HO TB BJ TL EW KP GMV). The funder had no role in study design, data collection and analysis, decision to publish, or preparation of the manuscript.

Competing Interests: The authors have declared that no competing interests exist.

* Email: charlotte.welinder@med.lu.se

Introduction

The latest epidemiological statistics position malignant melanoma (MM) as the third and most deadly type of skin cancer, while basal cell carcinoma is by far the most common type of skin cancer. Malignant melanoma develops in melanocytes, i.e., the pigment producing cells in the skin. Although MM accounts for only 4% of all skin cancers, it is more aggressive than the other types of skin cancer, and accounts for 80% of the mortality related to skin cancer [1].

Alpha-synuclein, encoded by the SNCA gene, is a protein with a yet unknown but complex mechanism of action in diseases. Alpha-synuclein is a synuclein protein with multiple functions, such as being involved in mitochondrial dysfunction, nuclear localization, vesicle trafficking etc [2]. Insights on altered mitochondrial function and dynamics in the pathogenesis of neurodegeneration [2–4] may help understand the role of alpha-synuclein in Parkinson's disease (PD). Alpha-synuclein is primarily found in neural tissue making up as much as 1% of all proteins in the cytosol, but also in melanoma and nevus tissues [5]. Mutation in

the alpha-synuclein gene (SNCA) as well as misfolding, and accumulation of the protein have been implicated in the development of PD [4,6]. Recently, Matsuo et al have shown that determination of alpha-synuclein protein expression could be useful also for the diagnosis of metastatic melanoma, although it cannot be used to distinguish between malignant and benign melanocytic skin lesions since melanosomes express alpha-synuclein [7]. In melanocytic cells, the protein expression of alpha-synuclein may be regulated by microphthalmia-associated transcription factor (MITF) [8]. MITF is a master regulator gene of melanocyte development and differentiation and is also associated with melanoma development and progression [9,10]. Lately, there has been growing evidence in the literature for mutual mechanisms between cancer and CNS disorders [11], and especially on shared risk and overlapping disease mechanisms in the development of PD and MM [12–15]. These findings suggested a link between MM and PD. Within the disease pathology of PD, alpha-synuclein is involved in a major pathway for protein aggregation. The monomeric protein form is natively unfolded, but will bind to membranes in an α-helical form. From this, unfolded monomers will aggregate first into small oligomeric species that can be stabilized by β-sheet interactions, and then into higher molecular weight insoluble fibrils [16]. Interaction with lipids is one of the ways aggregation occurs and promotes oligomer formation. The deposition of alpha-synuclein into pathological structures such as Lewy bodies is probably a late event that occurs and causes toxicity in neurons and neuronal cell death. Current hypotheses focus towards alpha-synuclein oligomers being the more toxic species [4,17,18].

The causal link between melanoma and PD may center on tyrosine metabolism [19]. Alpha-synuclein has been shown to negatively regulate the activity of tyrosine hydroxylase [20,21], the rate-limiting enzyme in the production of dopamine and melanin [22]. Additional data provide supporting evidence for the existence of a common, or at least related, pathogenic disease mechanism between MM and PD [13,23]. The interaction between alpha-synuclein and tyrosinase may occur more frequently in patients with PD who have shortage in dopamine levels, and the fibrillar forms of alpha-synuclein within PD disease may be responsible for impairments within the tyrosine pathway involved in melanogenesis, predisposing the individual to melanoma [19]. Notably however, there seem to be no direct positive correlation between melanin and alpha-synuclein expression in melanoma tissue cells. Thus, more than fifty percent of the cells highly expressing alpha-synuclein in melanoma were lacking melanin pigments [7].

Quantification of alpha-synuclein in human cerebrospinal fluid (CSF) has been suggested to serve as a biomarker candidate for PD [24]. However, recent works from several groups trying to quantify alpha-synuclein is inconsistent with the reported absolute concentrations of alpha-synuclein. Some studies found reduced concentration in CSF in PD relative to controls [25–29] and another study reported no change [30].

There are at least four isoforms of alpha-synuclein, which are produced through alternative splicing. The major form of the protein is the full length form with a 140 amino acids long transcript. The other isoforms are alpha-synuclein-126, where the exon 3 is lost and the protein lacks residues 41–54. The alpha-synuclein-112 lacks residues 103–130 due to loss of exon 5 [31]. Finally, alpha-synuclein-98 lacks exon 3 and 5. Alpha-synuclein structure also goes through post-translational modifications in a number of annotations, as well as alternative splicing as aggregation enhancers [31,32]. In addition, truncated forms as well as peptide cleavage products of alpha-synuclein has been shown to have chemotactic functions, in addition to a number of complexes occurring with other target proteins [33–38]. These highly complex and potentially disease driven alterations of alpha-synuclein makes it an enormous challenge trying to capture the full profile related to MM and to elucidate the various peptide functions. This includes what detailed mechanisms drug molecule(s) needs to be directed towards in order to reach efficacy, using conventional immunoassay technologies.

By using novel mass spectrometry technology a SRM assay was developed and validated in order to capture the interplay of alpha-synuclein peptides within MM disease mechanisms, thus being able to quantify the entire cascade of variants by nano LC separation interfaced to tandem mass spectrometry.

The expression level of alpha-synuclein was evaluated within metastatic tissue samples from patients diagnosed with stage III melanoma. The mRNA levels of the gene SNCA was analyzed by microarray in the same metastatic tissue samples.

Materials and Methods

2.1. Clinical Samples

Ten lymph node metastasis samples (Stage III) from MM cancer patients, archived in the local malignant melanoma biobank were obtained from Skåne University Hospital, Sweden. The clinical information on respective patients is summarized in Table 1. Ethical approval was granted by Central Ethical Review board at Lund University; approval number: DNR 191/2007, 101/2013. All patients within the study provided a written informed consent. The malignant melanoma biobank is located at Barngatan 2B, 221 85 Lund, Sweden. The biobank is called "Tissuebank for research on tumor diseases" (BD20).

2.2. Sample Preparation

Proteins and mRNAs were extracted from frozen melanoma tumor tissue. Tissues were carefully dissected during surgery and subdivided into 5–8 mm^2 fragments and placed into cryo-tubes. Tissue (15–20 mg), were processed by TissueLyser (Qiagen, Hilden, Germany) and AllPrep DNA/RNA/Protein Mini Kit (Qiagen Ltd, Crawley; UK), according to the manufacturer's-instructions. Extracted proteins were precipitated with ice-cold acetone to a final concentration of 80% acetone and incubated for 30 min at −20 C followed by centrifugation at 16000 g for 2 minutes. The supernatant was removed, and the protein pellets were allowed to air dry. The dried protein pellets were resolved in 8 M urea in 50 mM ammonium bicarbonate (pH 7.6). Protein concentration was determined by the BCA method (Pierce, Rockford, IL, USA). From the total protein fraction, 150 μg, was reduced with 10 mM DDT (1 h at 37 C) and alkylated using 40 mM iodoacetamide (30 min, kept dark at room temperature). Buffer were exchanged to 50 mM ammonium bicarbonate buffer (pH 7.6) using a 10 kDa cut-off spin filter (YM10 filter, AMICON). The samples were subsequently digested with sequencing grade trypsin (Promega, Madison, WI, USA) overnight at 37°C with a of ratio 1:120 w/w (trypsin:protein).

2.3. mRNA Analysis

SNCA mRNA levels for the ten melanoma tumors analyzed by the SRM assays were extracted from a whole genome gene expression assay (Illumina CA) on HT-12 v4 arrays (Cirenajwis et al. in preparation).

2.4 Haemoglobin Analysis – Western Blot

The ten tumor lysates, 20 μg, was diluted in NuPAGE LDS sample buffer (Invitrogen, California, US) with 50 mmol/L dithiotreitol (DTT) and incubated at 95°C for 10 min. Thirty μg of the protein was separated using 4–12% NuPAGE, Bis-Tris, 1 mm thick gels with 15 wells (Invitrogen, California, US) with SeeBlue Plus2 (Invitrogen, California, US) as molecular mass standard. Whole blood from a healthy volunteer (female) was used as a positive control for the rabbit anti-human haemoglobin (A0118, Dako). The electrophoresis was run in MES buffer at 180 V for 1 h and the proteins were then transferred to 0.2 μm PVDF membrane (Trans-Blot Turbo Transfer Pack, Mini format, Bio-Rad) at 25V for 30 min using Trans-Blot Turbo Transfer System (Bio-Rad). The membrane was blocked in 5% non-fat dry milk in 0.2% Tween-20, 150 mM NaCl and 20 mM Tris, pH 7.5 (TTBS) for 3 hours and incubated with rabbit anti-human haemoglobin (10 μg/mL) over night. The membrane was washed three times in TTBS, 10 min each. The membrane was incubated with FITC conjugated polyclonal swine anti-rabbit Immunoglobulins diluted 1:50, for 1 hour, washed three times, 10 min each, with TTBS and antibody binding was detected with Gel Doc EZ

Table 1. Clinical information of patient characteristics. Breslow thickness and Clarks refer to primary melanoma feature.

Tumor	Gender	Age at metastases	Age at primary	Breslow class	Clark	Stage	Status
MM35	Male	55	54	3	4	3	Alive
MM98	Male	75	73	4	4	3	Dead
MM504	Male	54	NA	NA	NA	NA	Dead
MM687	Male	74	72	1	2	3	Dead
MM787	Male	81	78	2	4	3	Dead
MM812	Male	51	NA	NA	NA	NA	Alive
MM813	Female	54	54	2	3	3	Alive
MM825	Female	66	64	2	4	3	Alive
MM829	Male	55	49	1	2	3	Alive
MM835	Female	36	32	3	3	3	Alive

NA-not available.

Imager (Bio-Rad). The staining density for each band was analysed with Image Lab Software (Bio-Rad).

2.5. In Silico Selection of Signature Peptides

The theoretical digestion of the neXtProt entry NX_P37840 was performed by the PeptideMass tool (available at the ExPASy Proteomics Server website, http://expasy.org/sprot/[39] using the following settings: iodoacetamide as alkylation agent with oxidation on methionine and no miss-cleavage. The resulted tryptic peptides were investigated for uniqueness by using Basic Local Alignment Search Tool (BLAST) [40]. Finally, two tryptic peptides were chosen for crude peptide synthesis with and without heavy isotope labeling.

2.6. SRM Assay Development

Crude peptides, both light and heavy, were supplied by Thermo Scientific (Ulm, Germany). The heavy peptides were isotopically labeled on the C-terminal-lysine residue (^{13}C, ^{15}N). A mixture was created for the two peptides (EQVTNVGGAVVTGVTAVAQK and TVEGAGSIAAATGFVK), corresponding to residues 61–80 and 81–96 of alpha-synuclein, respectively. Based on the total peptide content of the crude peptides, each peptide was diluted to an estimated concentration of 50 fmol/μL. The transition lists were created in Skyline v1.2 software [41] (MacCoss Lab Software, Seattle, WA). Primarily, high numbers of transitions, all possible y-ion series that matches the criteria (from $m/z >$ precursor-2 to last ion-2, precursor m/z exclusion window: 20 Th), were selected for each peptide at both 2+ and 3+ charge states. The peptide mixture was analyzed by nano LC-MS/MS using a TSQ Vantage triple quadrupole mass spectrometer equipped with an Easy n-LC II pump (Thermo Scientific, Waltham, MA). The samples were injected onto an Easy C18-A1 pre-column (2 cm, ID 100 μm with 5 μm particles) (Thermo Scientific, Waltham, MA), and following on-line desalting and concentration the tryptic peptides were separated on a 75 μm ×150 mm fused silica column packed with ReproSil C18 (3 μm, 120 Å from Dr. Maisch GmbH, Germany). Separations were performed in a 45-min linear gradient from 10 to 35% acetonitrile containing 0.1% formic acid; at a flow rate 300 nL/min. The MS analysis was conducted in positive ion mode with the spray voltage and declustering potential were set to 1750 V and 0, respectively. The transfer capillary temperature was set to 270°C and the amplitude of the S-lens was 143. SRM transitions were acquired in Q1 and Q3 operated at unit resolution (0.7 FWHM), the collision gas pressure in Q2 was set to 1.2 mTorr. The cycle time was 2.5 s and the dwell times were 0.11 and 0.10 for TVEGAGSIAAATGFVK and EQVTNVGGAVVTGVTAVAQK, respectively. The three best transitions per precursor were selected by manual inspection of the data in Skyline and scheduled transition lists were created for the final assays. Collision energies were optimized for each peptide. The collision energy was ramped round the predicted value in 3 steps on both sides, in 2V increments. The optimized collision energies were 21 and 29 for TVEGAGSIAAATGFVK and EQVTNVGGAVVTGVTAVAQK, respectively. The selected transitions were tested in real matrix also by spiking the heavy peptide mixtures into human MM tissue digests.

2.7. Standard Curves of Synthetic Peptides

A reverse standard curve approach was used for the calibration curve, thereby providing relative quantitative data to the melanoma patient samples [42]. A dilution series of heavy labeled synthetic peptide mixtures (in the estimated concentration range of 6.25 −200 fmol/μl) in tumor digest containing constant amount of non-labeled synthetic peptides (in the estimated concentration of 100 fmol/µl) for characterization of the assay linearity were analyzed in triplicates. Calibration curves were generated by linear regression analysis on the peak areas ratios (heavy/light) versus concentration ratios for the targeted peptides.

2.8. SRM Assay of Alpha-Synuclein

For relative quantification, the two heavy labeled peptides were spiked into ten tumor lysate digests at an estimated concentration of 6.25 fmol/µL and 50 fmol/µL for TVEGAGSIAAATGFVK and EQVTNVGGAVVTGVTAVAQK, respectively.

2.9. Data Analysis

Data sets were imported into Skyline (v1.2 http://proteome.gs. washington.edu/software/skyline) and peaks were automatically integrated. After automatic integration of the data sets, the data was also manually inspected. Integration of the peaks was adjusted when signals were not intense and the software could not reliably determine the peak. Interferences with the matrix, detector saturations and variable peak area ratios in replicate samples were also investigated. Data from the individual tumor lysates are presented as mean of triplicates measurements +/− standard deviation. Because originally mRNA expression index was reported after log2 transformation, for correlating mRNA and protein levels the mRNA expression index was transformed by raising 2 to the power equal to the expression index. Pearson correlation between mRNA and proteins levels was then calculated.

Results

In order to utilize genes and proteins as biomarkers for disease and/or treatments as for drug responders, more experience and Research & Development inputs are requested. For diagnostic quantification of protein(s) in clinical studies, there are currently no standard guidelines from the Food and Drug Administration (FDA) or European Medicines Agency (EMEA) that need to be met for approval. However, there are on-going projects concerning such biomarkers development between the FDA, academics projects and pharmaceutical industry that are investigating standardization procedures for future utilization. A guideline for industry has recently been introduced (http://www.fda.gov/downloads/RegulatoryInformation/Guidances/ucm126957.pdf). The initial step of a SRM assay development usually relates to apply an *in silico* step, where a selection of suitable peptides from the proteins are made, followed by BLAST searching in protein database, where identified peptides from the proteins can verify the utility of target peptides identified as candidates. Precursors and m/z of the peptide products are the keys to the assay development. Recently, a rapid assay development with SRM-MS instrumentations was presented [43]. In this respect, peptides libraries used as standards, are the most valuable tools in order to funnel the large number of peptide candidates in the in silico processing step that make judgments of the most useful target peptide candidates for the assay [44,45].

3.1. Selection of Transitions for SRM

By selection of the consensus sequence of alpha-synuclein (neXtProt entry NX_P37840) a theoretical tryptic peptide list was generated. Two sequences were identified; TVEGAG-SIAAATGFVK and EQVTNVGGAVVTGVTAVAQK. These peptides were chosen in order to encompass an assay that will provide quantitative data on all three alpha-synuclein isoforms. The peptides were evaluated for their uniqueness using BLAST [40]. The selected peptides were synthesized in both heavy labeled

and unlabeled forms in order to determine optimal SRM transitions detected by mass spectrometry. The three highest intensity fragment ions were selected for SRM transitions for each peptide (Table 2). Spiking heavy labeled peptides into the tumor lysate digest made experimental verification of the selected transitions. The relative signal intensities of the daughter ions generated from endogenous peptides were compared to those of the heavy labeled peptides. These experiments verified that the pattern of the daughter ions from the endogenous, and the heavy labeled peptides were the same. This experimental proof is shown in Figure 1, which ensures that the selected transitions are free from interference from matrix.

3.2. Quantification and Analytical Assay Performance

Clinical guidelines for creating MS-based assays require that heavy labeled synthetic peptides be added to samples at concentrations close to the mean concentration of the endogenous peptides to permit reproducible measurements [46]. The linearity and reproducibility of the SRM assay was investigated in 6 dilution steps using nano-LC separation. The tissue samples were spiked with different amount of heavy labeled synthetic peptides while keeping amounts of non-labeled synthetic peptides constant. The nano-LC provides improved sensitivity with an acceptable robustness, typically providing a relative standard deviation (RSD) <1% for the retention times.

The chromatographic separation conditions with the Internal Standards (IS), were chosen so that sufficient separation was achieved by the two IS (0.7 min peak-to-peak separation), and at the same time to have a fast turnaround in the cycle time (45 min).

Each analysis was repeated three times and the peak ratio of the internal standard (IS) peaks were plotted against their estimated concentrations (Figure 2). Corresponding calibration curves were generated by linear regression analysis of the peak areas (heavy/light) for both targeted peptides. The linear regression was found to be 0.99 (R^2-values) within the investigated concentration range (6.25–200 fmol/µl).

The LOQ have been found to be 75 attomole, defined by 10 times the RSD of the noise level in the assay, and the LOD was 23 attomole, defined as three times of the noise level.

By repeated series of analysis using this SRM assay variations (RSD) were found to be 9.3% and 7.9% for TVEGAG-SIAAATGFVK and EQVTNVGGAVVTGVTAVAQK, respectively.

In order to characterize the chromatographic separation of these alpha-synuclein peptides, their retention times were monitored (n = 33) with RSD values <1% for both peptides, TVEGAGSIAAATGFVK (Retention time = 19.8 min) and EQVTNVGGAVVTGVTAVAQK (Retention time = 20.5 min).

3.3. Haemoglobin Analysis in Tissue Samples

To understand the implication of red blood cells containing alpha-synuclein which may confound the SRM assay, we conducted haemoglobin analysis in the tumor lysates by Western blot. In most of the tumor lysates, low levels of haemoglobin were detected. Lysate from tumor MM825 seemed to contain higher levels indicating some hemolysis in that sample. Importantly, no correlation could be established between estimated levels of haemoglobin (used as a surrogate marker for red blood cells, Figure 3) and alpha-synuclein in any of the tissue samples (Figure 4), suggesting that the quantitation of alpha-synuclein in the tumor tissues by the SRM assay was not influenced by contaminating red blood cells.

TVEGAGSIAAATGFV**K**

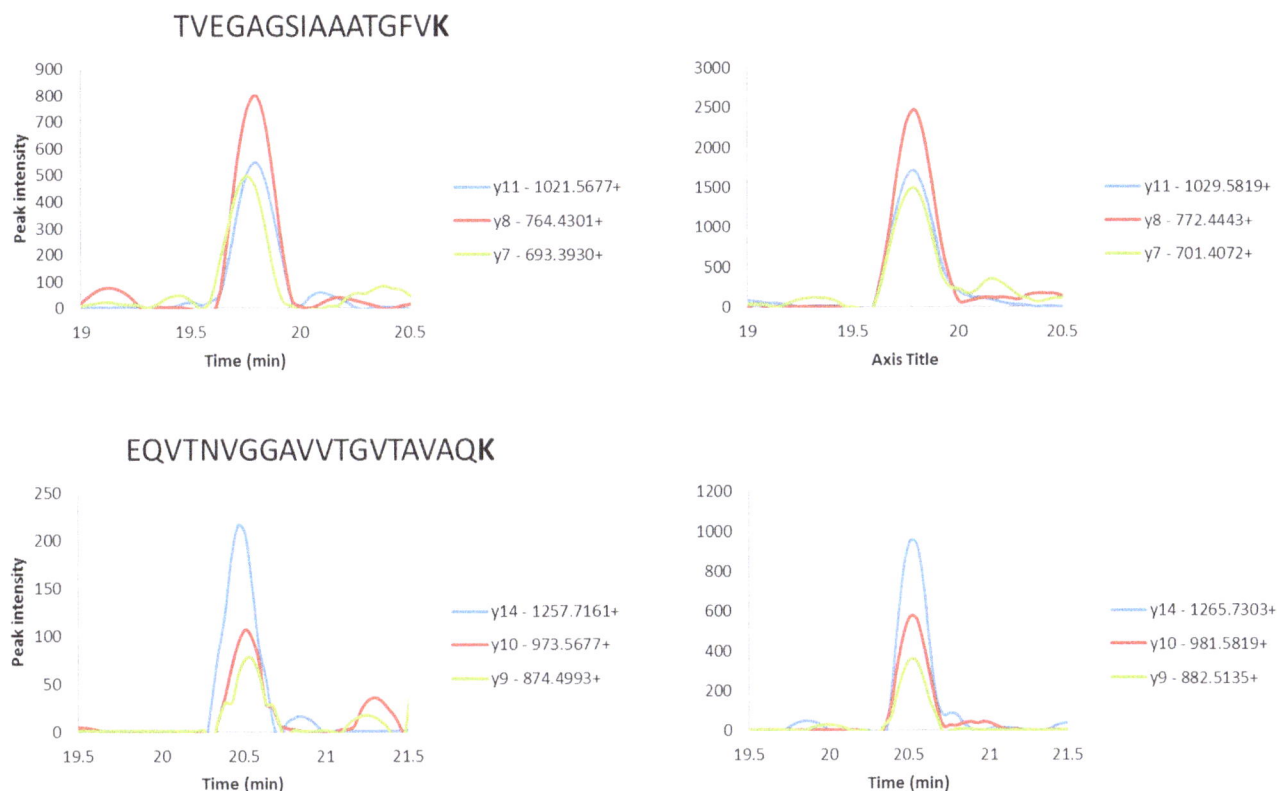

Legend (upper left):
- y11 - 1021.5677+
- y8 - 764.4301+
- y7 - 693.3930+

Legend (upper right):
- y11 - 1029.5819+
- y8 - 772.4443+
- y7 - 701.4072+

EQVTNVGGAVVTGVTAVAQ**K**

Legend (lower left):
- y14 - 1257.7161+
- y10 - 973.5677+
- y9 - 874.4993+

Legend (lower right):
- y14 - 1265.7303+
- y10 - 981.5819+
- y9 - 882.5135+

Figure 1. Extracted ion chromatograms show the 3 monitored transitions for the endogenous (left) and the heavy labeled peptides (right) for the two peptides.

3.4. Quantitative Expression Analysis in Biobank Tissue Samples

The tissue samples from ten MM patients were analyzed in triplicates. Reproducibility of quantification was achieved by spiking heavy isotope labeled peptides into the samples prior to LC-MS/MS analysis; resulting in a final protein amount of 1.25 μg. The amount of heavy labeled peptides was chosen in the linear range for each peptide (6.25 fmol/μL for TVEGAG-SIAAATGFVK and 50 fmol/μL for EQVTNVGGAVVTGV-

TAVAQK). The technical validation of the alpha-synuclein assay provided evidence of low RSD values, typically <3%. The selected alpha-synuclein peptide sequences were also found to perform analytically well, with respect to transition repeatability, and could be used throughout the entire patient samples investigated in this study. RSD values by running melanoma tissues were found to be similar to those reported on earlier [47], and typically; 15–20%. Figure 4 show the results of the SRM assay performed on individual tissue samples.

Table 2. Proteotypic peptides sequences and selected SRM transitions for the two peptides.

Accession no	Protein	Position	Peptide sequence	Q1	Q3
P37840	a- synuclein	81–96	TVEGAGSIAAATGFVK	739.9 (2+)	1021.6 (y11^{1+})
					764.4 (y8^{1+})
					693.4.4 (y7^{1+})
			TVEGAGSIAAATGFVK	743.9(2+)	1029.6 (y11^{1+})
					772.4 (y8^{1+})
					701.4 (y7^{1+})
		61–80	EQVTNVGGAVVTGVTAVAQK	964.5(2+)	1257.7 (y14^{1+})
					973.6 (y10^{1+})
					874.5 (y9^{1+})
			EQVTNVGGAVVTGVTAVAQK	968.5 (2+)	1265.7 (y14^{1+})
					981.6 (y10^{1+})
					882.5 (y9^{1+})

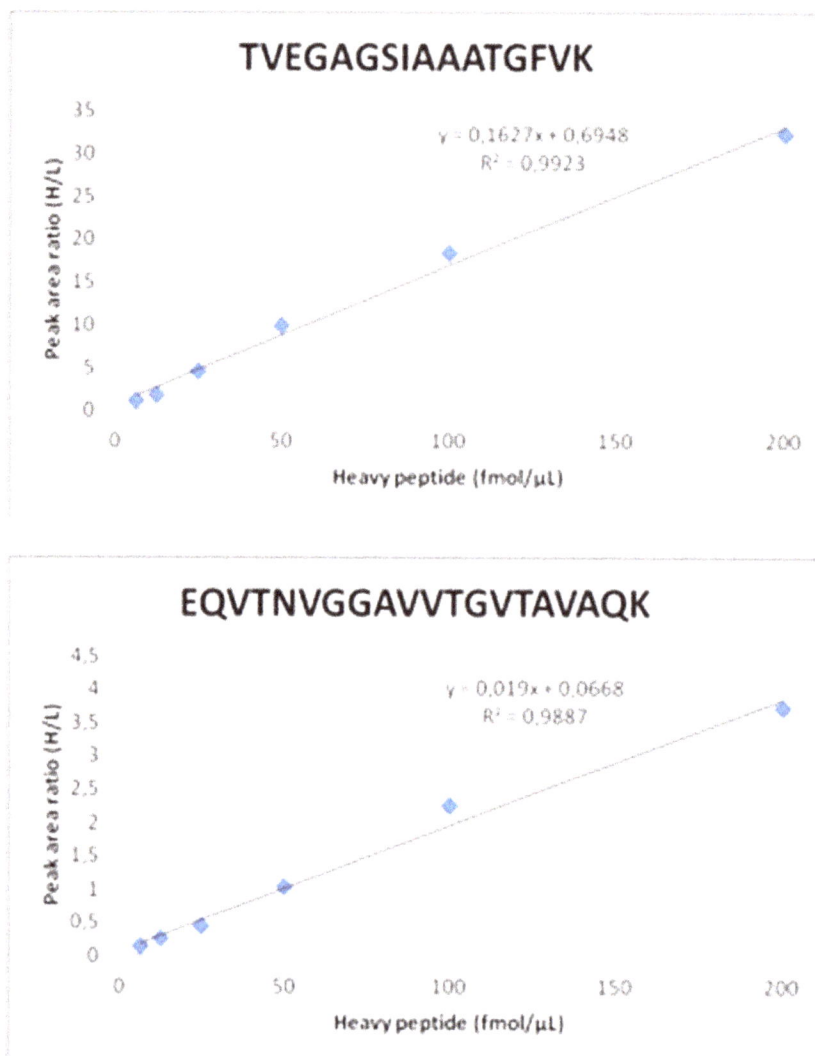

TVEGAGSIAAATGFVK

y = 0.1627x + 0.6948
R² = 0.9923

EQVTNVGGAVVTGVTAVAQK

y = 0.019x + 0.0668
R² = 0.9887

Figure 2. Linearity of the alpha-synuclein SRM assay determined by using heavy labeled peptides spiked into a pooled tissue sample at various estimated concentrations (6.25–200 fmol/μL).

Four tissue samples (MM35, MM504, MM687 and MM787) showed a higher protein expression of alpha-synuclein compared to the other six tumor samples. Comparing to the mRNA expression there was a strong correlation between variation in protein levels and mRNA in the sample group (Figure5). Correlation was indeed high, R = 0.79, and significant (p-value = 0.007).

The present data demonstrates heterogeneity of alpha-synuclein expression within MM patient lymph node metastases.

Conclusion and Future Perspectives

We have developed and applied a SRM assay for quantification of alpha-synuclein in MM tissue lysate using a stable isotope dilution strategy. Alpha-synuclein is expressed predominantly in the brain but also in melanocytic lesions. The relationship between PD and melanoma has been observed in a number of reports, for example a comprehensive study by Gao and colleagues, where an increased risk of PD was associated with a first-degree family history of melanoma (p = 0.004) [13]. Along with recently published studies where a higher prevalence of developing

melanoma in patients with PD was identified, this clearly supports the idea that melanoma and PD share common pathogenic pathways [12,14,48]. The disease relation between PD and MM indicates a possible link between functional effects of CNS disorder and cancer disease mechanisms.

Our current generic SRM assay opens up a new opportunity to quantify any alpha-synuclein form that may occur in MM, including monomer forms, truncated ones, peptides thereof, as well as oligomers, polymers and genetic variants. Specifically, the target peptides chosen in this study will enable simultaneous analysis of the four alpha-synuclein isoforms. Thus, the amount of alpha-synuclein protein detected in this study represents a summary of the four isoforms (140 kDa, 126 kDa, 112 kDa and 98 kDa). Data suggest however that specific quantification of the individual isoforms may have disease relevance and that possibly an imbalance occurs between the levels of the major form (140 kDa) and the minor forms [49] in PD. Such analysis will then require identification of new target peptides. Several general aspects for optimal selection of target peptides need to be taken into consideration [45]. Notably, these peptides should be unique to the protein to be analyzed and easily detected by mass

Western blot - Haemoglobin

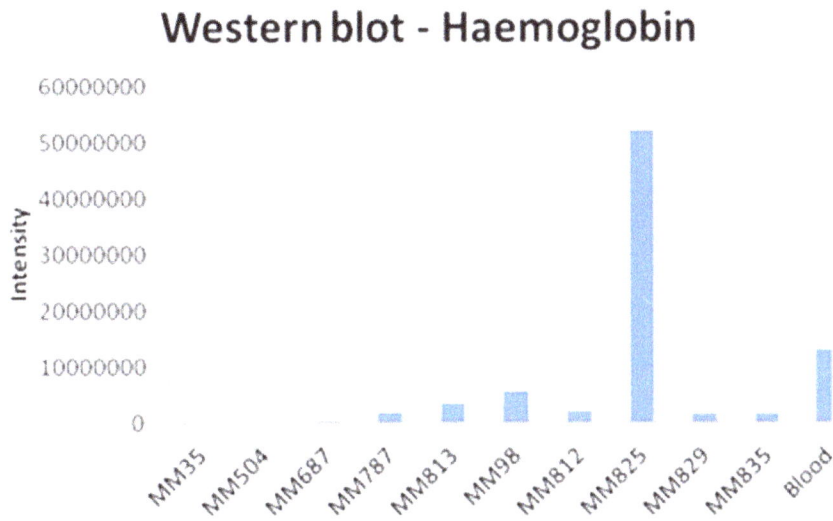

Figure 3. Haemoglobin expression levels in ten individual patient tissue samples and positive control (positive control corresponding to 1 µL human blood).

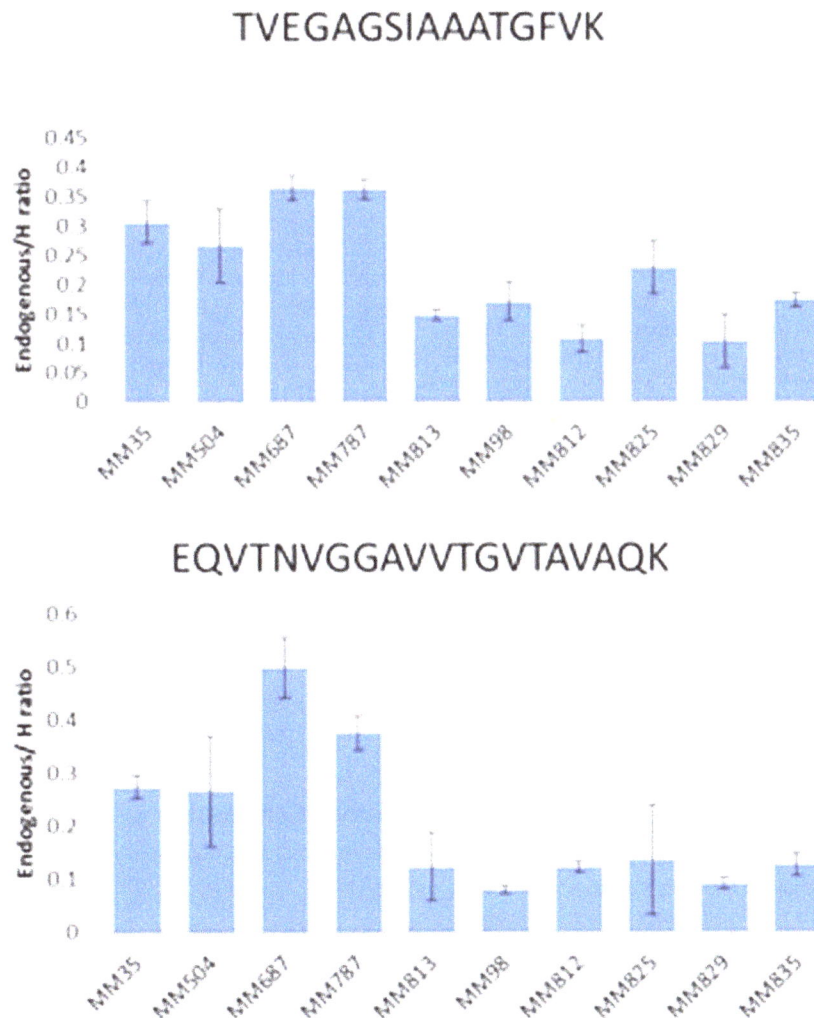

TVEGAGSIAAATGFVK

EQVTNVGGAVVTGVTAVAQK

Figure 4. Alpha-synuclein expression levels of the two selected peptides in ten individual patient tissue samples as determined by TVEGAGSIAAATGFVK and EQVTNVGGAVVTGVTAVAQK, run in triplicates.

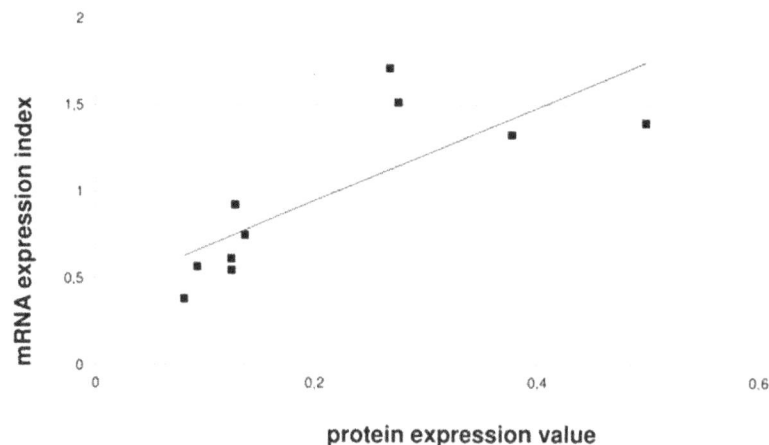

Figure 5. Correlation chart showing mRNA and protein expression for the 10 individual patient tissue samples. The protein expression value was calculated from the EQVTNVGGAVVTGVTAVAQK peptide as the average of total transitions. For mRNAexpression, the expression value was calculated as 2 raised to the power of expression index. Average values for triplicate measurements shown. Units are arbitrary units. Linear trend line is shown.

spectrometry. When studying PTM of alpha-synuclein, adequate standards may be less accessible. However, recently chemical and semisynthetic methods to introduce sites-specific PTMs have been described [24]. Important PTM that can occur is phosphorylation at amino acid residues serine-87, serine-129, tyrosine-125, tyrosine-133 and tyrosine-136. In addition to phosphorylation, alpha-synuclein can be modified by glycosylation of serine-129 [50] and nitration at amino acid residues; tyrosine-125, tyrosine-133 and tyrosine-136 following both oxidative and nitrative stress as well as C-terminal truncation [51]. In vitro and in vivo data suggest several of these to be involved in disease pathology [3]. As a common disease presentation, somatic mutations appear in melanoma patients in both lymph nodes as well as other vital organs. These DNA mutations, resulting in amino acid shifts, have proven to be key genetic factors in cancer diseases, which can efficiently be monitored by the SRM assay format. A future mutation-SRM assay will target every single nucleotide polymorphism (SNP) that may to be considered. For melanoma, learning's can be made from studies in PD were altered aggregation and misfolding of alpha-synuclein has been implicated. In particular mutations A53T, E46K and A60P has been linked to familial PD and may induce altered aggregation properties resulting in neurotoxicity [3]. The SRM assay format enables specific study of such mutations in tissues and body fluids by the use of specific target peptides.

Alpha-synuclein protein can be identified in various cells and compartments within melanoma patients. Vascularization and angiogenesis within melanoma is a vital part of the pathology and disease development. It is well known that in blood, erythrocytes and to minor extent platelets contain alpha-synuclein [27,52,53], why the origin and localization of this target protein is multifold. When developing a biomarker assay it is important to understand confounding factors like blood contamination or hemolysis effects

in order for correct data interpretation. In a previous paper we presented histology images providing tissue characteristics to the tumor samples used in the present study. These images do not suggest a large variation in vascularization between the samples [54]. Furthermore, by investigating haemoglobin levels in the tumor lysates (as surrogate marker for red blood cells) no correlation could be established between haemoglobin expression and alpha-synuclein levels which verifies that the amount of alpha-synuclein detected from the tissues was not influenced by contaminating red blood cells.

The presented data demonstrates strong correlation between protein and mRNA expression levels of alpha-synuclein in metastatic tumor lysate from MM patients. With the disease relation of PD to MM this study enables investigation of the expression levels of this target protein in various forms within different MM phenotypes. The future opportunity to correlate alpha-synuclein tumor tissue levels and specific forms of the protein within different MM patient subtypes warrants further investigations.

Supporting Information

Table S1 SNCA mRNA levels of the ten metastases. Gene expression was analyzed using Illumina arrays (HT12v4) and data normalized using the cubic spine method.

Author Contributions

Conceived and designed the experiments: CW MR GMV. Performed the experiments: CW GBJ. Analyzed the data: TB KP GBJ. Contributed reagents/materials/analysis tools: CI LL BB HO GMV. Contributed to the writing of the manuscript: CW BJ TEF TL EW KP GMV.

References

1. Miller AJ, Mihm MC Jr (2006) Melanoma. N Engl J Med 355: 51–65.
2. McInnes J (2013) Insights on altered mitochondrial function and dynamics in the pathogenesis of neurodegeneration. Transl Neurodegener 2: 12.

3. Rochet JC, Hay BA, Guo M (2012) Molecular insights into Parkinson's disease. Prog Mol Biol Transl Sci 107: 125–188.

4. Waxman EA, Giasson BI (2009) Molecular mechanisms of α-synuclein neurodegeneration. Biochimica et Biophysica Acta (BBA) - Molecular Basis of Disease 1792: 616–624.

5. Iwai A, Masliah E, Yoshimoto M, Ge N, Flanagan L, et al. (1995) The precursor protein of non-A beta component of Alzheimer's disease amyloid is a presynaptic protein of the central nervous system. Neuron 14: 467–475.

6. George S, Rey NL, Reichenbach N, Steiner JA, Brundin P (2013) α-Synuclein: The Long Distance Runner. Brain Pathology 23: 350–357.

7. Matsuo Y, Kamitani T (2010) Parkinson's disease-related protein, alpha-synuclein, in malignant melanoma. PLoS One 5: e10481.

8. Hoek KS, Schlegel NC, Eichhoff OM, Widmer DS, Praetorius C, et al. (2008) Novel MITF targets identified using a two-step DNA microarray strategy. Pigment Cell Melanoma Res 21: 665–676.

9. Yajima I, Kumasaka MY, Thang ND, Goto Y, Takeda K, et al. (2011) Molecular Network Associated with MITF in Skin Melanoma Development and Progression. J Skin Cancer 2011: 730170.

10. Garraway LA, Widlund HR, Rubin MA, Getz G, Berger AJ, et al. (2005) Integrative genomic analyses identify MITF as a lineage survival oncogene amplified in malignant melanoma. Nature 436: 117–122.

11. Tabares-Seisdedos R, Rubenstein JL (2013) Inverse cancer comorbidity: a serendipitous opportunity to gain insight into CNS disorders. Nat Rev Neurosci 14: 293–304.

12. Bertoni JM, Arlette JP, Fernandez HH, Fitzer-Attas C, Frei K, et al. (2010) Increased melanoma risk in Parkinson disease: a prospective clinicopathological study. Arch Neurol 67: 347–352.

13. Gao X, Simon KC, Han J, Schwarzschild MA, Ascherio A (2009) Family history of melanoma and Parkinson disease risk. Neurology 73: 1286–1291.

14. Olsen JH, Friis S, Frederiksen K (2006) Malignant melanoma and other types of cancer preceding Parkinson disease. Epidemiology 17: 582–587.

15. Olsen JH, Friis S, Frederiksen K, McLaughlin JK, Mellemkjaer L, et al. (2005) Atypical cancer pattern in patients with Parkinson's disease. Br J Cancer 92: 201–205.

16. Cookson MR (2009) alpha-Synuclein and neuronal cell death. Mol Neurodegener 4: 9.

17. Kalia LV, Kalia SK, McLean PJ, Lozano AM, Lang AE (2013) α-Synuclein oligomers and clinical implications for Parkinson disease. Annals of Neurology 73: 155–169.

18. McGlinchey RP, Yap TL, Lee JC (2011) The yin and yang of amyloid: insights from [small alpha]-synuclein and repeat domain of Pmel17. Physical Chemistry Chemical Physics 13: 20066–20075.

19. Paisan-Ruiz C, Houlden H (2010) Common pathogenic pathways in melanoma and Parkinson disease. Neurology 75: 1653–1655.

20. Perez RG, Waymire JC, Lin E, Liu JJ, Guo F, et al. (2002) A role for alpha-synuclein in the regulation of dopamine biosynthesis. J Neurosci 22: 3090–3099.

21. Peng X, Tehranian R, Dietrich P, Stefanis L, Perez RG (2005) Alpha-synuclein activation of protein phosphatase 2A reduces tyrosine hydroxylase phosphorylation in dopaminergic cells. J Cell Sci 118: 3523–3530.

22. Slominski A, Tobin DJ, Shibahara S, Wortsman J (2004) Melanin pigmentation in mammalian skin and its hormonal regulation. Physiol Rev 84: 1155–1228.

23. Olsen JH, Tangerud K, Wermuth L, Frederiksen K, Friis S (2007) Treatment with levodopa and risk for malignant melanoma. Mov Disord 22: 1252–1257.

24. Schmid AW, Fauvet B, Moniatte M, Lashuel HA (2013) Alpha-synuclein post-translational modifications as potential biomarkers for Parkinson's disease and other synucleinopathies. Mol Cell Proteomics.

25. Mollenhauer B, Cullen V, Kahn I, Krastins B, Outeiro TF, et al. (2008) Direct quantification of CSF alpha-synuclein by ELISA and first cross-sectional study in patients with neurodegeneration. Exp Neurol 213: 315–325.

26. Mollenhauer B, Locascio JJ, Schulz-Schaeffer W, Sixel-Doring F, Trenkwalder C, et al. (2011) alpha-Synuclein and tau concentrations in cerebrospinal fluid of patients presenting with parkinsonism: a cohort study. Lancet Neurol 10: 230–240.

27. Hong Z, Shi M, Chung KA, Quinn JF, Peskind ER, et al. (2010) DJ-1 and alpha-synuclein in human cerebrospinal fluid as biomarkers of Parkinson's disease. Brain 133: 713–726.

28. Tokuda T, Salem SA, Allsop D, Mizuno T, Nakagawa M, et al. (2006) Decreased alpha-synuclein in cerebrospinal fluid of aged individuals and subjects with Parkinson's disease. Biochem Biophys Res Commun 349: 162–166.

29. Tokuda T, Qureshi MM, Ardah MT, Varghese S, Shehab SA, et al. (2010) Detection of elevated levels of alpha-synuclein oligomers in CSF from patients with Parkinson disease. Neurology 75: 1766–1772.

30. Ohrfelt A, Grognet P, Andreasen N, Wallin A, Vanmechelen E, et al. (2009) Cerebrospinal fluid alpha-synuclein in neurodegenerative disorders-a marker of synapse loss? Neurosci Lett 450: 332–335.

31. Beyer K (2006) Alpha-synuclein structure, posttranslational modification and alternative splicing as aggregation enhancers. Acta Neuropathol 112: 237–251.

32. Schmid AW, Fauvet B, Moniatte M, Lashuel HA (2013) Alpha-synuclein post-translational modifications as potential biomarkers for Parkinson's disease and other synucleinopathies. Molecular & Cellular Proteomics.

33. Bonini NM, Giasson BI (2005) Snaring the function of alpha-synuclein. Cell 123: 359–361.

34. Chandra S, Gallardo G, Fernandez-Chacon R, Schluter OM, Sudhof TC (2005) Alpha-synuclein cooperates with CSPalpha in preventing neurodegeneration. Cell 123: 383–396.

35. Burre J, Sharma M, Tsetsenis T, Buchman V, Etherton MR, et al. (2010) Alpha-synuclein promotes SNARE-complex assembly in vivo and in vitro. Science 329: 1663–1667.

36. Cooper AA, Gitler AD, Cashikar A, Haynes CM, Hill KJ, et al. (2006) Alpha-synuclein blocks ER-Golgi traffic and Rab1 rescues neuron loss in Parkinson's models. Science 313: 324–328.

37. Alim MA, Hossain MS, Arima K, Takeda K, Izumiyama Y, et al. (2002) Tubulin seeds alpha-synuclein fibril formation. J Biol Chem 277: 2112–2117.

38. Alim MA, Ma QL, Takeda K, Aizawa T, Matsubara M, et al. (2004) Demonstration of a role for alpha-synuclein as a functional microtubule-associated protein. J Alzheimers Dis 6: 435–442; discussion 443–439.

39. Artimo P, Jonnalagedda M, Arnold K, Baratin D, Csardi G, et al. (2012) ExPASy: SIB bioinformatics resource portal. Nucleic Acids Res 40: W597–603.

40. Altschul SF, Gish W, Miller W, Myers EW, Lipman DJ (1990) Basic local alignment search tool. Journal of Molecular Biology 215.

41. MacLean B, Tomazela DM, Shulman N, Chambers M, Finney GL, et al. (2010) Skyline: an open source document editor for creating and analyzing targeted proteomics experiments. Bioinformatics 26: 966–968.

42. Campbell J, Rezai T, Prakash A, Krastins B, Dayon L, et al. (2011) Evaluation of absolute peptide quantitation strategies using selected reaction monitoring. Proteomics 11: 1148–1152.

43. Picotti P, Rinner O, Stallmach R, Dautel F, Farrah T, et al. (2010) High-throughput generation of selected reaction-monitoring assays for proteins and proteomes. Nat Methods 7: 43–46.

44. Deutsch EW, Lam H, Aebersold R (2008) PeptideAtlas: a resource for target selection for emerging targeted proteomics workflows. EMBO Rep 9: 429–434.

45. Picotti P, Aebersold R (2012) Selected reaction monitoring-based proteomics: workflows, potential, pitfalls and future directions. Nat Methods 9: 555–566.

46. Chace DH, Barr JR, Duncan MW, Matern D, Morris MR, et al. (2006) Mass Spectrometry in the Clinical Laboratory: General Principles and Guidance; Approved Guideline. Clinical and Laboratory Standards Institute, Wayne, PA.

47. Kitchen S, Jennings I, Woods T, Kitchen D, Walker I, et al. (2006) Laboratory tests for measurement of von Willebrand factor show poor agreement among different centers: results from the United Kingdom National External Quality Assessment Scheme for Blood Coagulation. Semin Thromb Hemost 32: 492–498.

48. Wirdefeldt K, Weibull CE, Chen H, Kamel F, Lundholm C, et al. (2013) Parkinson's Disease and Cancer: A Register-based Family Study. American Journal of Epidemiology.

49. McLean JR, Hallett PJ, Cooper O, Stanley M, Isacson O (2012) Transcript expression levels of full-length alpha-synuclein and its three alternatively spliced variants in Parkinson's disease brain regions and in a transgenic mouse model of alpha-synuclein overexpression. Mol Cell Neurosci 49: 230–239.

50. McLean PJ, Hyman BT (2002) An alternatively spliced form of rodent alpha-synuclein forms intracellular inclusions in vitro: role of the carboxy-terminus in alpha-synuclein aggregation. Neurosci Lett 323: 219–223.

51. Takahashi T, Yamashita H, Nakamura T, Nagano Y, Nakamura S (2002) Tyrosine 125 of alpha-synuclein plays a critical role for dimerization following nitrative stress. Brain Res 938: 73–80.

52. Barbour R, Kling K, Anderson JP, Banducci K, Cole T, et al. (2008) Red blood cells are the major source of alpha-synuclein in blood. Neurodegener Dis 5: 55–59.

53. Kasuga K, Nishizawa M, Ikeuchi T (2012) alpha-Synuclein as CSF and Blood Biomarker of Dementia with Lewy Bodies. Int J Alzheimers Dis 2012: 437025.

54. Welinder C, Jonsson G, Ingvar C, Lundgren L, Baldetorp B, et al. (2014) Feasibility study on measuring selected proteins in malignant melanoma tissue by SRM quantification. J Proteome Res 13: 1315–1326.

18

Identification of Reference Proteins for Western Blot Analyses in Mouse Model Systems of 2,3,7,8-Tetrachlorodibenzo-P-Dioxin (TCDD) Toxicity

Stephenie D. Prokopec[1], John D. Watson[1], Raimo Pohjanvirta[2,3], Paul C. Boutros[1,4,5]*

1 Informatics and Bio-computing Program, Ontario Institute for Cancer Research, Toronto, Ontario, Canada, 2 Laboratory of Toxicology, National Institute for Health and Welfare, Kuopio, Finland, 3 Department of Food Hygiene and Environmental Health, University of Helsinki, Helsinki, Finland, 4 Department of Medical Biophysics, University of Toronto, Toronto, Ontario, Canada, 5 Department of Pharmacology & Toxicology, University of Toronto, Toronto, Ontario, Canada

Abstract

Western blotting is a well-established, inexpensive and accurate way of measuring protein content. Because of technical variation between wells, normalization is required for valid interpretation of results across multiple samples. Typically this involves the use of one or more endogenous controls to adjust the measured levels of experimental molecules. Although some endogenous controls are widely used, validation is required for each experimental system. This is critical when studying transcriptional-modulators, such as toxicants like 2,3,7,8-tetrachlorodibenzo-p-dioxin (TCDD).To address this issue, we examined hepatic tissue from 192 mice representing 47 unique combinations of strain, sex, Ahr-genotype, TCDD dose and treatment time. We examined 7 candidate reference proteins in each animal and assessed consistency of protein abundance through: 1) TCDD-induced fold-difference in protein content from basal levels, 2) inter- and intra- animal stability, and 3) the ability of each candidate to reduce instability of the other candidates. Univariate analyses identified HPRT as the most stable protein. Multivariate analysis indicated that stability generally increased with the number of proteins used, but gains from using >3 proteins were small. Lastly, by comparing these new data to our previous studies of mRNA controls on the same animals, we were able to show that the ideal mRNA and protein control-genes are distinct, and use of only 2–3 proteins provides strong stability, unlike in mRNA studies in the same cohort, where larger control-gene batteries were needed.

Editor: Xuejiang Guo, Nanjing Medical University, China

Funding: This study was conducted with the support of the Academy of Finland (grant nos. 123345 and 261232 to RP), the Canadian Institutes of Health Research (grant no. MOP-57903 to ABO and PCB), and the Ontario Institute for Cancer Research to PCB through funding provided by the Government of Ontario. Dr. Boutros was supported by a Terry Fox Research Institute New Investigator Award and a CIHR New Investigator Salary Award. The above funding sources had no involvement in the study design, in the collection, analysis and interpretation of data, in the writing of the document, or in the decision to submit the work for publication.

Competing Interests: The authors have declared that no competing interests exist.

* Email: Paul.Boutros@oicr.on.ca

Introduction

2,3,7,8-tetrachlorodibenzo-*p*-dioxin (TCDD) is a member of a class of environmental contaminants, known as dioxins, and is primarily produced through industrial processes including incineration and manufacture of herbicides and pesticides [1,2] as well as electronics recycling [3]. Exposure to TCDD evokes a wide range of toxicities in laboratory animals, including wasting syndrome and death [4]. In humans, short-term exposure to high levels of TCDD often presents as liver damage and chloracne, while low-dose long-term exposure has been linked to immune deficiency [5], diabetes [6], and various cancer types [2,7].

TCDD is an exogenous ligand for the aryl hydrocarbon receptor (AHR) [8]. Upon cell entry, TCDD binds cytoplasmic AHR, leading to the formation of a ligand-receptor complex which translocates into the nucleus, dimerizes with the AHR nuclear translocator (ARNT) and binds to DNA to regulate transcription of target genes [9]. Previous studies have shown that TCDD exposure results in the dysregulation of hundreds of genes in numerous models [10,11,12,13,14]. While specific

changes to the transcriptome resulting from TCDD-mediated regulation have been identified across a wide range of experimental models, downstream effects on the proteome which may prove causative of toxicities, remain unclear. Complete examination of various –omics data will be required to identify the specific molecules responsible for the severe toxic effects induced by TCDD.

Animal models have been, and will continue to be, crucial to understanding the mechanisms described above [15]. In particular, the varying sensitivities to TCDD of different species and strains of rodent greatly contribute to our understanding of TCDD-mediated toxicities. For example, the Long-Evans rat strain (*Turku/AB*; L-E) displays a very low tolerance for TCDD (LD$_{50}$ = 10 μg/kg) while the Han/Wistar rat (*Kuopio*; H/W) is resistant to TCDD-induced lethality (LD$_{50}$>9600 μg/kg) [16]. This difference in sensitivity is caused by a point mutation in the H/W *Ahr*, resulting in expression of multiple isoforms of the AHR [17], leading to differential regulation of a subset of genes in H/W animals [18]. These differentially abundant transcripts, and any ensuing changes to the proteome, may lead to strain-specific TCDD toxicities. Similarly, in mice, both the C57BL/6 and DBA/2 strains exhibit TCDD-mediated toxic effects, however DBA/2 mice are much more resistant (approximately 10 to 20 times) than the C57BL/6 strain [19]. This resistance is caused by a point mutation within the ligand binding domain of the *Ahr* in the DBA/2 mice [20]. TCDD-toxicity also varies between male and female animals within a species. Female rats are more sensitive to TCDD-lethality than male rats, while in mice this relationship is reversed [21].

Analysis of protein content is the general end-point for many biological experiments. While mass spectrophotometry is a highly sensitive and specific technique, both the data generation and analysis steps are highly complex [22]. As such, western blot has become the standard method of use, as it allows for the sensitive and specific detection of target proteins with accurate relative quantitation of protein content in a relatively simple and inexpensive manner [23]. However, as in transcriptomic studies, accurate assessment of protein abundance by western blot requires thorough normalization of the data prior to the interpretation of results. This normalization typically involves the use of total protein or one or more endogenous loading controls in order to account for technical variability and to determine relative target abundance, thereby allowing multiple samples to be compared. While measurement of total protein is a relatively simple approach, it leads to complications down-stream [24]. Specifically, coomassie stained gels cannot be transferred to membrane for subsequent analysis and thereby requires the assumption that simultaneously run gels are loaded with identical amounts of protein [25]. The use of endogenous controls bypasses the need for additional steps, thereby reducing the number of gels and amount of sample used. Ideal endogenous control proteins maintain consistent levels of abundance regardless of environmental conditions, and thus often perform functions essential for cell survival [26]. Glyceraldehyde-3-phosphate dehydrogenase (GAPDH) and beta-actin (ACTB) have frequently been used as reference genes for both

mRNA expression measured by qPCR [26,27] and western blot analyses of protein content [28]. However, studies have shown that the stability of these widely used reference genes is not always consistent under different experimental conditions [29,30]. Factors such as tissue-type [30], organism (between and within species) [31], experimental manipulation [32] and even reagents used [33] can affect the abundance of candidate reference molecules. For these reasons, it is essential that endogenous reference proteins be thoroughly evaluated prior to experimental use.

Investigations into TCDD-induced proteomic changes are necessary to further our understanding of dioxin toxicity. Before these studies can proceed, candidate reference proteins must be carefully validated for use in western blot within the model systems used. Several reference genes have been previously validated for use in transcriptomic studies in rat [34] and mouse models [31] of TCDD toxicity. Currently, reference proteins for use in proteomic studies within these animal models have yet to undergo thorough validation. Since the transcriptomic responses differ dramatically across animal models [14,35], it is unclear whether these validated transcriptomic reference genes will translate to proteomic studies in either species. While it is not necessary to use the same controls for assessments of both gene and protein abundance, it is generally accepted that stably expressed genes may result in consistent abundance of protein [36,37]. We therefore chose to examine those genes previously identified as suitable references for transcriptomic studies of TCDD-toxicity [31], in addition to ACTB, to determine their validity for proteomic studies. Seven candidate proteins (*i.e.* ACTB, EEF1A1, GAPDH, HPRT, PGK1, PPIA and SDHA) were tested in hepatic tissue from multiple mouse models of TCDD-toxicity. This allows us to experimentally verify the idea that similar controls can be used at the RNA and protein levels, which would reduce the workload inherent in establishing controls.

Table 1. Experimental Design.

Study	Strain	Sex	Genotype	Treatment (TCDD µg/kg)	Time of tissue harvest (hours)	Number of animals
1	C57BL/6	Male	WT	0, 500	6	4, 5
	C57BL/6	Female	WT	0, 500	6	4, 5
2	C57BL/6	Male	rWT	0, 5, 500	19	4, 4, 4
	DBA/2J	Male	Ala375Val	0, 5, 500	19	4, 4, 4
3	C57BL/6	Male	WT	0, 500	24	4, 5
	C57BL/6	Female	WT	0, 500	24	3, 5
4	C57BL/6	Male	WT	0, 500	72	4, 5
	C57BL/6	Female	WT	0, 500	72	4, 5
5	C57BL/6	Male	WT	0, 500	144	3, 4
	C57BL/6	Female	WT	0, 500	144	3, 5
6	C57BL/6	Male	WT	0, 125, 250, 500, 1000	96	4, 4, 4, 4, 4
7	C57BL/6	Male	DEL	0, 125, 250, 500, 1000	96	5, 4, 3, 3, 4
8	C57BL/6	Male	INS	0, 125, 250, 500, 1000	96	5, 4, 4, 4, 5
9	C57BL/6	Male	rWT	0, 125, 250, 500, 1000	96	5, 3, 1, 4, 3
10	C57BL/6	Female	WT	0, 125, 250, 500, 1000	96	5, 5, 4, 4, 5

Animals analyzed (n = 192) varied in strain, sex, *Ahr*-allele, TCDD-treatment and time-point at which tissue was collected.

Table 2. Summary of analysis methods.

| | Student's t-test | NormFinder | | Normalization Method |
		Training	Validation	
ACTB	6/28	0.092	0.060	996.59
EEF1A1	11/28	0.112	0.050	278.40
GAPDH	5/31	0.072	0.077	316.07
HPRT	**1/31**	0.078	**0.046**	306.46
PGK1	6/29	0.144	0.081	**259.58**
PPIA	8/31	0.140	0.066	366.06
SDHA	10/26	**0.071**	0.056	286.62

Three analysis methods were used to evaluate the abundance consistency of each individual candidate protein; values in bold indicate the top ranked score for each method. 1) The difference between treated and untreated animals for each experimental condition was assessed by Student's t-tests; a p-value <0.05 was deemed significant. 2) The variation of each candidate was assessed using the NormFinder algorithm in two separate cohorts; a lower score indicates greater stability. 3) The comparative normalization method was used to evaluate the ability of each candidate to remove variation from a dataset; the average standard deviation for each pairwise comparison is reported.

Methods

Ethics Statement

All study plans were approved by the Finnish National Animal Experiment Board (Eläinkoelautakunta, ELLA; permit code: ESLH-2008-07223/Ym-23).

Animal Handling

Animal models and handling have been described previously [31]. Briefly, mouse colonies were maintained at the National Public Health Institute (today National Institute for Health and Welfare), Division of Environmental Health, Kuopio, Finland. Male and female C57BL/6 wild-type mice [21], male transgenic mice [38] and male DBA/2J mice [21] were studied. Wild-type animals were 12–15 weeks old and transgenic mice ranged up to 23 weeks. Animals were housed singly to avoid aggressive social behaviour, with environmental conditions maintained at $21 \pm 1°C$ with a relative humidity of $50 \pm 10\%$ on a 12 hour light cycle (12 hours of light followed by 12 hours of dark). Housing consisted of suspended, wire-mesh stainless-steel cages or Makrolon cages with aspen chip bedding (Tapvei Oy, Kaavi, Finland) and animals were provided with Altromin 1314 pellet feed (Altromin Spezialfutter GmbH & Co. KG, Lage, Germany) and water available *ad libitum*. The microbiological status of the animal facilities was regularly monitored in compliance with the recommendations of the Federation of European Laboratory Animal Science (FELASA), but individual mice were not tested in this regard. All experimental animals were drug and test naïve. Initial body weights for each animal are provided in Table S6.

Animals were stratified according to age such that groups contained a similar age-range, followed by randomization into experimental groups. Mice were treated in a group-wise manner, starting with the control in order to minimize the chance of human error. In most cases, the administration for a group was accomplished within an hour. Mice were treated with TCDD or corn oil vehicle alone and assessed following both timecourse and dose-response studies as described previously [31]. A total of 192 mice were used distributed across 47 separate experimental conditions (Table 1, Figure S1). TCDD was dissolved in corn oil and administered by oral gavage (10 mL/kg). Mice treated with corn oil alone acted as controls in each experiment.

Briefly, animals in the timecourse study were treated with a single dose of TCDD (500 µg/kg) or corn oil alone at time zero, followed by euthanasia at different time points (animals with tissue collected at the 19 hour time point received either 0, 5 or 500 µg/kg TCDD). Animals in the dose-response study received a single dose of 0, 125, 250, 500 or 1000 µg/kg TCDD followed by euthanasia 96 hours post-treatment. Although some of these doses were above the LD_{50} level of the exposed animals, the exposure time was in all cases maximally about 50% of the shortest time-to-death for these strains and genetic models as recorded in previous studies [21,33], and no mortality was therefore expected. However, all animals were carefully observed at least twice daily throughout the experimental period and, should signs consistent with severe suffering have been detected, those animals would have been euthanized immediately, as per the approved animal study plans.

Mouse livers were excised and snap-frozen in liquid nitrogen following euthanasia by carbon dioxide exposure. Tissue was shipped on dry ice to the analytical laboratory and stored at $-80°C$ or colder. All animal handling and reporting comply with ARRIVE guidelines [39].

Western analysis

Protein levels for candidate genes were determined by quantitative western blot. Each experiment was assessed on a single western blot to ensure identical analysis conditions between treated and control animals. Total protein was isolated from mouse liver using Tissue Extraction Reagent I (Life Technologies, Burlington, ON) supplemented with cOmplete protease inhibitor cocktail (Roche, Laval, QC). Protein extract, diluted 1/10 and 1/20 with 1XPBS, was quantified by Bradford assay and diluted to a final concentration of 10 µg/µL. A total of 65 µg protein [40,41] was loaded into each well of a Novex 4–12% Bis-Tris midi-gel system to ensure sufficient material would be available for the detection of low abundance targets [42]. Prepared gels were then electrophoresed for 40 minutes at 200V with MES running buffer (Life Technologies). Protein was transferred to PVDF membrane with the iBlot system using program P0 for 7 minutes (Life Technologies). The Colloidal Blue Staining Kit (Life Technologies) was used to observe total protein before and after electrophoresis and Ponceau staining was performed on the transferred membrane to ensure sufficient protein transfer (Figure S4). While there is some variation between samples, protein transfer appears consistent. Primary antibodies were purchased from Santa Cruz (Santa Cruz Biotechnology Inc.,

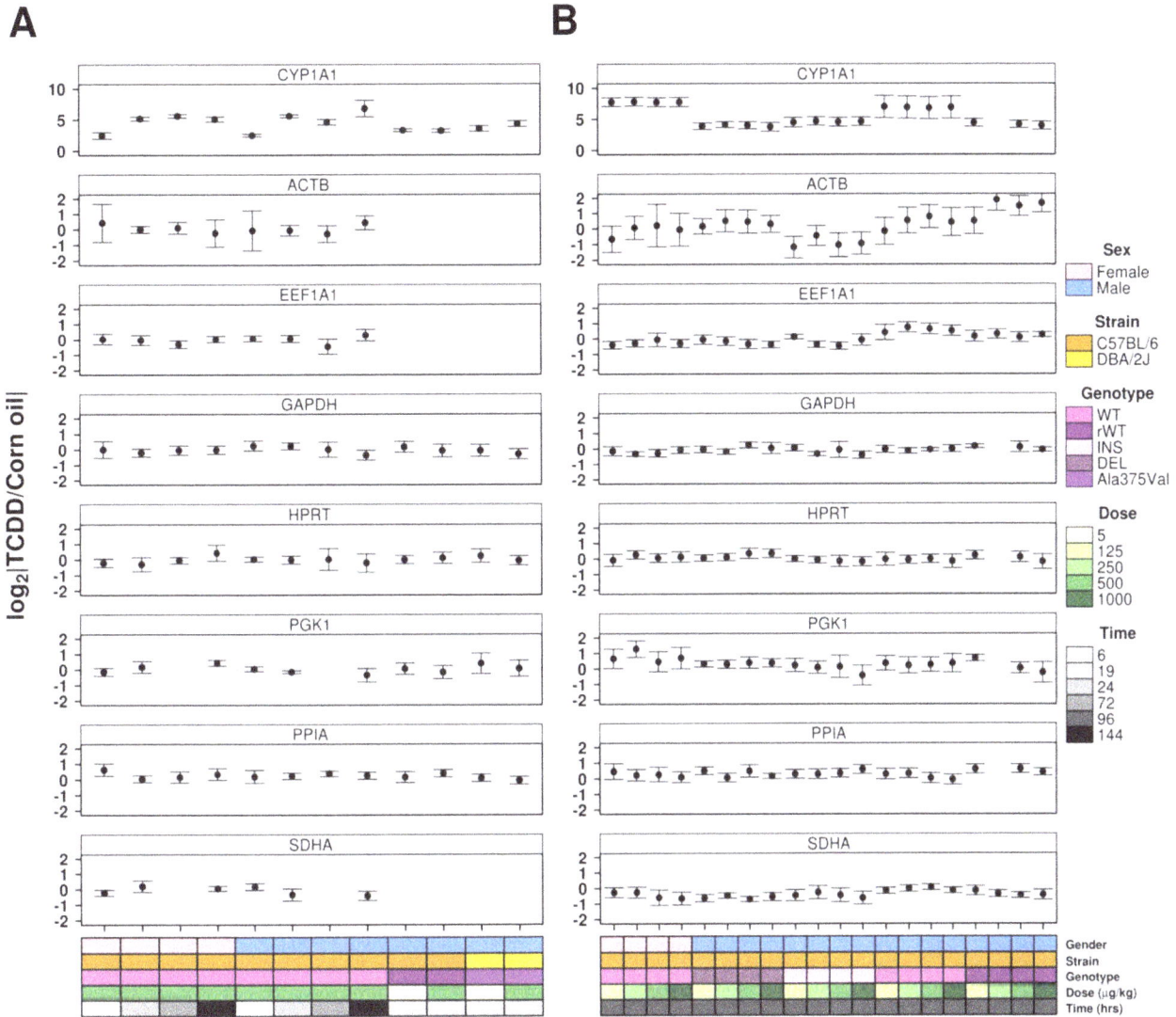

Figure 1. Timecourse and Dose-response by Treatment Group. The fold-differences in protein abundance between treated and control animals were calculated and results compared across all conditions. (**A**) Timecourse and (**B**) dose-response studies were visualized. Points represent the fold change in abundance (in log_2 space) and error bars indicate the standard deviation for each experimentally unique group.

Dallas, TX) or Abcam (Abcam Inc., Toronto, ON) and were diluted at the recommended concentrations in Li-Cor blocking buffer supplemented with 0.1% Tween-20, with overnight incubation at 4°C. Blots were washed three times with PBS supplemented with 0.1% Tween-20 at room temperature for 5 minutes each. The Li-Cor IRDye-labelled secondary antibodies (Mandel Scientific, Guelph, ON) were used at a dilution of 1:10,000 in Li-Cor blocking buffer supplemented as above with 0.01% SDS and incubated at room temperature for 1 hour (ordering information and optimal dilutions for all antibodies are provided in Table S1). After washing as described, blots were scanned and analyzed with the Odyssey quantitative western blot near-infrared system (Li-Cor Biosciences, Lincoln, NE, USA) using default settings. Antibodies were initially tested individually and then grouped based on banding patterns in order to reduce the number of blots required [43]. Average band intensities were normalized by subtraction of background levels. Background normalized values are provided in Table S2 and scanned images in Figure S2. Primary and secondary antibodies were initially

tested individually to identify optimal concentrations for the reduction of nonspecific banding patterns. Antibodies were then grouped where possible such that desired bands did not overlap.

Statistical Analyses and Visualization

Data were loaded in the R statistical environment (v3.0.3) for all analyses. Protein content was aggregated across biological replicates to obtain a mean abundance with standard for each candidate protein. Aggregation into biological replicates resulted in 47 separate experimental conditions. The ratio between treated and control abundances provided the fold-difference (M) in expression. Individual proteins and all possible combinations of multiple proteins were assessed. Visualizations were produced using the lattice (v0.20–29) and latticeExtra (v0.6–26) R packages.

Protein content was assessed across timecourse and dose-response studies. Animals treated with TCDD were compared to control animals of the same experimental group resulting in 26–31 comparisons (some comparisons were not done due to unsatisfactory loading patterns and/or lack of sufficient sample). Differential

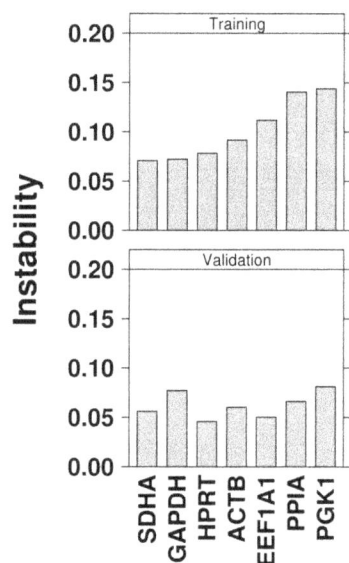

Figure 2. Univariate Analysis of Candidate Stability. Animals were separated into training and validation cohorts based on experiment, ensuring similar treatment conditions and animal numbers appeared in both sets. Within each cohort, animals were categorized as either TCDD-treated or control. Candidate proteins were analyzed using the NormFinder algorithm to determine stability across all treatment groups. A lower value indicates less variance across all experimental conditions.

abundance resulting from exposure to TCDD was evaluated for each candidate using an unpaired, two-tailed Student's t-test with Welch's adjustment for heteroscedasticity. Results were visualized as $M \pm$ standard-deviation for all experimental conditions.

Protein stability was evaluated using the NormFinder algorithm, which estimates the overall variation of a dataset by analysing its variance both within an experimental group and across experimental conditions [44]. Prior to analysis, animals were categorized into one of two groups (TCDD-treated or control) to estimate variance within experimental groups. Experiments were then split into 2 cohorts, labelled training (including experiments 1, 4, 6, 8 and 9) and validation (consisting of experiments 2, 3, 5, 7 and 10), such that each cohort contained similar types and number of animals and each cohort was analysed independently of the other. For each combination of candidates, the geometric mean of the background-normalized protein levels was calculated for each animal. For interpretation, a lower score indicates higher consistency of input across experimental groups signifying a potentially good loading control. Stability scores are available in Table S3. Linear modelling was performed to identify the contribution of each candidate protein [$Y_{OS} = \alpha_{ACTB} + \alpha_{EEF1A1} + \alpha_{GAPDH} + \alpha_{HPRT} + \alpha_{PGK1} + \alpha_{PPIA} + \alpha_{SDHA} + \varepsilon$] where Y_{OS} represents the overall stability of each combination of candidates and each protein is a Boolean variable indicating presence/absence in the combination while epsilon represents any error in the observations not explained by the model.

The comparative normalization method was used to contrast abundance levels between pairs of candidate molecules for each sample (adapted for use with protein abundance data from the comparative ΔC_q method [45]). The ability of each candidate to remove variability from other proteins was assessed and the mean standard deviation across comparisons provided a measure of stability.

mRNA analysis of candidate reference genes was reported previously for these animals and C_q data were downloaded (Supplementary Table 2, [31]); protein abundance and mean C_q data are provided in Table S4 for each animal. The correlation between protein levels and mean C_q values for each gene was assessed using Spearman's correlation using the AS89 method to assess statistical significance. NormFinder-generated stability scores were compared using the Spearman's correlation metric as the ordering of the scores is more meaningful than the magnitude (data available in Table S5).

Results

Quantitation of protein abundance by western blot is an essential technique widely used in the scientific community. In the past, this was typically performed using chemiluminesence. However, the Odyssey Infrared Imaging System is a well-documented alternative that provides many benefits over earlier methods, including an enhanced dynamic range of detection. Additionally, this system has the capacity for multiplexed reactions; specifically, antibodies are conjugated to IR fluorophores that can be detected at different wavelengths. As such, this system is ideal for detecting multiple targets [46].

Univariate Analysis

A good reference gene is one whose abundance is consistent across a wide range of conditions. This is most easily detected through analysis of the fold-difference (M) in expression from basal levels across specific treatment conditions. Candidate abundance was compared across conditions. Moderate correlations were observed between HPRT and PGK1 (Pearson's correlation, $R = 0.6$) as well as EEF1A1 and SDHA ($R = 0.49$), while the remaining candidates were weakly correlated (Figure S3).

To better understand this variation, each experimental group was examined individually (Figure 1). Of the 31 different experiment groups and 192 animals for which protein data were obtained (and for which mRNA data were obtained previously), HPRT was significantly altered by TCDD in only one group and GAPDH (5/31 conditions significantly altered) was also consistent, while the remainder of targets displayed less consistency, with greater than 20% of conditions altered (Table 2). To verify our samples and approach, the prototypical *Ahr*-regulated gene, CYP1A1, was examined as above and was determined to be significantly altered by TCDD at the protein level across all 31 conditions, as expected (Figure 1).

As this evaluation of differences in TCDD-altered abundance only accounts for variation within a single treatment, individual candidate stability across all experimental conditions was assessed using the NormFinder algorithm [44]. Briefly, NormFinder estimates the overall numerical stability of a molecule based on variability within a single treatment condition, variation within and between multiple conditions and systemic variation between experimental runs. Lower stability scores indicate less variation while higher scores indicate greater instability across experiments. As with our previous analysis of reference genes for transcriptomic analysis [31], experiments were organized into training and validation sets, thereby evaluating protein stability in two independent cohorts (Figure 2, Table 2). Although the cohorts differed in the magnitudes of stability scores, HPRT and SDHA were consistently amongst the most stable of the candidates, while PGK1 and PPIA were consistently the most unstable of the proteins evaluated.

To ensure that our results are not confounded by a shift in abundance caused by technical variation and independent of

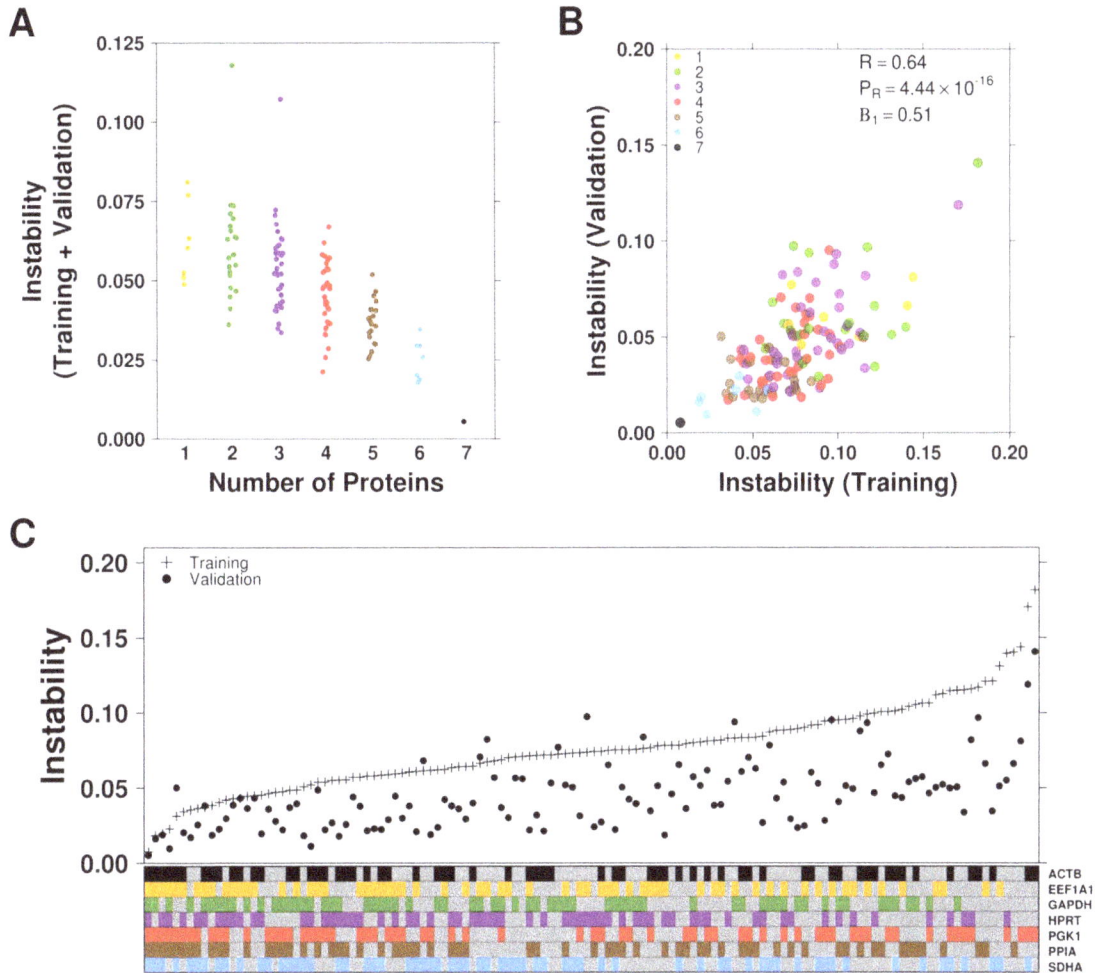

Figure 3. Multivariate Analysis of Candidate Stability. Animals were categorized as either TCDD-treated or control and separated into training and validation cohorts. All possible combinations of candidates were analyzed using the NormFinder algorithm. A lower value indicates less variance across all experimental conditions. (**A**) Combinations of candidates were organized according to the number of proteins included, in order to determine the optimal number of proteins used. (**B**) Stability results for each combination of candidates were compared between the training and validation sets to assess concordance. (**C**) Results for each combination of gene(s) were plotted for both the training (+) and validation (●) cohorts. Combinations are organized according to performance in the training set.

TCDD-treatment, we applied an alternate univariate analysis technique. Under typical experimental settings, it would be the purpose of the reference gene to normalize abundance levels for this shift. To this end, abundances of 6 proteins from each animal were normalized using the 7th, and the variance across technical replicates evaluated. This process was repeated using each protein as the normalization candidate. Using this approach, a lower score indicated greater stability across a dataset resulting from normalization with the given candidate protein (Table 2). By this method, PGK1 and EEF1A1 were determined to be the most stable of candidate proteins while ACTB was responsible for increased variation, likely due to the difference in magnitude of the intensity values between targets (intensity values for ACTB are significantly higher than for other candidates). Surprisingly, while PGK1 was identified as one of the more variable candidates both by analysis of fold-differences and the NormFinder algorithm, it was among the most stable candidates by this normalization method.

Multivariate Analysis

It has previously been shown that the use of multiple reference genes can improve normalization [31,47]. Although this generally applies to more high-throughput technologies capable of analyzing a large number of genes simultaneously, we evaluated the usefulness of utilizing multiple controls for western blot studies. The normalization capabilities of each possible combination of our candidate proteins were tested using the NormFinder algorithm, as described above. In general, including more control genes improved stability; however, specific pairs of candidates, and even some individual candidates, showed greater stability than some larger combinations (Figure 3A). Within each subset of samples, candidate combinations generally performed similarly; however, the training cohort demonstrated slightly more variance among samples (Pearson's correlation = 0.64) (Figure 3B). Despite this, the combination of all 7 candidates displayed the greatest stability in both cohorts (Figure 3C).

As a greater instability score appeared to primarily result from the inclusion of select candidates, linear modeling was performed to examine the contribution of each candidate to overall stability.

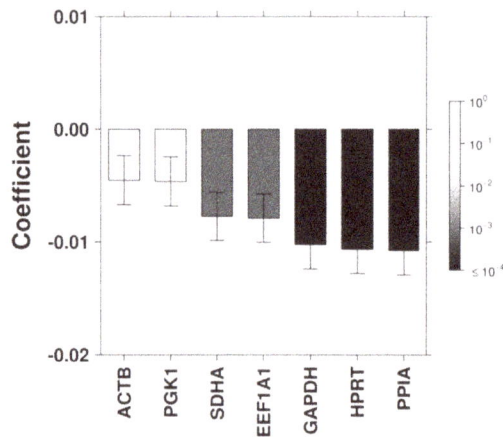

Figure 4. Linear Modelling of Multivariate Results. Linear modelling was performed to identify the contribution of each candidate to stability as determined by NormFinder across the complete dataset bars are coloured according to FDR-corrected p-value; error bars indicate standard error within the model; negative values are representative of decreased variation (increased stability).

ACTB and PGK1 decreased stability while GAPDH, HPRT and PPIA significantly increased stability (Figure 4).

Comparison with mRNA

As a similar analysis on the mRNA abundance of these candidates had been previously conducted in the liver of these animals, we thus compared the mRNA and protein abundances for each candidate. Spearman's correlation was used to determine whether protein abundance was concordant with mRNA levels. In general, there was little to no correlation between these molecules, possibly indicating differential regulation of translational mechanisms or variation in stability of the protein (Table 3, Figure 5A). To verify this, stability scores for each dataset generated by

NormFinder were combined, and the overlapping gene combinations compared (Figure 5B). Interestingly, while the abundance patterns of these candidates varied, combinations of candidates generally demonstrated similar stability (Spearman's correlation $= 0.5$, $p = 3.65 \times 10^{-5}$). Among the candidates (independently or in combination) that overlapped between studies, HPRT was among the most stable individual genes while the partnership of HPRT and GAPDH was consistently the most stable pair of candidates. Beyond this, the order of combination stability varied, sometimes dramatically, between data types. For example, the combination of EEF1A1, GAPDH and PPIA proved highly stable within the mRNA data, but was among the most unstable within the protein dataset. Alternatively, the pair-wise combination of EEF1A1 and PGK1 was among the most stable within the protein data and among the least stable in the mRNA data (Table S5).

Discussion

Thorough validation of reference genes is essential prior to any quantitative experimentation. Whether for evaluation of mRNA or protein abundance, all experimental methods are prone to some variation; the general rule is that each step in a process will introduce some error. This error may not be noticeable throughout the process, and only becomes apparent in downstream analyses, such as molecule quantitation. To ensure accurate interpretation, it is imperative to account for this technical variation. Estimation of target values relative to a reference molecule, whether internal or exogenous spike-in control is a proven method across technologies [48,49]. In the case of an endogenous molecule reference, careful validation must first occur as it has been shown that even classically-used controls can differ in abundance across different sample types or even by sample handling methods. For example, *Gapdh* was found to be less stable over time in FFPE breast tumour samples by qRT-PCR [50] whereas it was deemed a suitable reference gene for use in lung tumour FFPE samples [51]. In a proteomic analysis, multiple species of GAPDH were identified within human platelet samples; of these, the most abundant of species was highly variable across

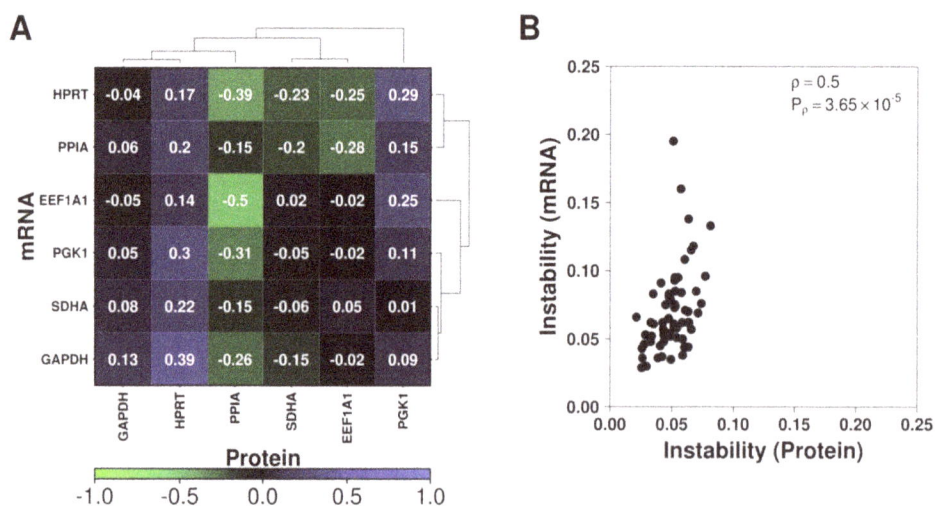

Figure 5. Comparison of candidate mRNA and protein abundances. mRNA and protein abundances as determined by qPCR and quantitative western blot were compared for each candidate. (**A**) Spearman's correlation was used to compare mean C_q values across technical replicates for qPCR and protein intensity for candidate genes and visualized in a heatmap organized using divisive clustering: blue indicates perfect correlation, green indicates inverse correlation and black indicates little or no correlation. Note that an increasing mRNA abundance results in a lower C_q; hence an inverse correlation indicates similarity between molecule abundances. (**B**) Spearman's correlation was used to assess similarity in candidate combination stability calculated by NormFinder for each data type.

Table 3. Comparison between mRNA and protein abundances.

	Spearman's correlation	
	ρ	p-value
EEF1A1	−0.02	0.79
GAPDH	0.13	0.08
HPRT	0.17	0.02
PGK1	0.11	0.14
PPIA	−0.15	0.04
SDHA	−0.06	0.46

Spearman's correlation was used to evaluate concordance between mRNA and protein abundances as determined by qPCR (mean C_q of technical replicates) and western blot ($\log_2$ of the protein intensity). Note that an increasing mRNA abundance results in a lower C_q; hence an inverse correlation indicates similarity between molecule abundance.

both age and sex [52]. This indicates that particular effort must be made when validating loading controls for western blot, as different antibodies may target different species.

Exposure to TCDD has been shown to have a dramatically different effect on transcriptomic regulation across various animal models. This has been shown to result from ligand activation of the AHR by TCDD-binding [8] while the degree of toxicity is directly related to the *Ahr*-genotype within rodents. While studies into the specific transcriptomic changes responsible for overall toxicity are still ongoing, progress has been made in the identification of candidate lists within various animal models, including strains of rats [53,54] and mice [55]. However, as toxicity likely results from subsequent changes in the proteome, further studies are required to verify which of these candidate genes are concomitantly altered at the protein level. While validation of reference genes for RNA quantitation in various mouse models has been completed [31], there is no reason to expect similar results to be obtained at the level of the proteome.

Here, we have evaluated the protein abundance of 7 reference genes for use in toxico-proteomic analyses of TCDD-induced toxicity within a wide range of mouse models. In particular, we have assessed the effect of TCDD exposure on protein abundance within mouse models of various strains, *Ahr*-genotype and sex across both a timecourse and dose-response approach. Protein abundance was assessed by quantitative western blot analysis and each candidate's suitability as a reference control was evaluated using 3 analysis methods: 1) the fold-difference in protein content from basal levels, 2) the NormFinder algorithm [44], which is an assessment of target stability and 3) the ability of each candidate to reduce instability of the others [45].

As TCDD is known to have a significant impact on transcriptional regulation, and has been shown to affect the proteome [56], the protein abundance of our candidates was first assessed using biologically similar animals that were treated with either TCDD (at various doses) or corn oil alone. HPRT was identified as the protein least affected by TCDD while EEF1A1 and SDHA showed significant variability across multiple experimental conditions (Figure 1, Table 2). The suitability of this method is proven through the evaluation of CYP1A1; a protein involved in the detoxification of xenobiotics known to be significantly induced by TCDD. As well, since data for both treated and control animals were generated on a single western blot (experiencing identical experimental settings), this metric was arguably the most appropriate for our goals. Next, as the purpose of a reference gene is to efficiently remove technical variation from the quantified results, we sought to characterize the residual

variability among the remaining proteins after normalization with each candidate. An assumption of this method is that all candidate proteins demonstrate consistent expression over experimental conditions and that increased variation indicates decreased stability of the candidate in question [47]. Here we identified EEF1A1 and PGK1 as the most consistently expressed candidate genes while PPIA was again determined to be the least stable candidate (Table 2). The high instability of ACTB should be interpreted with caution as it does not follow the above assumption. One limitation of this approach is its disregard for technical considerations; since each western blot contained a separate experiment, and were performed one at a time, some technical variation would be inherent across the entire study. Finally, unlike the above comparative method, the NormFinder algorithm considers variation both within and between experiments in its assessment of candidate stability [44]. While the specific order of stability varied, NormFinder analysis identified HPRT, ACTB and SDHA as the most stable candidates in all cohorts examined (training, validation and overall). Similarly, PGK1 and PPIA were always deemed the most unstable candidates. The consistency in stability scores for each candidate protein verifies that NormFinder is a robust and reproducible method for identifying good reference proteins.

A major finding of our previous study of reference gene stability in qPCR studies was that greater stability was obtained through increasing the number of reference genes used. This finding was consistent with other reference gene validation studies [47,57]. In order to determine whether this finding was consistent with proteomic analysis, NormFinder analysis was applied as above. In general, the trend of increasing stability was consistent with the inclusion of an increasing number of candidates (Figure 3). However, due to the low-throughput nature of any western blot analysis, increasing the number of reference proteins is largely impractical. Therefore, careful selection of 2 or 3 candidates with good stability would prove ideal. In some cases, even a single reference gene could provide a more stable normalization factor than a larger, less consistently expressed group of candidates. To this effect, linear modelling of the multivariate analysis indicated that 2 of the 3 most stable candidates identified in the univariate analysis (HPRT and SDHA) each contributed significantly to increased stability when included in combinations of any number of candidates (Figure 4) while PGK1 contributed less.

The availability of both mRNA and protein abundances collected from the same 192 animals presented an interesting opportunity, as an in-depth comparison of these molecules for these candidate genes across such a wide range of conditions has yet to be performed. We sought to determine whether targets selected as optimal reference genes at the level of mRNA would be suitable for normalization of protein abundance data. A comparison of abundance levels suggested little or no correlation between molecules (Table 3). The largest correlation coefficient, though showing an inverse relationship in abundance, was observed for HPRT. While analysis of the fold-changes identified HPRT as most stable univariate candidate at the protein level, it was much less stable at the level of mRNA abundance. However, it consistently ranked among the most stable genes across all analysis methods in each study. Alternatively, the least stable gene identified in the RNA study, *Sdha*, ranked among the most stable in the current protein analysis and did not show correlation between molecules. As such, the optimal reference gene for studies of mRNA abundance may not be optimal for studies of protein abundance and should be validated prior to use. Conversely, multivariate analysis by NormFinder generated stability scores that were moderately correlated between data types and, in general, these scores improved with the addition of an increased number of genes. Even so, the practicality of using a larger number of genes is limited by the technology used and must be taken into consideration. As such, while using a larger number of genes is encouraged for studies easily multiplexed (such as qPCR), careful selection of fewer genes is required for low-throughput methods such as western blot.

For any type of quantitative analysis, data must be thoroughly normalized in order to account for the technical variation inherent in any experiment and to ensure reliable and reproducible results. The use of multiple controls is ideal for generation of a normalization factor; however, a carefully selected group of fewer candidates can prove sufficient when larger numbers are impractical. Here we have identified and suggested specific combinations of loading controls, such as HPRT alone or combined with ACTB or GAPDH, for use in western blot analysis of various mouse models of TCDD toxicity.

Supporting Information

Figure S1   Experimental Design. Mice were treated with either 0, 5, 125, 250, 500 or 1000 μg/kg TCDD dissolved in corn oil vehicle and euthanized at 6, 19, 24, 72, 96 or 144 hours post-exposure. Timecourse experiments followed male (blue) and female (pink) wild-type C57BL/6 mice treated with 500 μg/kg TCDD. Male DBA/2J and ratonized-WT mice were collected at 19 hours post-exposure following treatment with either 5 or 500 μg/kg TCDD. Dose-response experiments followed male (blue) wild-type or ratonized mice and female (pink) wild-type mice treated with a single dose of TCDD and euthanized 96 hours following exposure.

Figure S2   Western blots. Western blots were scanned and analyzed with the Odyssey quantitative western blot near-infrared

system using default settings. Each blot was scanned twice as two groups of antibodies were used. Wells with unusual loading patterns (noted by the *) were not used in the downstream analysis.

Figure S3   Correlation of Candidate Proteins. The fold-difference in abundance between treated and control groups were calculated for each experimental condition and Pearson's correlations applied. Correlation results were visualized using a heatmap and organized by divisive clustering. Blue indicates perfect correlation; green represents inverse correlations while black indicates little or no correlation. Pearson's correlations are shown in white for each pair-wise comparison.

Figure S4   Ponceau Stain. Total protein abundance was assessed in a representative gel using Colloidal Blue Stain pre- (A) and post-transfer (B). Total protein was quantified and background-normalized intensity values were visualized for both gels (C). Transferred protein was also visualized on the membrane (D) using Ponceau stain. Lanes labelled in black indicate untreated samples, while blue labels are TCDD-treated (500 μg/kg) samples. The first four lanes show increasing amounts of loaded protein.

Table S1   Antibody Information.

Table S2   Protein Abundances.

Table S3   NormFinder Stability Scores.

Table S4   Comparison of mRNA and Protein Abundances.

Table S5   Comparison of mRNA and Protein Stability Scores.

Table S6   Animal Information.

Acknowledgments

The authors thank Arja Moilanen, Virpi Tiihonen, Janne Korkalainen and Susanna Lukkarinen for performing all animal experiments and sample preparation, as well as Herman Cheung for invaluable technical assistance.

Author Contributions

Conceived and designed the experiments: RP PCB. Performed the experiments: SDP. Analyzed the data: SDP. Contributed reagents/materials/analysis tools: SDP. Wrote the paper: SDP. Sample preparation: SDP JW.

References

1. Schecter A, Birnbaum L, Ryan JJ, Constable JD (2006) Dioxins: an overview. Environ Res 101: 419–428.
2. Marinkovic N, Pasalic D, Ferencak G, Grskovic B, Stavljenic Rukavina A (2010) Dioxins and human toxicity. Arh Hig Rada Toksikol 61: 445–453.
3. Shen C, Chen Y, Huang S, Wang Z, Yu C, et al. (2009) Dioxin-like compounds in agricultural soils near e-waste recycling sites from Taizhou area, China: chemical and bioanalytical characterization. Environ Int 35: 50–55.
4. Seefeld MD, Corbett SW, Keesey RE, Peterson RE (1984) Characterization of the wasting syndrome in rats treated with 2,3,7,8-tetrachlorodibenzo-p-dioxin. Toxicol Appl Pharmacol 73: 311–322.
5. Weisglas-Kuperus N, Patandin S, Berbers GA, Sas TC, Mulder PG, et al. (2000) Immunologic effects of background exposure to polychlorinated biphenyls and dioxins in Dutch preschool children. Environ Health Perspect 108: 1203–1207.
6. Longnecker MP, Michalek JE (2000) Serum dioxin level in relation to diabetes mellitus among Air Force veterans with background levels of exposure. Epidemiology 11: 44–48.
7. Bertazzi PA, Zocchetti C, Guercilena S, Consonni D, Tironi A, et al. (1997) Dioxin exposure and cancer risk: a 15-year mortality study after the "Seveso accident". Epidemiology 8: 646–652.
8. Okey AB, Riddick DS, Harper PA (1994) The Ah receptor: mediator of the toxicity of 2,3,7,8-tetrachlorodibenzo-p-dioxin (TCDD) and related compounds. Toxicol Lett 70: 1–22.

9. Okey AB (2007) An aryl hydrocarbon receptor odyssey to the shores of toxicology: the Deichmann Lecture, International Congress of Toxicology-XI. Toxicol Sci 98: 5–38.

10. Boutros PC, Yao CQ, Watson JD, Wu AH, Moffat ID, et al. (2011) Hepatic transcriptomic responses to TCDD in dioxin-sensitive and dioxin-resistant rats during the onset of toxicity. Toxicol Appl Pharmacol 251: 119–129.

11. Puga A, Sartor MA, Huang MY, Kerzee JK, Wei YD, et al. (2004) Gene expression profiles of mouse aorta and cultured vascular smooth muscle cells differ widely, yet show common responses to dioxin exposure. Cardiovasc Toxicol 4: 385–404.

12. Hanlon PR, Zheng W, Ko AY, Jefcoate CR (2005) Identification of novel TCDD-regulated genes by microarray analysis. Toxicol Appl Pharmacol 202: 215–228.

13. Kim S, Dere E, Burgoon LD, Chang CC, Zacharewski TR (2009) Comparative analysis of AhR-mediated TCDD-elicited gene expression in human liver adult stem cells. Toxicol Sci 112: 229–244.

14. Boverhof DR, Burgoon LD, Tashiro C, Sharratt B, Chittim B, et al. (2006) Comparative toxicogenomic analysis of the hepatotoxic effects of TCDD in Sprague Dawley rats and C57BL/6 mice. Toxicol Sci 94: 398–416.

15. Pohjanvirta R, Korkalainen M, Moffat ID, Boutros PC, Okey AB (2011) Role of the AHR and its Structure in TCDD Toxicity. In: Pohjanvirta R, editor. The AH Receptor in Biology and Toxicology. Hoboken, NJ, USA: John Wiley & Sons, Inc.

16. Pohjanvirta R, Unkila M, Tuomisto J (1993) Comparative acute lethality of 2,3,7,8-tetrachlorodibenzo-p-dioxin (TCDD), 1,2,3,7,8-pentachlorodibenzo-p-dioxin and 1,2,3,4,7,8-hexachlorodibenzo-p-dioxin in the most TCDD-susceptible and the most TCDD-resistant rat strain. Pharmacol Toxicol 73: 52–56.

17. Pohjanvirta R, Wong JM, Li W, Harper PA, Tuomisto J, et al. (1998) Point mutation in intron sequence causes altered carboxyl-terminal structure in the aryl hydrocarbon receptor of the most 2,3,7,8-tetrachlorodibenzo-p-dioxin-resistant rat strain. Mol Pharmacol 54: 86–93.

18. Franc MA, Moffat ID, Boutros PC, Tuomisto JT, Tuomisto J, et al. (2008) Patterns of dioxin-altered mRNA expression in livers of dioxin-sensitive versus dioxin-resistant rats. Arch Toxicol 82: 809–830.

19. Chapman DE, Schiller CM (1985) Dose-related effects of 2,3,7,8-tetrachlorodibenzo-p-dioxin (TCDD) in C57BL/6J and DBA/2J mice. Toxicol Appl Pharmacol 78: 147–157.

20. Poland A, Palen D, Glover E (1994) Analysis of the four alleles of the murine aryl hydrocarbon receptor. Mol Pharmacol 46: 915–921.

21. Pohjanvirta R, Miettinen H, Sankari S, Hegde N, Linden J (2012) Unexpected gender difference in sensitivity to the acute toxicity of dioxin in mice. Toxicol Appl Pharmacol 262: 167–176.

22. Kislinger T, Gramolini AO, MacLennan DH, Emili A (2005) Multidimensional protein identification technology (MudPIT): technical overview of a profiling method optimized for the comprehensive proteomic investigation of normal and diseased heart tissue. J Am Soc Mass Spectrom 16: 1207–1220.

23. Gerk PM (2011) Quantitative immunofluorescent blotting of the multidrug resistance-associated protein 2 (MRP2). J Pharmacol Toxicol Methods 63: 279–282.

24. Zeng L, Guo J, Xu HB, Huang R, Shao W, et al. (2013) Direct Blue 71 staining as a destaining-free alternative loading control method for Western blotting. Electrophoresis 34: 2234–2239.

25. Eaton SL, Roche SL, Llavero Hurtado M, Oldknow KJ, Farquharson C, et al. (2013) Total protein analysis as a reliable loading control for quantitative fluorescent Western blotting. PLoS One 8: e72457.

26. Li X, Bai H, Wang X, Li L, Cao Y, et al. (2011) Identification and validation of rice reference proteins for western blotting. J Exp Bot 62: 4763–4772.

27. Suzuki T, Higgins PJ, Crawford DR (2000) Control selection for RNA quantitation. Biotechniques 29: 332–337.

28. Weldon S, Ambroz K, Schutz-Geschwender A, Olive DM (2008) Near-infrared fluorescence detection permits accurate imaging of loading controls for Western blot analysis. Anal Biochem 375: 156–158.

29. Deindl E, Boengler K, van Royen N, Schaper W (2002) Differential expression of GAPDH and beta3-actin in growing collateral arteries. Mol Cell Biochem 236: 139–146.

30. Ferguson RE, Carroll HP, Harris A, Maher ER, Selby PJ, et al. (2005) Housekeeping proteins: a preliminary study illustrating some limitations as useful references in protein expression studies. Proteomics 5: 566–571.

31. Prokopec SD, Buchner NB, Fox NS, Chong LC, Mak DY, et al. (2013) Validating reference genes within a mouse model system of 2,3,7,8-tetrachlorodibenzo-p-dioxin (TCDD) toxicity. Chem Biol Interact 205: 63–71.

32. Greer S, Honeywell R, Geletu M, Arulanandam R, Raptis L (2010) Housekeeping genes; expression levels may change with density of cultured cells. J Immunol Methods 355: 76–79.

33. Linden J, Ranta J, Pohjanvirta R (2012) Bayesian modeling of reproducibility and robustness of RNA reverse transcription and quantitative real-time polymerase chain reaction. Anal Biochem 428: 81–91.

34. Pohjanvirta R, Niittynen M, Linden J, Boutros PC, Moffat ID, et al. (2006) Evaluation of various housekeeping genes for their applicability for normalization of mRNA expression in dioxin-treated rats. Chem Biol Interact 160: 134–149.

35. Boutros PC, Yan R, Moffat ID, Pohjanvirta R, Okey AB (2008) Transcriptomic responses to 2,3,7,8-tetrachlorodibenzo-p-dioxin (TCDD) in liver: comparison of rat and mouse. BMC Genomics 9: 419.

36. Kislinger T, Cox B, Kannan A, Chung C, Hu P, et al. (2006) Global survey of organ and organelle protein expression in mouse: combined proteomic and transcriptomic profiling. Cell 125: 173–186.

37. Gygi SP, Rochon Y, Franza BR, Aebersold R (1999) Correlation between protein and mRNA abundance in yeast. Mol Cell Biol 19: 1720–1730.

38. Pohjanvirta R (2009) Transgenic mouse lines expressing rat AH receptor variants – a new animal model for research on AH receptor function and dioxin toxicity mechanisms. Toxicol Appl Pharmacol 236: 166–182.

39. Kilkenny C, Browne WJ, Cuthill IC, Emerson M, Altman DG (2010) Improving bioscience research reporting: the ARRIVE guidelines for reporting animal research. PLoS Biol 8: e1000412.

40. Zhu M, Yu P, Jiang B, Gu Y (2012) Investigation of the influence of Arg555Trp and Thr538Pro TGFBI mutations on C-terminal cleavage and cell endoplasmic reticulum stress. Mol Vis 18: 1156–1164.

41. Hoene V, Fischer M, Ivanova A, Wallach T, Berthold F, et al. (2009) GATA factors in human neuroblastoma: distinctive expression patterns in clinical subtypes. Br J Cancer 101: 1481–1489.

42. Hinson JA, Michael SL, Ault SG, Pumford NR (2000) Western blot analysis for nitrotyrosine protein adducts in livers of saline-treated and acetaminophen-treated mice. Toxicol Sci 53: 467–473.

43. Anderson LV, Davison K (1999) Multiplex Western blotting system for the analysis of muscular dystrophy proteins. Am J Pathol 154: 1017–1022.

44. Andersen CL, Jensen JL, Orntoft TF (2004) Normalization of real-time quantitative reverse transcription-PCR data: a model-based variance estimation approach to identify genes suited for normalization, applied to bladder and colon cancer data sets. Cancer Res 64: 5245–5250.

45. Silver N, Best S, Jiang J, Thein SL (2006) Selection of housekeeping genes for gene expression studies in human reticulocytes using real-time PCR. BMC Mol Biol 7: 33.

46. Schutz-Geschwender A, Zhang Y, Holt T, McDermitt D, Olive DM (2004) Quantitative, Two-Color Western Blot Detection With Infrared Fluorescence. LI-COR Biosciences.

47. Vandesompele J, De Preter K, Pattyn F, Poppe B, Van Roy N, et al. (2002) Accurate normalization of real-time quantitative RT-PCR data by geometric averaging of multiple internal control genes. Genome Biol 3: RESEARCH0034.

48. Karge WH 3rd, Schaefer EJ, Ordovas JM (1998) Quantification of mRNA by polymerase chain reaction (PCR) using an internal standard and a nonradioactive detection method. Methods Mol Biol 110: 43–61.

49. Geiss GK, Bumgarner RE, Birditt B, Dahl T, Dowidar N, et al. (2008) Direct multiplexed measurement of gene expression with color-coded probe pairs. Nat Biotechnol 26: 317–325.

50. Tramm T, Sörensen BS, Overgaard J, Alsner J (2013) Optimal reference genes for normalization of qRT-PCR data from archival formalin-fixed, paraffin-embedded breast tumors controlling for tumor cell content and decay of mRNA. Diagn Mol Pathol 22: 181–187.

51. Walter RF, Mairinger FD, Wohlschlaeger J, Worm K, Ting S, et al. (2013) FFPE tissue as a feasible source for gene expression analysis – A comparison of three reference genes and one tumor marker. Pathol Res Pract.

52. Baumgartner R, Umlauf E, Veitinger M, Guterres S, Rappold E, et al. (2013) Identification and validation of platelet low biological variation proteins, superior to GAPDH, actin and tubulin, as tools in clinical proteomics. Journal of Proteomics 94: 540–551.

53. Yao CQ, Prokopec SD, Watson JD, Pang R, P'ng C, et al. (2012) Inter-strain heterogeneity in rat hepatic transcriptomic responses to 2,3,7,8-tetrachlorodibenzo-p-dioxin (TCDD). Toxicol Appl Pharmacol 260: 135–145.

54. Watson JD, Prokopec SD, Smith AB, Okey AB, Pohjanvirta R, et al. (2013) TCDD dysregulation of 13 AHR-target genes in rat liver. Toxicol Appl Pharmacol.

55. Boverhof DR, Burgoon LD, Tashiro C, Chittim B, Harkema JR, et al. (2005) Temporal and dose-dependent hepatic gene expression patterns in mice provide new insights into TCDD-Mediated hepatotoxicity. Toxicol Sci 85: 1048–1063.

56. Pastorelli R, Carpi D, Campagna R, Airoldi L, Pohjanvirta R, et al. (2006) Differential expression profiling of the hepatic proteome in a rat model of dioxin resistance: correlation with genomic and transcriptomic analyses. Mol Cell Proteomics 5: 882–894.

57. Teste MA, Duquenne M, Francois JM, Parrou JL (2009) Validation of reference genes for quantitative expression analysis by real-time RT-PCR in Saccharomyces cerevisiae. BMC Mol Biol 10: 99.

19

Detection of Serum Antibodies Cross-Reacting with *Mycobacterium avium* Subspecies *paratuberculosis* and Beta-Cell Antigen Zinc Transporter 8 Homologous Peptides in Patients with High-Risk Proliferative Diabetic Retinopathy

Antonio Pinna[1,3]*, **Speranza Masala**[2], **Francesco Blasetti**[1], **Irene Maiore**[1], **Davide Cossu**[2], **Daniela Paccagnini**[2], **Giuseppe Mameli**[2], **Leonardo A. Sechi**[2]

1 Department of Surgical, Microsurgical and Medical Sciences, Section of Ophthalmology, University of Sassari, Sassari, Italy, **2** Department of Biomedical Sciences, Section of Experimental and Clinical Microbiology, University of Sassari, Sassari, Italy, **3** Azienda Ospedaliero-Universitaria di Sassari, Sassari, Italy

Abstract

Purpose: MAP3865c, a *Mycobacterium avium* subspecies *paratuberculosis* (MAP) cell membrane protein, has a relevant sequence homology with zinc transporter 8 (ZnT8), a beta-cell membrane protein involved in Zn++ transportation. Recently, antibodies recognizing MAP3865c epitopes have been shown to cross-react with ZnT8 in type 1 diabetes patients. The purpose of this study was to detect antibodies against MAP3865c peptides in patients with high-risk proliferative diabetic retinopathy and speculate on whether they may somehow be involved in the pathogenesis of this severe retinal disorder.

Methods: Blood samples were obtained from 62 type 1 and 80 type 2 diabetes patients with high-risk proliferative diabetic retinopathy and 81 healthy controls. Antibodies against 6 highly immunogenic MAP3865c peptides were detected by indirect ELISA.

Results: Type 1 diabetes patients had significantly higher rates of positive antibodies than controls. Conversely, no statistically significant differences were found between type 2 diabetes patients and controls. After categorization of type 1 diabetes patients into two groups, one with positive, the other with negative antibodies, we found that they had similar mean visual acuity ($\sim$0.6) and identical rates of vitreous hemorrhage (28.6%). Conversely, Hashimoto's thyroiditis prevalence was 4/13 (30.7%) in the positive antibody group and 1/49 (2%) in the negative antibody group, a statistically significant difference ($P = 0.016$).

Conclusions: This study confirmed that type 1 diabetes patients have significantly higher rates of positive antibodies against MAP/ZnT8 peptides, but failed to find a correlation between the presence of these antibodies and the severity degree of high-risk proliferative diabetic retinopathy. The significantly higher prevalence of Hashimoto's disease among type 1 diabetes patients with positive antibodies might suggest a possible common environmental trigger for these conditions.

Editor: Demetrios Vavvas, Massachusetts Eye & Ear Infirmary, Harvard Medical School, United States of America

Funding: The authors have no support or funding to report.

* Email: apinna@uniss.it

Introduction

Diabetes mellitus is the most common endocrine disorder in industrialized countries. Two main forms are recognized [1]. Type 1 diabetes (T1D) is due to a deficiency in endogenous insulin secretion secondary to destruction of insulin-producing beta cells in the pancreas. Although T1D does have a peak incidence around the time of puberty, approximately 25% of cases present after 35 years of age. Type 2 diabetes (T2D) is characterized by insulin resistance with an insulin secretory defect leading to relative insulin deficiency. This group accounts for 90–95% of patients with diabetes and also has a strong genetic predisposition. T2D patients are usually, but not always, older than age 40 at presentation. Obesity is a frequent finding and, in the United States, is present in 80–90% of these patients.

Diabetic retinopathy is the leading cause of new cases of legal blindness among working-age people in developed countries. Proliferative diabetic retinopathy (PDR), the most vision-threatening form of diabetic retinopathy, has been reported to be present in approximately 50% of T1D patients with 25 years' duration of the disease and in 25% of patients who have had T2D for 25 years or more [2,3].

Zinc transporter 8 (ZnT8), a pancreas beta-cell membrane protein involved in Zn++ transportation, may act as a major autoantigen in T1D [4]. Auto-antibodies against ZnT8 have been found in 60–80% of newly-diagnosed cases of T1D [5]. Furthermore, dysregulation in Zn++ homeostasis caused by ZnT8 downregulation has been implicated in the pathogenesis of ischemic retinopathy [6]. *Mycobacterium avium* subspecies *paratuberculosis* (MAP) is transmitted from dairy herds to humans through food contamination. MAP causes an asymptomatic infection that is highly prevalent in patients with T1D, compared to those with T2D and healthy controls [7,8]. MAP3865c, a MAP cell membrane protein, has been shown to display a relevant sequence homology with ZnT8 [4,9]. Moreover, antibodies recognizing MAP3865c epitopes have been found to cross-react with ZnT8 in T1D patients [4,9,10].

We are unaware of any former study investigating a possible role of auto-antibodies against MAP/ZnT8 epitopes in the pathogenesis of PDR. The purpose of this study was to detect antibodies against 6 highly immunogenic MAP3865c peptides in patients with high-risk PDR, the most severe form of PDR, and in healthy controls and speculate on whether, or not, these antibodies may somehow be involved in the pathogenesis of PDR.

Methods

Patients and Controls

The present study used a case-control design, recruiting 62 T1D and 80 T2D patients with high-risk PDR and 81 healthy controls, all accrued between January and December 2013.

The inclusion criteria for the case group were diagnosis of T1D or T2D with high-risk PDR and age >18 years. Both newly-diagnosed cases of high-risk PDR and well-established cases, already treated with retinal laser photocoagulation, were included. According to the Early Treatment of Diabetic Retinopathy Study (ETDRS) classification, the diagnosis of high-risk PDR was made by the detection of new vessels on or within one disc diameter of the optic disc equaling or exceeding standard photograph 10A (about 1/4 to 1/3 disc area), with or without vitreous or preretinal hemorrhage; or vitreous and/or preretinal hemorrhage accompanied by new vessels either on the optic disc less than standard photograph 10A or new vessels elsewhere equaling or exceeding 1/4 disc area on ophthalmoscopic examination and fluorescein angiography [11,12]. Plasma glucose, creatinine, and glycated hemoglobin (HbA1c), and medical conditions, including body mass index (BMI), systemic hypertension, hypercholesterolemia, diabetic nephropathy, peripheral neuropathy, and cardio- and cerebrovascular status were recorded. All diabetic patients underwent a full ophthalmic evaluation, including best corrected visual acuity (BCVA), slit-lamp examination, applanation tonometry, fundus biomicroscopy, and fluorescein angiography. Exclusion criteria included any level of non-Sardinian ancestry and evidence of any other retinal vascular disorder.

Apparently healthy subjects, recruited from accompanying relatives or friends of patients or from hospital personnel, were used as controls. Exclusion criteria included clinical/laboratory evidence of diabetes mellitus, age <18 years, any level of non-Sardinian ancestry, and previous history of retinal artery occlusion, retinal vein occlusion, or anterior ischemic optic neuropathy. All controls underwent standard ophthalmic evaluation, including BCVA, slit-lamp examination, applanation tonometry, and fundus examination. Plasma glucose, systolic and diastolic blood pressure, and medical conditions were also recorded.

Subjects were classified as diabetic if they were under treatment for T1D or T2D or if they had a fasting plasma glucose level of ≥ 126 mg/dL and/or a plasma glucose level of ≥200 mg/dL 2 hours after a 75-g oral glucose load in a glucose tolerance test (as defined by the WHO). Subjects were considered to have hypertension if they were receiving treatment with anti-hypertension drugs or if their blood pressure was >140 mm Hg systolic or >90 mm Hg diastolic (as defined by the WHO/International Society of Hypertension). Hypercholesterolemia was defined by a fasting plasma cholesterol level of ≥200 mg/dL or the intake of lipid-lowering drugs.

Approval from the Ethics Committee/Institutional Review Board of the Department of Surgical, Microsurgical, and Medical Sciences, University of Sassari, Sassari, Italy, was obtained and the study was conducted in full accord with the tenets of the Declaration of Helsinki. Each participant received detailed information and provided written informed consent before inclusion.

Eight percent of the cases and 10% of the controls who were eligible for the study declined to participate. The major reason was "not interested".

Categorical values were compared by Chi-square test. The differences between cases and controls for quantitative variables were analyzed by Student's *t* test or Mann-Whitney test, when appropriate.

Blood collection

A blood sample was taken from each participant. Five ml of peripheral blood was collected in Vacutainer Serum tubes, transported at room temperature, left to clot for 15–30 minutes, and finally centrifuged at 2000×g for 10 minutes for plasma and serum separation. The supernatant serum was removed, stored at −80°C, and analyzed within 6 months.

Peptides

Six MAP/ZnT8 peptides were synthesized at >90% purity (LifeTein, South Plainfield, NJ 07080, USA). Four fell within the transmembrane region [$MAP3865c_{125-133}$ (MIAVALAGL), $MAP3865c_{133-141}$ (LAANFVVAL), $MAP3865c_{246-252}$ (LSPGKDM), $MAP3865c_{256-262}$ (HLISTGD)] and two were homologues to the human C-terminal ZnT8 immunogenic region [$MAP3865c_{261-267}$ (GDSARVL) and $MAP3865c_{281-287}$ (HATVQID)].

Enzyme-Linked Immunosorbent Assay (ELISA)

Antibodies specific for MAP3865c peptides were detected by indirect ELISA, as described previously [9]. Receiver operating characteristic (ROC) curves were used to calculate the cut-off values for positivity. The specificity was set at 93.8% (i.e. Ab + healthy controls <6.2%). The Area Under ROC Curve (AUC) and the percent fraction of antibody + sera were determined. Results were assessed by Fisher exact test and P values ≤0.05 were considered to be statistically significant. Statistical analysis was performed with commercial software (Graphpad Prism 6.0). ELISA results were normalized to a strongly positive control serum tested in all experiments, whose reactivity was set at 10.000 arbitrary units (AU)/ml. ELISA precision was determined by calculating both the inter- and intra-assay coefficients of variation (CV).

Results

The study group consisted of 62 T1D patients (mean age: 48.6±13.5 years; mean diabetes duration: 32.1±9.9 years) and 80 T2D patients (mean age: 66.6±7.4 years; mean diabetes duration: 15.1±8.5 years), all with bilateral high-risk PDR. The systemic characteristics of T1D and T2D patients are reported in Table 1. Both had similar median values of plasma glucose, creatinine, and glycated hemoglobin and similar rates of systemic hypertension, hypercholesterolemia, diabetic nephropathy, peripheral neuropathy, and Hashimoto's thyroiditis. Conversely, mean BMI was significantly higher in T2D patients, thus confirming that obesity is a frequent finding in T2D. Median glycated hemoglobin was 7.7 (61 mmol/mol) in T1D patients and 7.2 (55 mmol/l) in T2D patients, a result indicating poor glycemic control.

The control group consisted of 81 healthy subjects (39 men, 42 women; mean age: 48±12 years).

ELISA intra-assay CV ranged from 5.3% to 6.8%, whereas inter-assay CV ranged from 7.5% to 8.1%. These low CV values confirm that the results were consistent throughout the study.

Results on the detection of antibodies against MAP/ZnT8 peptides in T1D and T2D patients compared to healthy controls are summarized in Figures 1, 2, 3. All six peptides were highly recognized and showed detectable reactivity. T1D patients had significantly higher rates of positive antibodies than the control subjects for 5 out of the 6 MAP/ZnT8 peptides tested. On the other hand, no statistically significant differences were found between T2D patients and healthy controls.

In Figures 4 and 5, results on antibodies against MAP/ZnT8 peptides in T1D and T2D patients are compared. T1D patients showed significantly higher rates of positive antibodies against MAP3865c$_{125-133}$, MAP3865c$_{133-141}$, and MAP3865c$_{246-252}$.

After categorization of T1D patients into two groups, one with positive, the other with negative antibodies against MAP/ZnT8 peptides, we found that both had similar systemic characteristics, with the exception of Hashimoto's thyroiditis, whose prevalence was 4/13 (30.7%) in the positive antibody group and 1/49 (2%) in the negative antibody group, a statistically significant difference ($P = 0.016$). No antibodies against MAP/ZnT8 peptides were found in the only T2D patient with Hashimoto's disease.

Mean visual acuity was 0.65±0.36 (range: 0–1) in T1D patients and 0.57±0.36 (range: 0–1) in T2D patients, a not statistically significant difference. Furthermore, T1D and T2D patients also had almost identical mean intraocular pressure values (14.9±2.4 mm Hg and 15.3±4.1 mm Hg, respectively). Twelve (19.4%) T1D patients and 20 T2D patients (25%) had a posterior chamber intraocular lens. Vitreous hemorrhage was observed in 18 (29%) T1D and in 20 (25%) T2D patients. After categorization of T1D patients into two groups, one with positive, the other with negative antibodies against MAP/ZnT8 peptides, we found that they had similar mean BCVA (0.67±0.36 and 0.61±0.35, respectively) and identical rates of vitreous hemorrhage (28.6%).

Discussion

T1D is one of the most common autoimmune disorders, characterized by a T cell-mediated destruction of insulin-producing beta cells in the pancreas. This condition is believed to result from the interaction between multiple genetic and environmental factors, many of which are still poorly understood [13]. Former studies have shown that auto-antibodies against ZnT8 are common in newly-diagnosed cases of T1D [5], thus suggesting that this beta-cell membrane protein may act as a major autoantigen in T1D [4]. More recent investigations have found that antibodies against MAP3865c epitopes cross-react with ZnT8 epitopes [4,9,10], a result which could imply molecular mimicry between mycobacterial and pancreas beta-cell epitopes and raises the interesting question of whether MAP infection may be a potential environmental trigger for T1D. This finding, originally seen in T1D patients from Sardinia [4,9,10], an Italian island in

Table 1. Systemic characteristics of T1D and T2D patients with high-risk proliferative diabetic retinopathy (PDR).

	T1D patients (n = 62)	T2D patients (n = 80)	P T1D vs. T2D
Age, years, mean ± S.D.	8.6±13.5	66.6±7.4	<0.0001[a]
Gender			
- men, n. (%)	31 (50)	51 (63.8)	0.5[b]
- women, n. (%)	31 (50)	29 (36.2)	0.4[b]
Body mass index, mean ± S.D.	25.4±4	28.5±4.8	0.0001[a]
Diabetes duration, years, mean ± S.D.	32.1±9.9	15.1±8.5	<0.0001[a]
Plasma glucose, mg/dL, median (95% C.I.)	155 (150–171)	149 (140–160)	0.288[c]
Plasma creatinine, mg/dL, median (95% C.I.)	0.85 (0.79–0.91)	0.85 (0.76–0.9)	0.822[c]
HbA$_{1c}$, %, median (95% C.I.)	7.7 (7.4–8)	7.2 (6.8–8.2)	0.074[c]
HbA$_{1c}$, mmol/mol, median (95% C.I.)	61 (57–64)	55 (51–66)	0.074[c]
Systemic hypertension, n. (%)	32 (51.6)	58 (72.5)	0.25[b]
Hypercholesterolemia, n. (%)	26 (41.9)	40 (50)	0.63[b]
Diabetic nephropathy, n. (%)	1 (1.6)	5 (6.3)	0.36[b]
Peripheral neuropathy, n. (%)	3 (4.8)	0 (0)	0.18[b]
Hashimoto's thyroiditis, n. (%)	5 (8)	1 (1.3)	0.14[b]

[a]Student's t test
[b]Chi-square test
[c]Mann-Whitney test

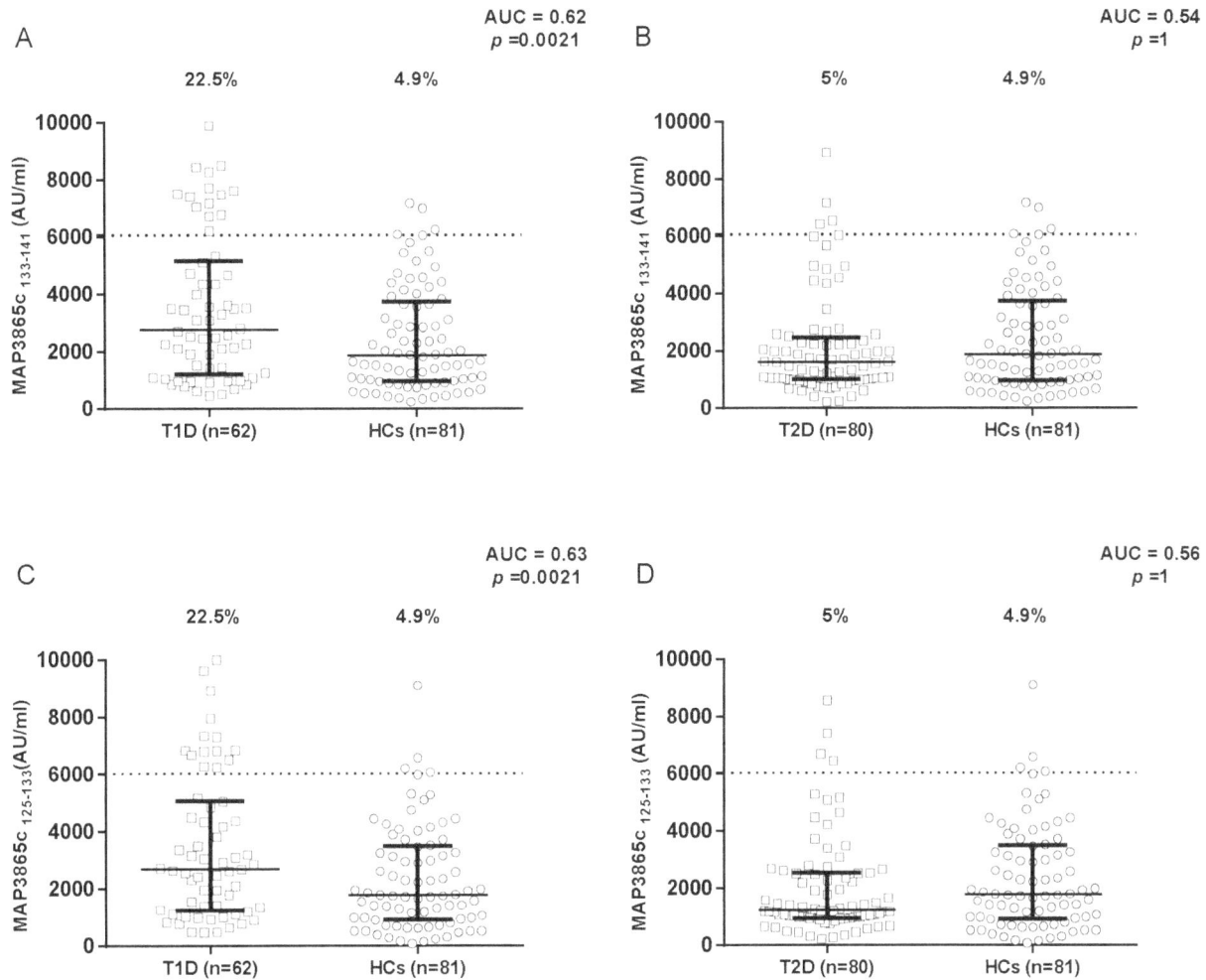

Figure 1. Prevalence of anti-MAP3865c antibodies in T1D and T2D patients with high-risk PDR and in healthy controls. Sera were tested for their reactivity against plate-coated MAP3865c$_{133-141}$ (A and B) and MAP3865c$_{125-133}$ (C and D) peptides. Antibody distribution is shown for T1D (A and C) and T2D (B and D) patients compared to controls. In each essay, the dotted line indicates the cut-off for positivity, as calculated by ROC analysis. The percent fraction of antibodies + sera is indicated on the top of each distribution, while bars indicate the corresponding median-interquartile range. AUC and p values determined by Fisher exact test are shown in the top right corner.

the middle of the Mediterranean Sea, was later also observed in T1D patients from mainland Italy [8,10].

There is a lot of experimental evidence indicating that Zn++ is abundant in the retina, where it plays an essential role in cell survival and in the normal functioning of antioxidant enzymes [6,14–16]. Recently, dysregulation in Zn++ homeostasis caused by ZnT8 downregulation has been implicated in the pathogenesis of ischemic retinopathy [6]. Therefore, it is theoretically possible that ZnT8 downregulation might also be involved in the pathogenesis of PDR, because retinal ischemia caused by capillary occlusion is an essential step in the development of this vision-threatening vascular disorder. The detection of antibodies against MAP3865c epitopes cross-reacting with ZnT8 epitopes, associated with a more severe form of PDR with lower visual acuity and higher prevalence of vitreous hemorrhage, might provide an indirect sign of a possible correlation between ZnT8 dysregulation and DR. In this survey, in order to investigate this hypothesis, serum antibodies against 6 MAP/ZnT8 peptides were detected in high-risk PDR patients and healthy controls. We found that T1D patients had significantly higher rates of positive antibodies than the control subjects, a result consistent with former studies

[4,9,10]. Similarly, T1D patients showed significantly higher rates of positive antibodies than T2D patients. On the other hand, no statistically significant differences were found between T2D patients and controls. Overall, these results confirm the association between T1D and the detection of antibodies against MAP/ZnT8 peptides, thus corroborating the hypothesis that MAP infection may be a potential environmental trigger for T1D.

In this study, T1D and T2D patients with high-risk PDR showed similar mean values of BCVA and had similar rates of vitreous hemorrhage. Likewise, after categorization of T1D patients into two groups, one with positive, the other with negative antibodies against MAP/ZnT8 peptides, we found that they had similar mean BCVA and identical rates of vitreous hemorrhage. On the whole, the absence of a more severe form of high-risk PDR in patients with positive antibodies against MAP/ZnT8 peptides might imply that auto-immune ZnT8 dysregulation plays a minor or no role in the pathogenesis of high-risk PDR.

One could argue that our choice of using BCVA and vitreous hemorrhage as ways of categorizing the severity degree of high-risk PDR might be questionable. Indeed, BCVA may be affected by concomitant macular edema and/or ischemia and cataract, all of

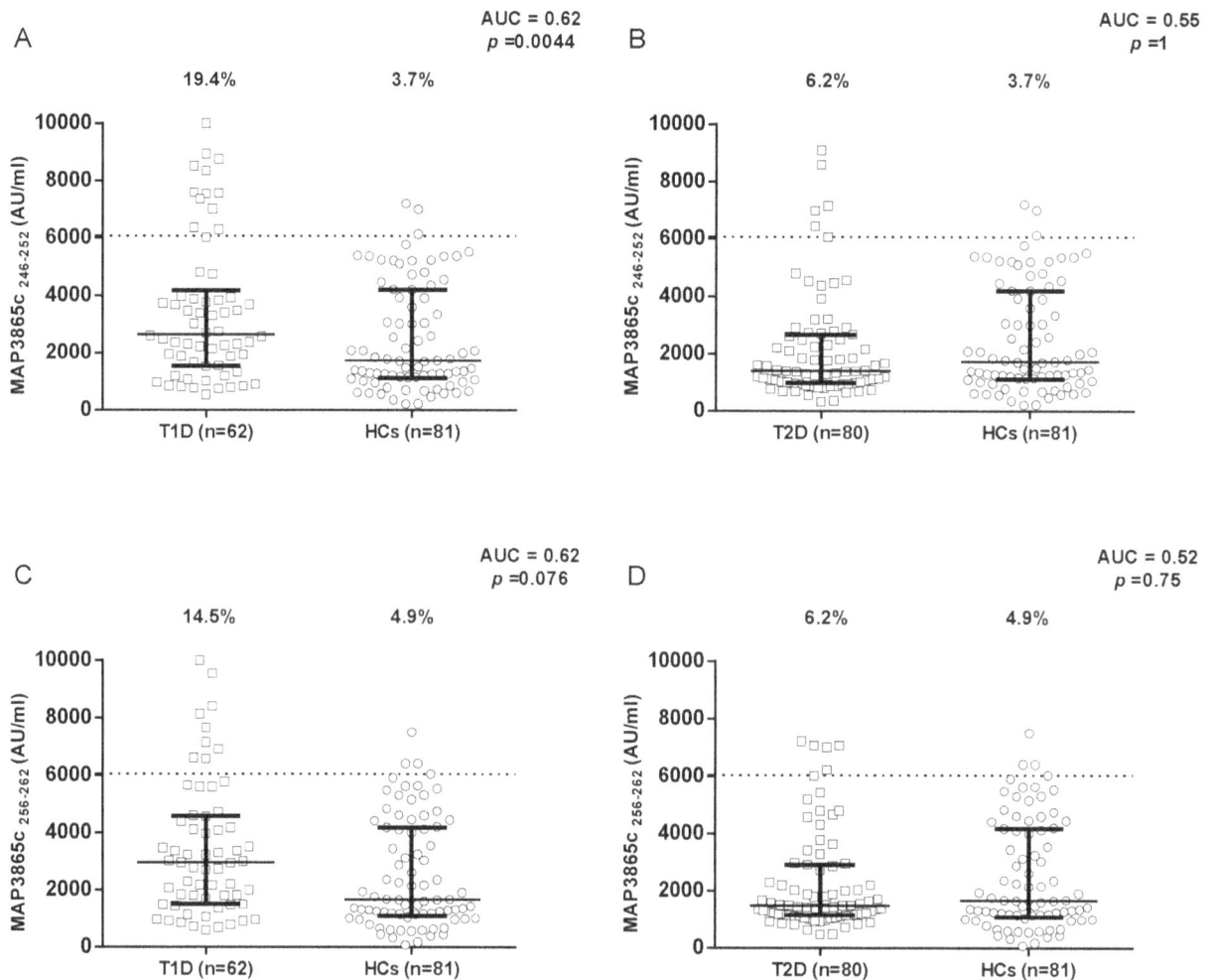

Figure 2. Prevalence of anti-MAP3865c antibodies in T1D and T2D patients with high-risk PDR and in healthy controls. Sera were tested for their reactivity against plate-coated MAP3865c$_{246-252}$ (A and B) and MAP3865c$_{256-262}$ (C and D) peptides. Antibody distribution is shown for T1D (A and C) and T2D (B and D) patients compared to controls. Data representation is the same as in Fig. 1.

which are common in patients with diabetic retinopathy. However, there is little doubt that the more serious forms of high-risk PDR are associated with reduced BCVA and vitreous hemorrhage. Therefore, we feel strongly that these parameters can provide a measure of the severity of high-risk PDR.

An unexpected result of post-hoc analysis was the significantly higher prevalence of Hashimoto's thyroiditis in the group of T1D patients with positive antibodies against MAP/ZnT8 peptides. To the best of our knowledge, we are unaware of any published study reporting a similar association. This finding is intriguing and raises the interesting questions of whether T1D and Hashimoto's disease share a common pathogenic mechanism and whether T1D patients with positive antibodies against MAP/ZnT8 peptides might have a higher risk of developing auto-immune thyroiditis.

In a recent population-based study exploring the prevalence and comorbidity of 12 auto-immune diseases in Sardinia, Sardu et al. disclosed that Hashimoto's thyroiditis is the most common autoimmune disease and that individuals affected by one autoimmune disorder are more likely to develop a second one [17]. Furthermore, experimental and clinical evidence indicates that MAP may be a potential environmental trigger of Hashomoto's thyroiditis, as suggested by the detection of viable MAP organism and serum antibodies against MAP/ZnT8

peptides in patients suffering from this auto-immune thyroid disease [18–20].

In ruminants, MAP specifically colonizes the mucosa-associated lymphoid tissue (MALT) of the small intestine, where this organism grows and multiplies within the intraepithelial macrophages, thus causing Johne's disease (JD) [21]. In humans, MAP is transmitted by the ingestion of contaminated dairy products and causes an asymptomatic infection, which has been associated not only with T1D and Hashimotos's thyroiditis, but also with Crohn's disease [22] and multiple sclerosis [23]. Overall, all these data support the idea that a common pathogenic mechanism may be responsible for these different autoimmune diseases.

Recent research has shown increased serum concentrations of CXCL10 chemokine in T1D children with antibodies against MAP3738c protein [24]. Furthermore, high circulating CXCL10 levels associated with T-helper (Th) 1 autoimmune response have been observed in new onset T1D and Hashimoto's disease [25,26]. Therefore, the significantly higher prevalence of Hashimoto's thyroiditis among T1D patients with positive antibodies against MAP/ZnT8 might in part be due to a common immuno-pathogenic mechanism involving Th 1 immune response and chemokines.

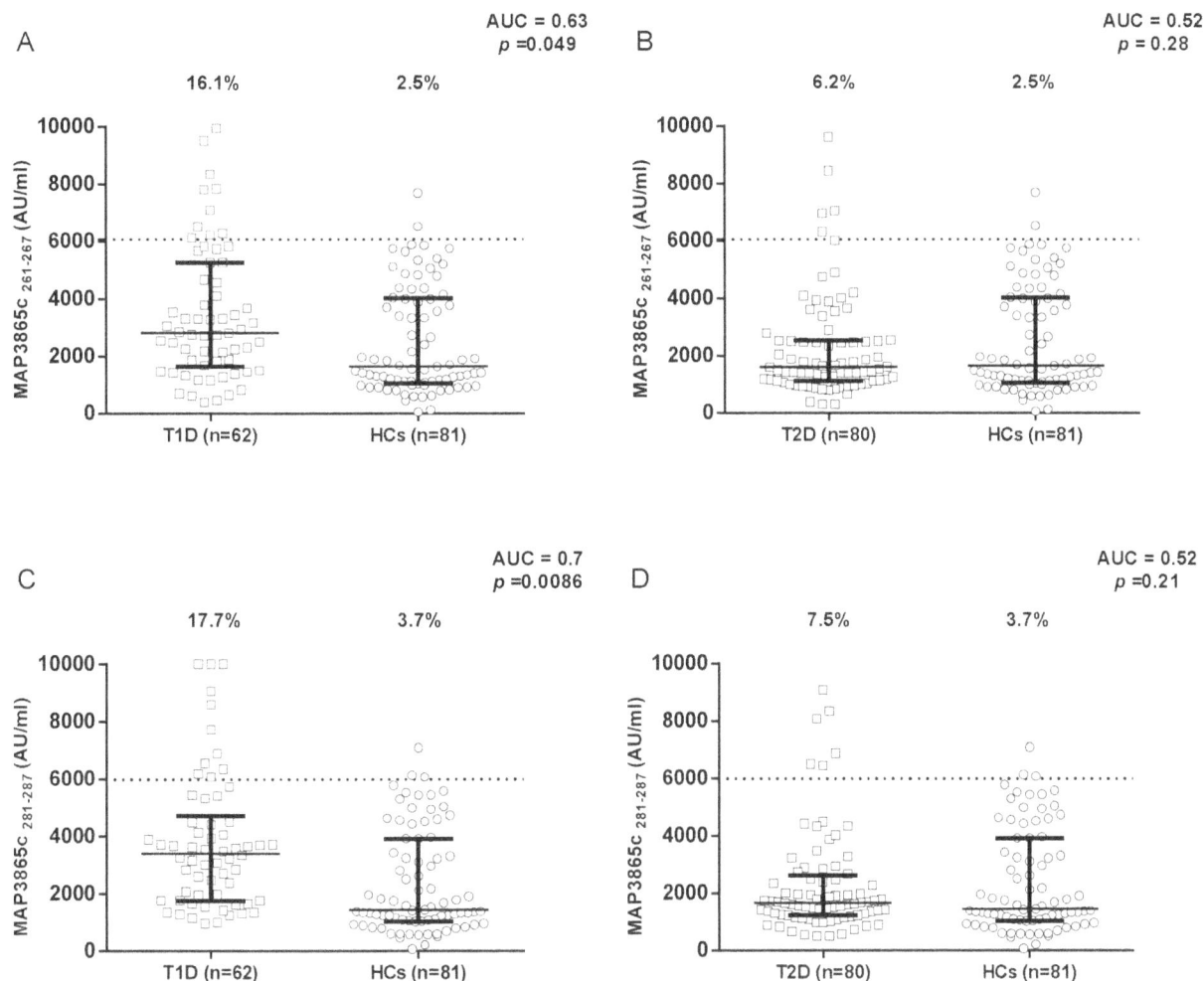

Figure 3. Prevalence of anti-MAP3865c antibodies in T1D and T2D patients with high-risk PDR and in healthy controls. Sera were tested for their reactivity against plate-coated MAP3865c$_{261-267}$ (A and B) and MAP3865c$_{281-287}$ (C and D) peptides. Antibody distribution is shown for T1D (A and C) and T2D (B and D) patients compared to controls. Data representation is the same as in Fig. 1.

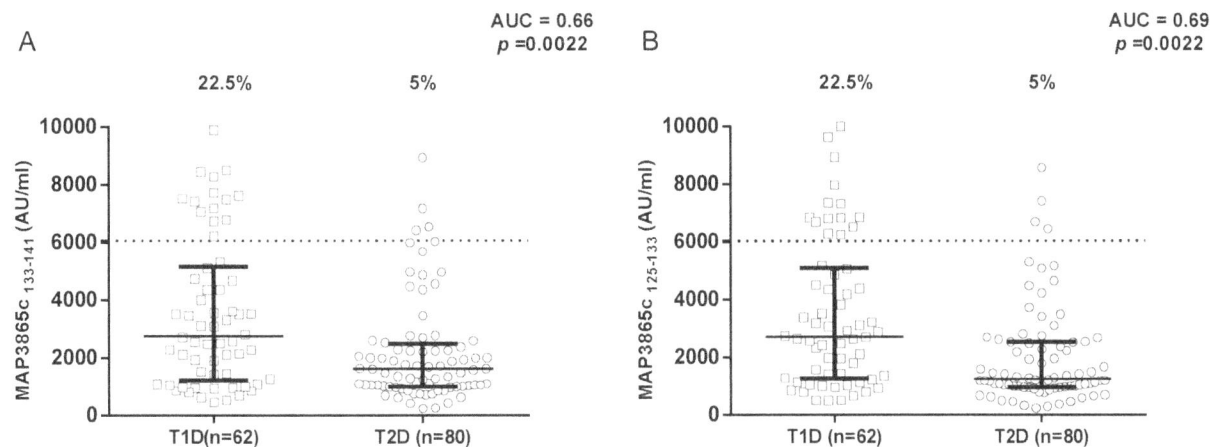

Figure 4. Prevalence of anti-MAP3865c antibodies in T1D and T2D patients with high-risk PDR. Sera were tested for their reactivity against plate-coated MAP3865c$_{133-141}$ (A) and MAP3865c$_{125-133}$ (B) peptides. Antibody distribution is compared between T1D and T2D patients. Data representation is the same as in Fig. 1.

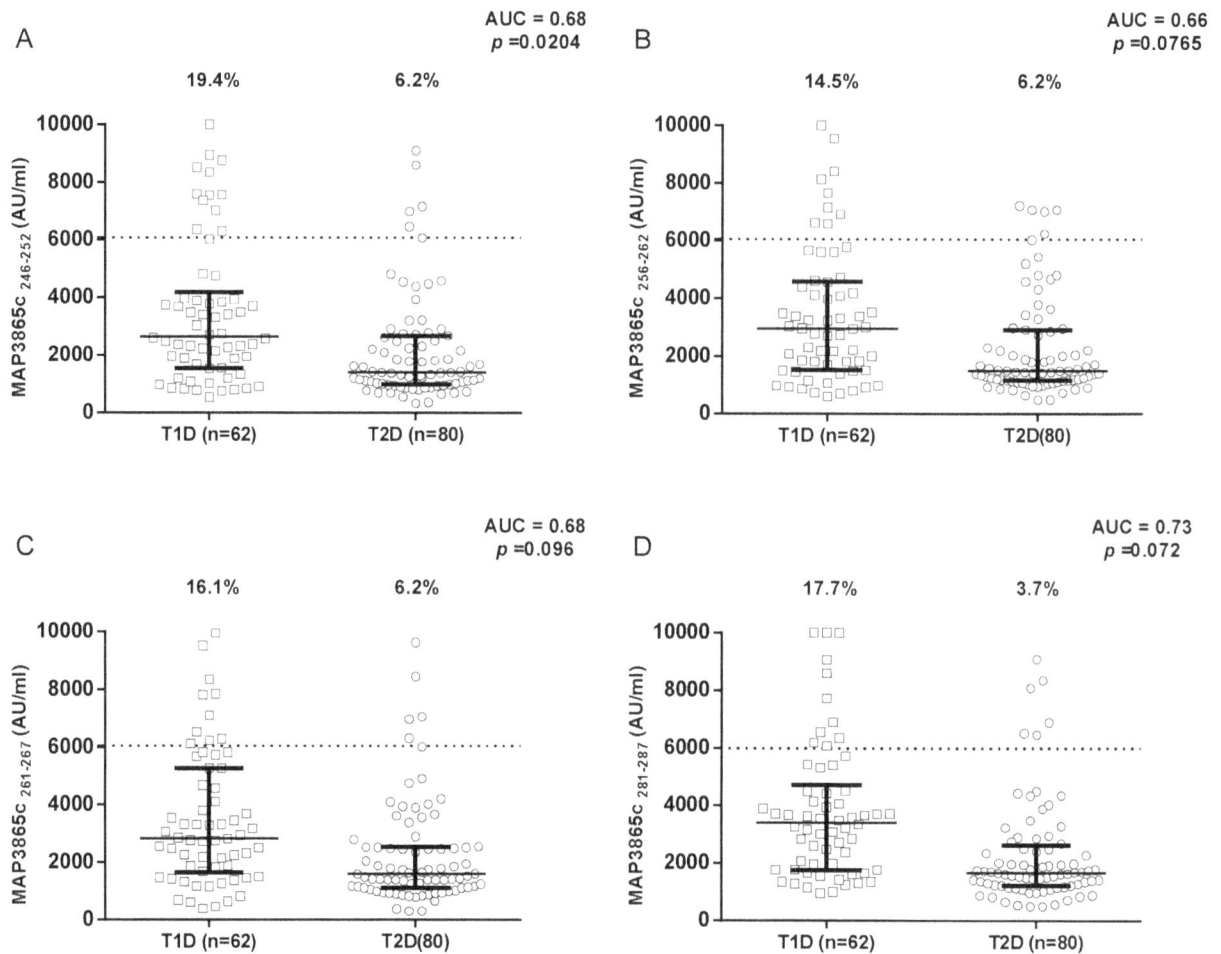

Figure 5. Prevalence of anti-MAP3865c antibodies in T1D and T2D patients with high-risk PDR. Sera were tested for their reactivity against plate-coated MAP3865c$_{246-252}$ (A), MAP3865c$_{256-262}$ (B), MAP3865c$_{261-267}$ (C), and MAP3865c$_{281-287}$ (D) peptides. Antibody distribution is compared between T1D and T2D patients. Data representation is the same as in Fig. 1.

Our study has several important limitations. First, it was restricted to a limited, genetically homogeneous group of patients (i.e. those of Sardinian ancestry); as a result, our findings may not be applicable to diabetic patients of non-Sardinian ancestry. Second, we analyzed a relatively small number of subjects. Third, we do not know whether T1D patients with positive antibodies against MAP/ZnT8 peptides developed high-risk PDR earlier than those with negative antibodies. Fourth, in patients suffering from Hashimoto's disease, we originally did not collect sufficiently detailed information about the thyroid condition, as the primary goal was to investigate the role of antibodies against MAP/ZnT8 peptides in the pathogenesis of high-risk PDR. Last, but not least, we compared T1D and T2D patients with high-risk PDR to normal subjects, rather than to diabetic controls without diabetic retinopathy. This second, potentially more informative comparison will be the subject of further investigation.

In conclusion, this study confirmed that T1D patients have significantly higher rates of positive antibodies against MAP/ZnT8 peptides, but failed to find a correlation between their presence in the patients' serum and the degree of severity of high-risk PDR.

On the other hand, we found a significantly higher prevalence of Hashimoto's thyroiditis in the group of T1D patients with positive antibodies. At this stage, no definitive conclusion can be drawn from our preliminary results. Further and larger clinical and experimental trials are required to clarify the exact role of antibodies against MAP/ZnT8 peptides in the pathogenesis of PDR in T1D and to establish whether, or not, T1D patients are more susceptible to Hashimoto's disease.

Acknowledgments

This manuscript was in part presented as a Scientific Poster at the 2013 Annual Meeting of the European Association for Vision and Eye (EVER), September 18–21, 2013, Nice, France

Author Contributions

Conceived and designed the experiments: AP SM FB IM DC DP GM LAS. Performed the experiments: AP SM FB IM DC DP GM LAS. Analyzed the data: AP SM FB LAS. Contributed to the writing of the manuscript: AP.

References

1. Diagnosis and Classification of Diabetes Mellitus (2011) American Diabetes Association. Diabetes Care 34 (Suppl 1): S62–S69.

2. Klein R, Klein BE, Moss SE, Davis MD, DeMets DL (1984) The Wisconsin epidemiologic study of diabetic retinopathy. II. Prevalence and risk of diabetic

retinopathy when age at diagnosis is less than 30 years. Arch Ophthalmol 102: 520–526.

3. Klein R, Klein BE, Moss SE, Davis MD, DeMets DL (1984) The Wisconsin epidemiologic study of diabetic retinopathy. III. Prevalence and risk of diabetic retinopathy when age at diagnosis is 30 or more years. Arch Ophthalmol 102: 527–532.

4. Masala S, Zedda MA, Cossu D, Ripoli C, Palermo M, et al. (2013) Zinc transporter 8 and MAP3865c homologous epitopes are recognized at T1D onset in Sardinian children. PLoS One 17;8(5): e63371.

5. Scotto M, Afonso G, Larger E, Raverdy C, Lemonnier FA, et al. (2012) Zinc transporter (ZnT)8(186–194) is an immunodominant CD8+ T cell epitope in HLA-A2+ type 1 diabetic patients. Diabetologia 55: 2026–2031.

6. Deniro M, Al-Mohanna FA (2012) Zinc transporter 8 (ZnT8) expression is reduced by ischemic insults: a potential therapeutic target to prevent ischemic retinopathy. PLoS One 7(11): e50360.

7. Paccagnini D, Sieswerda L, Rosu V, Masala S, Pacifico A, et al. (2009) Linking chronic infection and autoimmune diseases: Mycobacterium avium subspecies paratuberculosis, SLC11A1 polymorphisms and type-1 diabetes mellitus. PLoS One 21;4(9): e7109.

8. Bitti ML, Masala S, Capasso F, Rapini N, Piccinini S, et al. (2012) Mycobacterium avium subsp. paratuberculosis in an Italian cohort of type 1 diabetes pediatric patients. Clin Dev Immunol 2012: 785262.

9. Masala S, Paccagnini D, Cossu D, Brezar V, Pacifico A, et al. (2011) Antibodies recognizing Mycobacterium avium paratuberculosis epitopes cross-react with the beta-cell antigen ZnT8 in Sardinian type 1 diabetic patients. PLoS One 6(10): e26931.

10. Masala S, Cossu D, Piccinini S, Rapini N, Massimi A, et al. (2014) Recognition of zinc transporter 8 and MAP3865c homologous epitopes by new-onset type 1 diabetes children from continental Italy. Acta Diabetol 51: 577–585.

11. Early Treatment Diabetic Retinopathy Study Research Group (1991) Early Treatment Diabetic Retinopathy Study design and baseline patient characteristics. ETDRS report number 7. Ophthalmology 98(5 Suppl): 741–756.

12. Early Treatment Diabetic Retinopathy Study Research Group (1991) Grading diabetic retinopathy from stereoscopic color fundus photographs – an extension of the modified Airlie House classification. ETDRS report number 10. Ophthalmology 98(5 Suppl): 786–806.

13. La Torre D (2012) Immunobiology of beta-cell destruction. Adv Exp Med Biol. 771: 194–218.

14. Karcioglu ZA (1982) Zinc in the eye. Surv Ophthalmol 27: 114–122.

15. Grahn BH, Paterson PG, Gottschall-Pass KT, Zhang Z (2001) Zinc and the eye. J Am Coll Nutr 20(2 Suppl): 106–118.

16. Wills NK, Ramanujam VM, Kalariya N, Lewis JR, van Kuijk FJ (2008) Copper and zinc distribution in the human retina: relationship to cadmium accumulation, age, and gender. Exp Eye Res 87: 80–88.

17. Sardu C, Cocco E, Mereu A, Massa R, Cuccu A, et al. (2012) Population based study of 12 autoimmune diseases in Sardinia, Italy: prevalence and comorbidity. PLoS One 7(3): e32487.

18. Sisto M, Cucci L, D'Amore M, Dow TC, Mitolo V, et al. (2010) Proposing a relationship between Mycobacterium avium subspecies paratuberculosis infection and Hashimoto's thyroiditis. Scand J Infect Dis 42: 787–790.

19. D'Amore M, Lisi S, Sisto M, Cucci L, Dow CT (2010) Molecular identification of Mycobacterium avium subspecies paratuberculosis in an Italian patient with Hashimoto's thyroiditis and Melkersson-Rosenthal syndrome. J Med Microbiol 59: 137–139.

20. Masala S, Cossu D, Palermo M, Sechi LA (1014) Recognition of Zinc transporter 8 and MAP3865c homologue epitopes by Hashimoto's thyroiditis subjects from Sardinia: A common target with Type 1 Diabetes? PLoS One 9(5): e97621.

21. Pozzato N, Capello K, Comin A, Toft N, Nielsen SS, et al. (2011) Prevalence of paratuberculosis infection in dairy cattle in Northern Italy. Prev Vet Med 102: 83–86.

22. Sechi LA, Scanu AM, Molicotti P, Cannas S, Mura M, et al. (2005) Detection and isolation of Mycobacterium avium subspecies paratuberculosis from intestinal mucosal biopsies of patients with and without Crohn's disease in Sardinia. Am J Gastroenterol 100: 1529–1536.

23. Cossu D, Masala S, Sechi LA. (2013) A Sardinian map for multiple sclerosis. Future Microbiol 8: 223–232.

24. Cossu A, Ferrannini E, Fallahi P, Antonelli A, Sechi LA (2013) Antibodies recognizing specific Mycobacterium avium subsp. Paratuberculosis's MAP3738c protein in type 1 diabetes mellitus children are associated with serum Th1 (CXCL10) chemokine. Cytokine 61: 337–339.

25. Antonelli A, Ferrari SM, Giuggioli D, Ferrannini E, Ferri C, et al. (2014) Chemokine (C-X-C motif) ligand (CXCL)10 in autoimmune diseases. Autoimmun Rev 13: 272–280.

26. Antonelli A, Ferrari SM, Corrado A, Ferrannini E, Fallahi P (2014) CXCR3, CXCL10 and type 1 diabetes. Cytokine Growth Factor Rev 25: 57–65.

20

Development of a Mimotope Vaccine Targeting the *Staphylococcus aureus* Quorum Sensing Pathway

John P. O'Rourke[1]*[꠹], Seth M. Daly[2]꠹, Kathleen D. Triplett[2], David Peabody[1], Bryce Chackerian[1], Pamela R. Hall[2]*꠹

1 Department of Molecular Genetics and Microbiology, University of New Mexico School of Medicine, Albuquerque, NM, United States of America, **2** Department of Pharmaceutical Sciences, University of New Mexico School of Medicine, Albuquerque, NM United States of America

Abstract

A major hurdle in vaccine development is the difficulty in identifying relevant target epitopes and then presenting them to the immune system in a context that mimics their native conformation. We have engineered novel virus-like-particle (VLP) technology that is able to display complex libraries of random peptide sequences on a surface-exposed loop in the coat protein without disruption of protein folding or VLP assembly. This technology allows us to use the same VLP particle for both affinity selection and immunization, integrating the power of epitope discovery and epitope mimicry of traditional phage display with the high immunogenicity of VLPs. Previously, we showed that using affinity selection with our VLP platform identifies linear epitopes of monoclonal antibodies and subsequent immunization generates the proper antibody response. To test if our technology could identify immunologic mimotopes, we used affinity selection on a monoclonal antibody (AP4-24H11) that recognizes the *Staphylococcus aureus* autoinducing peptide 4 (AIP4). AIP4 is a secreted eight amino acid, cyclized peptide produced from the *S. aureus* accessory gene regulator (*agr*IV) quorum-sensing operon. The *agr* system coordinates density dependent changes in gene expression, leading to the upregulation of a host of virulence factors, and passive transfer of AP4-24H11 protects against *S. aureus* *agr*IV-dependent pathogenicity. In this report, we identified a set of peptides displayed on VLPs that bound with high specificity to AP4-24H11. Importantly, similar to passive transfer with AP4-24H11, immunization with a subset of these VLPs protected against pathogenicity in a mouse model of *S. aureus* dermonecrosis. These data are proof of principle that by performing affinity selection on neutralizing antibodies, our VLP technology can identify peptide mimics of non-linear epitopes and that these mimotope based VLP vaccines provide protection against pathogens in relevant animal models.

Editor: Nicholas J. Mantis, New York State Dept. Health, United States of America

Funding: JPO and BC were supported by NIH grant R01 AI083305. SMD was supported by the University of New Mexico Biology of Infectious Diseases and Inflammation Training Grant (T32-AI007538). PH was supported by NIH grant R01 AI091917. The funders had no role in study design, data collection and analysis, decision to publish, or preparation of the manuscript.

Competing Interests: The authors have declared that no competing interests exist.

* Email: jorourkejr@salud.unm.edu (JPO); phall@salud.unm.edu (PH)

꠹ These authors contributed equally to this work.

Introduction

The small particulate nature and multivalent structure of virus-like particles cause them to provoke strong immune responses and make them effective scaffolds for displaying heterologous antigens in a highly immunogenic format. Peptide-based vaccines are typically poorly immunogenic, however, peptides displayed on the surface of VLPs elicit high-titer and long-lasting antibody responses [1–5]. Although VLPs can be utilized to increase the immunogenicity of peptides, identifying relevant target epitopes and then presenting them to the immune system in a highly immunogenic context that mimics their native conformation, has largely been an unpredictable process of trial-and-error. The most widely used method for epitope identification is through affinity selection using peptide libraries displayed on a filamentous phage. This technology has identified the epitopes of many monoclonal antibodies (mAbs), and is a powerful technique for mapping linear epitopes and discovering peptide mimics of conformational and non-peptide epitopes. Nevertheless, peptides displayed on a filamentous phage are typically poorly immunogenic due to the low valency display of peptides on the phage surface. Thus, epitopes identified by phage display must be produced synthetically, linked to a carrier, and displayed in a structural context unrelated to the selected phage. Often, in this new conformation the peptides have vastly decreased affinity for the selecting molecule and frequently lose the ability to induce antibodies that mimic the selecting antibody.

VLP technology has not previously been adapted for use in epitope identification because recombinant VLPs are not well-suited for the construction of diverse peptide libraries. Insertion of heterologous peptides into viral structural proteins often result in protein folding and VLP assembly defects. [6–8]. To overcome these limitations, we engineered a version of the bacteriophage MS2 coat protein whose folding and assembly is highly tolerant of short peptide insertions [7]. This system has allowed us to generate large, complex libraries of VLPs displaying random peptide sequences. Because VLPs encapsidate the mRNA that encodes coat protein and its peptide [7,9], VLPs with specific binding characteristics can be affinity selected and then the nucleic acid encoding the selected peptide can be recovered by RT-PCR. Most importantly, the same VLP can be used for both affinity selection and immunization. Thus, this system integrates the power of epitope/mimotope discovery of traditional phage display with the high immunogenicity of VLPs. We recently showed the utility of this VLP technology to identify linear epitopes and to elicit the proper antibody response by performing affinity selection using a set of well-characterized mAbs [10].

In this study we used this VLP vaccine discovery platform to identify immunogenic mimics of a quorum-sensing peptide from the Gram-positive pathogen *Staphylococcus aureus*. *S. aureus* is the leading cause of skin and soft tissue infections (SSTI) presenting to emergency departments in the USA [11]. The *S. aureus* accessory gene regulator (*agr*) quorum-sensing system coordinates a density dependent switch in gene expression that includes upregulation of virulence factors critical for invasive SSTI [12–15]. The *agr* system signals through the use of a secreted thiolactone-cyclized autoinducing peptide (AIP) which, upon binding to its cognate surface receptor AgrC, initiates a regulatory cascade leading to changes in transcription of more than 200 genes [16,17]. Among the upregulated genes are those encoding secreted virulence factors essential for invasive skin infection, including upregulation of the pore-forming toxin alpha-hemolysin (Hla). Infection with *agr* or *hla* deletion mutants, loss of the Hla receptor ADAM10, or neutralization of Hla significantly attenuates virulence in mouse models of SSTI [13,17–21]. Furthermore, we and others have shown that host innate effectors which disrupt *agr*-signaling also provide defense against *S. aureus* infection [22–26]. These results suggest that a VLP-based epitope identification approach to vaccine development targeted towards disruption of *agr* signaling would be efficacious against *S. aureus* SSTI.

Among *S. aureus* strains there are four *agr* alleles (*agr*I to *agr*IV) and strains from a given allele secrete a unique thiolactone cyclized AIP (AIPI to AIPIV) ranging from seven to nine amino acids in length. Due to their size, these peptides are inherently non-immunogenic. To overcome this, Park et al. described the production of a monoclonal antibody (AP4-24H11) against a synthetic AIP4 hapten that binds with nM affinity to AIP4 and that largely did not bind to other AIP family members, including AIP1 (μM affinity), which differs by a single amino acid [15,27]. The crystal structure of AP4-24H11 bound to AIP4 reveals that the antibody recognizes the characteristic AIP thiolactone ring conformation, but does not interact with the N-terminal, linear region [28]. Importantly, passive transfer of AP4-24H11 protected against *S. aureus* pathogenicity in a mouse model of dermonecrosis and against a lethal intraperitoneal *S. aureus* challenge. The protection afforded by AP4-24H11 administration occurred without affecting normal bacterial growth, confirming that the AP4-24H11 mechanism of action was specific to inhibiting *S. aureus* virulence. Therefore, this work provided proof of principle that antibodies targeting AIP could be efficacious against *S. aureus* SSTIs [15].

We aimed to develop an active vaccine to provide protection against *S. aureus agr*-mediated pathogenesis. Traditionally, subunit vaccines utilize whole proteins, domains or epitopes conjugated to a carrier. We initially produced Qß VLPs with many copies of chemically conjugated AIP1 peptide, but they failed to elicit a protective response in the dermonecrosis mouse model (unpublished data). This failure may have resulted from potential instability of the native AIP molecule [15]. For example,

the conformational restraint imposed by the AIP thiolactone bond is necessary for binding AgrC and induction of *agr*-signaling, as linearization of the AIPs results in loss of function. Furthermore, oxidation of the C-terminal methionine of AIP1 or AIP4 by host-generated reactive oxygen species is sufficient to inactivate the peptides. Thus, if the VLP-linked AIPs became linearized or oxidized during the conjugation or vaccination process, they would no longer be presented to the immune system as an authentic antigenic target. Therefore, we pursued the novel approach reported here using our affinity selection technology and the previously reported AP4-24H11 mAb targeting *S. aureus* AIP4.

Our approach was to identify peptides that immunologically mimic AIP4. Starting with random sequence peptide libraries on MS2 VLPs, we conducted biopanning on the AP4-24H11 mAb and identified 8 different VLPs displaying peptides that specifically bound the antibody. Vaccination with two of these VLPs elicited an immune response that protected in a *S. aureus* mouse model of dermonecrosis. These data demonstrate the feasibility of our VLP technology to identify immunologic peptide mimics of conformational epitopes and the potential to develop efficacious vaccines against otherwise non-immunogenic, conformationally constrained peptides such as those regulating *S. aureus agr*-dependent virulence. To our knowledge, this is the first report of an efficacious active vaccine targeting the secreted autoinducing peptides of the *S. aureus agr* quorum-sensing system.

Materials and Methods

Plasmid construction and random peptide libraries

The plasmids pDSP62 and pDSP62(am) were previously described [10]. Briefly, pDSP62 expresses the single chain dimer of the MS2 coat protein under the control of the inducible T7 promoter. VLPs produced using pDSP62 contain 90 copies of the displayed peptide per VLP. pDSP62(am) is the same construct except it contains an amber stop codon at the junction of the two coat protein monomers in the single chain dimer. The pDSP62(am) vector produces VLPs that display peptides at low valency (~3 copies of the peptide per VLP) when expressed in a *E. coli* strain containing pNMsupA a plasmid that expresses an alanine-inserting, amber suppressing tRNA under the control of the lac promoter. The suppressor mediates occasional read through of the stop codon, so that pDSP62(am) produces a mixture of wild-type coat protein and the peptide-displaying single-chain dimer, which then coassemble into a mosaic VLP.

We have previously constructed random peptide plasmid libraries for use in our VLP affinity selection protocol that display peptides in the downstream AB loop of the MS2 single chain dimer coat protein [10]. Briefly, oligonucleotides were synthesized with 6, 7, 8 or 10 NNS codons, where N represents an equimolar mixture of all four nucleotides and S is an equal mixture of C and G. NNS codons encode all 20 amino acids and only a single stop codon. Using the Kunkel method, we produced plasmid libraries of at least 10^{10} individual transformants for each peptide library. All plasmids in this study were isolated using Qiagen Qiafilter or minipreps kits (Qiagen, Valencia CA).

VLP production and purification

Plasmid libraries from affinity selection or single plasmids containing defined sequences that bound to AP4-24H11 were electroporated into the *E. coli* T7 expression strain C41(DE3) (Lucigen, Middleton WI) and grown to mid-log phase. Coat protein expression was induced by the addition of IPTG (1 mM, Sigma-Aldrich, St. Louis MO) for three hours and bacteria were collected by centrifugation and the pellet was stored at –20°C overnight. Bacteria were lysed in SCB buffer (50 mM Tris, pH 7.5, 100 mM NaCl) by addition of 10 µg/ml of lysozyme, sonicated and purified from bacterial debris by centrifugation. The supernatant was treated with 10 units/mL of DNaseI (Sigma-Aldrich, St. Louis MO) and the VLPs were purified away from contaminating bacterial proteins by size exclusion chromatography using sepharose CL-4B resin (Sigma-Aldrich, St. Louis MO). Fractions that contained VLPs were combined and precipitated by the addition of ammonium sulfate at 50% saturation. Precipitated VLPs were collected by centrifugation, solubilized in SCB buffer and dialyzed in SCB overnight (Slide-a-lyzer cassettes 20 K MWCO, Millipore, Billerica MA). Purified VLPs were quantitated by Bradford assay (Biorad, Hercules CA) and analyzed by agarose gel electrophoresis and by SDS gel electrophoresis.

Affinity selections

VLP affinity selection was performed on the neutralizing AIP4 monoclonal antibody AP4-24H11 (a generous gift from Gunnar Kaufmann and Kim Janda, Scripps Research Institute). VLP affinity selections were performed as previously described [10]. Briefly, for the initial round of selection, we coated Nunc MaxiSorp ELISA plates (eBiosciences, San Diego, CA) with 250 ng of AP4-24H11 in PBS overnight at 4°C. After washing, wells were blocked with 0.5% nonfat dry milk in PBS and the four VLP-peptide libraries (2.5 µg each of VLPs displaying 6, 7, 8 and 10 mers) were applied to the blocked wells for 2 hours (10 µg total VLP/well). After extensive washing (PBS), bound VLPs were eluted with 0.1 M glycine, pH 2.7 and immediately neutralized by the addition of 1/10 volume of 1 M Tris, pH 9. To make enriched VLP libraries for subsequent rounds of affinity selection, RNA from the eluted VLP were reverse transcribed and the RT products (containing the downstream coat protein and AB loop peptide) were amplified by PCR, digested with Bam HI and Sal I, and ligated into the pDSP62(am) vector. Ligation products were electroporated into the *E. coli* 10 G bacterial strain (Lucigen, Middleton WI), with a one-hour outgrowth and then immediately placed into 100 mL of LB media containing 60 µg/mL of kanamycin. After overnight growth, plasmids were isolated (Qiafilter Midi kit, Valencia CA) and used for VLP production for the next round of affinity selection. All plasmid libraries constructed after affinity selection contained at least 10^6 individual transformants. Two additional rounds of affinity selection were performed with low valency peptide display (~3 peptides/VLP); one using 250 ng of mAb per well and the final round used 50 ng of mAb per well.

Identification and characterization of VLPs

After the final round of affinity selection, plasmid libraries enriched for VLPs displaying peptides that bound to the AIP4 mAb were isolated as described above. 1 pg of each library was electroporated into the C41(DE3) *E. coli* strain to ensure single transformants and bacteria were plated on agar plates containing kanamycin. The next day, single colonies were grown in 1 mL LB to an A_{600} of 0.6 and induced for VLP production with the addition of IPTG for 3 hours. Before induction, a 100 µL aliquot was removed for subsequent plasmid isolation. VLPs were isolated by sonication in SCB buffer and genomic DNA digested by incubation with 10 units of DNaseI. Crude VLP preps were assayed for binding to AP4-24H11 or an unconjugated control mouse IgG (Jackson ImmunoResearch Laboratories, West Grove PA) by ELISA. VLPs that bound to AP4-24H11 at least 5-fold higher then IgG control were further analyzed.

Plasmids were isolated from bacteria that produced VLPs that specifically bound to AP4-24H11, but not control antibody. Nucleotide sequences encoding the various peptides were determined (Eurofins Genomics, Huntsville AL) and plasmids encoding unique peptide sequences were electroporated into C41(DE3) for large scale VLP isolation as described above.

SDS PAGE and agarose gel electrophoresis was used to assess the purity and characterize the isolated VLPs. We ran 2 µg of total protein on a 10% NuPAGE SDS gel (Life Technologies, Grand Island NY) and stained the gel for total protein using Coomassie Blue (BioRad, Hercules CA). Since VLPs encapsulate their RNA we could also characterize VLP selectant particles by electrophoresis on an agarose gel. 10 µg of VLPs were run on a 1% TBE agarose gel containing ethidium bromide.

ELISA

ELISA was used to assess relative binding of VLP selectants to AP4-24H11. Briefly, wells were coated with 250 ng of purified VLPs in PBS and incubated overnight at 4°C. Wells were washed 3 times with PBS and blocked for 1 hour using 3% BSA. Different concentrations of AP4-24H11 in 3% BSA were applied to each well, and incubated at room temperature for 1 hour. Unbound antibody was removed by washing with PBS. Goat anti-Mouse IgG HRP conjugated antibody (Jackson ImmunoResearch Laboratories, West Grove PA) was diluted 1:5000 in 3% BSA and incubated for 1 hour at room temperature. ABTS solution (EMD Millipore, Billerica MA) was used to detect bound HRP antibody and color change was measured by absorbance at 405 nm (Opsys Plate Reader, Thermo Scientific, Waltham MA).

For competition ELISAs, plates were prepared as above. AP4-24H11 (100 ng/well) was mixed with different concentration of the cyclical, bioactive AIP4 peptide (10 µM–0.1 µM) for 10 minutes prior to incubation with VLPs. Secondary antibody and detection was the same as above. As control peptides, we used a linear form of AIP4 or a cyclical peptide from the L2 protein of human papillomavirus 16.

Immunization and skin infection model

Animal studies were carried out in accordance with the recommendation in the Guide for the Care and Use of Laboratory Animals, the Animal Welfare Act, and U.S. federal law. The protocol was approved by the Institutional Animal Care and Use Committee (IACUC) of the University of New Mexico Health Sciences Center. Four to six week old female Balb/c mice (Harlan Laboratories, South Easton MA) were immunized with 10 µg of VLPs in PBS (50 µL total) without the addition of adjuvant by intramuscular injection into the caudal thigh muscle. The initial immunization was followed with 2 boosts each 2 weeks apart. The initial experiment used 3 mice for each VLP and subsequent experiments used 5 mice per vaccine candidate. As negative controls, 2 groups of mice were immunized with either PBS alone or a control VLP (displays no peptide).

The dermonecrosis model of mouse skin infection was previously described and performed with minor modifications [29]. Briefly, one week after the final vaccine boost, mice were anesthetized with isoflurane and inoculated subcutaneously with 1×10^8 CFU of early-exponential phase *S. aureus agr*IV (AH1872–MN TG; generously provided by Dr. Alex Horswill, University of Iowa [30]). Animals were monitored for weight loss and lesion formation for three days post-infection. Lesion formation was assessed by measuring the maximal width and length of the abscess and necrotic ulcer with calipers. Area of the abscess was determined using the equation $A = (\pi/2) \times L \times W$, while the necrotic ulcer area was determined using the equation

$A = L \times W$ [20]. On day three post-infection animals were euthanized by CO_2 asphyxiation and abscesses (2.25 cm^2 area) were collected. The tissue was homogenized and serially diluted for CFU enumeration.

Cytokine analysis of abscess tissues

Abscess homogenates were stored at $-80°C$ until cytokine analysis. Homogenates were rapidly defrosted at $37°C$ and clarified by centrifugation at $12,500 \times g$ for 10 minutes. Cytokine concentrations in clarified supernatants were determined using a custom designed multiplex assay performed as per manufacturer's recommendations (EMD Millipore, Billerica, MA). The assay was read on a Bio-Plex 200 instrument and data analyzed using the Bio-Plex Manager Software (Bio-Rad, Hercules, CA). Statistical analysis of all in vitro data was performed using the two-tailed Student's t-test.

Alpha-hemolysin Western blot

Frozen homogenates were thawed and clarified as described above. Briefly, clarified tissue homogenate was separated by SDS-PAGE on a 16% Tris-glycine gel (Life Technologies, Grand Island, NY) before transfer to a polyvinylidene fluoride membrane. After blocking using TBST (20 mM Tris pH 7.5, 150 mM NaCl, 0.1% Tween 20) with 5% non-fat dry milk, immunodetection was performed using an anti-Staphylococcus alpha hemolysin antibody (Abcam, Cambridge, MA). Immunoreactive band intensity was determined using a FluorChem R System and AlphaView software (proteinsimple, Santa Clara, CA). Relative intensity is the ratio of measured intensity divided by the total protein concentration based on absorbance at 280 nm.

Statistical analysis

Statistical significance was determined using GraphPad Prism v.5.04. The two-tailed Student's t-test was used for analysis of in vitro data, and in vivo data were analyzed by the Mann-Whitney U test for non-parametrics. Results were considered significantly different at p<0.05.

Results

In order to identify mimotopes of the AIP4 mAb AP4-24H11 epitope, we performed affinity selection on AP4-24H11 using our random sequence peptide libraries displayed on MS2 VLPs. The basic methodology is found in Figure S1. We used a mixture of four libraries, each displaying 6-, 7-, 8- or 10-amino acid inserts, with each library containing more than 10^{10} transformants [7,10]. We find that many antibodies have strong preferences for peptide sequences of specific lengths, and using a mixture increases the probability of finding optimal binding peptides. Three iterative rounds of affinity selection were performed, each at increasing stringency; in round 2 we increased the stringency by reducing the display valency from 90 to about 3 peptides per particle. In round 3 stringency was further increased by reducing the amount of antibody 5-fold to 50 ng. Reaction of the selectant population with AP4-24H11 was monitored by ELISA after each round. By the end of round 3 binding was elevated more than 200-fold (data not shown).

Ten cloned round 3 selectants were subjected to sequence analysis, which identified 8 different peptide sequences (Figure 1A). The peptide sequence SGIMPH was found in 3/10 selectants, while the other seven clones had unique sequences. When performing affinity selections on antibodies with linear epitopes, families of related sequences are often encountered. However, there was little primary sequence homology amongst the

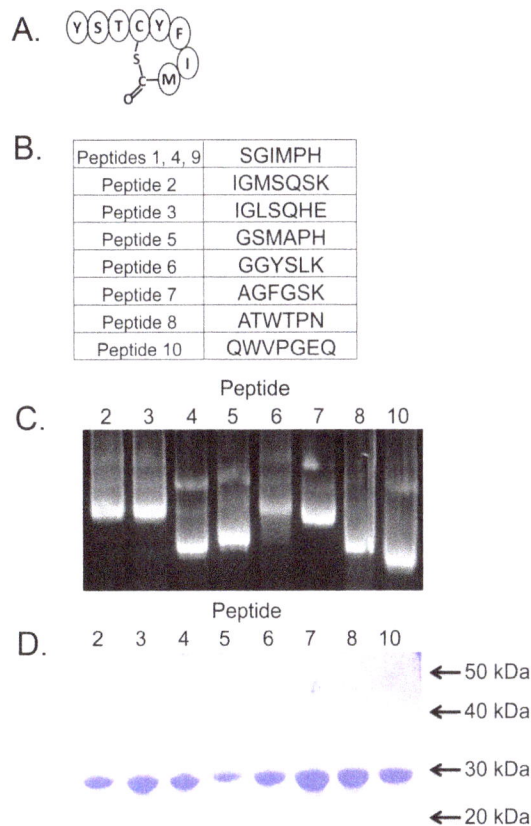

Figure 1. Identification and purification of affinity selected VLP displayed peptides. (A) The sequence and structure of the AIP4 peptide. (B) 10 VLP clones were sequenced after three rounds of VLP affinity selection. Peptide inserts are shown and demonstrate little primary sequence homology to the native AIP4 peptide. (C) Agarose gel analysis of purified VLPs. RNA staining is indicative of intact VLPs (which encapsulate their RNAs) and differences in VLP mobility are mostly due to differences in the charges of the peptides displayed on the VLP surface. (D) SDS-PAGE. Two μg of purified VLP protein were run of a 10% SDS gel and protein was detected by staining the gel using Coomassie Blue.

different peptides, with the exception of peptide 2 and peptide 3. Furthermore, none showed sequence identity to AIP4 (Figure 1A, right) suggesting that these peptides somehow structurally mimic AIP4, or that they bind an antibody paratope distinct from that occupied by AIP4.

The AP4-24H11 selectant VLPs were expressed in *E. coli* and purified by procedures we have described elsewhere [7]. Their elution behavior from the gel filtration matrix Sepharose CL-4B shows that each assembles into a particle the size expected of the VLP (not shown). Figure 1B shows the electrophoretic behavior of each VLP in agarose. Since only intact particles contain RNA, their staining with ethidium bromide verifies their intactness. Differences in electromobility are due mostly to charge differences imparted by the presence of the peptides on the VLP surface. To assess the purity of the VLP preparation prior to binding assays and immunizations, we analyzed protein content by SDS PAGE followed by Coommassie blue staining. VLPs (single chain dimer ~28 kD) were effectively purified away from contaminating bacterial protein (Figure 1C).

A direct ELISA was used to confirm that the affinity selected VLPs bound specifically to AP4-24H11. VLPs were used as the coating antigen (250 ng/well) and various amounts (0.1 to 500 ng) of AP4-24H11 were added to each well. AP4-24H11 bound to all of the VLPs displaying the selected peptides, whereas little to no binding of AP4-24H11 was observed with a control VLP (Figure 2A). The selected VLPs demonstrated a range of binding to AP4-24H11 with a ~3.5-fold difference between the strongest binder (VLP displaying peptide 5) and the lowest (VLP displaying peptide 8). As an additional control, VLP coated wells were incubated with a mouse IgG control antibody using the same dilutions as used with the AIP4 mAb. Little or no binding was detected with the control antibody for any VLP samples (data not shown).

To ensure that selecting VLPs are binding to the antigen-binding site of AP4-24H11 rather than the Fc region, we investigated the ability of bioactive AIP4 peptide to compete with VLPs for antibody binding. VLPs were the coating antigen, and prior to the addition of AP4-24H11 (100 ng/well) various amounts of bioactive AIP4 peptide (10 μM–0.1 μM) were incubated with the mAb. The peptide/antibody mixture was added to the VLP coated wells and incubated for 1 hour. Similar to the results shown in Figure 2A, there was a range of peptide concentrations required to inhibit antibody binding (Figure 2B). Importantly, all VLPs were competed off the antibody by bioactive AIP4, suggesting that the selected VLPs are interacting with the antigen binding site of AP4-24H11. These data demonstrate that affinity selection can identify a population of VLPs displaying peptides that bind specifically to the mAb AP4-24H11.

Next, we tested whether any of the selected VLPs could serve as an immunologic mimic of AIP4. Secreted virulence factors regulated by *agr*, such as Hla, mediate dermonecrosis, suggesting that vaccination with VLPs presenting immunologic mimics of the AP4-24H11 epitope would elicit protection against *agr*IV-mediated dermonecrosis. To test this, we vaccinated groups of 3 mice with VLPs presenting AP4-24H11-selected peptides or a VLP control, and challenged the mice by subcutaneous injection with *S. aureus agr*IV isolate AH1872 [24] (Figure 3A). We observed the

mice for three days post-infection as we typically see maximum ulcer development by this time point followed by resolution over approximately the next seven days [31,32]. Compared to VLP control vaccinated mice, mice vaccinated with peptide 4 VLPs showed significantly reduced abscess area on days 1 and 3 post-infection (Figure 3B). In addition, mice vaccinated with either peptide 2 or peptide 4 VLPs showed a trend toward reduced dermonecrosis (ulcer area) on days 2 and 3 post-infection, although this did not reach statistical significance (Figure 3C). Vaccination with VLPs displaying peptide 3 and peptides 5–10 were included in pilot testing but no protection was observed, therefore these VLPs were not included in further studies (data not shown).

Based on the reduced ulcer area in mice vaccinated with peptide 2 or 4 VLPs, we asked whether vaccination with a combination of peptide 2 and peptide 4 VLPs (peptide 2/4-VLPs) would result in a significant reduction in dermonecrosis following *S. aureus* challenge. To address this, mice were vaccinated with a mixture (1:1) of VLPs displaying peptide 2 and 4, control VLPs or PBS alone, and then challenged by subcutaneous injection with *S. aureus* AH1872. Whereas reductions in ulcer area in mice immunized with either peptide 2- or peptide 4-VLPs alone failed to reach statistical significance, mice immunized with combined peptide 2/4-VLPs had significantly reduced dermonecrosis compared to control vaccinated mice on day three post-infection (Figure 4A). Decreased ulcer area in peptide 2/4-VLP vaccinated mice was not due to a reduction in bacterial burden at the site of infection (Figure 4B) suggesting the decreased dermonecrosis resulted from inhibition of *agr* signaling. In support of this view, decreased dermonecrosis in the peptide 2/4-VLP vaccinated mice was associated with a significant decrease in local IL-1β levels, but not decreases in IL-6 or keratinocyte-derived chemokine (KC), compared to control vaccinated mice (Figure 4C). Such a decrease in local IL-1β levels is consistent with reduced *agr*-mediated expression of Hla, which causes pore-formation in host cells leading to NLRP3 inflammasome activation and IL-1β production

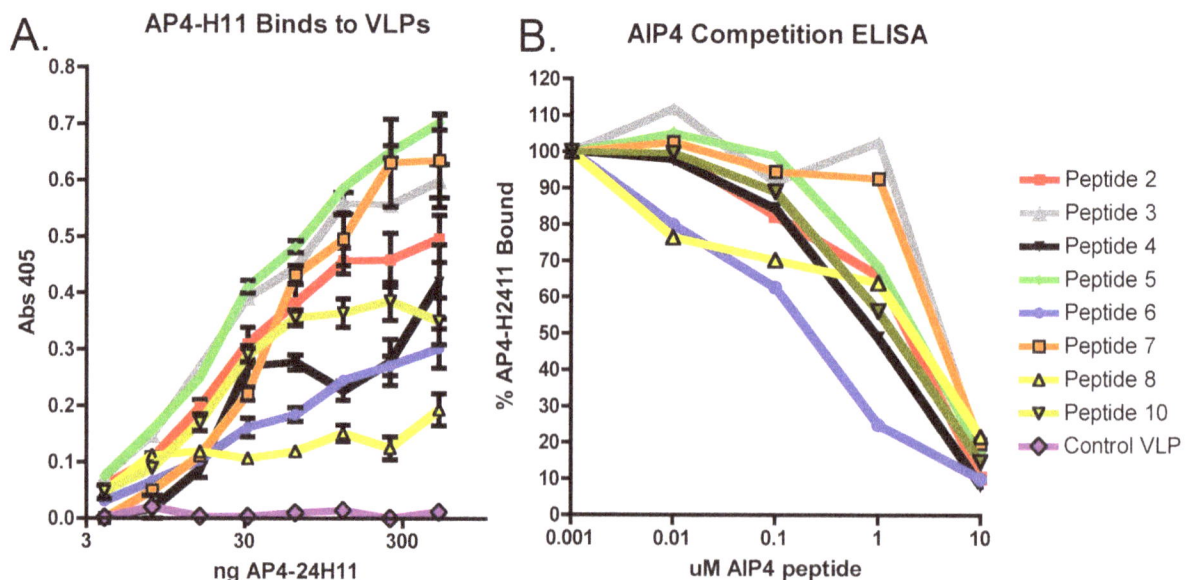

Figure 2. Affinity selected VLPs bind to and occupy the antigen binding site of mAb AP4-24H11. (A) ELISA was used to assess specific binding of VLP selectants to AP4-24H11. Wells were coated with the indicted VLPs and different concentrations of AP4-24H11 were applied. Error bars represent standard error of the mean. (B) For competition ELISAs, wells were coated with the indicated VLPs and AP4-24H11 was mixed with different concentration of the cyclical, bioactive AIP4 peptide prior to incubation with VLPs. Secondary antibody and detection was the same as above. As a control peptide, we used a linear form of AIP4. Results are representative of an experiment performed twice.

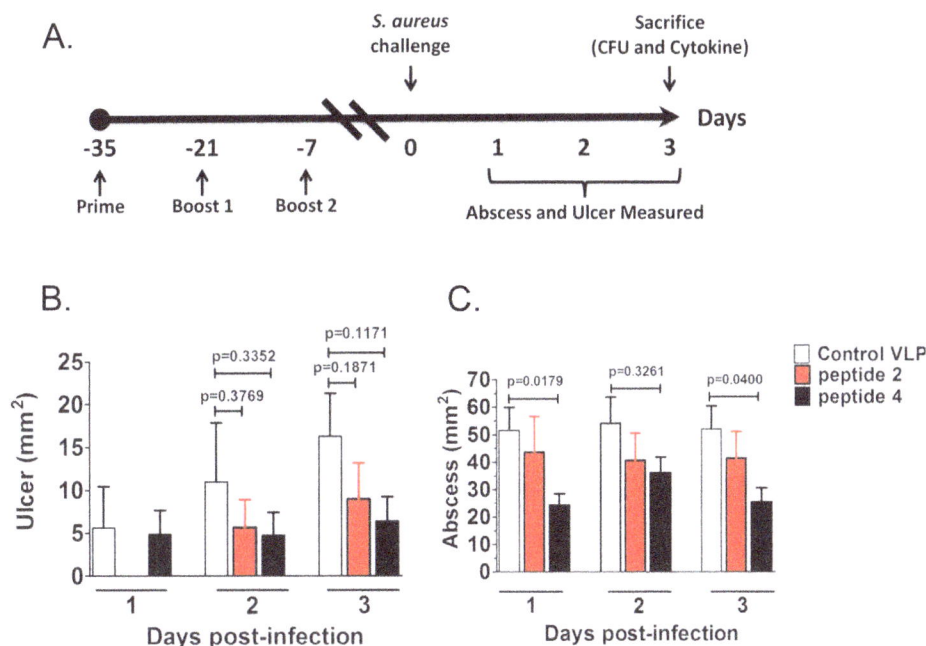

Figure 3. Vaccination with VLP mimotopes of AIP4 is efficacious in a mouse model of *S. aureus* **SSTI.** (A) Mice were vaccinated three times at two week intervals with 10 μg of VLP. One week after the final vaccination mice were inoculated subcutaneously with 1×10^8 CFU of *S. aureus agr*-IV isolate AH1872, and abscess (B) and ulcer (C) areas were measured over the course of 72 hours. Data are shown as the mean ± SEM of at least two independent experments totalling 8–10 mice per group.

[33–39]. To demonstrate that reduced IL-1β at the site of infection in peptide 2/4-VLP vaccinated mice was associated with decreased *agr*-mediated virulence factor expression, we measured Hla in tissue homogenate by Western blot analysis. As expected, peptide 2/4-VLP vaccinated mice had significantly less Hla at the site of infection compared to controls (Figure 4D). Together, these data suggest that peptides 2 and 4 identified by VLP affinity selection can serve as immunologic mimics of the *S. aureus* AIP4 mAb AP4-24H11 epitope and provide protection against *agr*-mediated dermonecrosis.

Discussion

Recent technological advances have resulted in the isolation and characterization of a host of broadly neutralizing monoclonal antibodies having prophylactic and therapeutic effects against a variety of pathogens. We have developed a novel vaccine technology that takes advantage of these newly identified antibodies that allows for epitope discovery and mimicry on a highly immunogenic platform. We recently reported the use of this VLP selection platform to identify epitopes for several previously characterized mAbs that recognize linear epitopes [10]. In this paper, we extend these observations by identifying peptide mimics of the conformational epitope from the *S. aureus* AIP4 mAb AP4-24H11. Of critical importance is that compared to controls, co-immunization with two of the selected VLP candidates limited *agr*-signaling and pathogenesis during *S. aureus* SSTI as indicated by (1) decreased expression of the *agr*-regulated virulence factor Hla, (2) reduced local levels of the inflammatory cytokine IL-1 and (3) reduced dermonecrosis. Furthermore, although immunization did not impact bacterial burden at the time point evaluated, we and others have shown that, along with preventing or limiting dermonecrosis, disruption of *agr*-signaling or neutralization of Hla leads to increased bacterial clearance during resolution of infection [18,31,40]. This suggests the potential for vaccination with VLP-based AIP mimotopes to not only limit *agr*-dependent pathogenesis, but also to eventually contribute to host-mediated bacterial clearance. Importantly, these data provide proof-of-principle that our VLP technology provides a background upon which to develop efficacious vaccines against otherwise non-immunogenic, conformationally constrained epitopes. Antibodies are by nature polyspecific. In the universe of all possible short peptide sequences, an antibody may be capable of binding a number of them. Therefore, it is possible that only some affinity-selected peptides will bind the antigen-combining site through interactions mimicking those of the authentic antigen. It is well known, for example, that M13 phage display frequently finds so-called functional mimics, peptides that bind the antibody at paratopes distinct from the antigen itself. When utilized as immunogens, functional mimics fail to elicit antibodies with the desired specificity against the original antigen. We suspect functional mimics are especially readily encountered with antibodies like AP4-24H11 whose binding sites have not been optimized for binding to a simple linear peptide epitope. Immunogenic mimics, on the other hand, form molecular contacts with the selecting antibody similar to those that engage the antigen itself, and are therefore more likely to provoke antibodies that bind the antigen. Even in these cases, without detailed structural analysis it may be impossible to discern any obvious structural similarity between the original epitope and its affinity-selected immunogenic mimic. Of the 10 peptides we characterized here, 8 are apparently in the functional mimic category; they bind the antibody but fail to elicit antibodies with specificity for AIP4. However, two peptides elicited antibodies that served as immunogenic mimics as determined by their ability to provoke an immune response that protected against *S. aureus*-mediated dermonecrosis.

To date, no anti-*Staphylococcus aureus* vaccine has succeeded in Phase III clinical trials [41]. Such vaccines have primarily relied on immunization with *S. aureus* surface protein antigens,

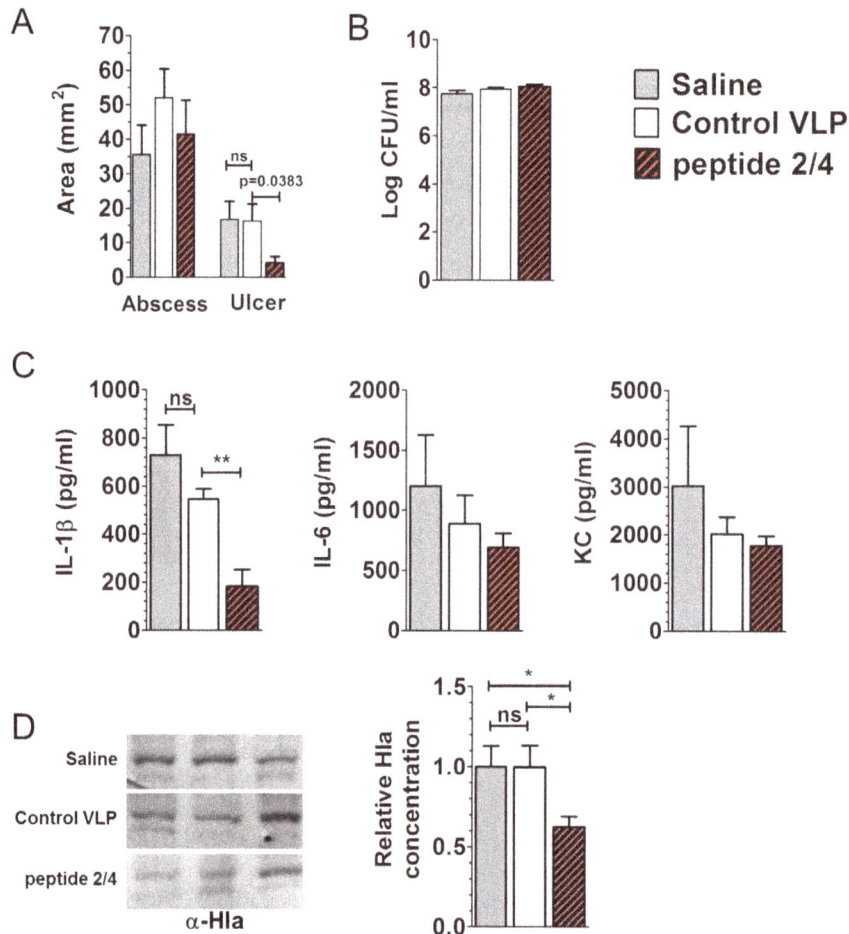

Figure 4. A combination vaccine of two VLP mimotopes limits pathogenesis in a mouse model of *S. aureus* dermonecrosis. Mice were vaccinated with 10 μg of a 1:1 suspension of peptide 2 and peptide 4 and inoculated with 1×10^8 CFU of AH1872 as described previously. At the apex of infection (day 3) (A) abscess and ulcer area were measured and (B) bacterial burden at the site of infection was determined, and (C) local cytokine and chemokine levels were determined. (D) Western blot showing relative HLA levels in tissue homogenate of vaccinated and challenged mice. Quantification based on Western blot band intensity relative to total protein concentration. Data are shown as the mean ± SEM of 6–10 mice per group. ns, not significant; *, $p < 0.05$; **, $p < 0.01$.

suggesting that strategies aimed at inducing opsonophagocytic antibodies are not sufficient to prevent disease by this pathogen. We are not alone in recognizing the possible utility of vaccines that target secreted virulence factors [42,43]. For example, neutralization of the secreted virulence factor Hla using active vaccination or passive transfer of neutralizing antibodies has proven efficacious in several *S. aureus* infection models [20,21,38,39,44–46] and a vaccine targeting recombinant Hla was part of a recent clinical trial (NCT01011335). The previous example is one of several strategies based on inhibition of a single secreted virulence factor; however, immune-based approaches to inhibit *S. aureus* virulence factor expression on a global level have been limited (reviewed in [42]).

Park et al. recently demonstrated in vivo protection via passive administration of a mAb, directed against a synthetic AIP4 hapten, which prevents global virulence factor expression by inhibiting *agr* quorum sensing [15]. However, the availability of both prophylactic vaccines and therapeutic mAbs targeting *S. aureus agr*-regulated virulence would significantly increase the translational spectrum of this anti-staphylococcal virulence approach. Herein, we expanded on the work of Park et al. using MS2 VLP libraries and affinity selection to develop a vaccine strategy for *agr*-

inhibition. Using this technique against the mAb AP4-24H11, we identified candidate vaccines with in vivo efficacy in a mouse model of *S. aureus* SSTI.

Supporting Information

Figure S1   VLP affinity selection to identify mimotopes of Mab AP4-24H11. Wells of an ELISA plate were coated with the Mab AP4-24H11 and were incubated with VLP libraries displaying random peptides. RNA sequences from bound VLPs were recovered by RT-PCR and re-cloned into VLP expression constructs and VLP libraries enriched for peptides binding to AP4-24H11 were produced. Three rounds of biopanning were used and clones of the resulting VLPs were sequence for peptide identification and subsequent functional analysis.

Author Contributions

Conceived and designed the experiments: JPO DP BC PH. Performed the experiments: JPO SD KDT. Analyzed the data: JPO SD PH. Contributed reagents/materials/analysis tools: DP PH. Contributed to the writing of the manuscript: JPO SD PH.

References

1. Chackerian B (2007) Virus-like particles: flexible platforms for vaccine development. Expert Rev Vaccines 6: 381–390.
2. Tumban E, Peabody J, Peabody DS, Chackerian B (2011) A pan-HPV vaccine based on bacteriophage PP7 VLPs displaying broadly cross-neutralizing epitopes from the HPV minor capsid protein, L2. PLoS One 6: e23310.
3. Tissot AC, Renhofa R, Schmitz N, Cielens I, Meijerink E, et al. (2010) Versatile virus-like particle carrier for epitope based vaccines. PLoS ONE 5: e9809.
4. Jennings GT, Bachmann MF (2008) The coming of age of virus-like particle vaccines. Biol Chem.
5. Mihailova M, Boos M, Petrovskis I, Ose V, Skrastina D, et al. (2006) Recombinant virus-like particles as a carrier of B- and T-cell epitopes of hepatitis C virus (HCV). Vaccine 24: 4369–4377.
6. Caldeira JC, Peabody DS (2011) Thermal stability of RNA phage virus-like particles displaying foreign peptides. J Nanobiotechnology 9: 22.
7. Peabody DS, Manifold-Wheeler B, Medford A, Jordan SK, do Carmo Caldeira J, et al. (2008) Immunogenic display of diverse peptides on virus-like particles of RNA phage MS2. J Mol Biol 380: 252–263.
8. Caldeira JC, Peabody DS (2007) Stability and assembly in vitro of bacteriophage PP7 virus-like particles. J Nanobiotechnology 5: 10.
9. Caldeira Jdo C, Medford A, Kines RC, Lino CA, Schiller JT, et al. (2010) Immunogenic display of diverse peptides, including a broadly cross-type neutralizing human papillomavirus L2 epitope, on virus-like particles of the RNA bacteriophage PP7. Vaccine 28: 4384–4393.
10. Chackerian B, Caldeira Jdo C, Peabody J, Peabody DS (2011) Peptide epitope identification by affinity selection on bacteriophage MS2 virus-like particles. J Mol Biol 409: 225–237.
11. Moran GJ, Krishnadasan A, Gorwitz RJ, Fosheim GE, McDougal LK, et al. (2006) Methicillin-resistant S. aureus infections among patients in the emergency department. N Engl J Med 355: 666–674.
12. Wright JS 3rd, Jin R, Novick RP (2005) Transient interference with staphylococcal quorum sensing blocks abscess formation. Proc Natl Acad Sci U S A 102: 1691–1696.
13. Montgomery CP, Boyle-Vavra S, Daum RS (2010) Importance of the global regulators Agr and SaeRS in the pathogenesis of CA-MRSA USA300 infection. PLoS One 5: e15177.
14. Wang R, Braughton KR, Kretschmer D, Bach TH, Queck SY, et al. (2007) Identification of novel cytolytic peptides as key virulence determinants for community-associated MRSA. Nat Med 13: 1510–1514.
15. Park J, Jagasia R, Kaufmann GF, Mathison JC, Ruiz DI, et al. (2007) Infection control by antibody disruption of bacterial quorum sensing signaling. Chem Biol 14: 1119–1127.
16. Novick RP, Geisinger E (2008) Quorum sensing in staphylococci. Annu Rev Genet 42: 541–564.
17. Cheung GY, Wang R, Khan BA, Sturdevant DE, Otto M (2011) Role of the accessory gene regulator agr in community-associated methicillin-resistant Staphylococcus aureus pathogenesis. Infect Immun 79: 1927–1935.
18. Kobayashi SD, Malachowa N, Whitney AR, Braughton KR, Gardner DJ, et al. (2011) Comparative analysis of USA300 virulence determinants in a rabbit model of skin and soft tissue infection. J Infect Dis 204: 937–941.
19. Inoshima N, Wang Y, Bubeck Wardenburg J (2012) Genetic requirement for ADAM10 in severe Staphylococcus aureus skin infection. J Invest Dermatol 132: 1513–1516.
20. Kennedy AD, Bubeck Wardenburg J, Gardner DJ, Long D, Whitney AR, et al. (2010) Targeting of alpha-hemolysin by active or passive immunization decreases severity of USA300 skin infection in a mouse model. J Infect Dis 202: 1050–1058.
21. Tkaczyk C, Hua L, Varkey R, Shi Y, Dettinger L, et al. (2012) Identification of anti-alpha toxin monoclonal antibodies that reduce the severity of Staphylococcus aureus dermonecrosis and exhibit a correlation between affinity and potency. Clin Vaccine Immunol 19: 377–385.
22. Rothfork JM, Timmins GS, Harris MN, Chen X, Lusis AJ, et al. (2004) Inactivation of a bacterial virulence pheromone by phagocyte-derived oxidants: new role for the NADPH oxidase in host defense. Proc Natl Acad Sci U S A 101: 13867–13872.
23. Peterson MM, Mack JL, Hall PR, Alsup AA, Alexander SM, et al. (2008) Apolipoprotein B Is an innate barrier against invasive Staphylococcus aureus infection. Cell Host Microbe 4: 555–566.
24. Hall PR, Elmore BO, Spang CH, Alexander SM, Manifold-Wheeler BC, et al. (2013) Nox2 modification of LDL is essential for optimal apolipoprotein B-mediated control of agr type III Staphylococcus aureus quorum-sensing. PLoS Pathog 9: e1003166.
25. Schlievert PM, Case LC, Nemeth KA, Davis CC, Sun Y, et al. (2007) Alpha and beta chains of hemoglobin inhibit production of Staphylococcus aureus exotoxins. Biochemistry 46: 14349–14358.
26. Pynnonen M, Stephenson RE, Schwartz K, Hernandez M, Boles BR (2011) Hemoglobin promotes Staphylococcus aureus nasal colonization. PLoS Pathog 7: e1002104.
27. Kaufmann GF, Park J, Mayorov AV, Kubitz DM, Janda KD (2011) Generation of quorum quenching antibodies. Methods Mol Biol 692: 299–311.
28. Kirchdoerfer RN, Garner AL, Flack CE, Mee JM, Horswill AR, et al. (2011) Structural basis for ligand recognition and discrimination of a quorum-quenching antibody. J Biol Chem 286: 17351–17358.
29. Malachowa N, Kobayashi SD, Braughton KR, DeLeo FR (2013) Mouse model of Staphylococcus aureus skin infection. Methods Mol Biol 1031: 109–116.
30. Malone CL, Boles BR, Lauderdale KJ, Thoendel M, Kavanaugh JS, et al. (2009) Fluorescent reporters for Staphylococcus aureus. J Microbiol Methods 77: 251–260.
31. Sully EK, Malachowa N, Elmore BO, Alexander SM, Femling JK, et al. (2014) Selective chemical inhibition of agr quorum sensing in Staphylococcus aureus promotes host defense with minimal impact on resistance. PLoS Pathog 10: e1004174.
32. Bose JL, Daly SM, Hall PR, Bayles KW (2014) Identification of the Staphylococcus aureus vfrAB operon, a novel virulence factor regulatory locus. Infect Immun 82: 1813–1822.
33. Munoz-Planillo R, Franchi L, Miller LS, Nunez G (2009) A critical role for hemolysins and bacterial lipoproteins in Staphylococcus aureus-induced activation of the Nlrp3 inflammasome. J Immunol 183: 3942–3948.
34. Craven RR, Gao X, Allen IC, Gris D, Bubeck Wardenburg J, et al. (2009) Staphylococcus aureus alpha-hemolysin activates the NLRP3-inflammasome in human and mouse monocytic cells. PLoS One 4: e7446.
35. Berube BJ, Bubeck Wardenburg J (2013) Staphylococcus aureus alpha-toxin: nearly a century of intrigue. Toxins (Basel) 5: 1140–1166.
36. Cho JS, Guo Y, Ramos RI, Hebroni F, Plaisier SB, et al. (2012) Neutrophil-derived IL-1beta is sufficient for abscess formation in immunity against Staphylococcus aureus in mice. PLoS Pathog 8: e1003047.
37. Bhakdi S, Muhly M, Korom S, Hugo F (1989) Release of interleukin-1 beta associated with potent cytocidal action of staphylococcal alpha-toxin on human monocytes. Infect Immun 57: 3512–3519.
38. Bubeck Wardenburg J, Schneewind O (2008) Vaccine protection against Staphylococcus aureus pneumonia. J Exp Med 205: 287–294.
39. Hua L, Hilliard JJ, Shi Y, Tkaczyk C, Cheng LI, et al. (2014) Assessment of an anti-alpha-toxin monoclonal antibody for prevention and treatment of Staphylococcus aureus-induced pneumonia. Antimicrob Agents Chemother 58: 1108–1117.
40. Brady CS, Bartholomew JS, Burt DJ, Duggan-Keen MF, Glenville S, et al. (2000) Multiple mechanisms underlie HLA dysregulation in cervical cancer. Tissue Antigens 55: 401–411.
41. Fowler VG Jr, Proctor RA (2014) Where does a Staphylococcus aureus vaccine stand? Clin Microbiol Infect 20 Suppl 5: 66–75.
42. Kaufmann GF, Park J, Janda KD (2008) Bacterial quorum sensing: a new target for anti-infective immunotherapy. Expert Opin Biol Ther 8: 719–724.
43. Cheung GY, Otto M (2012) The potential use of toxin antibodies as a strategy for controlling acute Staphylococcus aureus infections. Expert Opin Ther Targets 16: 601–612.
44. Ragle BE, Bubeck Wardenburg J (2009) Anti-alpha-hemolysin monoclonal antibodies mediate protection against Staphylococcus aureus pneumonia. Infect Immun 77: 2712–2718.
45. Adhikari RP, Karauzum H, Sarwar J, Abaandou L, Mahmoudieh M, et al. (2012) Novel structurally designed vaccine for S. aureus alpha-hemolysin: protection against bacteremia and pneumonia. PLoS One 7: e38567.
46. Foletti D, Strop P, Shaughnessy L, Hasa-Moreno A, Casas MG, et al. (2013) Mechanism of action and in vivo efficacy of a human-derived antibody against Staphylococcus aureus alpha-hemolysin. J Mol Biol 425: 1641–1654.

21

Quantitative Proteomic Analysis of Serum from Pregnant Women Carrying a Fetus with Conotruncal Heart Defect Using Isobaric Tags for Relative and Absolute Quantitation (iTRAQ) Labeling

Ying Zhang[1], Yuan Kang[1], Qiongjie Zhou[1], Jizi Zhou[1], Huijun Wang[2], Hong Jin[3,4], Xiaohui Liu[3], Duan Ma[5]*, Xiaotian Li[1,5,6]*

1 Obstetrics and Gynecology Hospital, Fudan University, Shanghai, China, 2 Children's Hospital, Fudan University, Shanghai, China, 3 Department of Chemistry, Fudan University, Shanghai, China, 4 Institute of Biomedicine, Fudan University, Shanghai, China, 5 Key Laboratory of Molecular Medicine, Ministry of Education, Department of Biochemistry and Molecular Biology, Institute of Biomedical Sciences, Shanghai Medical College, Fudan University, Shanghai, China, 6 Shanghai Key Laboratory of Female Reproductive Endocrine Related Diseases, Shanghai, China

Abstract

Objective: To identify differentially expressed proteins from serum of pregnant women carrying a conotruncal heart defects (CTD) fetus, using proteomic analysis.

Methods: The study was conducted using a nested case-control design. The 5473 maternal serum samples were collected at 14–18 weeks of gestation. The serum from 9 pregnant women carrying a CTD fetus, 10 with another CHD (ACHD) fetus, and 11 with a normal fetus were selected from the above samples, and analyzed by using isobaric tags for relative and absolute quantitation (iTRAQ) coupled with two-dimensional liquid chromatography-tandem mass spectrometry(2D LC-MS/MS). The differentially expressed proteins identified by iTRAQ were further validated with Western blot.

Results: A total of 105 unique proteins present in the three groups were identified, and relative expression data were obtained for 92 of them with high confidence by employing the iTRAQ-based experiments. The downregulation of gelsolin in maternal serum of fetus with CTD was further verified by Western blot.

Conclusions: The identification of differentially expressed protein gelsolin in the serum of the pregnant women carrying a CTD fetus by using proteomic technology may be able to serve as a foundation to further explore the biomarker for detection of CTD fetus from the maternal serum.

Editor: Philippe Rouet, I2MC INSERM UMR U1048, France

Funding: This project was supported by Shanghai Municipal Natural Science Foundation (10ZRI404900), Health industry special funds for Public Benefit Research Foundation from the Ministry of Health, People's Republic of China (201002013), the National Science Fund of China (81270712), Program of Shanghai Subject Chief Scientist (12XD1401300), Program of Shanghai Leading Talent (2012), Shanghai Municipal Health Bureau (12GWZX0301), and National Key Basic Research Plan of China (973 Plan) (2010CB529500). The funders had no role in study design, data collection and analysis, decision to publish, or preparation of the manuscript.

Competing Interests: The authors have declared that no competing interests exist.

* Email: xiaotianli555@163.com (XL); duanma@fudan.edu.cn (DM)

Introduction

Congenital heart defects (CHDs) comprise the most common type of human birth defects, occurring in approximately one in 100 live births [1,2,3]. CHDs can be attributed to chromosomal and genetic abnormalities [4,5], exposure to teratogens [6], maternal diabetes [7], maternal folate status, and folate-related genes [8]. Conotruncal heart defects (CTDs) account for 20–30% of CHDs [9,10,11], and affect the ventricular outflow tract and the arterial pole of the heart [12,13,14]. Only 20–25% of CTDs can be attributed to the above risk factors [15], most cases are nonsyndromic, with little known about their cause and risk.

Currently there are no effective strategies for reducing the occurrence of CTDs, and no methods of early detection.

Serum protein screening is an important diagnostic tool, with a rich source, good sensitivity and simplicity. Proteins originating from the placenta, amniotic fluid or fetus circulation may cross the placenta barrier and exist in maternal serum. Currently, non-invasive procedures based on protein screening from maternal serum have been applied in the early screening of Down's syndrome [16,17] by using the following serum markers: human chorionic gonadotropin (hCG), a-fetoprotein(AFP) [18], pregnancy-associated plasma protein (PAPP-A), unconjugated estriol (uE3) and inhibin-A [19]. An effective prenatal screening for CTDs is, however, lacking.

ITRAQ coupled with 2D LC-MS/MS appears a powerful technique in proteomics for identification of the protein quantitative changes caused by exposure or disease processes in cells, tissues or biological fluids [20,21]. Tandem mass spectrometric analysis allows for the identification of multiple peptides per protein, providing increased confidence in both the identification and quantification of dysregulated protein. Recently, in many pathological pregnancies, proteomic technology has been used for the identification of differentially expressed proteins in amniotic fluid or maternal serum/plasma. These include screening for fetus' with abnormal karyotypes such as trisomy 21 [16,22,23], trisomy 18 [24], Turner syndrome [25,26], Klinefelter syndrome (47, XXY karyotype) [27], intra-uterine growth restriction [28,29], preeclampsia [29,30], spontaneous preterm birth [31,32] and intra-amniotic infection [33]. To the best of our knowledge, there are currently no reports on the application of proteomics to characterize differentially expressed proteins in maternal serum/plasma with CTD fetus.

In this study, we performed a relative quantitative comparison of differentially expressed proteins from the sera of women carrying a CTD fetus using iTRAQ combined with 2D LC-MS/MS, in order to explore potential screening markers of CTD fetus with sufficient sensitivity and specificity for clinical applications.

Materials and Methods

1. Study population and design

This nested case-control study was carried out in the Obstetrics and Gynecology Hospital affiliated with Fudan University (Shanghai, China) between August 2009 and July 2010. 5437 pregnant women were enrolled, maternal peripheral venous blood samples were collected at 14–18 weeks of gestation, serum was isolated at 4°C and stored at −80°C for use. Prenatal ultrasound examinations were performed on all fetuses to screen for developmental abnormalities at 20–24 weeks of gestation; CTD, ACHD fetuses and normal controls were confirmed by prenatal and postnatal echocardiography or autopsy. Fetuses with chromosome abnormalities and multi-malformation, pregnant women with multiple pregnancies, pregnancy-related complications, abnormalities of cardiac structure and function or other comorbidities were excluded. Clinical information was collected for the established cases and controls. The controls were matched for (i) maternal age, (ii) gestational time at serum sample collection, (iii) gestational age at diagnosis by routine obstetric ultrasound, (iv) Number of pregnancies, and (v) parity (Table 1). Additionally, four heart tissues of CTD fetuses from the above cases and four heart tissues of normal controls from induced abortions because of unintended pregnancies were collected at 23–25 weeks of gestation and frozen at −80°C. The study was approved by the Research Ethics Board of Obstetrics and Gynecology Hospital affiliated with

Fudan University, and participants gave signed written informed consent.

2. Depletion of high abundance proteins from maternal serum

Total protein content was determined in each serum sample, and samples from the same disease states were pooled with an equal protein amount to limit variability. Highly-abundant proteins were depleted by the Multiple Affinity Removal LC Column (4.6*50mm, Agilent Technologies, Inc, Palo Alto, CA) as per manufacturer's instructions. In each pooled sample, the high abundance proteins included albumin, immunoglobulin (Ig) A, IgG, IgM, transferrin, R1-acid glycoprotein, fibrinogen, α2-Macroglobulin, a1-antitrypsin, haptoglobin, apolipoprotein A-I, and apolipoprotein A-II. The depleted samples were then concentrated using centrifugal filter units, and the protein concentration was determined by the Bradford Protein Assay using a bovine serum albumin standard curve.

3. Protein digestion and iTRAQ labeling

Depleted protein samples were digested by using Acetone precipitation Ready Prep 2-D Cleanup Kit (Bio-Rad, Inc., CA). 100 μg of protein from each sample was dissolved in iTRAQ dissolution buffer according to the manufacturer's instructions (Applied Biosystems, Foster City, CA), reduced by 2 μl reducing reagent, incubated at 60°C for 1 h, alkylated in 1μl cysteine blocking reagent for 10 min at room-temperature and digested with trypsin at a ratio of 1:20 overnight at 37°C. Digested samples were labeled with iTRAQ reagents following the recommended protocol. Isopropanaol was used to solubilize the iTRAQ isobaric tagging reagents. CTD group, ACHD group, and normal control were amino-labeled by iTRAQ reagents-118, 119, 121 respectively. The three samples were pooled, and the mixture of the trypsin-digested and iTRAQ-labeled samples was evaporated to dryness under vacuum and resuspended in acetonitrile (ACN) for the following MS analysis.

4. 2D LC–MS/MS analysis for protein identification and relative quantification

The labeled peptide mixture was fractionated by strong cation exchange (SCX) chromatography on an ultimate high-performance liquid chromatography (HPLC) system (Shimadzu, Kyoto, Japan) with a SCX column (2.1mm*100mm, 5μm, 200A, The Nest Group, Inc., MA). The mixed sample was suspended in buffer A (defined below) and loaded onto the column. Peptides were separated with a linear gradient of 0–80% buffer B in buffer A at a constant flow rate of 200 μl/min for 60 min. Buffer A consisted of 10 mM KH_2PO_4 and 25% ACN, pH 2.6 and Buffer B contained 10 mM KH_2PO_4, 25% ACN, 350mM KCl, pH 2.6.

Table 1. Clinical characteristics of the pregnant women with a CTD, ACHD or normal fetus.

	CTD (n = 9)	ACHD (n = 10)	Control (n = 11)	p
Maternal age(years)	28.7±3.1	28.0±3.5	27. 0±3.2	NS
Gestational age for collection of serum(weeks)	16.9±1.1	16.1±1.5	16.4±1.1	NS
Gestational age at diagnosis by routine obstetric ultrasound(weeks)	22.5±1.0	22.3±1.3	22.5±1.3	NS
Number of pregnancies	2.0±1.3	2.2 ±1.	1.9±1.	NS
parity	1.1±0.4	1.2±0.	1.0±0.	NS

Statistical significant difference was not observed (one way-ANOVA) in three groups.

The chromatogram was monitored at 214 nm and 280 nm. 8 SCX fractions were collected along the gradient, and dried using a rotary vacuum concentrator (Christ RVC 2–25,Christ,Germany). The dried SCX peptides were dissolved in buffer C [5% ACN, 0.1% formic acid (FA)], prior to analysis on a QSTAR XL system (Applied Biosystems, Concord, ON, Canada). Peptides were loaded on a ZORBAX 300SB-C18 column (5 μm, 300 Å, 0.1×150 mm, Microm, USA) combined with the HPLC system at a flow rate of 0.3 μl/min for 90 min, and separated by C18 chromatography (75 μm ID, 100 Å) PepMap100 analytical column. The HPLC gradient ramped from 5% to 80% buffer D (95% ACN, 0.1% FA) in buffer C. The mass spectrometer data was acquired in information-dependent acquisition (IDA) mode with the m/z range set at 400–1800, and the four most intense peaks of were MS/MS scanned from m/z 100–2000. Data was acquired and collected with Analyst QS software (Version 1.1, service pack 8, AB/SCIEX).

5. Protein identification and quantitation

Both protein identification and quantification were performed by using the Paragon Algorithm in ProteinPilot software (version 3.0, Applied Biosystems). The settings were as follows: Sample Type, iTRAQ 8-plex (Peptide Labeled), Cys alkylation: methyl methane thiosulfonate (MMTS), digestion: trypsin, Instrument: QSTARESI, search effort: through ID, ID focus: biological modification, Database: non-redundant international protein index (IPI version 3.45 human). Protein ratio >1.40 or <0.71 was considered as potential differentially expressed proteins between diseased cases and controls for further investigation [34].

6. Western blotting

To validate the iTRAQ results, a total of 20 serum samples were screened by Western blot, including 9 CTDs and 11 normal controls, as previously described. The serum protein concentration was determined by a Bradford protein assay, with a bovine serum albumin (BSA) standard curve [35]. 30 μg of non-depleted serum proteins were loaded on a single lane and separated with 10% (or 12%) sodium dodecyl sulfate polyacrylamide gel electrophoresis (SDS-PAGE), with molecular weight standards run in parallel. The separated proteins were electrotransferred to nitrocellulose membranes at 4°C. The membranes were blocked for 2 h at room temperature with 5% non-fat milk in Tris-Buffered-Saline with Tween(TBST) and incubated with primary antibodies overnight at 4°C; the primary antibodies were rabbit polyclonal gelsolin (GSN) antibody (1:75, Thermo, USA), mouse monoclonal ceruloplasmin (CERU) antibody (1:1000, Abcam, Hong Kong), mouse mono-clonal Attractin(ATRN) antibody (1:1000, Abcam, Hong Kong), goat monoclonal Alpha-2-macroglobulin(A2M) antibody (1:5000, Thermo, USA), goat polyclonal Pregnancy zone protein(PZP) antibody(1:150 SANTA CRUZ, USA), respectively. After washing with TBST in triplicate, the blots were incubated in HRP-conjugated secondary antibody for 1 h at room temperature. Immunoreactive proteins were detected with SuperSigal West FemtoMaximun Sensitivity Substrate (Thermo, USA), and then exposed to X-ray film. The densitometry of the bands was estimated by Quantity One.

The GSN expression in the four heart tissues of CTD fetus and four normal controls were also analyzed using Western blot. The tissues were directly lysed with a fixed volume of cell lysis buffer (50mM Tris-HCl, pH 7.4, 150mM NaCl, 1% Triton-100), followed by vortexing for 10 minutes. Protein inhibitor was added during cell lysis. 20 μg of heart proteins were loaded on a single lane and separated with 10% SDS-PAGE and electrotransferred to nitrocellulose membranes. Blots were incubated with primary GSN antibody (1:75, Thermo, USA), followed by incubation with secondary antibody. GAPDH was used as a loading control. Triplicate blots were carried out for each tissue sample to ensure robustness of the data generated.

7. Immunohistochemistry

Fetal heart tissues were examined by immunohistochemistry as described previously. Sections of heart tissue were prepared from paraffin embedded fresh tissues, blocked for 1 h at room temperature with 1% bovine serum albumin in PBS and incubated with primary rabbit polyclonal GSN antibody (1:100, Thermo) overnight at 4°C and goat anti-rabbit HRP-conjugated secondary antibody for 1 h at room temperature. Pictures were analyzed under a phase microscope at a 400× magnification.

8. Statistical analysis

Data were expressed as mean ± SD. Mann–Whitney U-tests or one-way ANOVA was used to determine the significant difference of variables between two groups or three groups. Statistical analyses were performed using the SPSS19.0 software (SPSS Inc. Chicago, IL, USA). $p < 0.05$ was considered statistically significant.

Results

1. Derivation of the study population

Figure 1 displays the division of the analytical data set. 12 fetuses were diagnosed with isolated CTD, 15 fetuses with isolated ACHD and 7 with multi-malformation by ultrasound examination at 20-24 weeks of gestation. Termination of pregnancy was chosen in 7 of CTD cases after diagnosis, and CTD was verified as an isolated defect by autopsy. Other CTDs, ACHDs and normal controls were confirmed by prenatal and postnatal echocardiography. 3 cases of fetal 21-trisomy, 7 cases of fetal multi-malformation, 19 cases of fetal CHD which was diagnosed at birth, and 1 case combined with pregnancy–induced hypertension and fetal malformation were excluded. After exclusion of these women and those whose serum samples were lost, 9 participants with CTD fetus, 10 participants with ACHD fetus, 11 participants with normal fetus were recruited. Table 2 shows the types of prenatally diagnosed CTDs and ACHDs from the collected samples.

2. The identification of differentially expressed maternal serum proteins in women carrying a CTD fetus and ACHD fetus by iTRAQ and 2D LC/MS/MS analysis

A total of 105 unique proteins present in the three groups were identified, and relative expression data were obtained for 92 of them with high confidence by employing the iTRAQ combined with 2D LC/MS/MS technologies. The information of identification and quantification is provided in Table S1. The detailed information for one differentially expressed protein-gelsolin (GSN) is shown in Figure 2A.

The criteria of a differential abundance of 1.40 and 0.71 fold changes were applied to the 92 identified proteins, and a total of 13 proteins showed significant changes between the two CHD groups and control group. Compared to the normal controls, 11 proteins in CTD group were dysregulated, with 3 upregulated and 8 downregulated proteins. Meanwhile, 7 proteins in the ACHD group were dysregulated, with 2 elevated and 5 depleted proteins. Of the dysregulated proteins, 5 proteins were shared between the CTD and ACHD groups, and 6 proteins were differentially expressed only in the CTD groups, including Ceruloplasmin(-CERU), Complement factor B(CFAB), Pregnancy zone pro-

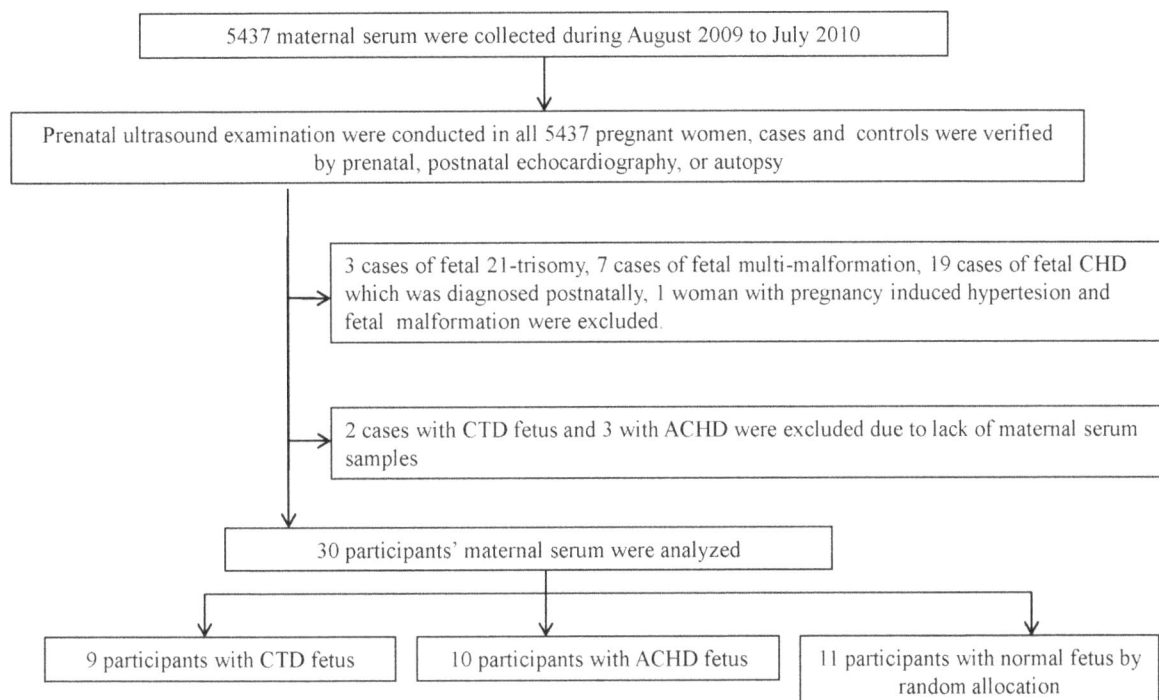

Figure 1. Flow diagram describing the selection of the analytical data set used for iTRAQ analysis.

tein(PZP), Gelsolin (GSN), Alpha-2-macroglobulin (A2M), Attractin (ATRN). The full list of differentially expressed proteins can be found in Table 3.

3. Western blot validation of the protein expression in maternal serum

Except of the protein CFAB whose antibody cannot be available, the other five proteins, including CERU, GSN, A2M, ATRN, PZP were verified using Western blot (Figure S1). Figure 2B displays a summary of the protein expression from the maternal sera of 9 women carrying CTD fetus in comparison to that of 11 women carrying normal fetus, detected by Western blot analysis. GSN expression was downregulated in both iTRAQ and Western blot, in the maternal sera of women carrying a CTD

fetus compared to those of the normal controls ($p = 0.008$). A2M, CERU, ATRN, PZP were not significantly differentially expressed ($p > 0.05$).

4. Western blot and IHC validation of the gelsolin protein expression in fetal heart tissue

Having validated that GSN protein was downregulated in maternal serum with a CTD fetus, we also examined the expression of GSN in 4 gestation week-paired CTD fetal hearts (23–25 gestational weeks), and 4 normal controls by Western blot and IHC. Figure 3A demonstrates that the GSN expression in heart tissues of fetuses with CTD in comparison with normal tissue, detected by Western blot. The expression of GSN was downregulated in CTD fetuses in comparison with normal

Table 2. Types of congenital heart defects included in this study.

Types of congenital heart defects	Number
CTD	
Tetralogy of Fallot (TOF)	4
Truncus arteriosus TA	1
Transposition of the great arteries (TGA)	2
Double-outlet right ventricle (DORV)	1
Pulmonary atresia (PA)	1
ACHD	
Ventricular septal defect (VSD)	6
Atrial septal defect (ASD)	1
Complete atrioventricular septal defect (CAVSD)	2
Complete endocardial cushion defect with uniatrium	1

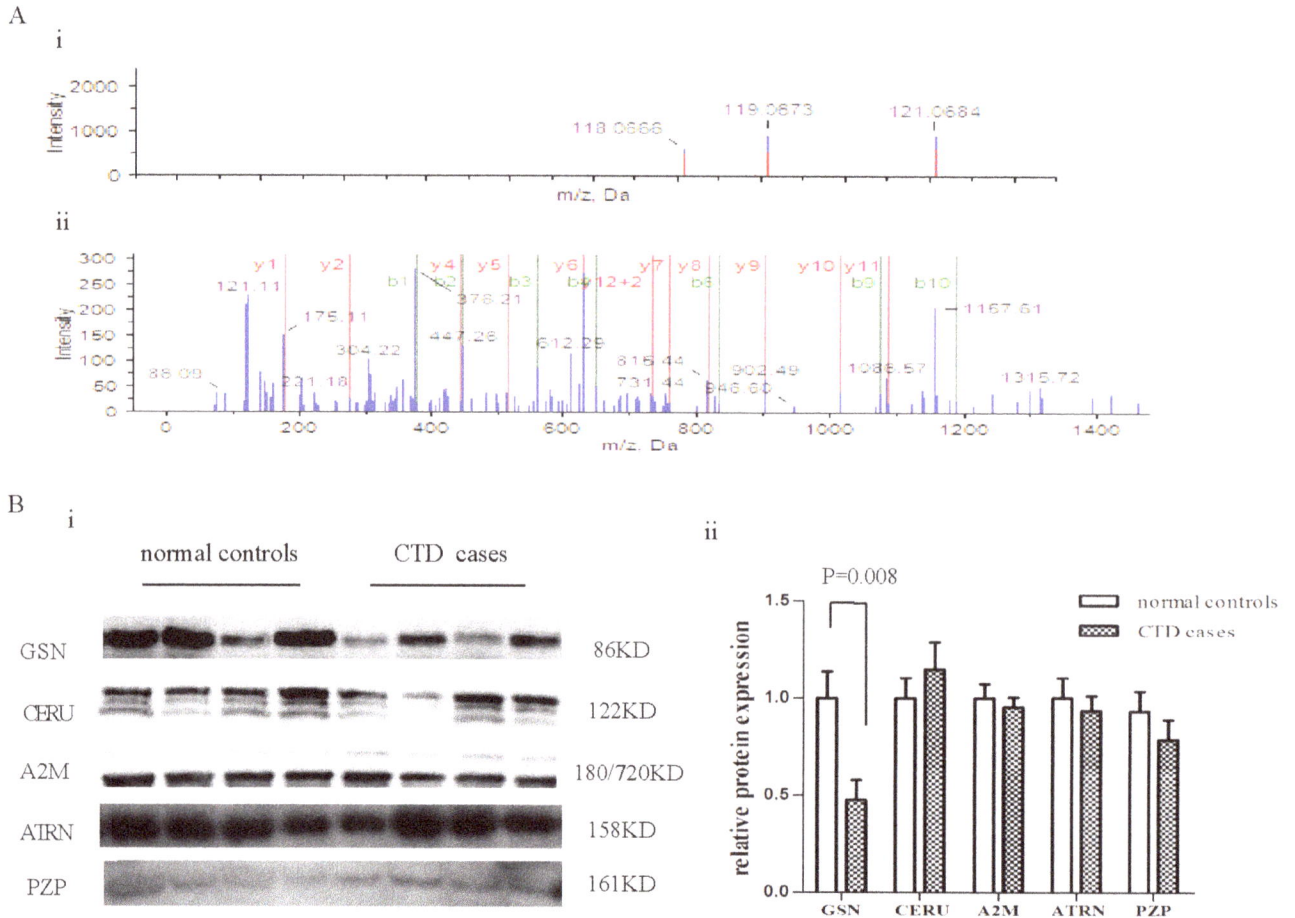

Figure 2. Identification of the protein gelsolin as being differentially expressed in maternal sera of pregnant women carrying a CTD fetus. (A) Identification of the protein gelsolin in maternal serum by iTRAQ and 2D LC-MS/MS. (i) The peptide quantitation information of gelsolin was derived from the intensities of three iTRAQ reporter ions. Reporter ions 118,119 were used to label sera from pregnant women carrying CTD and ACHD fetuses, respectively. Reporter ion 121 was used to label normal controls. (ii)The representative MS/MS spectrum of a peptide from gelsolin. (B). The Western blot validated the relative expression of GSN, A2M, CERU, ATRN, PZP in maternal serum samples(i), and confirmed relative decreased level of maternal serum gelsolin in CTD group compared with normal controls (ii). n = 9 in CTD group, n = 11 in normal control.*p = 0.008.

controls. Figure 3B shows the expression level and localization of GSN in heart tissues of the representative cases by IHC. GSN protein is expressed in the cytoplasm of myocardiocytes. The dark brown immunostaining is weaker in the CTD heart tissues than the normal controls, further verifying the downregulation of GSN in CTD fetus' heart (i, normal control, 23 gestational weeks, ii, CTD 23 gestational weeks; iii normal control, 25 gestational weeks, iv, CTD 25 gestational weeks).

Discussion

In this nested-control study, we identified the protein gelsolin as being differentially expressed in the sera of pregnant women carrying a CTD fetus at 14–18 gestational weeks. In comparison with sera from women carrying a normal fetus, gelsolin was significantly downregulated in the CTD group, as validated by proteomic analysis and Western blot. Moreover, decreased level of gelsolin in CTD fetal heart tissue was confirmed. It provided a valuable clue to search for the potential pathogenesis of CTD and develop a predictor of CTD.

As the prognosis of CTD infants is much poorer than that of other CHD(such as ventricular septal defect, etc) infants, we specially focused on the differentially expressed proteins of CTD in

this study. Our study used pooled samples after depletion of the highly abundant proteins for iTRAQ labeling, which worked effectively for detecting low abundance proteins and avoiding clogging to the LC columns by the unprocessed serum [31]. However, pooled sample beforehand could mask the inter-sample variability, so we validated each sample by Western blot. Abnormal cardiac development usually occurs before 8 weeks of gestation, which is much earlier than ultrasound detection of CTD at around 24 gestational weeks. The environmental pathogenic factors for early cardiac development in maternal serum may appear earlier than abnormities can be detected by ultrasound, therefore, we collected the maternal serum at 14–18 gestational weeks, as near to the onset of CTD as possible. Currently, some markers (such as brain natriuretic peptide (BNP) and N-terminal pro-brain natriuretic peptide (NT-proBNP)) have been proven to be useful for the diagnosis of complex and significant congenital heart disease in neonates, children and adults [36,37,38], however, biomarkers for detection complex congenital heart disease in the fetus have not been reported.

Gelsolin is a Ca^{2+}-dependent actin filament severing and capping protein which is involved in multiple important biological and clinical functions. Low expression of gelsolin has been observed in many cancers, including ovarian [39], breast [40],

Table 3. Maternal serum proteins showed differently expression in CTD group and ACHD group compared to normal control (CON).

Accession number	Gene	Protein	Function	CTD/CON		ACHD/CON	
				Ratio	p	Ratio	p
POCOL5	CO4B	Complement C4-B	Complement activity	0.3565	0.0005	0.4875	0.0001
P00450	CERU	Ceruloplasmin	oxidoreductase activity	0.6546	0.0371	0.8630	0.6704
P02751	FINC	Fibronectin	intracellular protein transport	0.9204	0.9958	0.5754	0.0078
P00751	CFAB	Complement factor B	Unknown	0.6792	0.0010	0.7178	0.0155
P19827	ITIH1	Inter-alpha-trypsin inhibitor heavy chain H1	serine protease inhibitor	0.6546	0.0582	0.6427	0.0265
Q14624	ITIH4	Inter-alpha-trypsin inhibitor heavy chain H4	serine protease inhibitor	0.5297	0.0310	0.6026	0.0143
P20742	PZP	Pregnancy zone protein	cytokine activity	0.2884	0.0021	1.3062	0.1249
P06727	APOA4	Apolipoprotein A-IV	lipid transporter activity	0.4529	0.0195	0.6792	0.0130
P06396	GSN	Gelsolin	structural constituent of cytoskeleton	0.4656	0.0009	0.8630	0.3166
P01023	A2MG	Alpha-2-macroglobulin	cytokine activity	0.6138	0.0300	0.7586	0.5209
P43652	AFAM	Afamin	transport	2.7542	0.0378	2.6546	0.0431
P02787	TRFE	Serotransferrin	serine-type peptidase activity	3.9084	0.0228	11.1686	0.0008
O75882	ATRN	Attractin	oxidoreductase activity	1.4723	0.0493	0.9727	0.9742

A

B

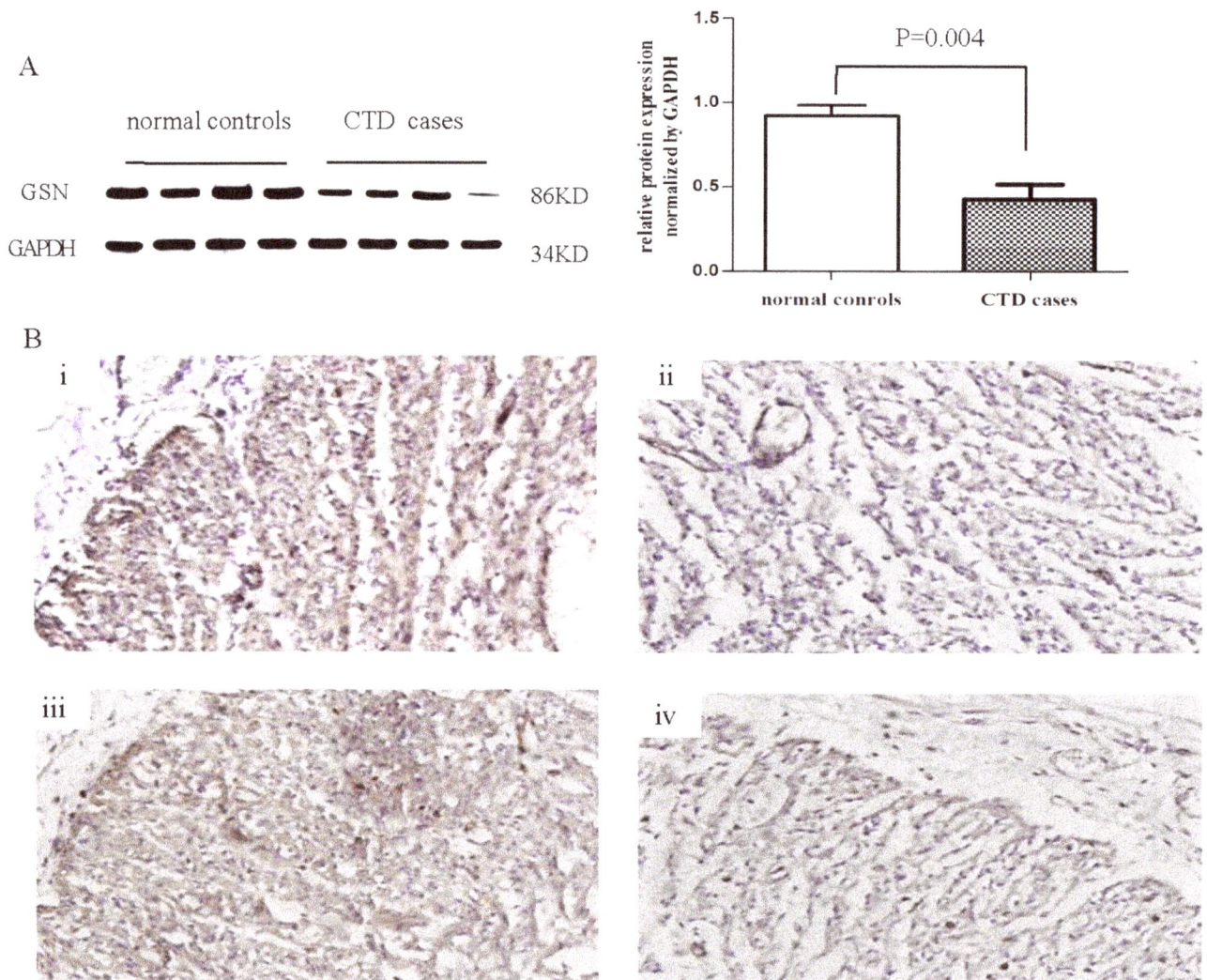

Figure 3. Western blot and IHC analyses confirmed the relative expression of GSN in fetal heart tissues with CTD. (A) Western blot analysis examined the relative expression of GSN, and confirmed that the GSN protein was downregulated in heart tissues of CTD fetuses compared with normal controls. n = 4, respectively. *p = 0.004. (B) IHC demonstrated the dark brown immunostaining was weaker in the CTD heart tissues than the normal controls (i, normal control, 23 gestational weeks, ii, CTD 23 gestational weeks; iii normal control, 25 gestational weeks, iv, CTD 25 gestational weeks). n = 4, respectively. Magnification was set at 400×.

colon [41] and prostate [42], cervical [43]. Plasma gelsolin level also decreases in many critically ill patients such as burn. [44], sepsis [45], traumatic brain injury [46,47]. The reduced level in plasma gelsolin is positively associated with the severity of these diseases, which can be used to predict the prognosis of diseases. The expression change of gelsolin in heart tissues is closely related to cardiac injury. The expression of gelsolin is upregulated in failing human hearts [48], pressure overload myocardium [49], and the gelsolin overexpression could induce cardiac hypertrophy [50]. The downregulation of gelsolin is a prosurvival factor for inhibition of heart failure progression after myocardial infarct. $Gsn^{-/-}$ mice had a lower mortality, reduced hypertrophy, less interstitial fibrosis and improved cardiac function compared with $Gsn^{+/+}$ mice after they were subjected left anterior descending coronary artery ligation. Gelsolin exists in midtrimester amniotic fluid of normal pregnancy and inhibits the proinflammatory immune response [51]; and serum gelsolin concentration was decreased with gestational age during normal pregnancy [52]. The expression of gelsolin was found to be downregulated in amniotic

fluid supernatants from Klinefelter syndrome fetuses [27], and in maternal plasma of fetuses with Down Syndrome [23], preeclampsia and intra-uterine growth restriction [29]. However, the validations of proteomic results have not yet been reported for these pathological pregnancies. In this study, the decreased gelsolin level in the sera of women carrying a CTD fetus was verified by Western blot for the first time. But, the exact mechanism for this result is still unclear, which is probably of maternal origin itself, or may be caused by reduced secretion from placental and amniotic fluid. The plasma levels of gelsolin in tetralogy newborns were significantly lower than those in normal controls [53], so it is also plausible that the decreased gelsolin in the CTD fetus' circulation is a potential source of that in maternal serum. Interestingly, we found that gelsolin was also downregulated in the heart tissue of CTD fetuses by Western blot and IHC. Gelsolin knockout mice (Gsn-/-) are not lethal, and gelsolin is expressed in the myocardium of the atria and left ventricle from E10.5 days which is a critical time point in heart chamber morphogenesis and early septal development instead of outflow tract [54], thus, decreased expressed of gelsolin may not directly be a risk factor of CTDs. Further studies are needed to explore the reason of the decreased gelsolin level in the sera of women carrying a CTD fetus and the effect of low gelsolin level in maternal serum on development of embryo.

Currently, ultrasound scan and echocardiography are the major tools for CTD fetal prenatal diagnosis [11,55], however, the first ultrasound examination is usually beyond 20 gestational weeks. The accurate diagnosis of such conditions requires special facilities, highly qualified operators, and the correct prenatal assessment of the ventricular septum or the distal aortic arch is difficult [55,56]. Furthermore, many hospitals in China can not develop the routine fetal ultrasound scan and echocardiography. There is therefore an increasing demand for the development of a new method for earlier screening of fetal CTDs. Although the reason for decreased expression of gelsolin in maternal serum with CTD fetus is not still understood, the detection of gelsolin from maternal serum is non-invasive, requires small amounts of sample, and we believe that this study may provide an independent, low cost and complementary strategy for screening of fetal CTD. However, the confirmation of gelsolin protein as a potential screening biomarker requires validation of this protein level in all kinds of pregnancies, with a larger sample size. Furthermore, it is necessary to set a large–scale, systematic, prospective experiments for elucidating the sensitivity and specificity for population screening.

In summary, we successfully detected decreased levels of the protein gelsolin in the maternal sera of pregnant women carrying a CTD fetus in comparison with normal controls using iTRAQ labeling combined with 2D LC–MS/MS. The downregulation of gelsolin associated with CTD fetuses was confirmed by Western blot, both in sera and fetal heart tissues with CTD. Future studies should assess the quality of gelsolin as a maternal serum biomarker of fetuses with CTD.

Supporting Information

Figure S1 The Western blot confirmed relative decreased level of maternal serum gelsolin in CTD group compared with normal controls. n = 9 in CTD group, n = 11 in normal control. *$p = 0.008$.

Table S1 The information of identification for 105 proteins and relative quantification for 92 proteins present in the three groups obtained by iTRAQ combined with 2D LC/MS/MS. 118, 119, 121 were used to represent CTD group, ACHD group, and normal control, respectively.

Acknowledgements

We thank all of the pregnant women for their participation in this study.

Author Contributions

Conceived and designed the experiments: YZ YK DM X. Li. Performed the experiments: YZ X. Liu. Analyzed the data: YZ YK. Contributed reagents/materials/analysis tools: YZ JZ HW HJ. Wrote the paper: YZ QZ X. Li.

References

1. Shaw GM, Carmichael SL, Yang W, Lammer EJ (2010) Periconceptional nutrient intakes and risks of conotruncal heart defects. Birth Defects Res A Clin Mol Teratol 88: 144–151.
2. Botto LD, Correa A, Erickson JD (2001) Racial and temporal variations in the prevalence of heart defects. Pediatrics 107: E32.
3. Centers for Disease Control and Prevention (CDC) (1998) Trends in infant mortality attributable to birth defects—United States, 1980–1995. MMWR Morb Mortal Wkly Rep 47: 773–778.
4. Benson DW, Basson CT, MacRae CA (1996) New understandings in the genetics of congenital heart disease. Curr Opin Pediatr 8: 505–511.
5. Pierpont ME, Basson CT, Benson DW Jr., Gelb BD, Giglia TM, et al. (2007) Genetic basis for congenital heart defects: current knowledge: a scientific statement from the American Heart Association Congenital Cardiac Defects Committee, Council on Cardiovascular Disease in the Young: endorsed by the American Academy of Pediatrics. Circulation 115: 3015–3038.
6. Ratajska A, Ciszek B, Zajaczkowska A, Jablonska A, Juszynski M (2009) Angioarchitecture of the venous and capillary system in heart defects induced by retinoic acid in mice. Birth Defects Res A Clin Mol Teratol 85: 599–610.
7. Lisowski LA, Verheijen PM, Copel JA, Kleinman CS, Wassink S, et al. (2010) Congenital heart disease in pregnancies complicated by maternal diabetes mellitus. An international clinical collaboration, literature review, and meta-analysis. Herz 35: 19–26.
8. van Beynum IM, Kapusta L, den Heijer M, Vermeulen SH, Kouwenberg M, et al. (2006) Maternal MTHFR 677C >T is a risk factor for congenital heart defects: effect modification by periconceptional folate supplementation. Eur Heart J 27: 981–987.
9. Debrus S, Berger G, de Meeus A, Sauer U, Guillaumont S, et al. (1996) Familial non-syndromic conotruncal defects are not associated with a 22q11 microdeletion. Hum Genet 97: 138–144.
10. Hoffman JI, Kaplan S (2002) The incidence of congenital heart disease. J Am Coll Cardiol 39: 1890–1900.
11. Galindo A, Mendoza A, Arbues J, Graneras A, Escribano D, et al. (2009) Conotruncal anomalies in fetal life: accuracy of diagnosis, associated defects and outcome. Eur J Obstet Gynecol Reprod Biol 146: 55–60.
12. Hutson MR, Kirby ML (2003) Neural crest and cardiovascular development: a 20-year perspective. Birth Defects Res C Embryo Today 69: 2–13.
13. Kirby ML, Waldo KL (1995) Neural crest and cardiovascular patterning. Circ Res 77: 211–215.
14. Clark EB (1996) Pathogenetic mechanisms of congenital cardiovascular malformations revisited. Semin Perinatol 20: 465–472.
15. Long J, Ramadhani T, Mitchell LE (2010) Epidemiology of nonsyndromic conotruncal heart defects in Texas, 1999–2004. Birth Defects Res A Clin Mol Teratol 88: 971–979.
16. Kolialexi A, Tsangaris GT, Papantoniou N, Anagnostopoulos AK, Vougas K, et al. (2008) Application of proteomics for the identification of differentially expressed protein markers for Down syndrome in maternal plasma. Prenat Diagn 28: 691–698.
17. Park SJ, Yoon WG, Song JS, Jung HS, Kim CJ, et al. (2006) Proteome analysis of human amnion and amniotic fluid by two-dimensional electrophoresis and matrix-assisted laser desorption/ionization time-of-flight mass spectrometry. Proteomics 6: 349–363.
18. Merkatz IR, Nitowsky HM, Macri JN, Johnson WE (1984) An association between low maternal serum alpha-fetoprotein and fetal chromosomal abnormalities. Am J Obstet Gynecol 148: 886–894.
19. Wald NJ, Kennard A, Hackshaw A, McGuire A (1997) Antenatal screening for Down's syndrome. J Med Screen 4: 181–246.
20. Zhang X, Yin X, Yu H, Liu X, Yang F, et al. (2012) Quantitative proteomic analysis of serum proteins in patients with Parkinson's disease using an isobaric tag for relative and absolute quantification labeling, two-dimensional liquid chromatography, and tandem mass spectrometry. Analyst 137: 490–495.
21. Huang Z, Wang H, Huang H, Xia L, Chen C, et al. (2012) iTRAQ-based proteomic profiling of human serum reveals down-regulation of platelet basic protein and apolipoprotein B100 in patients with hematotoxicity induced by chronic occupational benzene exposure. Toxicology 291: 56–64.
22. Kang Y, Dong X, Zhou Q, Zhang Y, Cheng Y, et al. (2012) Identification of novel candidate maternal serum protein markers for Down syndrome by integrated proteomic and bioinformatic analysis. Prenat Diagn 32: 284–292.
23. Kolla V, Jeno P, Moes S, Tercanli S, Lapaire O, et al. (2010) Quantitative proteomics analysis of maternal plasma in Down syndrome pregnancies using isobaric tagging reagent (iTRAQ). J Biomed Biotechnol 2010: 952047.
24. Wang TH, Chao AS, Chen JK, Chao A, Chang YL, et al. (2009) Network analyses of differentially expressed proteins in amniotic fluid supernatant associated with abnormal human karyotypes. Fertil Steril 92: 96–107.
25. Kolialexi A, Anagnostopoulos AK, Papantoniou N, Vougas K, Antsaklis A, et al. (2010) Potential biomarkers for Turner in maternal plasma: possibility for noninvasive prenatal diagnosis. J Proteome Res 9: 5164–5170.
26. Mavrou A, Anagnostopoulos AK, Kolialexi A, Vougas K, Papantoniou N, et al. (2008) Proteomic analysis of amniotic fluid in pregnancies with Turner syndrome fetuses. J Proteome Res 7: 1862–1866.
27. Anagnostopoulos AK, Kolialexi A, Mavrou A, Vougas K, Papantoniou N, et al. (2010) Proteomic analysis of amniotic fluid in pregnancies with Klinefelter syndrome foetuses. J Proteomics 73: 943–950.

28. Cecconi D, Lonardoni F, Favretto D, Cosmi E, Tucci M, et al. (2011) Changes in amniotic fluid and umbilical cord serum proteomic profiles of foetuses with intrauterine growth retardation. Electrophoresis 32: 3630–3637.

29. Auer J, Camoin L, Guillonneau F, Rigourd V, Chelbi ST, et al. (2010) Serum profile in preeclampsia and intra-uterine growth restriction revealed by iTRAQ technology. J Proteomics 73: 1004–1017.

30. Park J, Cha DH, Lee SJ, Kim YN, Kim YH, et al. (2011) Discovery of the serum biomarker proteins in severe preeclampsia by proteomic analysis. Exp Mol Med 43: 427–435.

31. Esplin MS, Merrell K, Goldenberg R, Lai Y, Iams JD, et al. (2011) Proteomic identification of serum peptides predicting subsequent spontaneous preterm birth. Am J Obstet Gynecol 204: 391 e391–398.

32. Gercel-Taylor C, Taylor DD, Rai SN (2012) Proteomic identification of serum peptides predicting subsequent spontaneous preterm birth. Am J Obstet Gynecol 206: e3; author reply e3–4.

33. Gravett MG, Novy MJ, Rosenfeld RG, Reddy AP, Jacob T, et al. (2004) Diagnosis of intra-amniotic infection by proteomic profiling and identification of novel biomarkers. JAMA 292: 462–469.

34. Datta A, Qian J, Chong R, Kalaria RN, Francis P, et al. (2014) Novel pathophysiological markers are revealed by iTRAQ-based quantitative clinical proteomics approach in vascular dementia. J Proteomics 99: 54–67.

35. Bradford MM (1976) A rapid and sensitive method for the quantitation of microgram quantities of protein utilizing the principle of protein-dye binding. Anal Biochem 72: 248–254.

36. Eindhoven JA, van den Bosch AE, Jansen PR, Boersma E, Roos-Hesselink JW (2012) The usefulness of brain natriuretic peptide in complex congenital heart disease: a systematic review. J Am Coll Cardiol 60: 2140–2149.

37. Cantinotti M, Giovannini S, Murzi B, Clerico A (2011) Diagnostic, prognostic and therapeutic relevance of B-type natriuretic hormone and related peptides in children with congenital heart diseases. Clin Chem Lab Med 49: 567–580.

38. Nir A, Luchner A, Rein AJ (2012) The natriuretic peptides as biomarkers for adults with congenital heart disease. Biomark Med 6: 827–837.

39. Noske A, Denkert C, Schober H, Sers C, Zhumabayeva B, et al. (2005) Loss of Gelsolin expression in human ovarian carcinomas. Eur J Cancer 41: 461–469.

40. Winston JS, Asch HL, Zhang PJ, Edge SB, Hyland A, et al. (2001) Downregulation of gelsolin correlates with the progression to breast carcinoma. Breast Cancer Res Treat 65: 11–21.

41. Gay F, Estornes Y, Saurin JC, Joly-Pharaboz MO, Friederich E, et al. (2008) In colon carcinogenesis, the cytoskeletal protein gelsolin is down-regulated during the transition from adenoma to carcinoma. Hum Pathol 39: 1420–1430.

42. Lee HK, Driscoll D, Asch H, Asch B, Zhang PJ (1999) Downregulated gelsolin expression in hyperplastic and neoplastic lesions of the prostate. Prostate 40: 14–19.

43. Liao CJ, Wu TI, Huang YH, Chang TC, Wang CS, et al. (2011) Overexpression of gelsolin in human cervical carcinoma and its clinicopathological significance. Gynecol Oncol 120: 135–144.

44. Xianhui L, Pinglian L, Xiaojuan W, Wei C, Yong Y, et al. (2014) The association between plasma gelsolin level and prognosis of burn patients. Burns.

45. Cohen TS, Bucki R, Byfield FJ, Ciccarelli NJ, Rosenberg B, et al. (2011) Therapeutic potential of plasma gelsolin administration in a rat model of sepsis. Cytokine 54: 235–238.

46. Jin Y, Li BY, Qiu LL, Ling YR, Bai ZQ (2012) Decreased plasma gelsolin is associated with 1-year outcome in patients with traumatic brain injury. J Crit Care 27: 527 e521–526.

47. Xu JF, Liu WG, Dong XQ, Yang SB, Fan J (2012) Change in plasma gelsolin level after traumatic brain injury. J Trauma Acute Care Surg 72: 491–496.

48. Yang J, Moravec CS, Sussman MA, DiPaola NR, Fu D, et al. (2000) Decreased SLIM1 expression and increased gelsolin expression in failing human hearts measured by high-density oligonucleotide arrays. Circulation 102: 3046–3052.

49. Mani SK, Shiraishi H, Balasubramanian S, Yamane K, Chellaiah M, et al. (2008) In vivo administration of calpeptin attenuates calpain activation and cardiomyocyte loss in pressure-overloaded feline myocardium. Am J Physiol Heart Circ Physiol 295: H314–326.

50. Hu WS, Ho TJ, Pai P, Chung LC, Kuo CH, et al. (2014) Gelsolin (GSN) induces cardiomyocyte hypertrophy and BNP expression via p38 signaling and GATA-4 transcriptional factor activation. Mol Cell Biochem 390: 263–270.

51. Sezen D, Bongiovanni AM, Gelber S, Perni U, Hutson JM, et al. (2009) Gelsolin down-regulates lipopolysaccharide-induced intraamniotic tumor necrosis factor-alpha production in the midtrimester of pregnancy. Am J Obstet Gynecol 200: 191 e191–194.

52. Scholl PF, Cole RN, Ruczinski I, Gucek M, Diez R, et al. (2012) Maternal serum proteome changes between the first and third trimester of pregnancy in rural southern Nepal. Placenta 33: 424–432.

53. Xuan C, Gao G, Yang Q, Wang XL, Liu ZG, et al. (2014) Proteomic study reveals plasma protein changes in congenital heart diseases. Ann Thorac Surg 97: 1414–1419.

54. Plageman TF, Jr., Yutzey KE (2006) Microarray analysis of Tbx5-induced genes expressed in the developing heart. Dev Dyn 235: 2868–2880.

55. Gardiner HM (2001) Fetal echocardiography: 20 years of progress. Heart 86 Suppl 2: II12–22.

56. Buskens E, Stewart PA, Hess J, Grobbee DE, Wladimiroff JW (1996) Efficacy of fetal echocardiography and yield by risk category. Obstet Gynecol 87: 423–428.

PERMISSIONS

All chapters in this book were first published in PLOS ONE, by The Public Library of Science; hereby published with permission under the Creative Commons Attribution License or equivalent. Every chapter published in this book has been scrutinized by our experts. Their significance has been extensively debated. The topics covered herein carry significant findings which will fuel the growth of the discipline. They may even be implemented as practical applications or may be referred to as a beginning point for another development.

The contributors of this book come from diverse backgrounds, making this book a truly international effort. This book will bring forth new frontiers with its revolutionizing research information and detailed analysis of the nascent developments around the world.

We would like to thank all the contributing authors for lending their expertise to make the book truly unique. They have played a crucial role in the development of this book. Without their invaluable contributions this book wouldn't have been possible. They have made vital efforts to compile up to date information on the varied aspects of this subject to make this book a valuable addition to the collection of many professionals and students.

This book was conceptualized with the vision of imparting up-to-date information and advanced data in this field. To ensure the same, a matchless editorial board was set up. Every individual on the board went through rigorous rounds of assessment to prove their worth. After which they invested a large part of their time researching and compiling the most relevant data for our readers.

The editorial board has been involved in producing this book since its inception. They have spent rigorous hours researching and exploring the diverse topics which have resulted in the successful publishing of this book. They have passed on their knowledge of decades through this book. To expedite this challenging task, the publisher supported the team at every step. A small team of assistant editors was also appointed to further simplify the editing procedure and attain best results for the readers.

Apart from the editorial board, the designing team has also invested a significant amount of their time in understanding the subject and creating the most relevant covers. They scrutinized every image to scout for the most suitable representation of the subject and create an appropriate cover for the book.

The publishing team has been an ardent support to the editorial, designing and production team. Their endless efforts to recruit the best for this project, has resulted in the accomplishment of this book. They are a veteran in the field of academics and their pool of knowledge is as vast as their experience in printing. Their expertise and guidance has proved useful at every step. Their uncompromising quality standards have made this book an exceptional effort. Their encouragement from time to time has been an inspiration for everyone.

The publisher and the editorial board hope that this book will prove to be a valuable piece of knowledge for researchers, students, practitioners and scholars across the globe.

LIST OF CONTRIBUTORS

Youhui Si, Jianguo Li, Yuqiang Niu, Xiuying Liu, Lili Ren, Li Guo, Min Cheng, Hongli Zhou, Jianwei Wang, Qi Jin and Wei Yang
Ministry of Health Key Laboratory of Systems Biology of Pathogens, Institute of Pathogen Biology, Chinese Academy of Medical Sciences & Peking Union Medical College, Beijing, China

Dongliang Liu, Yang Li, Jing Zhao, Fuchun Zhang, Yongxing Wang, Ji Ma, Surong Sun
Xiaomei Duan, Chun Kou, Ting Wu and Yijie Li
Xinjiang Key Laboratory of Biological Resources and Genetic Engineering, College of Life Science and Technology, Xinjiang University, Urumqi, Xinjiang, China

Fei Deng and Zhihong Hu
State Key Laboratory of Virology, Chinese Academy of Sciences, Wuhan, Hubei, China

Jianhua Yang
Xinjiang Key Laboratory of Biological Resources and Genetic Engineering, College of Life Science and Technology, Xinjiang University, Urumqi, Xinjiang, China
Texas Children's Cancer Center, Department of Pediatrics, Dan L. Duncan Cancer Center, Baylor College of Medicine, Houston, Texas, United States of America

Yujiang Zhang
Center for Disease Control and Prevention of the Xinjiang Uyghur Autonomous Region, Urumqi, Xinjiang, China

Hao Chen and Yunjie Zhao
Department of Physics, The George Washington University, Washington, D. C., United States of America

Haotian Li, Dongyan Zhang, Yanzhao Huang and Shiyong Liu
Department of Physics, Huazhong University of Science and Technology, Wuhan, Hubei, China

Qi Shen
BNLMS, Center for Quantitative Biology, Peking University, Beijing, China

Fatah Kashanchi
George Mason University, National Center for Biodefense & Infectious Diseases, Manassas, Virginia, United States of America

Rachel Van Duyne
George Mason University, National Center for Biodefense & Infectious Diseases, Manassas, Virginia, United States of America
The George Washington University Medical Center, Department of Microbiology, Immunology, and Tropical Medicine, Washington, D. C., United States of America

Chen Zeng
Department of Physics, Huazhong University of Science and Technology, Wuhan, Hubei, China
Department of Physics, The George Washington University, Washington, D. C., United States of America

Harmeet Singh, Guiying Nie and Yao Wang
MIMR-PHI Institute of Medical Research, Clayton, Victoria, Australia, 2 Monash University, Clayton, Victoria, Australia

Tracy L. Nero
ACRF Rational Drug Discovery Centre, St Vincent's Institute of Medical Research, Fitzroy, Victoria, Australia

Michael W. Parker
ACRF Rational Drug Discovery Centre, St Vincent's Institute of Medical Research, Fitzroy, Victoria, Australia
Department of Biochemistry and Molecular Biology, Bio21 Molecular Science and Biotechnology Institute, the University of Melbourne, Parkville, Victoria, Australia

Stefan J. Kempf, Soile Tapio and Simone Moertl
Institute of Radiation Biology, Helmholtz Zentrum München, German Research Center for Environmental Health GmbH, Neuherberg, Germany

Sonja Buratovic and Per Eriksson
Department of Environmental Toxicology, Uppsala University, Uppsala, Sweden

Christine von Toerne and Stefanie M. Hauck
Research Unit Protein Science, Helmholtz Zentrum München, German Research Center for Environmental Health GmbH, Neuherberg, Germany

Michael J. Atkinson
Chair of Radiation Biology, Technical University Munich, Munich, Germany

Bo Stenerlöw
Division of Biomedical Radiation Sciences, Rudbeck Laboratory, Uppsala University, Uppsala, Sweden

Jianjun Cheng, Mingwei Leng, Longjie Li, Hanhai Zhou and Xiaoyun Chen
School of Information Science and Engineering, Lanzhou University, Lanzhou, Gansu Province, China

Benjamin Y. S. Li, King T. Ko and Lam F. Yeung
Department of Electronic Engineering, City University of Hong Kong, Hong Kong, Hong Kong

Choujun Zhan
Department of Electronic and Information Engineering, The Hong Kong Polytechnic University, Hong Kong, Hong Kong

Genke Yang
Department of Automation, Shanghai Jiao Tong University, Shanghai, China

Santiago Marfà, Vedrana Reichenbach, Gregori Casals and Manuel Morales-Ruiz
Biochemistry and Molecular Genetics Service, Centro de Investigación Biomédica en Red de Enfermedades Hepáticas y Digestivas (CIBEREHD), Hospital Clínic, Institut d'Investigacions Biomèdiques August Pi i Sunyer (IDIBAPS), University of Barcelona, Barcelona, Spain

Gonzalo Crespo, Xavier Forns and Miquel Navasa
Liver Unit, Centro de Investigación Biomédica en Red de Enfermedades Hepáticas y Digestivas (CIBEREHD), Hospital Clínic, Institut d'Investigacions Biomèdiques August Pi i Sunyer (IDIBAPS), University of Barcelona, Barcelona, Spain

Wladimiro Jiménez
Biochemistry and Molecular Genetics Service, Centro de Investigación Biomédica en Red de Enfermedades Hepáticas y Digestivas (CIBEREHD), Hospital Clínic, Institut d'Investigacions Biomèdiques August Pi i Sunyer (IDIBAPS), University of Barcelona, Barcelona, Spain
Departament de Ciencies Fisiologiques I, Centro de Investigación Biomédica en Red de Enfermedades Hepáticas y Digestivas (CIBEREHD), Hospital Clínic, Institut d'Investigacions Biomèdiques August Pi i Sunyer (IDIBAPS), University of Barcelona, Barcelona, Spain

Suyu Mei
Software College, Shenyang Normal University, Shenyang, China
Bioinformatics Section, School of Basic Medical Sciences, Southern Medical University, Guangzhou, China

Hao Zhu
Bioinformatics Section, School of Basic Medical Sciences, Southern Medical University, Guangzhou, China

Yeng Chen
Department of Oral Biology & Biomedical Sciences, Faculty of Dentistry, University of Malaya, Kuala Lumpur, Malaysia
Oral Cancer Research and Coordinating Center, Faculty of Dentistry, University of Malaya, Kuala Lumpur, Malaysia

Siti Nuraishah Azman
Institute for Research in Molecular Medicine, Universiti Sains Malaysia, Georgetown, Penang, Malaysia

Jesinda P. Kerishnan, Yin-Ling Wong and Subash C. B. Gopinath
Department of Oral Biology & Biomedical Sciences, Faculty of Dentistry, University of Malaya, Kuala Lumpur, Malaysia

Rosnah Binti Zain
Oral Cancer Research and Coordinating Center, Faculty of Dentistry, University of Malaya, Kuala Lumpur, Malaysia
Department of Oro-Maxillofacial and Medical Science, Faculty of Dentistry, University of Malaya, Kuala Lumpur, Malaysia

Yu Nieng Chen
Chen Dental Specialist Clinic, Kueh Hock Kui Commercial Centre, Jalan Tun Ahmad Zaidi Adruce, Kuching, Sarawak, Malaysia

Yun-Zi Dong, Li-Juan Zhang, Zi-Mei Wu, Ling Gao, Yi-Sang Yao and Ning-Zhi Tan
Pharmanex Beijing Clinical Pharmacology Center, Beijing, China

Jian-Yong Wu
Department of Applied Biology and Chemistry Technology, Hong Kong Polytechnic University, Hung Hom, Kowloon, Hong Kong
Shenzhen TCM Pharmacy and Molecular Pharmacology Kay Laboratory, Hong Kong Polytechnic University, Shenzhen, Guangdong, China

Luqun Ni
Department of Mechanical and Aerospace Engineering, University of California San Diego, La Jolla, CA, United States of America

Jia-Shi Zhu
Department of Applied Biology and Chemistry Technology, Hong Kong Polytechnic University, Hung Hom, Kowloon, Hong Kong

Shenzhen TCM Pharmacy and Molecular Pharmacology Kay Laboratory, Hong Kong Polytechnic University, Shenzhen, Guangdong, China
NS Center for Anti-Aging Research, Provo, UT, United States of America

Liang-Hui Chu and Esak Lee
Department of Biomedical Engineering, School of Medicine, Johns Hopkins University, Baltimore, Maryland, United States of America

Joel S. Bader
Department of Biomedical Engineering, School of Medicine, Johns Hopkins University, Baltimore, Maryland, United States of America
High-Throughput Biology Center, Johns Hopkins University, Baltimore, Maryland, United States of America

Aleksander S. Popel
Department of Biomedical Engineering, School of Medicine, Johns Hopkins University, Baltimore, Maryland, United States of America
Department of Oncology and Sidney Kimmel Comprehensive Cancer Center, School of Medicine, Johns Hopkins University, Baltimore, Maryland, United States of America

Masamoto Murakami, Xiuju Dai, Yasushi Hanakawa, Mikiko Tohyama, Koji Sayama and Hidenori Okazaki
Department of Dermatology, Ehime University Graduate School of Medicine, Ehime, Japan

Takaaki Kaneko and Akemi Ishida-Yamamoto
Department of Dermatology, Asahikawa Medical College, Asahikawa,Japan

Teruaki Nakatsuji
Division of Dermatology, University of California San Diego, and VA San Diego Healthcare Center, San Diego, California, United States of America

Kenji Kameda
Integrated Center for Science, Ehime University Graduate School of Medicine, Ehime, Japan

Ruili Ma
College of Veterinary Medicine, Northwest Agriculture & Forestry University, Yangling, Shaanxi, China
College of Life Sciences, Northwest Agriculture & Forestry University, Yangling, Shaanxi, China

Yanming Zhang and Pengbo Ning
College of Veterinary Medicine, Northwest Agriculture & Forestry University, Yangling, Shaanxi, China

Haiquan Liu
School of Computer Science and Engineering, Xi'an Technological University, Xi'an, Shaanxi, China

Wentao Hu, Ganesh Anand, J. Sivaraman and Yu-Keung Mok
Department of Biological Sciences, 14 Science Drive 4, National University of Singapore, Singapore, Singapore 117543

Ka Yin Leung
Department of Biology, Faculty of Natural and Applied Sciences, Trinity Western University, Langley, British Columbia, Canada V2Y 1Y1
State Key Laboratory of Bioreactor Engineering, East China University of Science and Technology, Shanghai, China 200237

KiYoung Lee
Department of Medical Informatics, School of Medicine, Ajou University, Yeongtong-gu, Suwon si, Republic of Korea

Jae W. Lee
Neuroscience Section, Papé Family Pediatric Research Institute, Department of Pediatrics, Oregon Health and Science University, Portland, Oregon, United States of America

Sunjae Lee, Doheon Lee and Seyeol Yoon
Department of Bio and Brain Engineering, KAIST, Yuseong-gu, Daejeon, Republic of Korea

Charlotte Welinder and Thomas E. Fehniger
Division of Oncology and Pathology, Clinical Sciences, Lund University, Lund, Sweden
Centre of Excellence in Biological and Medical Mass Spectrometry, Lund University, Lund, Sweden

Göran B. Jönsson, Bo Baldetorp, Thomas Breslin, Bo Jansson and Elisabet Wieslander
Division of Oncology and Pathology, Clinical Sciences, Lund University, Lund, Sweden

Christian Ingvar
Skåne University Hospital, Lund, Sweden
Dept. of Surgery, Clinical Sciences, Lund University, Skåne University Hospital, Lund, Sweden

Lotta Lundgren
Division of Oncology and Pathology, Clinical Sciences, Lund University, Lund, Sweden
Skåne University Hospital, Lund, Sweden

Håkan Olsson
Division of Oncology and Pathology, Clinical Sciences, Lund University, Lund, Sweden
Skåne University Hospital, Lund, Sweden

Dept. of Cancer Epidemiology, Clinical Sciences, Lund University, Lund, Sweden

Melinda Rezeli
Clinical Protein Science & Imaging, Biomedical Center, Biomedical Engineering, Lund University, Lund, Sweden

Thomas Laurell
Centre of Excellence in Biological and Medical Mass Spectrometry, Lund University, Lund, Sweden
Clinical Protein Science & Imaging, Biomedical Center, Biomedical Engineering, Lund University, Lund, Sweden

Krzysztof Pawlowski
Division of Oncology and Pathology, Clinical Sciences, Lund University, Lund, Sweden
Dept. of Experimental Design and Bioinformatics, Faculty of Agriculture and Biology, Warsaw University of Life Sciences, Warszawa, Poland

György Marko-Varga
Centre of Excellence in Biological and Medical Mass Spectrometry, Lund University, Lund, Sweden
Clinical Protein Science & Imaging, Biomedical Center, Biomedical Engineering, Lund
University, Lund, Sweden
First Department of Surgery, Tokyo Medical University, Tokyo, Japan

Stephenie D. Prokopec and John D. Watson
Informatics and Bio-computing Program, Ontario Institute for Cancer Research, Toronto, Ontario, Canada

Raimo Pohjanvirta
Laboratory of Toxicology, National Institute for Health and Welfare, Kuopio, Finland
Department of Food Hygiene and Environmental Health, University of Helsinki, Helsinki, Finland

Paul C. Boutros
Informatics and Bio-computing Program, Ontario Institute for Cancer Research, Toronto, Ontario, Canada
Department of Medical Biophysics, University of Toronto, Toronto, Ontario, Canada
Department of Pharmacology & Toxicology, University of Toronto, Toronto, Ontario, Canada

Antonio Pinna
Department of Surgical, Microsurgical and Medical Sciences, Section of Ophthalmology,
University of Sassari, Sassari, Italy
Azienda Ospedaliero-Universitaria di Sassari, Sassari, Italy

Davide Cossu, Daniela Paccagnini, Giuseppe Mameli, Leonardo A. Sechi and Speranza Masala
Department of Biomedical Sciences, Section of Experimental and Clinical Microbiology, University of Sassari, Sassari, Italy

Francesco Blasetti and Irene Maiore
Department of Surgical, Microsurgical and Medical Sciences, Section of Ophthalmology, University of Sassari, Sassari, Italy

John P. O'Rourke, David Peabody and Bryce Chackerian
Department of Molecular Genetics and Microbiology, University of New Mexico School of Medicine, Albuquerque, NM, United States of America

Seth M. Daly, Kathleen D. Triplett and Pamela R. Hall
Department of Pharmaceutical Sciences, University of New Mexico School of Medicine, Albuquerque, NM United States of America

Ying Zhang, Yuan Kang, Qiongjie Zhou and Jizi Zhou
Obstetrics and Gynecology Hospital, Fudan University, Shanghai, China

Huijun Wang
Children's Hospital, Fudan University, Shanghai, China

Hong Jin
Department of Chemistry, Fudan University, Shanghai, China
Institute of Biomedicine, Fudan University, Shanghai, China

Xiaohui Liu
Department of Chemistry, Fudan University, Shanghai, China

Duan Ma
Key Laboratory of Molecular Medicine, Ministry of Education, Department of Biochemistry and Molecular Biology, Institute of Biomedical Sciences, Shanghai Medical College, Fudan University, Shanghai, China

Xiaotian Li
Obstetrics and Gynecology Hospital, Fudan University, Shanghai, China
Key Laboratory of Molecular Medicine, Ministry of Education, Department of Biochemistry and Molecular Biology, Institute of Biomedical Sciences, Shanghai Medical College, Fudan University, Shanghai, China
Shanghai Key Laboratory of Female Reproductive Endocrine Related Diseases, Shanghai, China

Index